Luigi Morino, Renzo Piva (Eds.)

Boundary Integral Methods

Theory and Applications

Proceedings of the IABEM Symposium
Rome, Italy, October 15-19, 1990

Springer-Verlag Berlin Heidelberg GmbH

Prof. Luigi Morino
Prof. Renzo Piva

University of Rome "La Sapienza"
Dip. di Meccanica e Aeronautica
Via Eudossiana 18
I-00184 Roma
Italy

ISBN 978-3-642-85465-1 ISBN 978-3-642-85463-7 (eBook)
DOI 10.1007/ 978-3-642-85463-7

Library of Congress Cataloging-in-Publication Data
IABEM (Organization). Symposium (1990: Rome, Italy)
Boundary integral methods: theory and applications
proceedings of the IABEM Symposium, Rome,Italy, October 15-19, 1990
L. Morino, R. Piva (eds.)
ISBN 0-387-53773-2 (acid-free-paper)
1. Boundary element methods-Congresses.
I. Morino, Luigi II. Piva, Renzo. III. Title.
TA347. B6912 1990
620'. 001' 51535-dc20 91-25894

Softcover reprint of the hardcover 1st edition 1991

Typesetting: Camera ready by authors
61/3020-543210 – Printed on acid-free paper

To Nancy and Marina

Foreword

This volume contains edited papers from IABEM-90, the 1990 Symposium of the International Association for Boundary Element Methods (IABEM). As stated in the By-Laws of the Association, the purposes of IABEM are:

1. to promote the international exchange of technical information related to the development and application of boundary-integral equation (BIE) formulations and their numerical implementation to problems in engineering and science, commonly referred to as the boundary element method (BEM);

2. to promote research and development activities for the advancement of boundary-integral equation methods and boundary element solution algorithms;

3. to foster closer personal relationships within the BEM community of researchers.

The objectives of the Symposium, in line with those of the Association, was to provide a forum where the two "souls" of the Association, *i.e.*, (*i*) mathematical foundations and numerical aspects, and (*ii*) engineering applications could be integrated. We believe that the first aspect has been neglected in too many of the BEM Symposia held in the past, which, with a few exceptions (notably, the IUTAM Symposia on the subject) have emphasized the practical aspects of the method. As a consequence, we have tried to give a stronger emphasis to the more theoretical issues: this is attested for instance, by the fact that the two general lectures were given by Prof. Gaetano Fichera, of the University of Rome "La Sapienza," and Prof. Wolfgang Wendland, of the University of Stuttgart – two mathematicians whose contribution to the theoretical and the numerical aspects of boundary integral equations need not to be emphasized. In addition, we have noticed that several papers dealt with sophisticated issues, such as theoretical analysis of first kind integral equations, mathematical treatment of hypersingular kernels, and coupled boundary-element finite-element analysis.

In order to underline that this symposium puts unusual emphasis on the theoretical issues, we have chosen to entitle this volume "Boundary Integral Methods: Theory and Applications", giving equal weight to theory and applications. These last include the fields of solid mechanics (e.g., crack propagation, elasto-plastic analysis, and optimal design), fluid mechanics (e.g., free-surface and free-wake analysis, and viscous flows), diffusion problems (heat conduction and ground flows), as well as acoustics, wave scattering, magnetohydrodynamics.

We acknowledge the financial support provided for the organization of the Symposium by the following groups: University of Rome "La Sapienza," CNR (Italian National Research Council), CIRA (Italian Center for Aerospace Research), Aeritalia, and Agusta.

We wish to thank Ms. Paola Agosti, Floriana Trollini, and Marisa Zaninotto for their invaluable contribution to the organization of the Symposium. Finally, we want to thank Nancy and Marina for their patience and understanding, which allowed us to devote all the attention that was necessary to the organization of the Symposium and the editing of this volume.

Luigi Morino and Renzo Piva

January, 1991
Università di Roma "La Sapienza"

IABEM-90

Symposium of the International Association for Boundary Element Methods

Scientific Committee

Contents

General Lectures

Contributions

Simple Layer Potentials for Elliptic Equations of Higher Order

G. FICHERA

University of Rome "La Sapienza"

1. *Preliminaries. Fundamental solution.*

Let Ω be a bounded domain of the x,y-plane with a smooth boundary Σ. For the sake of simplicity we assume that Ω is simply connected. Σ is a Liapounov closed contour.

The most classical method for solving the Dirichlet problem for harmonic functions in Ω, i.e. the problem

$$\Delta_2 u = 0 \quad \text{in } \Omega , \tag{1.1}$$

$$u|_{\Sigma} = f \tag{1.2}$$

f given function on Σ, is the classical "Fredholm method" which consists in representing u like a *double layer potential*

$$u(z) = \frac{1}{2\pi}\int_{\Sigma} \varphi(\zeta)\frac{\partial}{\partial \nu_{\zeta}} \log|z-\zeta|\, ds \qquad (z=x+iy,\ \zeta=\xi+i\eta) \tag{1.3}$$

$\frac{\partial}{\partial \nu_{\zeta}}$ denotes differentiation with respect to the inner normal ν_{ζ} in the point ζ of Σ.

By imposing (1.2) one gets for $z \in \Sigma$

$$f(z) = -\tfrac{1}{2}\varphi(z) + \frac{1}{2\pi}\int_{\Sigma} \varphi(\zeta)\frac{\partial}{\partial \nu_{\zeta}} \log|z-\zeta|\, ds_{\zeta} . \tag{1.4}$$

This is the classical Fredholm integral equation of the 2nd kind whose kernel $K(z,\zeta) = \frac{1}{2\pi}\frac{\partial}{\partial \nu_{\zeta}}\log|z-\zeta|$ has a weak singularity, i.e. $K(z,\zeta) = O(|z-\zeta|^{\alpha-1})$ where α $(0 < \alpha \leq 1)$ is the Hölder exponent of the boundary Σ.

Eq.(1.4) has, for any $f \in C^{\circ}(\Sigma)$, one and only one solution $\varphi \in C^{\circ}(\Sigma)$ and (1.3) gives the solution of (1.1), (1.2) which is continuous in $\bar{\Omega} = \Omega \cup \Sigma$.

A second method for solving (1.1), (1.2), via integral equations, consists in representing u by a *simple layer potential*

$$u(z) = \frac{1}{2\pi} \int_{\Sigma} \varphi(\zeta) \log|z - \zeta| ds_{\zeta} . \tag{1.5}$$

Condition (1.2) yields to the integral equation of the first kind

$$\frac{1}{2\pi} \int_{\Sigma} \varphi(\zeta) \log|z - \zeta| ds_{\zeta} = f(z) \tag{1.6}$$

which for $f \in C^{\circ}(\Sigma)$ is a ill-posed problem of analysis. However if we assume more smoothness about f, for instance suppose $\partial f/\partial s$ uniformly Hölder continuous on Σ, we obtain the singular integral equation

$$\frac{1}{2\pi} \int_{\Sigma}^{*} \varphi(\zeta) \frac{\partial}{\partial s_z} \log|z - \zeta| ds_{\zeta} = \frac{\partial}{\partial s_z} f(z) \tag{1.7}$$

where $\int_{\Sigma}^{*}$ means that the integral must be understood like a singular integral over Σ in the Cauchy sense.

This equation by using the Muskhelishvili theory of singular integral equation has a solution which by (1.5) determines u up to an additive constant and hence the solution of (1.1), (1.2). The solution given by (1.5) belongs to $C^{1+\lambda}(\bar{\Omega})$ for some λ $(0 < \lambda < 1)$. This second method requires more strict hypothesis on the datum f, but it gives a more regular solution.

Both methods of double and simple layer have been extended to an elliptic equation of order 2m in the papers [1] and [8] .

We shall not discuss here the method of Agmon [1] of *multiple layer potential* for the elliptic equation

$$Eu \equiv \sum_{k=0}^{2m} a_k \frac{\partial^{2m} u}{\partial x^k \partial y^{2m-k}} = 0 \tag{1.8}$$

where $m > 1$ and the a_k's are real constants such that the polynomial

$$L(w) = \sum_{k=0}^{2m} a_k w^k$$

has only complex zeroes.

This method highly interesting, permits to solve the Dirichlet problem for the elliptic P.D.E. (1.8) in the class of $C^{m-1+\lambda}(\bar{\Omega})$ functions $(0 < \lambda < 1)$ provided Σ and the boundary data satisfy suitable smoothness assumptions.

In the paper [8] a definition was given for potential of simple layer for a large class of elliptic equations of order 2m with variable coefficients.

However we restrict ourselves here to the case of Eq.(1.8). To this end let us consider the same *fundamental solution* for (1.8) used by Agmon [1]. Let Γ be a closed smooth Jordan contour of the plane of the complex variable w which belongs to the half plane $\mathrm{Im}\, w < 0$ and encloses in the interior of the bounded domain having Γ like boundary all the zeroes of L(w) with negative imaginary part. Set

$$P(z,\zeta) = \tag{1.9}$$

$$= \frac{-1}{2\pi^2(2m-2)!}\, \mathrm{Re} \int_{+\Sigma} \frac{[(x-\xi)w + (y-\eta)]^{2m-2} \log[(x-\xi)w + (y-\eta)]}{L(w)}\, dw\,.$$

We assume for $\log[(x-\xi)w + (y-\eta)]$ the *principal branch*, i.e. $-\pi < \arg[(x-\xi)w + (y-\eta)] \le \pi$, which for $x \neq \xi$ is holomorphic in the w-plane cut along the straight half line $\mathrm{Im}\, w = 0$, $\mathrm{Re}\, w \le -\frac{y-\eta}{x-\xi}$.

The function $P(z,\zeta)$ is a monodromic function of z and ζ defined for $z \neq \zeta$ which is analytic in x, y, ξ, η. All the derivatives of $P(z,\zeta)$ can be obtained by differentiating under the integral sign. If we set $p = (p_1, p_2)$, $q = (q_1, q_2)$, $|p| = p_1 + p_2$, $|q| = q_1 + q_2$

$$D_z^p = \frac{\partial^{|p|}}{\partial x^{p_1}\, \partial y^{p_2}}\,, \qquad D_\zeta^q = \frac{\partial^{|q|}}{\partial \xi^{q_1}\, \partial \eta^{q_2}}$$

we have

$$D_z^p D_\zeta^q P(z,\zeta) \begin{cases} = O(|z-\zeta|^{2m-2-|p|-|q|} \log|z-\zeta|), & 0 \le |p|+|q| \le 2m-2 \\ = O(|z-\zeta|^{-1}) & , |p|+|q| = 2m-1\,. \end{cases}$$

Moreover $E_z P(z,\zeta) = E_\zeta P(z,\zeta) = 0$ and for $g(z) \in C^{0+\mu}(\bar{\Omega})$ the function

$$v(z) = \iint_\Omega g(\zeta)\, P(z,\zeta)\, d\xi\, d\eta$$

belongs to $C^{2m+\lambda}(\bar{\Omega})$ for some λ $(0 < \lambda < 1)$ and $Ev = g$.

2. *A potential theoretic result.*

Let φ be a function uniformly Hölder continuous on Σ. We denote by $x = x(s)$ $y = y(s)$ the parametric representation of Σ, where s is the counter-clockwise increasing arc-length. By $\dot{x}$, $\dot{y}$ we denote differentiation with respect to s. Let $z = x + iy$ be any fixed point of Σ. The following limit relation holds (see [8] p.65)

$$\lim_{\tilde z \to z} \int_\Sigma \varphi(\zeta) \frac{\partial^{2m-1}}{\partial \tilde x^{2m-1-k}\, \partial \tilde y^k} P(\tilde z, \zeta) ds_\zeta =$$

$$= -\varphi(z) \frac{1}{2\pi} Im \int_{+\Gamma} \frac{w^{2m-1-k}}{L(w)(\dot x w + \dot y)} dw \tag{2.1}$$

$$-\frac{1}{2\pi^2} Re \int_\Sigma^* \varphi(\zeta)\, ds_\zeta \int_{+\Gamma} \frac{w^{2m-1-k}}{L(w)\ (x-\xi)w + (y-\eta)} dw$$

$$(0 \le k \le 2m-1)$$

when $\tilde z = \tilde x + i\tilde y$ tends to z remaining in the interior of Ω. This limit is uniform with respect to z.

The limit relation (2.1) contains all the limit relations connected with line potentials in the classical two-dimensional potential theory.

3. *Potential of simple layer for Eq.(1.6). Dirichlet problem.*

In [8] the *potential of simple layer* for Eq.(1.6)[(1)] is defined as follows

$$u(z) = \sum_{k=0}^{m-1} \int_\Sigma \varphi_k(\zeta) \frac{\partial^{m-1}}{\partial \xi^{m-1-k}\, \partial \eta^k} P(z,\zeta) ds_\zeta \qquad (z \in \Omega).$$

It is easily seen that for m=1 and $L(w) = w^2 + 1$ (3.1) essentially reduces to (1.5).

Suppose we wish to solve the Dirichlet problem

$$Eu = 0 \quad \text{in} \quad \Omega \tag{3.2}$$

$$\frac{\partial^{m-1} u}{\partial x^{m-1-h}\, \partial y^h} = f_h \quad \text{on } \Sigma \quad (h=0,..,m-1) . \tag{3.3}$$

We represent u by (3.1) and assume that $\frac{\partial}{\partial s} f_k$ is uniformly Hölder continuous on Σ . Let us impose the boundary conditions

$$\frac{\partial}{\partial s_z} \frac{\partial^{m-1} u}{\partial x^{m-1-h}\, \partial y^h} \frac{\partial f_h}{\partial s_z} = \Psi_h \quad (h=0,..,m-1) . \tag{3.4}$$

Using (2.1) we get

---

(1) Actually in [8] a much larger class of elliptic equations is considered.

$$(-1)^{m-1}\, 2\pi\ \Psi_h(z) = \sum_{k=0}^{m-1} \left\{ a_{hk}\, \varphi_k(z) + \frac{b_{hk}}{\pi i} \int^{*}_{+\Sigma} \frac{\varphi_k(\zeta)}{\zeta - z}\, d\zeta \right. \tag{3.5}$$
$$\left. + \int_{+\Sigma} \varphi_k(\zeta)\, M_{hk}(z,\zeta)\, d\zeta \right\} \quad (h=0,\ldots,m-1)$$

where

$$a_{hk} = -\,Im \int_{+\Gamma} \frac{w^{2m-2-(h+k)}}{L(w)}\, dw\ , \qquad b_{hk} = i\, Re \int_{+\Gamma} \frac{w^{2m-2-(h+k)}}{L(w)}\, dw$$

$$M_{hk}(z,\zeta) = O(|z-\zeta|^{\alpha-1}) \qquad (0 < \alpha \leq 1).$$

The system (3.5) is a system of singular integral equations of the following kind:

$$\Psi(z) = A(z)\, \Phi(z) + B(z)\, S\Phi + M\Phi \tag{3.6}$$

where $\Psi(z) \equiv \left[(-1)^{m-1}\, 2\pi\ \Psi_o(z), \ldots, (-1)^{m-1}\, 2\pi\ \Psi_{m-1}(z)\right]$, $A(z)$ and $B(z)$ are $m\times m$ matrices with uniformly Hölder continuous entries (actually, in the case of (3.5), A and B are constant) $S\Phi$ is the singular integral vector operator

$$S\Phi \equiv \left[\frac{1}{\pi i} \int^{*}_{+\Sigma} \frac{\varphi_o(\zeta)}{\zeta - z}\, d\zeta\, , \ldots, \frac{1}{\pi i} \int^{*}_{+\Sigma} \frac{\varphi_{m-1}(\zeta)}{\zeta - z}\, d\zeta\right]$$

and $M(\Phi)$ is a matrix Fredholm integral operator.

The system (3.6) is of *regular type*, according to the Muskhelishvili theory, when for any $z \in \Sigma$

$$\det(A - B)\, \det(A + B) \neq 0\, .$$

Let us denote by $K\Phi$ the operator on the right hand side of (3.6) and set for $U \equiv (u_o, \ldots, u_{m-1})$, $V \equiv (v_o, \ldots, v_{m-1})$

$$(U,V) = \sum_{k=0}^{m-1} u_k\, v_k\ .$$

Let $K^*\Phi$ be the operator such that for any Φ and Ψ

$$(K\Phi\, , \Psi) = (\Phi\, , K^*\Psi).$$

If (3.6) is of regular type then the following results hold (see [13]):

i) The homogeneous systems

$$K\Phi = 0\ , \tag{3.7}$$

$$K^* Z = 0 \tag{3.8}$$

have only finite sets of independent eigensolutions.

ii) System (3.6) has a solution when and only when $(\Psi, Z) = 0$ for any eigensolution Z of (3.8).

iii) The *index* $\varkappa$ of the system (3.6) i.e. the difference between the dimensions of the spaces of solutions of (3.7) and (3.8), respectively, is given by (Muskhelishvili's formula)

$$\varkappa = \frac{1}{2\pi i} \left[\log\left[\frac{\det A - B}{\det A + B}\right]\right]_{+\Sigma}$$

where $[\quad]_{+\Sigma}$ denotes the jump of the function between brackets after a counter-clockwise tour along Σ.

The main results of the theory hinge on the fact that the operator K is *reducible*. This means that an operator K' exists such that $K'K = I + T$, where I is the identity operator and T a compact operator (in this particular connection an integral Fredholm operator).

The particular system (3.5) is of regular type (see [8]) and the conditions ii) are satisfied when the f_h are the boundary values of the (m-1)th derivatives of a $C^m(\bar{\Omega})$ function. If Φ is any solution of (3.5) the simple layer potential (3.1) gives a solution of the problem (3.2),(3.4) determined up to an additive arbitrary polynomial in x and y of degree m-1. This solution belongs to $C^{m+\lambda}(\bar{\Omega})$ for some λ $(0 < \lambda < 1)$.

The method we have briefly described was introduced like a procedure of pure mathematical analysis and, except for a simple application, shown in [8], to plane elasticity, no connections were suspected with more general applications and with numerical analysis. However some researchers found out, later, that these connections exist and several papers were produced in this direction. Very active in this respect have been Robert Gilbert, George Hsiao and Wolfgang Wendland (2). The first paper on this subject was due to R.C. Maccamy [12] . Of main interest for connections with numerical analysis is the paper

---

(2) For complete bibliographies we refer to the papers of G.Hsiao and of W.Wendland in this Symposium .

[10] by G.Hsiao and R.C.Maccamy. These Authors consider the following elliptic equation

$$\Delta_2^m u - \sigma \Delta_2^{m-1} u = 0 \qquad \left(\Delta_2 = \frac{\partial^2}{\partial x^2} + \frac{\partial^2}{\partial y^2} \right)$$

and they cannot use the fundamental solution $P(z,\zeta)$ given by (1.9), but the one which is possible to construct by using Bessel functions.

Let us quote a few sentences from paper [10].

"*It has been known for some time that in two dimensions one can also solve the Dirichlet problem for Laplace's equation with a simple layer potential [13]. This procedure leads to singular integral equations of the first kind and hence has not been pursued. It was observed by Fichera [8], however, that this second method generalizes much more readily to higher order equations. It is the purpose of this paper to discuss Fichera's method. We present the method in a highly specialized situation which is nevertheless of quite wide physical interest. This permits a semplification of the method and seems to illuminate both its advantages and disadvantages as compared to the method of Fredholm. For the fourth order problem we consider a version of the Fredholm method could be developed by using ideas of Agmon [1]. However, we believe that Fichera's method is simpler in these cases and it does have wider applicability... There are obvious disadvantages of Fichera's method.... it leads in fact to singular integral equations with Caucgy kernels....however, we show that in the applications here the Cauchy kernels can be eliminated and replaced with logorithmic ones....*"

An important remark made by Hsiao and Maccamy is the following:

"*A more serious drawback to Fichera's method is that it does not seem to generalize easily to boundary conditions other than those of Dirichlet type.*"

Paolo Emilio Ricci in his important paper [14] has shown how to overcome this difficulty and how to handle general boundary conditions by the simple layer potential approach.

4. *P.E. Ricci's results.*

Let us consider the boundary operators on Σ

$$B_h u = \sum_{j=0}^{m} b_h^j(z) \frac{\partial^m u}{\partial x^{m-j} \partial y^j} + \tilde{B}_h u$$

$$\tilde{B}_h u = \sum_{p=0}^{m=1} \sum_{q=0}^{m-1-p} b_h^{m-1-p,q}(z) \frac{\partial^{m-1-p} u}{\partial x^{m-1-p-q} \partial y^q}$$

$$(h=0,\dots,m-1)$$

where all the coefficients $b_h^j(z)$, $b_h^{m-1-p,q}(z)$ belong to $C^{o+\mu}(\Sigma)$. The boundary value problem considered by P.E.Ricci [14] is the following:

$$Eu = 0 \quad \text{in } \Omega , \tag{4.1}$$

$$B_h u = \psi_h \quad \text{on } \Sigma \quad (h=0,\dots,m-1) . \tag{4.2}$$

It is evident that problem (3.2),(3.4) is a very particular case of (4.1),(4.2). In handling with problem (4.1), (4.2) it is of fundamental interest the following

Lopatinski condition: Let $L^-(w)$ the polynomial of degree m whose zeroes are the zeroes of L(w) with negative imaginary part. Let us consider the polynomial

$$L_h(w,z) = \sum_{j=0}^{m} b_h^j(z) w^{m-j} \qquad (h=0,\dots,m-1).$$

We say that the operators B_h satisfy the Lopatinski condition with respect to E if for no $z \in \Sigma$ and for no choice of the complex coefficients $c_o,\dots,c_{m-1}$ ($|c_o| + \dots + |c_m| > 0$) the polynomial $c_o L_o(w,z) + \dots + c_{m-1} L_{m-1}(w,z)$ is divisible for $L^-(w)$.

As we know the Lopatinski condition, stated in a slightly different but equivalent algebraic form, plays an important role in the theory of boundary value problems for elliptic equations ([11],[2]).

By representing u by (3.1) and imposing the boundary conditions (4.1), one,by using (2.1), gets a system like (3.6) where $A(z) = ((a_{hk}(z)))$, $B(z)=((b_{hk}(z)))$ $(h,k = 0,\dots,m-1)$ and

$$a_{hk}(z) = - Im \int_{+\Gamma} \frac{L_h(w,z)\, w^{m-1-k}}{L(w)(xw+y)} dw , \tag{4.3}$$

$$b_{hk}(z) = i\,Re \int_{+\Gamma} \frac{L_h(w,z)\, w^{m-1-k}}{L(w)(xw+y)}\, dw \quad . \qquad (4.4)$$

P.E. Ricci proves the important theorem:

The system (3.6) with A and B given by (4.3),(4.6) is of regular type if and only if the boundary operators B_h *satisfy the Lopatinski condition with respect to E.*

This permits to apply the Muskhelishvili theory to the system (3.6). When the Ψ_h are uniformly Hölder continuous on Σ and satisfy the compatibility conditions provided by this theory, one gets solutions of (4.1),(4.2) belonging to $C^{m+\lambda}(\Sigma)$ $(0 < \lambda < 1)$.

It must be remarked that the theory has been fully considered for the equation

$$\Delta_2 u + b_1(z)\frac{\partial u}{\partial x} + b_2(z)\frac{\partial u}{\partial y} + c(z)u = 0$$

with a general first order boundary operator, under the very general conditions $b_1(z)$, $b_2(z)$, $c(z) \in C^{o+\lambda}(\bar{\Omega})$, by Alberto Cialdea (see [3],[4],[5],[6]).

5. *Simple layer potentials of order n.*

The concept of simple layer potential for the elliptic Eq.(1.8) was generalized in the paper [9] in order to handle boundary conditions expressed by differential operators of order m + n with an arbitrary n.

Set for any nonnegative integer n

$$P_n(z,\zeta) =$$

$$= \frac{-1}{2\pi^2(2m+n-2)!}\, Re \int_{+\Gamma} \frac{[(x-\xi)w + (y-\eta)]^{2m+n-2} \log[(x-\xi)w+(y-\eta)]}{L(w)}\, dw \, .$$

The same choice is made for the branch of $\log[(x-\xi)w + (y-\eta)]$ like for $P(z,\zeta) = P_o(z,\zeta)$.

$P_n(z,\zeta)$ is the *fundamental solution of order n* of Eq.(1.7). The function

$$u(z) = \sum_{k=0}^{m-1} \int_\Sigma \varphi_k(z) \frac{\partial^{m-1}}{\partial\xi^{m-1-k}\,\partial\eta^k} P_n(z,\zeta)\, ds_\zeta \qquad (5.1)$$

is defined like a *potential of simple layer of order n* for the Eq.(1.7).

If the φ_k's are uniformly Hölder continuous on Σ, u is a solution of (1.7)

belonging to $C^{m+n+\lambda}(\bar{\Omega})$ $(0 < \lambda < 1)$.
Let us consider the boundary operators

$$B_h^{(n)} u = \sum_{j=0}^{m+n} b_h^j(z) \frac{\partial^{m+n} u}{\partial x^{m+n-j} \partial y^j} + \tilde{B}_h^{(n)} u$$

$$\tilde{B}_h^{(n)}(u) = \sum_{p=0}^{m+n-1} \sum_{q=0}^{m+n-1-p} b_h^{m+n-1-p,q}(z) \frac{\partial^{m+n-1-p}}{\partial x^{m+n-1-p-q} \partial y^q}$$

$$(h=0,\ldots,m-1)$$

whose coefficients $b_h^j(z)$, $b_h^{m+n-1-p,q}(z)$ are uniformly Hölder continuous on Σ.
Consider the boundary value problem

$$E u = 0 \qquad (5.2)$$

$$B_h^{(n)} u = \Psi_h \qquad (h=0,\ldots,m-1) \qquad (5.3)$$

and suppose that the Ψ_h's are uniformly Hölder continous on Σ.
If we represent u by (5.1) we get a singular integral system like (3.6). The Ricci theorem can be extended to this more general case, i.e. the integral system (3.6) connected with the B.V.P. (5.2),(5.3) is of regular type if and only if the boundary operators $B_h^{(n)}$ satisfy the Lopatinski condition with respect to E. In other words, if we set

$$L_h^{(n)}(w,z) = \sum_{j=0}^{m+n} b_h^j(z)\, w^{m+n-j} ,$$

the relevant integral system is of regular type if and only if for no $z \in \Sigma$ and for no choice of the complex constants $c_0,\ldots,c_{m-1}$ the polynomial $c_0 L_0^{(n)}(w,z) + \ldots + c_{m-1} L_{m-1}^{(n)}(w,z)$ is divisible for $L^-(w)$.
Theoretical and applied opportunities offered by the method of simple layer potentials of order n have not been yet fully exploited.

6. *Extension to the space R^n. The research work of A.Cialdea.*

Extension of the theory summarized in the previous Section to elliptic problems in a domain of $\mathbb{R}^n$ space is anything but an easy task. Some important results have, however, been obtained by Alberto Cialdea in this connection.

Let us first consider the possibility of extending the simple layer potential approach for the operator Δ_2 to the space $\mathbb{R}^n$.

Let us consider the classical n dimensional potential of simple layer

$$u(x) = \frac{-1}{(n-2)\,\omega_n} \int_\Sigma \varphi(y) \frac{1}{|x-y|^{n-2}} \, d\sigma_y \tag{6.1}$$

where ω_n is the measure of the sphere $|x| = 1$ and Σ is the Liapounov (connected) boundary of the bounded domain Ω of $\mathbb{R}^n$ $(n>2)$.

If we wish to solve the n-dimensional Dirichlet problem (1.1),(1.2),in order to avoid the first kind integral equation analogous to (1.6) we cannot use (1.7). However we have this new kind of singular integral equation

$$\frac{-1}{(n-2)\,\omega_n} \int_\Sigma^* \varphi(y) d_x \frac{1}{|x-y|} \, d\sigma_y = d_x f \equiv \Psi(x) \tag{6.2}$$

where d_x denotes the differential of the relevant function of x on Σ . The idea is to bild a theory of singular integral equations such that while the "datum" is a differential form on Σ like $\Psi = d_x f$, the "unknown" is a scalar function φ . This is an old idea of the present writer who,with this in mind, in his lecture notes of an advanced course delivered in Rome in 1963,proposed an abstract theory of "reducible"linear operators which map a Banach space B_0 into a different one B_1 .

The application to the integral equation (6.2) has been adroitly carried out by Alberto Cialdea [7] .

He consider the operator given by the left hand side of (6.2); let us denote it by $S\varphi$. Set

$$S'\Psi = \circledast \frac{4}{(n-2)\,\omega_n} \int_\Sigma^* \Psi(x) \wedge \sum_{j_1<\cdots<j_{n-2}} \left[d_z |z-x|^{2-n} dz_{j_1} \ldots dz_{j_{n-2}} \right] dx_{j_1} \ldots dx_{j_{n-2}}$$

which maps the 1-form $\Psi(x)$ into the scalar $S'\Psi$; $\circledast$ is the *adjointness operator* for differential forms on Σ .

Cialdea proves that S' *reduces* S. In fact one has

$$S'S\varphi = \varphi(z) + \int_\Sigma \varphi(y) L(z,y) \, d\sigma_y$$

where L(z,y) is a *Fredholm kernel*, i.e. has a weak singularity for $z = y$. Hence equation $S\varphi = \psi$ has a solution if and only if ψ is orthogonal to any eigensolution of the equation $S^*\gamma = 0$. This condition is satisfied when and only when ψ is weakly homologous to zero on Σ. The theory can be carried out considering φ and ψ uniformly Hölder continuous on Σ or, alternatively, belonging to the relevant $L^p(\Sigma)$ spaces $(1 < p < \infty)$.

This permits to extend the simple layer potential approach to the Dirichlet problem for $\Delta_2 u = 0$ in $\mathbb{R}^n$.

As far as higher order equations are concerned, Cialdea considers the B.V.P.

$$\Delta_2^2 u = 0 \quad \text{in } \Omega , \tag{6.3}$$

$$\frac{\partial u}{\partial x_h} = f_h \quad \text{on } \Sigma \qquad (h=1,..,n) \tag{6.4}$$

and defines as potential of simple layer for the operator Δ_2^2 the following

$$u(x) = \sum_{k=1}^{n} \int_{\Sigma} \varphi_k(y) \frac{\partial}{\partial y_k} F(x,y) d\sigma_y \tag{6.5}$$

where

$$F(x,y) \begin{cases} = \dfrac{4}{(n-4)(n-2)\,\omega_n} |x - y|^{4-n} & n \neq 4 \\[2ex] = -\dfrac{2}{\omega_4} \log |x - y| & n = 4. \end{cases}$$

Eq. (6.4) lead him to the singular system

$$Q\varphi \equiv \sum_{k=1}^{n} \int_{\Sigma}^{*} \varphi_k(y)\, d_x \left[\frac{\partial^2}{\partial x_h\, \partial y_k} F(x,y)\right] d\sigma_y = d_x f_h .$$

The operator Q on the left hand side transforms the vector $\varphi \equiv (\varphi_1,\dots,\varphi_n)$ into an ordered set of n 1-forms. Unfortunately this operator is not *reducible* since its null-space is infinite-dimensional.[3]

However Cialdea succeeds in constructing a reducible operator H such that its range coincides with the range of S. From this he gets the Fredholm alternative for the equation $Q\varphi = \psi$ and a fully treatment of problem (6.3),(6.4) via the

---

(3) A necessary condition for an operator to be reducible is to have a finite-dimensional null-space.

simple layer potential (6.5). Cialdea results will be soon published. These outstanding results of Cialdea make to believe that the simple layer approach could be extended to general situations for elliptic B.V.P. in $\mathbb{R}^n$. This research field looks extremely appealing.

References.

1. Agmon, S.: Multiple Layer Potential and the Dirichlet Problem for Higher Order Elliptic Equations in the Plane. Comm.on Pure and Appl.Mathem.X,2 1957 179-239.

2. Agmon, S.; Douglis, A. ; Nirenberg, L. : Estimates Near the Boundary for Solution of Elliptic Partial Differential Equations Satisfying General Boundary Conditions I. Comm. on Pure and Appl. Mathem. XII,4 1959 623-727 and II, ibid. XVII,1 1964 35-92.

3. Cialdea, A. : L'equazione $\Delta_2 u + a_{10}(x,y)\frac{\partial u}{\partial x} + a_{01}(x,y)\frac{\partial u}{\partial y} + a_{00}(x,y)u = F(x,y)$. Teorema di esistenza per un generale problema al contorno. Rend. Acc. Naz. Lincei VIII,80,3 1986 89-99.

4. Cialdea, A. : L'equazione $\Delta_2 u + a_{10}(x,y)\frac{\partial u}{\partial x} + a_{01}(x,y)\frac{\partial u}{\partial y} + a_{00}(x,y)u = F(x,y)$. Calcolo dell'indice dei problemi al contorno e soluzioni deboli. Rend. Acc. Naz. Lincei VIII,80,4 1986 185-195.

5. Cialdea, A. : L'equazione $\Delta_2 u + a_{10}(x,y)\frac{\partial u}{\partial x} + a_{01}(x,y)\frac{\partial u}{\partial y} + a_{00}(x,y)u = F(x,y)$. Formole di maggiorazione relative a problemi al contorno. Rend. Acc. Naz. Lincei VIII,80 1986 510-524.

6. Cialdea, A. : L'equazione $\Delta_2 u + a_{10}(x,y)\frac{\partial u}{\partial x} + a_{01}(x,y)\frac{\partial u}{\partial y} + a_{00}(x,y)u = F(x,y)$. Teoremi di completezza. Rend.Acc.Naz. Lincei VIII,81 1987 245-257.

7. Cialdea, A. : Sul problema della derivata obliqua per le funzioni armoniche e questioni connesse. Rend. Acc. Scienze detta dei XL, 106, XII 1988 181-200.

8. Fichera, G. : Linear elliptic equations of higher order in two independent variables and singular integral equations, with applications to anisotropic inhomogeneous elasticity. Partial Diff. Equat. and Continuum Mech. edited by R.E.Langer, Madison The Univ. of Wisconsin Press 1961 55-80.

9. Fichera,G. ; Ricci, P.E. : The sigle layer potential approach in the theory of boundary value problems for elliptic equations. Function Theoretic Methods for Part. Diff. Equat. Darmstadt 1976, Lecture Notes in Mathem. 561 Springer Verlag Berlin Heidelberg New York 40-50.

10. Hsiao, G.; Maccamy, R.C. : Solution of Boundary Value Problems by Integral Equations of the First Kind. SIAM Review 15,1 1973 687-705.

11. Lopatinski, Y.B. : On a method of reducing boundary problems for a system of differential equations of elliptic type to regular equations.Ukrain Mat. Zurn. 5 1953 123-151.

12. Maccamy, R.C. : On a class of two-dimensional Stokes flows. Arch. for Rat. Mech. & Anal. 21 1966 256-258.

13. Muskhelishvili, N.I. : Singular Integral Equations. 2nd Ed. Transl. edit. by J.R.M. Radok, P.Noordhoff N.V. Groningen 1953.

14. Ricci, P.E. : Sui potenziali di semplice strato per le equazioni ellittiche di ordine superiore in due variabili. Rend. di Matem. e delle sue appl 7,VI 1974 1-39.

Variational Methods for BEM

W. L. Wendland

University of Stuttgart

1. *Introduction*

As we all know, variational methods and formulations are basic for a big variety of problems in mechanics [12]. Their exploitation in connection with finite element approximation has created some of the most powerful algorithms in computational mechanics, the finite element methods.

For boundary element methods, variational formulations and corresponding coerciveness properties are also the fundamental tools for successful algorithms and their analysis. This point of view is here not so obvious since one usually struggles first with the reduction to the boundary, with the Gauss-Green theorems and with potentials of boundary distributions whose integral kernels have singularities of various orders. Nevertheless, variational formulations reveal the boundary element approximations of boundary integral equations to be closely related to the principles of virtual work and least energy. This lecture is devoted to show some of these relations. In particular it is intended to show the unifying character of variational principles for boundary element methods which allow the formulation and analysis of several different aspects as coupling with finite elements, coercitivity properties in connection with the energy of the equilibrium states considered, domain decomposition and error estimates. We shall consider here only elliptic problems which correspond to stationary states and we shall concentrate on the general scheme. Therefore, many extremely important problems of boundary element methods are left out - as numerical integration, solution methods for the discrete equations, time dependent problems - to name just a few. Moreover, it is not claimed that anything in this lecture is really new. The relations between boundary value problems and boundary integral equations are known for more than 150 years; C.F. Gauss already used the boundary integral equation of the first kind for solving the Dirichlet problem for the Laplacian and since, the boundary integral equations of various types were used to solve and to analyze elliptic boundary value problems.

The reduction of general elliptic boundary value problems to boundary integral equations can be found in detail in [11], for higher dimensions we find many general results in Mikhlin's work, see [31], also for references.

Here we shall follow ideas which can be found in the work by Nedelec [32, 34] and Hsiao and Wendland [21] where the coercivity of the Dirichlet bilinear forms in the interior and the exterior was used for proving coerciveness of boundary integral equations of the first kind. Costabel and Wendland in [9] applied this approach to general even order regular elliptic boundary value problems.

For brevity and simplicity let us start with the simple Dirichlet problem for the Laplacian,

$$\begin{aligned} \Delta u &= 0 \quad \text{in} \quad \Omega\,, \\ u|_\Gamma &= \varphi \quad \text{on} \quad \Gamma = \partial\Omega \end{aligned}$$

where Ω denotes a given domain in $\mathbb{R}^n$, $n = 2, 3$, interior (or exterior) to a sufficiently smooth boundary Γ. Then with the well known fundamental solution $E(x, y)$ of the Laplacian in $\mathbb{R}^n$, the solution admits the Green representation for

$$x \in \Omega: \; u(x) = \int_\Gamma E(x,y)\lambda(y)ds_y - \int_\Gamma (\partial_{ny}E(x,y))^T u(y)ds_y \tag{1}$$

in terms of a single and a double layer potential whose boundary distributions $\lambda = \partial_n u|_\Gamma$ and $u|_\Gamma$ are the Cauchy data of the harmonic solution u. If we take the boundary values on both sides of (1) on Γ and also the normal derivative with respect to the exterior normal n of Γ then with corresponding jump relations we find for $x \in \Gamma$ the two equations

$$\begin{aligned} u &= V\lambda + \left(\frac{1}{2}I - K\right)u\,, \\ \lambda &= \left(\frac{1}{2}I + K'\right)\lambda + Du \end{aligned} \tag{2}$$

where

$$Du = -\partial_{nx}\int_\Gamma (\partial_{ny}E(x,y))^T u(y)ds_y\,. \tag{3}$$

For a solution u of $\Delta u = 0$ in Ω and $\lambda = \partial_n u|_\Gamma$, the operators on the right hand side of (2) reproduce $u|_\Gamma$ and λ and, hence, define a projection, the Calderon projection.

If $u|_\Gamma = \varphi$ is given then it is already sufficient to use only one of the two equations in (2) in order to find λ and with (1) the solution to the boundary value problem. Hence, the Dirichlet problem is reduced to the following:

For given φ, find λ such that

$$V\lambda = \frac{1}{2}\varphi + K\varphi\,. \tag{4}$$

This is a Fredholm integral equation of the first kind.

Instead of (4) we can also use

$$\frac{1}{2}\lambda - K'\lambda = D\varphi \tag{5}$$

which is a Fredholm integral equation of the second kind. Although in classical analysis the equation (5) was preferred over (4), the latter will show much closer relation to the Dirichlet principle, i.e. the variational formulation of the boundary value problems in Ω.

2. *Variational Formulation*

We begin with the weak formulation of boundary integral equations by multiplying equation (4) with test functions and integrating over Γ. The space of desired admissible boundary functions let us denote by S. Then the weak formulation of (4) reads as to find $\lambda \in S$ such that

$$\forall\, \chi \in C^\infty(\mathbb{R}^n): \quad \langle \chi, V\lambda\rangle_\Gamma \;=\; \ell_\varphi(\chi)\ . \tag{6}$$

With (6) there is associated the following bilinear form on $\chi \in C^\infty$ and $\lambda \in S$

$$b(\chi,\lambda) : = \langle \chi, V\lambda\rangle_\Gamma \;=\int_\Gamma \chi(x)\int_\Gamma E(x,y)\lambda(y)ds_y ds_x \tag{7}$$

and with given φ, the linear functional

$$\ell_\varphi(\chi) : = \langle \chi, \frac{1}{2}\varphi + K\varphi\rangle_\Gamma \;= \int_\Gamma \chi(x)\left\{\frac{1}{2}\varphi(x) + \int_{\Gamma\setminus\{x\}} (\partial_{n_y} E)^T \varphi(y) ds_y\right\} ds_x\ . \tag{8}$$

If both, the bilinear and the linear functional can be extended from $\chi \in C^\infty$ to the space S of admissible functions - preferably by continuity - then instead of (4) or (6) we may deal with the Variational Form:

Find $\qquad \lambda \in S\ \forall\, \chi \in S : b(\chi,\lambda) = \ell_\varphi(\lambda)\ .$

For a simple treatment of the variational form in the framework of the Lax-Milgram theorem we would like S to be a Hilbert space, b and ℓ_φ to be continuous on S, b to be symmetric,

$$b(\chi,\lambda) = b(\lambda,\chi) \tag{9}$$

such that the variational solution is equivalent to the minimum problem:

Find $\lambda \in S$ *with*

$$\frac{1}{2}b(\lambda,\lambda) - \ell_\varphi(\lambda) = \inf_{\chi\in S}\left\{\frac{1}{2}b(\chi,\chi) - \ell_\varphi(\chi)\right\}\ . \tag{10}$$

The Lax-Milgram theorem is valid, if e.g. in addition, the bilinear form is $\mathcal{S}$-elliptic, i.e. it satisfies the coerciveness inequality

$$b(\chi,\chi) \geq \gamma_o \|\chi\|_{\mathcal{S}}^2 \tag{11}$$

for all $\chi \in \mathcal{S}$ where $\gamma_o > 0$ is a fixed constant. Hence, we are primarily interested in specializing the Hilbert space $\mathcal{S}$ and in finding boundary integral equations which provide the simple variational formulation as above.

3. *Strongly Elliptic Transmission Problems and 1st Kind BIE*

In order to see a more systematic approach to the foregoing formulation we first consider the single layer potential

$$\Lambda(x) := \int_\Gamma E(x,y)\lambda(y)ds_y = \begin{cases} \Lambda_i & \text{in} \quad \Omega_i, \\ \Lambda_e & \text{in} \quad \Omega_e \end{cases} \tag{12}$$

which decays like $0\left(\frac{1}{|x|}\right)$ for $|x| \to \infty$ in the exterior domain Ω_e. If λ is smooth enough then the potential is continuous across Γ and

$$\Lambda_i = \Lambda_e = V\lambda \qquad \text{on} \qquad \Gamma\,. \tag{13}$$

The classical jump relations [25] for the normal derivatives from Ω_i and Ω_e in the direction of the exterior normal to Γ yield the relation

$$\lambda = \partial_n \Lambda_i - \partial_n \Lambda_e \qquad \text{on} \qquad \Gamma\,. \tag{14}$$

Hence, we find with Green's formula in Ω_i and in Ω_e that the bilinear form (7) associated with the single layer potential is already non-negative; for any λ smooth enough we have

$$\begin{aligned} b(\lambda,\lambda) &= \langle \partial_n \Lambda_i, \Lambda_i \rangle - \langle \partial_n \Lambda_e, \Lambda_e \rangle \\ &= \int_{\Omega_i} |\nabla \Lambda_i|^2 dx + \int_{\Omega_e} |\nabla \Lambda_e|^2 dx \geq 0\,. \end{aligned} \tag{15}$$

Moreover, if the two domain integrals could serve as the square of a norm, we would have positive definiteness right away. Therefore let us define the Hilbert space of functions being harmonic in Ω_i and Ω_e,

$$\begin{aligned} \mathcal{W} := \quad & \Big\{ \Lambda \mid \Lambda_{|\Omega_i} \in H^1(\Omega_i) \wedge \Lambda_{|\Omega_e} \in H^1_{\frac{1}{R}}(\Omega_e) \wedge \\ & \Lambda_i = \Lambda_e \text{ on } \Gamma \wedge \Delta\Lambda = 0 \text{ in } \mathbb{R}^3 \backslash \Gamma \wedge \\ & \infty > \|\Lambda\|_{\mathcal{W}}^2 := \int_{\Omega_i} \left(|\nabla\Lambda|^2 + |\Lambda|^2 \right) dx + \int_{\Omega_e} \left(|\nabla\Lambda|^2 + |\frac{\Lambda}{|x|}|^2 \right) dx \Big\}\,. \end{aligned} \tag{16}$$

From (15) then follows:

Lemma 1 *If $\Lambda \in \mathcal{W}$ and λ is given by* (14) *then*

$$b(\lambda,\lambda) \geq \gamma \|\Lambda\|^2_{\mathcal{W}} \geq \gamma' \|V\lambda\|^2_{H^{\frac{1}{2}}(\Gamma)} \, . \tag{17}$$

In fact, the first inequality in (17) corresponds to the famous Trefftz principle [49], whereas the right inequality is due to the trace theorem [29]. The norm on the trace space can be defined by

$$\|\chi\|^2_{H^{\frac{1}{2}}(\Gamma)} = \int_\Gamma |\chi|^2 ds + \int_\Gamma \int_\Gamma \frac{|\chi(x)-\chi(y)|^2}{|x-y|^n} ds_x ds_y \tag{18}$$

where

$$\Gamma \subset \mathbb{R}^n; \quad \dim\Gamma = n-1 \, . \tag{19}$$

The space for λ is given by the (distributional) normal derivatives of harmonic functions

$$\Lambda \in H^1_{loc}\left(\Omega_{i|e}\right) \wedge \Delta\Lambda = 0 \tag{20}$$

then

$$\partial_n \Lambda_i, \partial_n \Lambda_e \in H^{-\frac{1}{2}}(\Gamma) \tag{21}$$

where

$$\|\lambda\|_{H^{-\frac{1}{2}}(\Gamma)} = \sup_{\|\chi\|_{H^{\frac{1}{2}}(\Gamma)} \leq 1} |\langle \lambda, \chi \rangle_\Gamma| \, . \tag{22}$$

Then inequality (17) yields

Theorem 2: *For $n=2$ let us suppose* $\operatorname{diam}(\Omega_i) < 1$. *(For $n=3$ we do not need an additional assumption). Then the bilinear form* (7) *is $H^{-\frac{1}{2}}(\Gamma)$ – elliptic and continuous on $H^{-\frac{1}{2}}(\Gamma) \times H^{-\frac{1}{2}}(\Gamma)$: there exists $\gamma_0' > 0$ and*

$$b(\lambda,\lambda) \geq \gamma_o' \|\lambda\|^2_{H^{-\frac{1}{2}}(\Gamma)} \qquad \text{for all } \lambda \in \mathcal{S} = H^{-\frac{1}{2}}(\Gamma) \, . \tag{23}$$

The proof of Theorem 2 is due to Nedelec and Planchard [34] and Hsiao and Wendland [21]. Theorem 2 holds under rather weak assumptions [6]: Γ only needs to be a Lipschitz boundary.

A similar result holds for the hypersingular operator D in (3).

Theorem 3: *The bilinear form associated with the hypersingular operator D in* (3) *can be transformed as*

$$b_D(\Psi,\mu) := \langle \Psi, D\mu \rangle_\Gamma = \begin{cases} \left\langle \dfrac{d\Psi}{ds}, V\left(\dfrac{d\mu}{ds}\right) \right\rangle_\Gamma, & \text{for } n=2 \\ \left\langle \left(\vec{n} \times \nabla\Psi\right), V\left(\vec{n} \times \nabla\mu\right) \right\rangle_\Gamma, & \text{for } n=3. \end{cases} \tag{24}$$

Moreover, b_D is $H^{\frac{1}{2}}(\Gamma)/\mathbb{R}$ – elliptic:

$$b_D(\mu,\mu) \geq \gamma_o''\|\mu\|^2_{H^{\frac{1}{2}}(\Gamma)} \quad \text{for all} \quad \mu \in H^{\frac{1}{2}}(\Gamma) \quad \text{with} \int_\Gamma \mu ds = 0 \,. \tag{25}$$

γ_o'' is a positive constant depending on Γ.

The formula (24) was proved by Maue [30] whereas the coerciveness estimate (25) can be found in [33] and also holds for Lipschitz boundaries Γ [6].

As a consequence, both traditional boundary value problems for the Laplacian, the Dirichlet problem and the Neumann problem can be reduced to the solution of corresponding boundary integral equations of the first kind in variational form and the associated bilinear forms are S-elliptic with $S = H^{-\frac{1}{2}}(\Gamma)$ for the Dirichlet problem and $S = H^{\frac{1}{2}}(\Gamma)/\mathbb{R}$ for the Neumann problem.

General Boundary Value Problems

The foregoing reduction can also be performed for general elliptic boundary value problems. For two-dimensional problems, Fichera in [11] and Fichera and Ricci [14] developed a rather detailed analysis for this reduction in connection with Cauchy singular integral equations. Here we follow Costabel and Wendland [9], where the reduction to first kind equations for arbitrary n follows the lines of our previous presentation.

Let us consider a regular elliptic linear boundary value problem of $2m$-th order,

$$\begin{aligned} L^{(2m)}u &= 0 \quad \text{in} \quad \Omega_i \quad (\text{or } \Omega_e), \\ B\gamma u &= g \quad \text{on} \quad \Gamma \end{aligned} \tag{26}$$

where the boundary conditions are given by

$$B\gamma u = ((B_{jk}))\left(\partial_n^k u\right), \; 0 \leq j \leq m-1, \; 0 \leq k, \; \mu_j \leq 2m-1 \tag{27}$$

with tangential differential operators B_{jk} of orders $\mu_j - k$. Note that u can also be a vector-valued desired solution.

For the generalized representation formula we use the algebraic decomposition

$$L^{(2m)} = \sum_{j=0}^{2m} P_j \partial_n^j \quad \text{along} \quad \Gamma \,. \tag{28}$$

Let $E(x,y)$ be the fundamental solution of $L^{(2m)}$ in $\mathbb{R}^n$. Then any solution of $L^{(2m)}u = 0$ admits the Green representation

$$u(x) = (\overset{+}{-}) \sum_{k=0}^{2m-1} \sum_{l=0}^{2m-k-1} \int_\Gamma \left(\partial_{ny}^{\prime k} E(x,y)\right) P_{k+l+1} \partial_n^l u(y) ds_y \tag{29}$$

for $x \in \Omega_i$ (or Ω_e) (see [10]).

We further need that to B there exist complementary boundary differential operators S such that $M = \begin{pmatrix} B \\ S \end{pmatrix}$ is invertible; i.e.

$$(A_1) \qquad \begin{pmatrix} B \\ S \end{pmatrix} \gamma u = M\gamma u = \begin{pmatrix} g \\ \lambda \end{pmatrix} \; ; \; \gamma u = M^{-1}\begin{pmatrix} g \\ \lambda \end{pmatrix} \tag{30}$$

where g is the given boundary datum and λ is the yet unknown complementary Cauchy datum on Γ. Inserting (30) into (29) yields the two boundary integral equations

$$A\lambda(x) = \sum_{k=0}^{2m-1} \sum_{l=0}^{2m-k-1} B\gamma \int_\Gamma \left\{ \left(\partial_{ny}^{\prime k} E\right) P_{k+\ell+1} M_\ell^{-1} \right\} \begin{pmatrix} 0 \\ \lambda \end{pmatrix} ds_y = \Psi \tag{31}$$

and

$$\underset{\sim}{A}\lambda(x) = \lambda(x)(\pm) \sum_{k=0}^{2m-1} \sum_{l=0}^{2m-k-1} S\gamma \int_\Gamma \left\{ \left(\partial_{ny}^{\prime k} E\right) P_{k+\ell+1} M_\ell^{-1} \right\} \begin{pmatrix} 0 \\ \lambda \end{pmatrix} ds_y = \underset{\sim}{\Psi} \,. \tag{32}$$

Equation (31) corresponds to (4), a system of integral equation of the first kind. A is for $m > 1$ a matrix $((A_{jk}))$ of operators where each entry defines a pseudo-differential operator A_{jk} on Γ having the order

$$\alpha_{jk} = \mu_j + \mu_k + 1 - 2m \qquad \text{with} \qquad j,k = 0,\ldots,m-1 \,. \tag{33}$$

Consequently, A maps the boundary spaces

$$\mathcal{V}^s := \prod_{j=0}^{m-1} H^{-m+\mu_j+s_j+\frac{1}{2}}(\Gamma)$$

with $s = (s_0,\ldots,s_{m-1})$ continuously into $\mathcal{V}^{s-\alpha}$ with $s - \alpha = (s_j - \alpha_{jj})$:

$$A := \mathcal{V}^s \longrightarrow \mathcal{V}^{s-\alpha} \,.$$

Equation (32) corresponds to the boundary integral equation (5) of the second kind eith the + sign for the interior and the - sign for exterior problem, respectively. $\underset{\sim}{A} = \left((\underset{\sim}{A}_{jk}) \right)$ where $\underset{\sim}{A}_{jk}$ is a pseudo-differential operator of order $\alpha_{jk} = \mu_k - \mu_j$, $j,k = 0,\ldots,m-1$ and

$$A: \; \mathcal{V}^s \longrightarrow \mathcal{V}^s \tag{34}$$

is continuous.

Now let us assume that the original boundary value problem (26) admits an energy bilinear form $\mathcal{E}(u,u)$ such that with $\mathcal{E}$ there holds a Green identity satisfying

$$(A_2) \qquad Re\mathcal{E}(u,u) = Re\sum_{j=o}^{m-1} \int_\Gamma \{(B_j\gamma u_i)\cdot(S_j\gamma u_i) - (B_j\gamma u_e)\cdot(S_j\gamma u_e)\}\, ds \tag{35}$$

For the bilinear form $\mathcal{E}$ we now assume:

(A_3) For all u with $u|_{\Omega_i} \in C^\infty\left(\bar{\Omega}_i\right)$, $u|_{\Omega_e} \in C^\infty\left(\bar{\Omega}_e\right)$, $\operatorname{supp} u \subset K \subset\subset \mathbb{R}^n$,
$$B\gamma u_i = B\gamma u_e \tag{36}$$

there holds:

$$\begin{aligned} Re\mathcal{E}(u,u) \; &\geq \; \gamma_{1K} \cdot \left\{\|u_i\|^2_{H^m(\Omega_i)} + \|u_e\|^2_{H^m(\Omega_e)}\right\} \\ &\quad - c_K \left\{\|u_i\|^2_{H^{m-\epsilon}(\Omega_i)} + \|u_e\|^2_{H^{m-\epsilon}(\Omega_e)}\right\} \end{aligned} \tag{37}$$

Here γ_{1K}, ε and c_K are positive constants depending on the compact set K. Note that the required coerciveness (39) is required for functions satisfying the homogeneous transmission property in (36) and being defined in the whole space having a compact support in K but being perhaps discontinuous across Γ.

Now we return to the boundary integral equation (31) of the first kind. Associate with any boundary distribution λ the "single layer potentials"

$$u(x) := (\pm) \sum_{k=0}^{2m-1} \sum_{\ell=0}^{2m-k-1} \int_\Gamma \left\{\left(\partial^k_{ny} E\right) P_{k+2+1} M_\ell^{-1}\right\} \begin{pmatrix} 0 \\ \lambda \end{pmatrix} ds_y$$

with the + sign in Ω_i and - sign in Ω_e.

<u>**Theorem 4**</u>: *Let*

$$b(\lambda, \chi) := \left\langle A\lambda, \chi \right\rangle_\Gamma \tag{38}$$

be the boundary bilinear form associated with the first kind equation (31) *and let the assumptions* $(A_1) - (A_3)$ *be fulfilled. Then there holds the Gårding inequality*

$$Re b(\lambda, \lambda) = Re\mathcal{E}(u,u) \geq \gamma_0 \|\lambda\|^2_{\mathcal{V}} - c_0 \|\lambda\|^2_{\mathcal{V}^{-\epsilon}} \tag{39}$$

where

$$\mathcal{V} = \prod_{j=0}^{m-1} H^{-m+\mu_j+\frac{1}{2}}(\Gamma) \tag{40}$$

with positive γ_0 *and* ε *and with* $c_0 \geq 0$.

<u>**Remarks**</u>: Note that the Gårding inequality (39) is less strict than the coerciveness property (11). However, (39) is already sufficient for the classical Fredholm alternative to hold for (31).

<u>*Boundary integral equations of the 2nd kind*</u>

In case of differential equations of the second order, i.e. $m = 1$ in (26), one can also prove a Gårding inequality for the second kind operator $\underset{\sim}{A}$ in (32) provided $\underset{\sim}{A}$ is a strongly elliptic pseudo-differential operator. A special example is given by Equation (5) for the Laplacian,

$$\frac{1}{2}\lambda - K'\lambda = D\varphi\ . \tag{41}$$

Here,

$$b_2(\lambda,\mu) := \left\langle \left(\frac{1}{2}I - K'\right)\lambda, \mu \right\rangle_\Gamma = \left\langle \lambda, \frac{1}{2}\mu - K\mu \right\rangle_\Gamma \tag{42}$$

is a bilinear form with K', K being pseudo-differential operators of order -1. Hence,

$$\begin{aligned} Re\, b_2(\lambda,\lambda) &= \frac{1}{2}\|\lambda\|^2_{H^0(\Gamma)} - \langle K'\lambda, \lambda\rangle_\Gamma \\ &\geq \frac{1}{2}\|\lambda\|^2_{H^0(\Gamma)} - c\|\lambda\|^2_{H^{-\frac{1}{2}}(\Gamma)}\ . \end{aligned}$$

For the Navier system of elasticity, the Mikhlin symbol σ of the corresponding Cauchy singular integral equations of the second kind (see [27, 31]) satisfies the criterion of strong ellipticity [50],

$$Re\varsigma^T\sigma_0(x;\xi)\bar{\varsigma} \geq \gamma|\varsigma|^2 \qquad \forall\varsigma \in \mathbb{C}^n \wedge |\xi| = 1, \xi \in \mathbb{R}^{n-1} \tag{43}$$

where $\gamma > 0$.

As is explained in [50], the strong ellipticity condition (43) implies the Gårding inequality

$$b_2(\lambda,\lambda) \geq \gamma_0\|\lambda\|^2_{H^0(\Gamma)} - c\|\lambda\|^2_{H^{-\epsilon}(\Gamma)} \tag{44}$$

with positive γ_0 and ϵ and with $c \geq 0$.

Whereas for A in case $m = 1$ the inequality (39) remains valid even for Lipschitz domains, the inequality (44) for $\underset{\sim}{A}$ associated with the displacement and with the traction problems and a corner on Γ gets lost if the tangent direction discontinuity at the corner exceeds 87 degrees [15].

4. *Coupling of FEM and BEM*

The variational approach allows us also to formulate coupling of finite elements and boundary elements. Let us consider again as a sample model problem the Poisson equation for u,

$$\begin{aligned} \Delta u &= f \quad \text{in} \quad \Omega\ , \\ u &= \varphi \quad \text{on} \quad \Gamma\ . \end{aligned} \tag{45}$$

With the Dirichlet bilinear form

$$a(u,v) := \int_{\Omega} \nabla u \cdot \nabla v dx \ , \tag{46}$$

the Dirichlet problem (45) is equivalent to

Dirichlet's principle: *Find $u \in H^1(\Omega)$ with $u|_\Gamma = \varphi$ and*

$$\left\{\frac{1}{2}a(u,u) - (f,u)_\Omega\right\} = \inf_{v|_\Gamma = \varphi} \left\{\frac{1}{2}a(v,v) - (f,v)_\Omega\right\} \ . \tag{47}$$

As everybody knows, (47) in turn is equivalent to the bilinear variational formulation:

Find

$$u \in H^1(\Omega) \qquad \textit{with} \qquad u|_\Gamma = \varphi \tag{48}$$

such that

$$\forall v \in \overset{0}{H^1}(\Omega) : \qquad a(u,v) = \int_{\Omega} f v dx \ . \tag{49}$$

Since

$$a(v,v) \geq \gamma_0 \|v\|^2_{H^1(\Omega)} \qquad \text{for all } v \in \overset{0}{H^1}(\Omega) \ , \tag{50}$$

the well known theorem by Lax and Milgram provides us with a unique solution of (48), (49).

For the variational formulation of the coupling of finite elements and boundary elements let us decompose Ω into two subdomains Ω_F and Ω_B with corresponding boundaries $\Gamma_F = \partial\Omega_F$, $\Gamma_B = \partial\Omega_B$, the coupling boundary $\Gamma_C = \Gamma_B \cap \Gamma_F$ and $\Omega = \Omega_F \cup \Omega_B \cup \overset{0}{\Gamma}_C$. Let us assume that Γ_F and Γ_B are piecewise smooth. Then with $\lambda = \partial_n u|_{\Gamma_B}$ and the Gauss-Green formula we can write

$$a(u,v) = \int_{\Gamma_B} \lambda v ds + \int_{\Omega_F} \nabla u \cdot \nabla v dx \tag{51}$$

for any function u which is harmonic in Ω_B. With (4) and (5) for Ω_B we have in addition to (51) on Γ_B the two boundary integral equations

$$\mu = \left(\frac{1}{2}I - K_B\right)\mu + V_B\lambda \tag{52}$$

and

$$\lambda = D_B\mu + \left(\frac{1}{2}I + K_B'\right)\lambda \tag{53}$$

where

$$\mu = u|_{\Gamma_B}, \ \ \lambda = \partial_n u|_{\Gamma_B} \ . \tag{54}$$

Taking (51) and (52) and using (45) we may reformulate the original variational formulation (48), (49) ending up with

Unsymmetric coupling: Find

$$(u,\mu,\lambda) \in H^1(\Omega_F) \times H^{\frac{1}{2}}(\Gamma_B) \times H^{-\frac{1}{2}}(\Gamma_B) \tag{55}$$

with

$$u|_{\Gamma_B \cap \Gamma_F} = \mu|_{\Gamma_B \cap \Gamma_F}\,, \qquad u|_{\Gamma_D} = \varphi \tag{56}$$

such that $\forall (v,\chi) \in \overset{0}{H^1}(\Omega) \times H^{-\frac{1}{2}}(\Gamma_B)$:

$$\begin{aligned} b(u,\mu,\lambda; v, v|_{\Gamma_B},\chi) :&= \langle \lambda, v\rangle_{\Gamma_B} - \langle \chi,\mu\rangle_{\Gamma_B} - 2\langle \chi, K_B\mu\rangle_{\Gamma_B} \\ &+ 2\langle \chi, V_B\lambda\rangle_{\Gamma_B} + a_F(u,v) = (f,v)_\Omega\,. \end{aligned} \tag{57}$$

If Γ_B is chosen to be smooth, then K_B for the Laplacian is the double layer potential operator, a pseudo-differential operator of order -1 which is on Γ_B a compact perturbation of the identity in $L_2(\Gamma)$. In this case if K_B is of order $-2\varepsilon < 0$ then we find from (11) for V_B in (7) the Gårding inequality

$$\begin{aligned} b(v, v|_{\Gamma_B},\chi; v, v|_{\Gamma_B},\chi) &= -2\langle \chi, K_B v|_{\Gamma_B}\rangle + 2\langle \chi, V_B\chi\rangle + a_F(v,v) \\ &\geq \gamma_0 \left\{ \|v\|^2_{H^1(\Omega_F)} + \|v\|^2_{H^{\frac{1}{2}}(\Gamma_B)} + \|\chi\|^2_{H^{-\frac{1}{2}}(\Gamma_B)} \right\} \\ &\quad -c_0 \left\{ \|v\|_{H^{\frac{1}{2}-\epsilon}(\Gamma_B)} \|\chi\|_{H^{-\frac{1}{2}-\epsilon}(\Gamma_B)} \right\}\,. \end{aligned} \tag{58}$$

In this case the Gårding inequality can serve as a general coerciveness inequality which provides stability and convergence for a coupled finite element-boundary element approximation when $H^1(\Omega_F)$ and $H^{-\frac{1}{2}}(\Gamma_B)$ are replaced by the finite element subspaces $H_{h_1}(\Omega_F)$ and boundary element subspaces $H_{h_2}(\Gamma_B)$, respectively. That analysis is due to Johnson and Nedelec [23].

For the corresponding formulation in elasticity, however, the double layer potential operator is Cauchy singular and, hence, of order zero and (58) only holds with $\varepsilon = 0$. But if only the space $H^1(\Omega_F)$ would be replaced by finite elements and (52) was kept as a continuous boundary integral equation, then for this approximation coerciveness (50) would still hold on the subspaces. Therefore a finer mesh refinement of the boundary elements than of the finite elements still yields stable and convergent methods also for elasticity problems as was shown by Brezzi and Johnson [4] and can be used for Schnack's macro-elements, see [20]. By the way, a finer finite element refinement than boundary element refinement also defines stable and convergent methods [51].

Symmetric formulation of coupling

Another possibility of coupling formulations for non-compact K_B is using both equations (52) and (53) simultaneously. We begin with the variational formulations,

$$a(u,v) = \langle \lambda, v\rangle_{\Gamma_B} + a_F(u,v) = (f,v)_\Omega \ , \tag{59}$$

$$\langle V_B\lambda, \chi\rangle_{\Gamma_B} = \frac{1}{2}\langle u, \chi\rangle_{\Gamma_B} + \langle K_B u, \chi\rangle_{\Gamma_B} \ , \tag{60}$$

$$\frac{1}{2}\langle \lambda, v\rangle_{\Gamma_B} - \langle K'_B\lambda, v\rangle_{\Gamma_B} = \langle D_B u, v\rangle_{\Gamma_B} \ . \tag{61}$$

Inserting (60) and (61) into (59) yields the so called "symmetric formulation" [5, 8, 16, 17, 19, 35]:

Find

$$(u,\mu,\lambda) \in H^1(\Omega_F) \times H^{\frac{1}{2}}(\Gamma_B) \times H^{-\frac{1}{2}}(\Gamma_B) \tag{62}$$

with

$$u|_{\Gamma_B \cap \Gamma_F} = \mu|_{\Gamma_B\cap\Gamma_F}, \qquad u|_{\Gamma_D} = \varphi \tag{63}$$

such that $\forall (v,\chi) \in \overset{0}{H^1}(\Omega) \times H^{-\frac{1}{2}}(\Gamma_B)$:

$$\begin{aligned} b(u,\mu,\lambda; v, v|_{\Gamma_B}, \chi) : &= a_F(u,v) \\ &+ \frac{1}{2}\langle\lambda, v\rangle_{\Gamma_B} + \langle K'_B\lambda, v\rangle_{\Gamma_B} + \langle D_B u, v\rangle_{\Gamma_B} \\ &- \frac{1}{2}\langle u, \chi\rangle_{\Gamma_B} - \langle K_B u, \chi\rangle_{\Gamma_B} + \langle V_B\lambda, \chi\rangle_{\Gamma_B} \\ &= (f,v)_\Omega \ . \end{aligned} \tag{64}$$

If we identify $u = v$, $\mu = v|_B$ and $\lambda = \chi$, then we immediately find the coerciveness property

$$\begin{aligned} b(v,\mu,\chi; v,\mu,\chi) &= a_F(v,v) + \langle D_B\mu, \mu\rangle_{\Gamma_B} + \langle V_B\chi, \chi\rangle_{\Gamma_B} \\ &\geq \gamma_0 \left\{ \|v\|^2_{H^1(\Omega_F)} + \|\mu\|^2_{H^{\frac{1}{2}}(\Gamma_B)} + \|\chi\|^2_{H^{-\frac{1}{2}}(\Gamma_B)} \right\} \end{aligned} \tag{65}$$

for all $v \in H^1(\Omega)$, $v|_{\Gamma_B} = \mu \in H^{\frac{1}{2}}(\Gamma_B)$, $\chi \in H^{-\frac{1}{2}}(\Gamma_B)$.

Since (65) also holds on finite element-boundary element subspaces $v_{h_1} \in H_{h_1}(\Omega_F)$ with $\mu_{h_1} = v_{h_1}|_{\Gamma_B}$ and $\chi_{h_2} \in H_{h_2}(\Gamma_B)$ we obtain a stable and convergent coupling procedure which is still valid if K_B is only a pseudo-differential operator of order zero as in elasticity. This method also works and can be justified for a coercive bilinear form a_F corresponding to nonlinear problems as in visco-elasticity or obstacle problems with variational inequalities (see [8, 16, 35]).

5. *Domain Decomposition*

Variational formulations with boundary elements can also be used for domain decomposition techniques which can be dealt with using parallel computing. Let here

Ω be decomposed into a finite union of finite element subdomains Ω_{B_j}, $j = 1, \ldots, N$. Then (49) takes the form

$$a(u,v) = \sum_B \langle \lambda_B, v\rangle = (f,v)_\Omega \tag{66}$$

where the summation is taken over all the subdomains. On each of the subdomains Ω_B we introduce

$$\mu_B := u|_{\Gamma_B}, \quad \lambda_B = \partial_n u|_{\Gamma_B} \tag{67}$$

and resolve the boundary integral equation

$$\frac{1}{2}\mu_B + K_B\mu_B = V_B\lambda_B \qquad \text{on each of } \Gamma_{B_j} \tag{68}$$

with respect to

$$\lambda_B = V_B^{-1}\left(\frac{1}{2}I + K_B\right)\mu_B =: T_B\mu_B\ . \tag{69}$$

Note that T_B maps the boundary values μ_B of a solution of the differential equation (45) in Ω_B onto the values of $\lambda_B = \partial_n u|_{\Gamma_B}$; hence, T_B is the corresponding Steklov-Poincaré operator. For the boundary element domain decomposition method approximate μ_{B_j} and λ_{B_j} in (68) by boundary elements which creates an approximation of T_B (see [22]). For the corresponding algorithm one now needs a preconditioner for the discrete equations, then they can be solved iteratively with the preconditioned conjugate gradient method. Preconditioners have been constructed by Khoromsky, Mazurkewich, Zhidkov [26] and by Rjasanov [38]. If we denote by H the meshwidth parameter of the finite element subdomains and by h the meshwidth parameter of the individual boundary element approximations on Γ_{B_j} then one can prove the following asymptotic error estimate:

If c_0H is sufficiently small then one has

$$\left\{\sum_{j=1}^{N}\left(\|u|_{\Gamma_{B_j}} - \mu_{B_{jh}}\|^2_{H^{\frac{1}{2}}(\Gamma_{B_j})} + \|\partial_n u|_{\Gamma_{B_j}} - \lambda_{B_{jh}}\|^2_{H^{-\frac{1}{2}}(\Gamma_{B_j})}\right)\right\}^{\frac{1}{2}} \le c\,(c_0H)^{l-1}\,\|u\|_{H^l(\Omega)} \tag{70}$$

$$\text{for } 1 < l < d + \frac{3}{2}\ ;\ h \le c_0H\ .$$

For further details see [22].

6. *Signorini Problem*

The Signorini problem with the Laplacian

$$
\begin{aligned}
-\Delta u &= f \qquad \text{in} \qquad \Omega\,, \\
u|_\Gamma &= 0 \qquad \text{on} \qquad \Gamma_D
\end{aligned}
\tag{71}
$$

and

$$u\partial_n u = 0 \wedge u \geq 0 \wedge \partial_n u \geq 0 \;\text{ on } \Gamma_S \tag{72}$$

is a simple model problem for the contact problem of elasticity which also can be solved with appropriate boundary element variational methods. Let u_f be the solution of

$$\Delta u_f = f, \qquad u_{f_{|\Gamma}} = 0\,. \tag{73}$$

Then the solution of (71) can be represented by

$$u(x) = \int_\Gamma E(x,y)\lambda(y)ds_y - \int_\Gamma (\partial_{n_y} E(x,y))^T u(y) ds_y + u_f(x) \tag{74}$$

for any $x \in \Omega$. On the boundary we have the equation

$$\lambda = Tu := \; V^{-1}\left(\frac{1}{2}I + K\right)u \quad \text{on} \quad \Gamma \tag{75}$$

with the Steklov-Poincaré operator T. Let

$$\mathcal{V} := \; \left\{ v \in H^{\frac{1}{2}}(\Gamma) \mid v_{|\Gamma_D} = 0 \wedge v_{|\Gamma_S} \geq 0 \right\}$$

denote the convex closed set of admissible functions. Then the Signorini problem (71), (72) is equivalent to the variational inequality:

Find $\tau \in \mathcal{V}$ *such that*

$$\langle T\tau, v-\tau\rangle_\Gamma \geq -\langle \partial_n u_f, v-\tau\rangle_\Gamma \quad \text{for all} \quad v \in \mathcal{V}\,. \tag{76}$$

It is well known that this problem has exactly one solution (see [13]). for the boundary element treatment one can either use a boundary element subspace of $\mathcal{V}$ and solve (76) on the subspace as done by Spann in [47] or one uses further approximation by nonlinear boundary integral equations in connection with the Yosida approximation of (72) following Kawohl [24].

Yosida approximation

Find $\tau_\varepsilon \in H^{\frac{1}{2}}(\Gamma)$ such that

$$\langle T\tau_\varepsilon, v\rangle_\Gamma + \frac{1}{\varepsilon}\langle \tau_\varepsilon^-, v\rangle_\Gamma = -\langle \partial_n u_f, v\rangle_\Gamma \quad \text{for all} \quad v \in H^{\frac{1}{2}}(\Gamma) \,. \tag{77}$$

Here $\tau_\varepsilon^-(x) = \inf\{\tau_\varepsilon(x), 0\}$ for any measurable representative $\tau_\varepsilon(x)$ of $\tau_\varepsilon \in H^{\frac{1}{2}}(\Gamma)$. Schmitz and Schneider show in [41] (see also [42]) that the boundary element approximation of (70) with piecewise linear continuous polynomials on Γ converges and satisfies an a-priori asymptotic error estimate of the form

$$\|\tau_{\varepsilon_h} - \tau\|_{H^{\frac{1}{2}}(\Gamma)} \le c_1\varepsilon + c_2\left(1 + \frac{\sqrt{h}}{\varepsilon}\right) h \le c h^{\frac{1}{2}} \text{ for } \varepsilon \sim h^{\frac{1}{2}} \tag{78}$$

7. *Consequences of Variational Formulations with Bilinear Forms Satisfying Gårding Inequalities*

The variational formulation of boundary integral equations and corresponding boundary element methods has many consequences:

1. The mathematical foundation of boundary integral equations can be given in the framework of elliptic bilinear forms, see e.g. [48]. It provides the classical Fredholm alternative, and with uniqueness, the solvability, a-priori estimates and regularity of the solution.
2. Stability and quasi - optimal convergence of boundary element-Galërkin schemes can be shown along the well known lines of finite element analysis. (See e.g. [21, 32, 50]). The Sobolev space error estimates are of the form

$$\|\lambda - \lambda_h\|_{H^{\alpha-d-1}(\Gamma)} \le c h^{2d+1-\alpha}\|\lambda\|_{H^{d+1}(\Gamma)} \tag{79}$$

   if the boundary elements λ_h are given by piecewise polynomials of degree d. As before, α denotes the order of the boundary integral operator involved.
3. Also boundary element collocation and the recently developed quadrature based generalized collocation, the so-called *qualocation methods* [45, 46] can be put into the variational formulation based on the Galërkin-Petrov formulation:

   Find $\lambda_h \in H_h$ where H_h denotes the trial space, such that

$$\langle A\lambda_h, \chi_h\rangle_\Gamma = \langle \Psi, \chi_h\rangle_\Gamma \tag{80}$$

   holds for all $\chi_h \in \mathcal{T}_h$ where $\mathcal{T}_h$ us now the test space.

With boundary element trial spaces H_h, different choices of test spaces yield different discretization methods for the boundary integral equations. With

$$\mathcal{T}_h = \mathcal{L}\{\delta(x - x_k)\}$$

we have collocation where $\mathcal{L}$ denotes the span of the Dirac functionals to the collocation points $\{x_k\}$.

If we take for the k-th equation several points x_{k_j} associated with a specific quadrature rule related to the principal part of A, then with

$$\mathcal{T}_h = \mathcal{L}\left\{\sum_{j=1}^{M} c_{kj}\delta(x - x_{kj})\right\}$$

we have qualocation, where the coefficients c_{kj} are defined by a careful analysis. This is a very efficient method due to Sloan [45, 46]. All these different methods can be handled in the variational framework if there exists a linear mapping from the trial into the test space,

$$\mathcal{B}_h : H_h \longrightarrow \mathcal{T}_h .$$

Then we can write the corresponding method as a family of variational problems on $H_h \times H_h$:

Find $\lambda_h \in H_h$ such that

$$a_h(\lambda_h, \mu_h) := \langle A\lambda_h, \mathcal{B}_h\mu_h\rangle_\Gamma = \langle \mathcal{B}_h' A\lambda_h, \mu_h\rangle_\Gamma \tag{81}$$

holds for all $\mu_h \in H_h$.

Here $\mathcal{B}_h'$ denotes the adjoint of $\mathcal{B}$ with respect to the L^2-boundary scalar product.

<u>$n = 2$:</u> For two-dimensional problems with Γ a system of simple non intersecting curves, all functions on Γ can be identified with vector-valued periodic functions on the unit circle. If H_h is given by piecewise polynomial splines of odd degree d which are $d-1$ times continuously differentiable then Arnold and Wendland choose in [2]

$$\mathcal{B}_h\mu_h = \left(\frac{d}{ds}\right)^{d+1} \mu_h + \mathcal{J}_h\mu_h \tag{82}$$

where $\mathcal{J}_h\mu_h$ denotes the trapezoidal rule applied to μ_k with the break points of H_h as knots. With (82) inserted into (81), here one finds from strong ellipticity of A coerciveness properties of a_h in form of a Gårding inequality which yields stability and convergence results of point collocation.

Further analysis is based on the Fourier transform $\mathcal{F}$ which becomes Fourier series expansion on the periodic functions. With the Parseval equality we may rewrite (81) in the form

$$a_k(\lambda_k, \mu_k) = \left\langle \mathcal{F}\mathcal{B}_h A\mathcal{F}^{-1}(\mathcal{F}\lambda_h), \mathcal{F}\mu_k\right\rangle_{L^2} \tag{83}$$

defining a bilinear form on the space $\mathcal{F}H_h$.

Here techniques for convolutional operators and pseudo-differential operators provide stability and convergence results for spline collocation methods [3, 36, 40, 44] and for qualocation methods with piecewise polynomial splines H_h of even and odd degree d. For C^∞-data, the qualocation methods provide convergence of arbitrarily high orders [18, 39, 45, 46]. Using Fourier series approximation as H_h, one even gets spectral methods and exponential convergence rates [28].

$\underline{n = 3}$: In three-dimensional problems the analysis of Galërkin-Bubnov methods for solving (31) or (32) are the same as in two-dimensional problems. For collocation and qualocation, however, only a few recent results of error analysis are known [37, 43, 7]. There Γ is not anymore a closed surface but a compact subdomain of the plane or of a surface where extension by zero can be combined with the Fourier transform and (83) is analyzed on $\mathcal{F} H_h$.

Acknowledgements

I want to thank Prof. Luigi Morino and Prof. Renzo Piva for the great boundary element conference in Rome and for their warm hospitality and patience. I also want to thank them, their students and Mrs. Paola Agosti for typing the manuscript.

References

1. Agranovich M.S., 1977, "Spectral properties of diffraction problems. In: The General Method of Natural Vibrations in Diffraction Theory (Voitovic N.N., Katzenellenbaum B.Z., Sivov A.N., Eds.), Izdat. Nauka, Moscow, 289-416.

2. Arnold D.N., Wendland W.L., 1983, "On the asymptotic convergence of collocation methods". Math. Comp., 41, 349-381.

3. Arnold D.N., Wendland W.L., 1985, "The convergence of spline collocation for strongly elliptic equations on curves". Numer. Math., 47, 317-341.

4. Brezzi F., Johnson C., 1979, "On the coupling of boundary integral and finite element methods". Calcolo, 16, 189-201.

5. Costabel M., 1987, "Principles of boundary element methods". Computer Physics Report, 6, 243-274.

6. Costabel M., 1988, "Boundary integral operators on Lipschitz domains: Elementary results". SIAM J. Math. Anal., 19, 613-626.

7. Costabel M., Penzel F., Schneider R., 1990, "Error analysis of boundary element collocation for a screen problem in $\mathbb{R}^3$. (THD-Preprint, TH Darmstadt).

8. Costabel M., Stephan E.P., 1990, "Coupling of finite and boundary element methods for an elastoplastic interface problem". SIAM J. Numer. Anal., 27.

9. Costabel M., Wendland W.L., 1986, "Strong Ellipticity of boundary integral operators", J. für die reine und angewandte Mathematik 372, 34-63.

10. Dieudonné J., 1978, Elemént d'Analyse. VIII, Gauthier-Villars, Paris.

11. Fichera G., 1961, "Linear elliptic equations of higher order in two independent variables and singular integral equations". In: Proc. Conf. Partial Differential Equations in Continuum Mechanics (Langer R.E., Ed.), University Press, Wisconsin, Madison, 55-80.

12. Fichera G., 1972, "Existence theorems in elasticity". In: Handbuch der Physik, vol. VIa/2, Festkörpermechanik II, Springer-Verlag, Berlin, 347-389.

13. Fichera G., 1972, "Boundary value problems of elasticity with unilateral constraints". In: Handbuch der Physik, vol. VIa/2, Festkörpermechanik II, Springer-Verlag, Berlin.

14. Fichera G., Rizzi P., 1976, "The single layer potential approach of boundary value problems for elliptic equations". In: Lecutre Notes in Mathematics, 561, Springer-Verlag, Berlin, 39-50.

15. Flassig U., 1984, "Das 2. Fundamentalproblem der ebenen Elastizität". Diplom Thesis, Technical University Darmstadt.

16. Gatica G.N., Hsiao G.C., 1989, "The coupling of boundary element and finite element methods for a nonlinear exterior boundary value problem". Zeitschr. Analysis Anwendungen, 28, 377-387.

17. Grannell J.J., 1987, "On simplified hybrid methods for coupling of finite elements and boundary elements". In: Boundary Elements IX, vol. 1, (Brebbia C.A., Kuhn G., Wendland W.L., Eds.), Springer-Verlag, Berlin, 447-460.

18. Hagen R., Silbermann B., 1988, "On the stability of the qualocation method". Seminar Anal. Operator Equat. and Numer. Anal. Karl Weierstrass Inst. for Mathematics, Berlin, 43-52.

19. Houde Han, 1988, "A new class of variational formulations for the coupling of finite and boundary element methods". Preprint Univ. Maryland, College Park, USA.

20. Hsiao G.C., Schnack E., Wendland W.L., "A hybrid coupled finite-boundary element method". In preparation.

21. Hsiao G.C., Wendland W.L., 1977, "A finite element method for some integral equations of the first kind". J. Math. Anal. Appl., 58, 449-481.

22. Hsiao G.C., Wendland W.L., "Domain decomposition in boundary element methods". In: Proc. Domain Decomposition Conf., Moscow, 1990 SIAM. To appear.

23. Johnson C, Nedelec J.C., 1980, "On Coupling of boundary integral and finite element methods". Math. Comp., 35, 1063-1079.

24. Kawohl B., 1980, "On nonlinear mixed boundary value problems for second order elliptic differential equations on domains with corners". Proc. Royal Soc., Edinburgh, 87A, 35-51.

25. Kellog O.D., 1929, Foundations of Potential Theory, New York.

26. Khoromskij B.N., Mazurkevich G.E., Zhidkov E.P., 1990, "Domain decomposition method for magnetostatics nonlinear problems in combined formulations". Sov. J. Numer. Anal. Math. Modelling, 5, 121-165.

27. Kupradze V.D., Gegelia T.G., Basheleishvili M.O., Burchuladze T.V., 1979, Three-Dimensional Problems of the Mathematical Theory of elasticity and Thermoelasticity. North-Holland, Amsterdam.

28. Lamp U., Schleicher K.-T., Wendland W.L., 1985, "The fast Fourier transform and the numerical solution of one-dimensional boundary integral equations". Numer. Math., 47, 15-38.

29. Lions J.L., Magenes E., 1972, Non-Homogeneous Boundary Value Problems and Applications. Springer-Verlag, Berlin.

30. Maue A.W., 1949, "Über die Formulierung eines allgemeinen Diffraktionsproblems mit Hilfe einer Integralgleichung". Zeitschr. f. Physik, 126, 601-618.

31. Mikhlin S.G., Prössdorf S., 1986, Singular Integral Operators, Springer-Verlag, Berlin.

32. Nedelec J.C., 1977, "Approximation des équations intégrales en mécanique et en physique". Cours de l'Ecole d'Eté CEA-IRIA-EDF.

33. Nedelec J.C., 1978, "Approximation par double couche du probléme de Neumann extérieur", C.R.A.S., Sec. A, 286, 103-106.

34. Nedelec J.C., Planchard J., 1973, Une methode variationelle d'elelements finis pour la résolution numérique d'un problèm extérieur dans $\mathbb{R}^3$. RAIRO Anal. Numer., 7, 105-129.

35. Polizzotto C., "A symmetric-definite BEM formulation for the elasto-plastic rate problem". In: Boundary Elements IX, vol. 2, (Brebbia C.A., Kuhn G., Wendland W.L., Eds.), Springer-Verlag, Berlin, 315-337.

36. Prössdorf S., Schmidt G., 1981, "A finite element collocation method for singular integral equations". Math. Nachr., 100, 33-60.

37. Prössdorf S., Schneider R., "Spline approximation methods for multidimensional periodic pseudo-differential equations". Integral Equations and Operator Theory, to appear.

38. Rjasanov S., 1991, "Vorkonditionierte iterative Auflösung von Randelementgleichungen für die Dirichlet-Aufgabe". Doctoral B Thesis, Techn. Univ. Chemnitz.

39. Saranen J., 1988, "The convergence of even degree spline collocation solution for potential problems in smooth domains of the plane". Numer. Math. 53, 499-512.

40. Saranen J., Wendland W.L., 1985, "On the asymptotic convergence of collocation methods with spline functions of even dregree". Math. Comp., 45, 91-108.

41. Schmitz H., Schneider G., "Boundary element solution of the Dirchlet-Signorini problem by a penalty method". Applicable Analysis, to appear.

42. Schmitz H., Schneider G., Wendland W.L., "Boundary element methods for problems involving unilateral boundary conditions". In: Nonlinear Computational Mechanics, State of the Art. Springer-Verlag, Berlin, to appear.

43. Schneider R., 1989, "Stability of a spline collocation method for strongly elliptic multidimensional singular integral equations". Preprint 1272, Techn. Univ. Darmstadt, to appear.

44. Silbermann B., 1989, "Symbol constructions and numerical analysis", Preprint Techn. Univ. Chemnitz, to appear.

45. Sloan I.H., 1988, "A quadrature-based approach to improving the collocation method". Numer. Math., 54, 41-56.

46. Sloan I.H., Wendland W.L., 1989, "A quadrature-based approach to improving the collocation method for splines of even degree". Zeitschr. Analysis Anw., 8, 361-376.

47. Spann W., 1989, "Fehlerabschätzungen zur Randelementmethode beim Signorini-Problem für die Laplace-Gleichung". Doctoral Thesis, Univ. München.

48. Stummel F., 1969, "Rand-und Eigenwertaufgaben in Sobolewschen Räumen". Lecture Notes in Mathematics, 102, Springer-Verlag, Berlin.

49. Trefftz E., 1927, "Ein Gegenstück zum Ritzschen Verfahren". In: Verhandlungen d. 2. Intern. Kongress f. techn. Mechanik, Zürich, 1926, 131-138.

50. Wendland W.L., 1987, "Strongly elliptic boundary integral equations". In: The State of the Art in Numerical Analysis (Iserles A., Powell M., Eds.), 511-561, Clarendon Press, Oxford.

51. Wendland W.L., 1988, "On asymptotic error estimates for combined boundary element methods and finite element methods". In: Finite Element and Boundary Element Techniques from Mathematical and Engineering Point of View (Stein E., Wendland W.L., Eds.). CISM Courses and Lectures, 301, Springer-Verlag, Vienna.

The Use of Spline Approximated Particular Integrals in the Free Vibration Analysis of Membranes by BEM

C. Alessandri and A. Tralli

Dipartimento di Costruzioni, University of Florence, Italy.

Summary

Efficient methods for free vibration analyses via BEM have been proposed in recent years. They are based on the use of "localized particular solutions" which allow the transferal to the boundary of the domain integrals containing the inertial forces. In the present paper an alternative approach, leading to a different algebraic eigenvalue problem, is discussed with particular reference to membranes vibrations. This approach, quite general, is based on the use of basis functions on compact support (cubic splines and Hermite polynomials) and collocation methods.

Introduction

The Boundary Element Method has been applied successfully to the solution of dynamic problems and, in particular, to the free vibration analysis of elastic bodies since its first appearing [1-4]. Most of the early applications of BEM to free vibration problems required sweeping the frequency range of interest to obtain the dynamic response of the structural system at hand [5-6]. Such a research was needed because the fundamental solution of the governing differential equations (e.g. Hankel functions for membranes) was itself frequency dependent, producing therefore a non algebraic eigenvalue problem. An alternative approach was proposed by Nardini and Brebbia [7][8] in the early eighties and later on re-formulated and generalized by Banerjee and co-workers [9-11]. It differs from the previous approach for the use of the static - i.e. frequency independent - fundamental solution; therefore, the boundary integrals need to be computed only once, as they are frequency independent, and the free vibration problem is reduced to an algebraic non symmetric eigenvalue problem. In particular, in [7][8] a general technique (DRM) is introduced, which allows to transfer to the boundary the domain integrals containing the inertia terms (as well as other domain terms) and is based on the approximation of the effects of the domain sources by a series of localized particular solutions. Banerjee and co-workers [9-11], with the same purpose, suggested an effective technique which consists of decomposing the solution in a complementary part satisfying the homogeneous differential equation and in a particular integral, which , in the general case, is still approximated by a series of localized particular solutions. As a general remark, the DRM [7-8] and Banerjee's technique [9-11] may be interpreted as a way of approximating the particular integral by "global shape functions" or, in other words, by resorting to the old method of the undetermined coefficients through a truncated series of M terms which are defined by collocation in an arbitrary number of boundary and interior points. The reduction of the computational effort turns out to be considerable, but many questions still remain open, particularly those concerning the choice criterion for the collocation points or the shape functions to be assumed. Furthemore, such procedures require inevitably the inversion of a full matrix about which not much can be said and and for which double pivoting and scaling are required [11].

The aim of the present paper is to show how the main advantages of the quoted approach to free vibration analyses via BEM can be retained if the Authors' technique [12][13] for avoiding domain integrations (based on spline approximations of the particular integrals) is adopted. This technique, backed by a consistent mathematical theory [14], exhibits the further advantage of dealing with banded and well conditioned systems.

Vibration Analysis of Membranes

Let us consider, for the sake of simplicity, the simplest 2-D eigenvalue problem described by the well-known Helmholtz equation

$$\nabla^2 u(\underline{x}) + \frac{\rho}{T}\,\omega^2 u(\underline{x}) = 0 \qquad \underline{x} \in \overline{\Omega} \subset R^2 \tag{1}$$

where Ω is assumed to be an open, bounded, simply connected domain of the Euclidean space R^2 referred to an orthogonal carthesian reference system (O, x_i, i=1,2). Moreover $\overline{\Omega} = \Omega \cup \Gamma$;the boundary Γ is assumed to be sufficiently regular. Equation (1) describes the free vibrations of a homogeneous membrane having surface density ρ and stretched by a uniform tension T. The natural frequency ω is assumed as the problem unknown and the ratio $\rho\omega^2/T$ will be denoted in the sequel by μ^2. Moreover, if the membrane is supposed to be fixed along the boundary, the usual condition $u(\underline{x})=0$, $\underline{x} \in \Gamma$, holds on Γ. Since the problem is described by a linear ordinary differential operator, it is possible to apply the classical technique of decomposing the actual solution in a particular integral $u^p(\underline{x})$ and a complementary part $u^c(\underline{x})$ satisfying the homogeneous equation, as previously suggested by Banerjee and co-workers in the free vibration analysis by BEM [9-10]. That allows to formulate an algebraic eigenvalue problem and to avoid domain integrations. Therefore the following equations hold

$$u(\underline{x}) = u^c(\underline{x}) + u^p(\underline{x}) \qquad \text{in } \overline{\Omega} \tag{2}$$

$$\nabla^2 u^p(\underline{x}) + \mu^2 u(\underline{x}) = 0 \qquad \text{in } \Omega \tag{3}$$

$$\nabla^2 u^c(\underline{x}) = 0 \qquad \text{in } \Omega$$
$$u^c(\underline{x}) = - u^p(\underline{x}) \qquad \text{on } \Gamma \tag{$4_{1,2}$}$$

The starting point for the application of the BEM to the problem at hand is given by the following integral relationship, referred to the complementary part, which does not involve any domain integral.

$$c_i u_i^c = \int_\Gamma q^c u^* \, d\Gamma - \int_\Gamma u^c q^* \, d\Gamma \tag{5}$$

As usual, $q^*(\underline{x},\underline{\xi})$, $u^*(\underline{x},\underline{\xi})$ (with $\underline{\xi} \in \Gamma$) denote the fundamental solutions of the problem at hand [1-2], u_i represents the potential at a given point x_i , $q = \nabla \underline{u}.\underline{n}$ denotes the flux along the outward normal unit vector $\underline{n}$ and u*, q* are the fundamental solutions of the Laplace equation applied at "i". The first integral must be interpreted as a Cauchy principal value integral if $\underline{x} \in \Gamma$. It is worth noting that in Equation (3) the inertia term $\mu^2 u(\underline{x})$, which is required to be square integrable, is still undetermined. The current method involves considering the inertia forces as unknown loadings distributed transversally in the interior domain.

Discretization of the Inertia Forces and Spline Approximation of the Particular Integrals

The inertia force field, once it has been assumed as a problem unknown , can be approximated by means of functions defined all over the domain (for instance polynomials etc.) or, as usual in FEM, by compact support functions, i.e.

$$\mu^2 u(\underline{x}) \simeq \mu^2 \sum_{k=1}^{N} \Psi_k(\underline{x}) \, \hat{\underline{U}}_k \tag{6}$$

where $\Psi_k(\underline{x})$ denotes the basis functions assumed and vector $\hat{\underline{U}}$ collects the displacement values corresponding to the discretized inertia forces. The choice of the most suitable function is mainly a matter of computational efficiency. Since the inertia terms are required to be only square integrable, even piecewise unit constant functions can be employed, as is shown in the numerical applications. By replacing Equation (6) in Equation (3) one obtains

$$\nabla^2 \tilde{u}^P(\underline{x}) = - \mu^2 \sum_{k=1}^{N} \Psi_k(\underline{x}) \, \hat{\underline{U}}_k \tag{7}$$

where $\tilde{u}^P$ denotes the particular integral associated to the approximation of the inertia force distribution. The most widely used technique for obtaining a suitable approximation of the obviously not unique particular integral is the use of localized particular solutions [7-11] (for a critical discussion see [12-13]). When looking for an alternative technique to set up a suitable approximated particular integral for a generic source distribution, some aspects have to be considered. In order to retain the most significant advantages of the BEM solutions, the minimum amount of data inside the domain has to be provided. Therefore collocation based techniques have to be preferred to other well established methods such as FEM or FDM. It is worth noting that this way of proceeding is in agreement with the characteristics of the most widely diffused BEM solutions obtained by collocating at boundary nodes. Moreover, the use of basis functions on a compact support, such as splines, provides sparse banded coefficient matrices [12-14]. Because a particular integral, and not the solution of the problem, is looked for, some classical drawbacks of this approach (such as the difficulty of dealing with geometrically complex domains or with mixed inhomogeneous boundary conditions) turn out to be not important. As a matter of fact, only problems defined on simple domains (square, rectangular or circular) and with simple Dirichlet boundary conditions can be easily solved by these methods. Therefore, the approximation u^N of the unknown function u can be expressed as follows

$$u^N(x_1,x_2) = \sum_{i=1}^{2nx_1} \sum_{j=1}^{2nx_2} a^{ij} \phi_i(x_1) \phi_j(x_2) \tag{8}$$

where $\phi_i(x_1)$, $\phi_j(x_2)$ represent the standard Hermite polynomials and nx_1, nx_2 are the partitions along the coordinate axes x_1, x_2 respectively. The substitution of relationship (8) in the left hand side term of Equation (7) provides

$$\nabla^2 \left(\sum_{i=1}^{2nx_1} \sum_{j=1}^{2nx_2} a^{ij} \phi_i(x_1) \phi_j(x_2) \right) = - \mu^2 \sum_{k=1}^{N} \Psi_k(\underline{x}) \, \hat{\underline{U}}_k \tag{9}$$

After dividing the domain Ω into n rectangular subdomains, with $n = nx_1 * nx_2$ the coefficients α^{ij} are determined by solving the system of linear equations obtained by collocating Equation (9) at 4n Gaussian knots. The inversion of Equation (9) provides explicitely, once for all, the α^{ij} coefficients which, replaced in Equation (8) allow to generate the particular integral. In a matrix form, at any $\underline{x} \in \overline{\Omega}$ the particular solution can be cast as follows

$$\hat{u}^p(\underline{x}) = \mu^2 \underline{p}^T(\underline{x})\ \underline{\hat{U}} \tag{10}$$

where vector $\underline{p}$ collects the particular solution due to any unit $\hat{U}_k$.

Algebraic Model

The particular integral approximated as described in the previous Section and evaluated at the boundary is replaced in Equation (4.2) and in the integral equation (5) in order to provide a B.E. approximation of the complementary part of the solution. After subdividing the boundary Γ into elements [1-2] and after integrating over Γ, the following linear system of equations is obtained

$$\underline{\underline{H}}\ \underline{U}^c = \underline{\underline{G}}\ \underline{Q}^c \tag{11}$$

where $\underline{Q}^c$ contains the unknown complementary flux values and $\underline{U}^c$ denotes the opposite of the particular integral evaluated at the nodes; moreover, $\underline{\underline{H}}$ and $\underline{\underline{G}}$ are the coefficient matrices obtained by integrating q* and u* respectively over Γ . The solution of Equation (11) provides

$$\underline{Q}^c = \underline{\underline{G}}^{-1}\underline{\underline{H}}\ \underline{U}^c = \mu^2 \underline{\underline{G}}^{-1}\underline{\underline{H}}\ \underline{\underline{P}}_\Gamma \underline{\hat{U}} \tag{12}$$

where matrix $\underline{\underline{P}}_\Gamma$ collects the values of the particular integral at the boundary nodes due to a unit $\hat{U}_k$ and is obtained by assembling the $\underline{p}$ vectors. The complementary part of the solution at any internal point can be immediately evaluated by means of the integral equation (5) (c_i =1), once vector $\underline{Q}^c$ has been computed. Thus, Equation (5) can be re-written as follows

$$u^c(\underline{x}_i) = \sum_{j=1}^{N} [\ G_j(\underline{x}_i,\underline{\xi}_j)\ Q_j^c - H_j(\underline{x}_i,\underline{\xi}_j)\ U_j^c\ ] \tag{13}$$

where G_j (.,.) and H_i (.,.) have been evaluated by integrating over the boundary elements and by taking into account the shape function $N(\underline{\xi}_j)$. In a matrix form, at any $\underline{x} \in \overline{\Omega}$ the complementary solution can so be cast as follows

$$u^c(\underline{x}) = \mu^2 \underline{c}^T(\underline{x})\ \underline{\hat{U}} \tag{14}$$

where vector $\underline{c}$ collects the effect on the complementary solution due to a unit $\hat{U}_k$. Therefore, by taking into account Equation (2), at any $\underline{x} \in \bar{\Omega}$ one obtains

$$u(\underline{x}) \simeq \mu^2(\underline{p}^T(\underline{x}) + \underline{c}^T(\underline{x}))\ \hat{\underline{U}} \tag{15}$$

The unknown eigenvalue μ^2 and the associated eigenvector can be obtained by equating at N points Equation (15) with the approximate expression of the inertia terms (Equation (6)). When the inertia forces are approximated by means of piecewise constant functions ,as is done in the numerical examples, the collocation can be performed either in each Gaussian knot or in the barycentre of each ractangle containing the knot. By this way, the following linear algebraic eigenvalue problem with non symmetric coefficient matrix is obtained

$$\mu^2(\ \underline{\underline{P}} + \underline{\underline{C}}\ )\ \hat{\underline{U}} = \mu^2 \underline{\underline{F}}\ \hat{\underline{U}} = \underline{\underline{I}}\ \hat{\underline{U}} \tag{16}$$

where matrices $\underline{\underline{P}}$ and $\underline{\underline{C}}$ collect vectors $\underline{p}$ and $\underline{c}$ respectively and matrix $\underline{\underline{F}}$ has the obvious mechanical meaning of flexibility matrix. The non symmetry of the coefficient matrix is due to the fact that the influence coefficients at any point have been computed by means of an approximate procedure based on the solution of the Boundary Value Problem at hand via BEMs based on collocations.

Numerical Examples

Circular membrane: The free vibration analysis of a circular, homogeneous membrane with unit radius, unit ratio T / ρ and fixed along the boundary has been performed. In Table I some of the lowest eigenfrequencies obtained by using the procedure presented in the previous Sections are compared with the exact ones expressed, as is well known, in terms of Hankel functions. All the approximate results have been computed with reference to a boundary discretized by means of 16 standard linear boundary elements. A different number of piecewise constant functions, defined either on square domains (cases A,B) or on trapezoidal domains (case C) (Figure 1), has been employed to approximate the inertia forces. In obtaining the approximate particular integrals by means of cubic Hermite polynomials, the square domain circumscribing the circle has been divided into two (A) or four (B,C) equal partitions along each direction, so that 16 or 64 Gaussian points respectively have been employed as collocation points.

L-shaped membrane: The problem at hand has been used by many authors [15-17] as a significant case for testing numerical solutions. Although there exists no exact analytical solution of such a problem, very accurate values, at least for the first eigenfrequencies, have been computed by series [15], by B.Es [16] or by F.Es [17], which can be used for comparison (Table II). The B.E. solution has been obtained by using 32 standard linear elements, while the inertia force distribution has been approximated by 12 or 48 piecewise constant functions defined on square domains (Figure 2). Finally, a collocation at 16 or 64 Gauss points respectively is performed to approximate the particular integrals by means of cubic Hermite polynomials whereas the algebraic eigenvalue problem is formulated by collocating at 12 and 48 Gauss points. The numerical results for the non symmetric eigenvalue problem have been computed by using the EIGRF - IMSL routine in simple precision on an Olivetti OH5450 Computer. A more detailed numerical analysis is still necessary to test the accuracy of the method presented in this paper and to evaluate the dependence of the numerical accuracy on the number of the functions approximating the inertia forces (dimension of the eigenvalue problem) and on the approximation of the particular integrals (number of Gaussian points).

Concluding Remarks

A boundary element approach to the free vibration problem of membranes and a first set of numerical results have been presented. Actually, the proposed method is not a pure Boundary Element Method, because a domain discretization is necessary as well. However, since the main ingredient of the method is the establishment of Green's function using BEM, the method retains most of the advantages over the pure domain methods. Moreover, the method presented so far is general for not being bound to any particular structural model and reminds, in a certain way, some free vibration analyses carried out by BEM on membranes [16] and isotropic and orthotropic plates [18]. Yet, it definitely differs from these ones for being formulated without involving domain integrations.

References

[1] Banerjee P.K.; Butterfield R.: Boundary Element Methods in Engineering Sciences, McGraw-Hill 1981.
[2] Brebbia C.A.; Telles J.C.F.; Wrobel L.C.: Boundary Element Techniques, Springer-Verlag 1984.
[3] Kitahara M.: Boundary Integral Equation Methods in Eigenvalue Problems of Elastodynamics and Thin Plates, Elsevier 1985.
[4] Manolis G.D.; Beskos D.E.; Boundary Element Methods in Elastodynamics, Unwin Hyman 1988.
[5] Demey G.: Calculation of Eigenvalues of the Helmholtz Equation by an Integral Equation, Int. Jour. Num. Meth. in Engng, 10 (1976) 56-66.
[6] Hutchinson J.R.: An Alternative BEM Formulation Applied to Membrane Vibration, VII Int. BEM Conf., Italy, Springer-Verlag (1985) 6.13-6.25.
[7] Nardini D.; Brebbia C.A.: A New Approach to Free Vibration Analysis Using Boundary Elements, Appl. Math. Modelling, 7 (1983) 157-162.
[8] Partridge P.W.; Brebbia C.A.: The Dual Reciprocity B.E.M. for the Helmholtz Equation, Boundary Elements in Mech. and Electr. Engng, CM Publ. and Springer-Verlag (1990) 543-555.
[9] Ahmad S.; Banerjee P.K. : Free Vibration Analysis by BEM Using Particular Integrals, ASCE Jour. Engng Mech. 112, 7 (1986) 682-695.
[10] Wilson R.B.; Miller N.M.; Banerjee P.K.: Free Vibration Analysis of Three-Dimensional Solids by BEM, Int. Jour. Num. Meth. in Engng 29 (1990)1737-1757.
[11] Henry D.P. Jr.; Banerjee P.K.: A New Boundary Element Formulation for Two and Three - Dimensional Thermoelasticity Using Particular Integrals, Int. Jour. Num. Meth. in Engng 26 (1988) 2061-2077.
[12] Alessandri C.; Tralli A.: An Alternative Technique for Reducing Domain Integrals to the Boundary, Boundary Elements in Mech. and Electr. Engng CM Publ. and Springer-Verlag (1990) 517-529.
[13] Alessandri C.; Tralli A.: A Spline Based Approach for Avoiding Domain Integrations in BEM, submitted.
[14] Prenter P.M.: Splines and Variational Methods, Jhon Wiley & Sons 1975.
[15] Fox L.; Henrici P.; Moler C.: Approximations and Bounds for Eigenvalues of Elliptic Operators, SIAM Jour. Num. Anal. 4 (1967) 89-102.
[16] Katsikadelis J.T.; Sapountzakis E.J.: An Approach to the Vibration Problem of Homogeneous, Nonhomogeneous and Composite Membranes Based on the B.E.M., Int. Jour. Num. Meth. in Engng 26 (1988) 2439-2455.
[17] Ladeveze P.; Pelle J.P.: Accuracy in Finite Element Computation for Eigenfrequencies, Int. Jour. Num. Meth. in Engng 28 (1989) 1929-1949.
[18] Shi G.; Bezine G.: The Direct Boundary Integral Equation Method for the Free Vibration Analysis of Orthotropic Plates, Eur.Jour. Mech., A, 8, 4 (1989) 277-291.

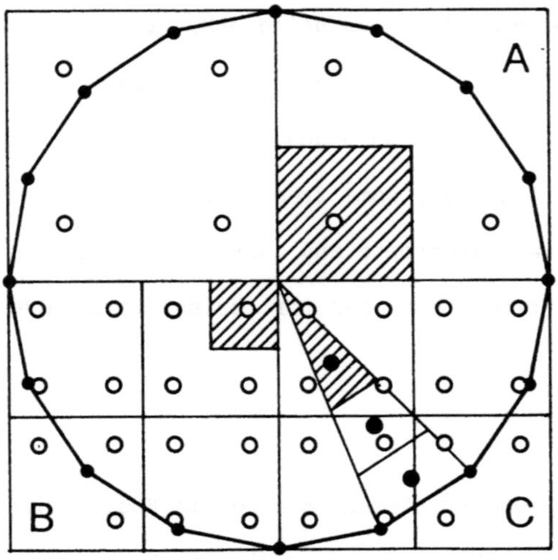

Fig. 1. Circular membrane

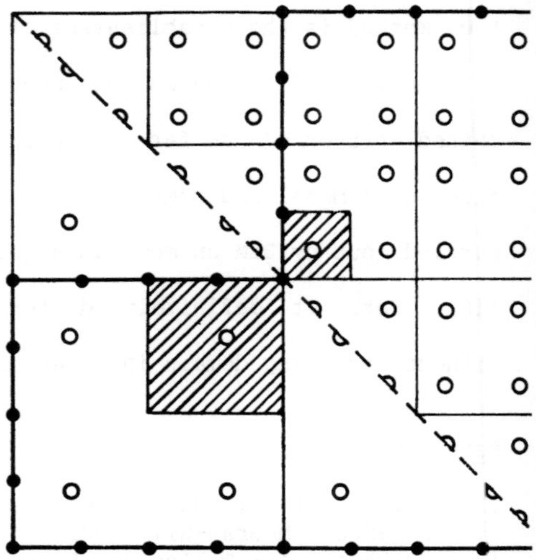

Fig. 2. L - shaped membrane

Table I Dimensionless eigenfrequencies for circular membrane.
m = number of nodal diameters
n = number of concentric nodal circles

$\Omega_{m,n}$	BEM: case A. 12 p.c. func. 16 G. points	BEM: case B. 52 p.c. func. 64 G. points	BEM: case C. 48 p.c. func. 64 G. points	Exact Solution (Hankel func.)
$\Omega_{0,1}$	2.388	2.405	2.061	2.405
$\Omega_{1,1}$	4.161	3.835	3.414	3.832
$\Omega_{0,2}$	6.033	5.487	5.271	5.520

Table II Values of the first four eigenfrequencies of an L-shaped membrane .
e(%) = discretization error computed for the F.E. mesh with 462 el.

Ω_n	BEM 12 p.c.f. 16 G. p.	BEM 48 p.c.f. 64 G. p.	[15]	[16]	[17] 96 F.Es	[17] 462 F.Es	e(%)
Ω_1	6.62	6.26	6.21	6.20	6.22	6.20	1.1
Ω_2	8.55	7.78	7.79	7.79	7.77	7.78	1.0
Ω_3	9.84	8.88	8.88	8.87	8.86	8.86	1.3
Ω_4	11.60	10.88	10.86	10.84	10.92	10.82	1.9

Two and Three Dimensional Crack Growth Using the Surface Integral and Finite Element Hybrid Method

Balkrishna S. Annigeri

United Technologies Research Center
East Hartford, Connecticut 06108
U.S.A.

William D. Keat

Massachusetts Institute of Technology
Cambridge, Massachusetts 02139
U.S.A.

SUMMARY

This paper discusses the recent developments in the modelling of 2D and 3D crack propagation using the Surface Integral and Finite Element Hybrid method. The hybrid formulation is based on the coupling of a singular integral equation representation of the fracture and a finite element representation of the uncracked body. As examples of 2D applications, the results for crack propagation in compressor and turbine blade attachments are presented. For the 3D application, the fatigue growth of a surface crack in a beam under tension is presented. Results obtained using the hybrid method agree very well with experimental data for both opening and shear modes.

INTRODUCTION

The capability of modelling evolution of fractures is very important for assessing the integrity of structural components. This need arises from the fact that materials that are used for the manufacture of structural components inherently contain flaws or defects such as inclusions, voids, porosities, microcracks etc. These flaws are usually the result of the processes used for manufacturing these components. The design of structural components thus has to be made taking into account the presence of flaws/crack like defects. These cracks can grow, under mechanical and thermal loading and also as a result of fatigue loading, and it is important to be able to predict the evolution of these flaws as a function of the loading. This capability is utilized for predicting the total useful life of structural components and for determining inspection intervals.

Several methods exist for modelling crack propagation such as the finite element, boundary element and surface integral methods. The surface integral method which is employed here is based on utilization of dislocations or force dipoles to generate a displacement discontinuity and is a very powerful technique for representing fractures. Further, this surface integral method has been combined with the finite element method resulting in the surface integral and finite element hybrid method. A brief derivation of the governing equations is given in the next section. The emphasis in this paper is on the modelling of 2D and 3D crack growth. The criteria used for propagating cracks will be presented and comparison of hybrid results with experimental data will be made to ascertain the accuracy of numerical predictions.

THE HYBRID FORMULATION

The development of the Surface Integral and Finite Element Hybrid Method [1-9] can be explained by considering the problem of a crack in a plate as shown in Fig. 1. Using superposition, the plate with a crack is equivalent to the sum of a plate without the crack and an infinite plate with a crack; ensuring that tractions and displacements are matched at the boundaries. The plate without the crack is modelled using a finite element model and the crack in an infinite medium is modelled using a surface integral model. Since, the crack is being modelled in an infinite plate, a correction has to be applied along the surface of the actual plate. Thus an appropriate boundary load correction vector R^c is computed along the surface of the finite plate using the surface integral model and applied to the finite element model to ensure that the correct load is applied to the boundary. Similarly, the traction vector T^c acting along the crack line which is the result of the forces $R - R^c$ acting on the uncracked body is computed using the finite element model and applied to the surface integral model. The governing equation for the hybrid method is obtained as follows. The finite element equation for the uncracked plate is given by:

$$[K]\,\{U^{FE}\} \;=\; \{R\} \;-\; \{R^c\} \tag{1}$$

$[K]$ = finite element stiffness matrix of the uncracked body; $\{U^{FE}\}$ = vector of unknown finite element nodal displacements for the uncracked body; $\{R\}$ = vector of applied loads; $\{R^c\}$ = load correction vector for the boundary due to the presence of the crack.

The crack in an infinite domain is represented by using a continuous distribution of dislocations for 2D applications leading to a singular integral equation formulation [2]. For 3D applications, the crack is represented by a continuous distribution of force dipoles [6]. The 2D integral equation for an infinite domain with a crack is given by:

$$t_i(\vec{x}_0) = \int_{S_c} n_j \Gamma^l_{ij}(\vec{x}, \vec{x}_0)\mu_l(\vec{x})\; dS_c \tag{2}$$

$t_i(\vec{x}_0)$ = i'th component of traction at $\vec{x}_0$ due to a distribution of dislocations along the crack surface; n_j = the j'th component of the unit normal to the crack surface at $\vec{x}_0$; $\Gamma^l_{ij}(\vec{x}, \vec{x}_0)$ = fundamental stress solution for a dislocation(kernel function); $\mu_l(\vec{x})$ = l'th component of the dislocation density at $\vec{x}$; S_c = surface area of the crack.

The 3D integral equation for an infinite domain is given by:

$$t_i(x_0) = \iint_{S_c} n_j \Gamma^l_{ij}(\vec{x}, \vec{x}_0)\; [\delta_l(\vec{x}) - \delta_l(\vec{x}_0)]\; dS_c \tag{3}$$

where $\delta_l(\vec{x})$ denotes the l'th component of displacement discontinuity on the crack surface at location $\vec{x}$. The singular integral equation is evaluated numerically in the Cauchy principal value sense and the discretized form of equations (2) and (3) are both represented by:

$$[C]\,\{F\} \;=\; \{T\} \;-\; \{T^c\} \tag{4}$$

$[C]$ = coefficient matrix for the singular integral equation; $\{F\}$ = vector of dislocation density amplitudes at the interpolation points along the crack surface (for 2D) or the vector of nodal displacement discontinuities (for 3D); $\{T\}$ = vector of applied tractions along the crack; $\{T^c\}$ = traction vector for the crack surfaces due to the applied load $R - R^c$ on the finite element model.

Equations (1) and (4) are fully coupled through the boundary force matrix G ($\{R^c\} = [G]\{F\}$) and the stress feedback matrix S ($\{T^c\} = [S]\{U^{FE}\}$) which leads to the following representation for the hybrid formulation:

$$\begin{bmatrix} K & G \\ S & C \end{bmatrix} \begin{Bmatrix} U^{FE} \\ F \end{Bmatrix} = \begin{Bmatrix} R \\ T \end{Bmatrix} \tag{5}$$

In equation (5), the variable U^{FE} represents the continuous displacement field due to only the finite element model. This has to be changed to the total displacement field by addition of the displacement in the surface integral model so that the displacement boundary conditions can be applied, as follows:

$$\{U\} \;=\; \{U^{FE}\} \;+\; \{U^{SI}\} \tag{6}$$

where $\{U^{SI}\}$ is the vector of unknown surface integral displacements occurring at the finite element nodal positions. $\{U^{SI}\}$ is obtained by evaluating the integral equation for displacements [2]:

$$\{U^{SI}\} \;=\; [L]\;\{F\} \tag{7}$$

where $[L]$ = is the displacement matrix. Thus equations(5), (6) and (7) can be combined to form the governing equation for the hybrid formulation:

$$\begin{bmatrix} K & G - KL \\ S & C - SL \end{bmatrix} \begin{Bmatrix} U \\ F \end{Bmatrix} = \begin{Bmatrix} R \\ T \end{Bmatrix} \tag{8}$$

Equation (8) allows imposition of arbitrary force, traction and displacement boundary conditions. The sparsity and symmetry of the stiffness matrix is taken full advantage of using a special solution scheme which has been reported in [3]. This scheme has been implemented in the SAFE(Surface-Integral and Finite Element) 2D and 3D Hybrid computer codes for fracture mechanics. Crack propagation theories and results are discussed in the next section.

CRACK PROPAGATION THEORY AND RESULTS

2D Crack Propagation.

The SAFE-2D code has been developed for modelling propagation of through cracks in structural components. There are several theories available for modelling crack growth such as the maximum circumferential stress theory, the minimum strain energy density theory, and the maximum energy release rate theory . In the maximum circumferential stress theory, the crack propagates perpendicular to the maximum circumferential tensile stress direction in the neighborhood of the crack tip. In the SAFE-2D code two criteria are available for propagating cracks. One is the maximum circumferential stress theory based on the stress at the existing crack tip ; the second is based on $K_{II} = 0$ at the end of the incremented crack tip. The first option is suitable for Mode I and Mode II cracks and involves less computations as the direction of crack growth is determined according to equations given below. The second option is suitable especially for modelling propagation of shear bands (non-opening cracks) and is computationally more expensive as a boundary value problem has to be solved at every iteration for determining the direction at which $K_{II} = 0$.

The Maximum Circumferential Stress Theory and Implementation.

This theory dictates that crack propagation will occur when the maximum circumferential stress evaluated near the crack tip reaches a critical value. The value of the maximum stress and the associated direction can be obtained as follows:

$$\frac{\sigma_{\theta\theta}\sqrt{2\pi r}}{K_{Ic}} = 1 = \frac{\cos\theta_0}{2}\left[\frac{K_I}{K_{Ic}}\cos^2\frac{\theta_0}{2} - 1.5\frac{K_{II}}{K_{Ic}}\sin\theta_0\right] \tag{9}$$

$$\sigma_{r\theta} = 0 = \frac{\cos\theta_0}{2}[K_I \sin\theta_0 + K_{II}(3\cos\theta_0 - 1)] \tag{10}$$

By defining $K_{II}/K_I = \alpha$ [3], it can be shown that θ_0 can be obtained from the equation given in the following:

$$\theta_0 = \arccos\left[\frac{6\alpha^2 \pm \sqrt{(4 + 32\alpha^2)}}{2(9\alpha^2 + 1)}\right] \tag{11}$$

where the crack is propagated in the θ_0 direction. The crack length increment can be chosen as a fixed length or a fixed percentage of the current crack length. These length parameters can be varied to determine convergence of the propagated path.

Results for 2D geometries.

Crack propagation in gas turbine engine compressor and turbine attachments have been modelled using the SAFE-2D code. The first example is that of a compressor blade attachment where the crack was initiated at a stress concentration location. The geometry of the blade attachment and the finite element model is as shown in Fig. 2. The SAFE-2D predicted crack propagation path is in reasonable agreement with the experimentally observed path. The modelling of the interface between the attachment and the disk lug is very important as friction can play an important role in the load redistribution and can affect the crack path. In the model shown, the interfaces are in perfect adhesion. Results have also been run allowing for complete slippage i.e. no friction. In reality, the situation is somewhere in between. Another example of crack growth in a multiple tooth turbine blade attachment is shown in Fig. 3. The experimental results for this geometry have not been obtained as of yet; the SAFE-2D code however provides capability for modelling growth of single and multiple cracks in complicated geometries.

3D Crack Propagation.

The postulation and verification of 3-D fracture criteria is only now entering the realm of practical investigation with the advent of new computational approaches and the general availability of more powerful computing resources. Among the future issues that must be considered is that of local versus global criteria. The presumption of a local criterion is that the advance of a local point on the fracture perimeter is governed solely by the state of stress at that point and thus occurs independently of all other points for small perturbations of the crack front. In contrast, a global criterion assumes an interdependence of all the local perturbations of the crack front. It is thus a characteristic of the use of a local criterion that the calculation of the updated crack front geometry is straightforwardly deterministic, whereas a global criterion will usually involve an iterative search for the most probable solution. However, our purpose here is not to evaluate the relative merits of the various criteria, but rather to demonstrate the application of the 3-D hybrid method to Mode I problems in crack propagation by choosing a representative criterion.

Two-parameter Fatigue Crack Growth Model.

In this initial crack propagation study we chose to constrain the fracture perimeter to assuming elliptical shapes. Cruse and Besuner[12] proposed two different criteria that might be used with this form of self-similar crack growth. Both are two parameter models, i.e. they reduce the search for the updated crack front to a two degree of freedom problem. The first is a local criterion based on advancing the elliptical crack front in accordance with the local values of K_I occuring on the major and minor axes of the ellipse. The second, which we employ here, is based on two average values of K_I, each found analytically by independently perturbing the major and minor axes from the current shape. The following formula is taken from [12]:

$$\overline{K}_x^2 = \frac{1}{\Delta A_x} \iint_{\Delta A_x} K^2(s)\ dA \qquad (12)$$

where $\overline{K}_x$ is the averaged value of stress intensity resulting from perturbing the axis (major or minor) which is parallel to the x-axis; ΔA_x is the area between the perturbed and unperturbed crack shapes; s is a parametric variable which is local to the crack front. Despite the averaging, this is still essentially a local criterion.

To apply the proposed two-parameter model to fatigue, we introduce the standard Paris law:

$$\frac{da}{dN} = C\,(\Delta K)^n \tag{13}$$

where for the plane strain or plane stress fracture, da is the crack extension with dN being the corresponding number of load cycles; C and n are material constants; ΔK is the stress intensity factor range. Following Cruse and Besuner, two independent equations each of the form of equation (13) are assumed to govern fatigue crack growth in the directions of the major and minor axes:

$$\frac{da_x}{dN} = C\,(\Delta \overline{K}_x)^n\,; \quad \frac{da_y}{dN} = C\,(\Delta \overline{K}_y)^n \tag{14}$$

If we then specify the maximum allowed crack extension, be it either da_x or da_y, then the corresponding increment in the complementary direction is found from the following expression which results from dividing the two equations in (14), i.e.:

$$\frac{da_x}{da_y} = \left(\frac{d\overline{K}_x}{d\overline{K}_y}\right)^n \tag{15}$$

Results for a 3D geometry.

The two-parameter model just described was implemented in the SAFE-3D code in order to model the fatigue growth of a surface crack in a thick plate under tension (see Fig. 4).

The accuracy of hybrid models akin to the one depicted in Fig. 4 was tested in [8] for various crack-depth to plate-thickness ratios and was found to lie consistently in close agreement with the finite element results of Raju and Newman [11]. The finite element component of the hybrid model was constructed of 48 eight-noded isoparametric brick elements. The surface integral model of the fracture employed a mix of interpolation functions consisting of discontinuous linear elements in the interior of the mesh, and tip elements which assume a $\rho^{1/2}$ variation of crack opening at the crack front, with ρ being the perpendicular distance from the crack front. These tip elements have also been equipped with a fifth geometric node, located at the midpoint of the side bordering on the fracture perimeter, so as to represent the crack front as a piecewise continuous series of quadratic segments.

A key feature of the hybrid model is the use of specialized dipole solutions which explicitly account for the free surface intersected by the fracture. This produces the relative independence of the finite element and surface integral meshes, as evidenced by the lack of any severe grading in either mesh. This strategy could have been taken a step further by introducing a second set of influence functions to represent the back face, but this was found to be unnecessary for the range of geometries examined.

The surface integral model of the initial precrack was incrementally advanced through the fixed finite element mesh using 17 load steps. The number of cycles corresponding to each load step can be found directly by substitution into either of equations (14). In

Fig. 5, these results have been superimposed onto the experimental results reproduced from [10]. Also appearing in the figure are the results of an alternative computational approach known as the weight-function method, which for this application was limited by the fact that it did not take account of the back face. Excellent correlation between the hybrid results and the test results can be observed.

A few additional comments regarding these results are in order. First, excellent correlation would have also been obtained had we simply applied the tabulated results of Raju and Newman in conjunction with the Besuner criterion. However, such will be the case only when we are fortunate enough to have the fracture grow as a sequence of ellipses. In fact, this is a rare event as exemplified by the case of a surface crack growing in a thick plate under bending, for which there is strong experimental evidence to suggest that the evolving shapes are nonelliptical. Thus it is noteworthy that the surface integral method is not, in any way, restricted to modelling elliptical crack shapes. Second, the importance of remeshing the fracture so as to maintain a high level of accuracy needs to be emphasized. In general one cannot always rely upon interactive means to accomplish this because of the potential use of remote supercomputing facilities. The automatic remeshing of the elliptical fractures used in this study was relatively straightforward. The capability to remesh fractures of arbitrary shape poses a greater challenge which is now being addressed by integrating cubic spline representations of the crack front, built-in rules governing crack element shape in the near-tip region, and automatic triangulation of the interior elements.

CONCLUSIONS

The purpose of this paper is to demonstrate the advantages of the Surface Integral and Finite Element Hybrid Method for modelling crack propagation. An important feature is that the fracture discretization is relatively independent of the finite element discretization of the uncracked body. In fact, the finite element discretization remains fixed as the crack advances. This feature simplifies the topological problem of remeshing the fracture and is very efficient computationally because a majority of the matrix coefficients can be retained between load steps.

ACKNOWLEDGEMENTS

The research performed in this paper was supported by the Pratt and Whitney Division of United Technologies Corporation and by a corporate sponsored research program by the United Technologies Research Center. Thanks are also due to the personnel of the illustrating group at the United Technologies Research Center for their assistance with the preparation of the figures.

REFERENCES

1. Annigeri, B.S.: Surface Integral Finite Element Hybrid Method For Localized Problems in Continuum Mechanics, Sc.D. Thesis, Department of Mechanical Engineering, M.I.T., 1984.
2. Annigeri, B.S.; Cleary M.P.: Surface integral finite element hybrid (SIFEH) method for fracture mechanics. International Journal for Numerical Methods in Engineering, Vol.20, 869-885, 1984.
3. Annigeri, B.S.; Cleary M.P.: Quasi-static fracture propagation using the surface integral finite element hybrid method. *ASME PVP Vol.85*, 1984.
4. Annigeri, B.S.: Effective modelling of stationary and propagating cracks using the surface integral and finite element hybrid method. *ASME AMD Vol.72*, 1985.
5. Annigeri, B.S.: Thermoelastic fracture analysis using the surface integral and finite element hybrid method. Presented at the ICES-88 Conference, Atlanta, Georgia, 1988.
6. Keat, W.D.; Annigeri B.S.; Cleary, M.P.: Surface integral and finite element hybrid method for two and three dimensional fracture mechanics analysis. International Journal of Fracture,Vol.36, 35-53, 1988.
7. Keat, W.D.: Surface Integral and Finite Element Hybrid Method For the Analysis of Three Dimensional Fractures, Ph.D. Thesis, Department of Mechanical Engineering, M.I.T., 1989.
8. Annigeri, B.S.; Keat, W.D.; Cleary, M.P.: Fracture mechanics research using the surface integral and finite element hybrid method. Proceedings of First Joint Japan/US Symposium on Boundary Element Methods, University of Tokyo, Pergamon Press, 191-202, 1988.
9. Keat, W.D.; Cleary, M.P.: Analysis of 3-D near-interface fractures in bounded, heterogeneous domains using the surface integral and finite element hybrid method. Proceedings of International Symposium on Boundary Element Methods, United Technologies Research Center, October 1989, Springer Verlag , in press.
10. Cruse, T.A.; Meyers, G.J.; Wilson, R.B.: Fatigue growth of surface cracks. ASTM STP 631, American Society for Testing Materials, 174-189, 1977.
11. Raju, I.S.; Newman J.C.: Stress-intensity factors for a wide range of semi-elliptical surface cracks in finite thickness plates. Engineering Fracture Mechanics, Vol.11, 817-829, 1979.
12. Cruse, T.A.; Besuner, P.M.: Residual life prediction for surface cracks in complex structural details. Journal of Aircraft, Vol.4, 369-375, 1975.

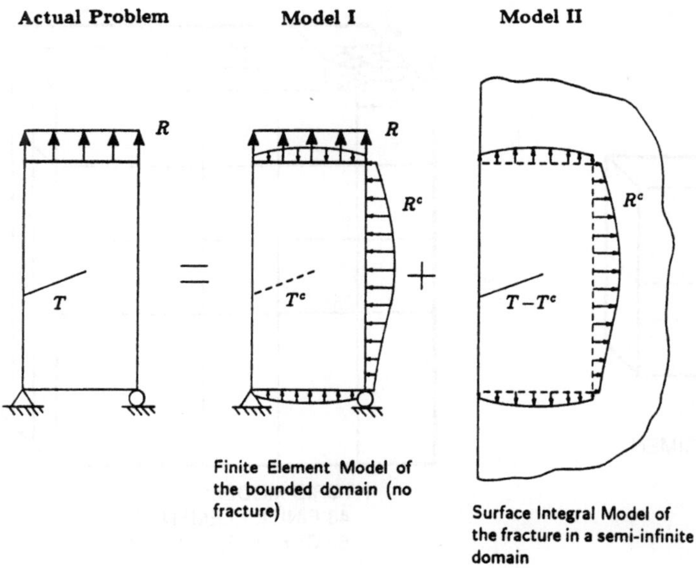

Fig.1. Superposition of the surface integral and finite element models

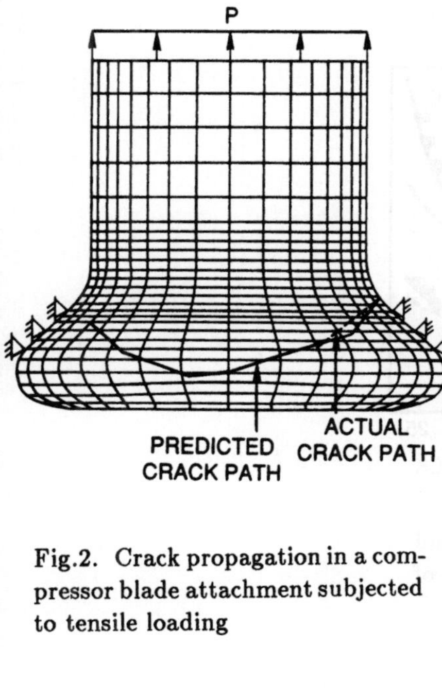

Fig.2. Crack propagation in a compressor blade attachment subjected to tensile loading

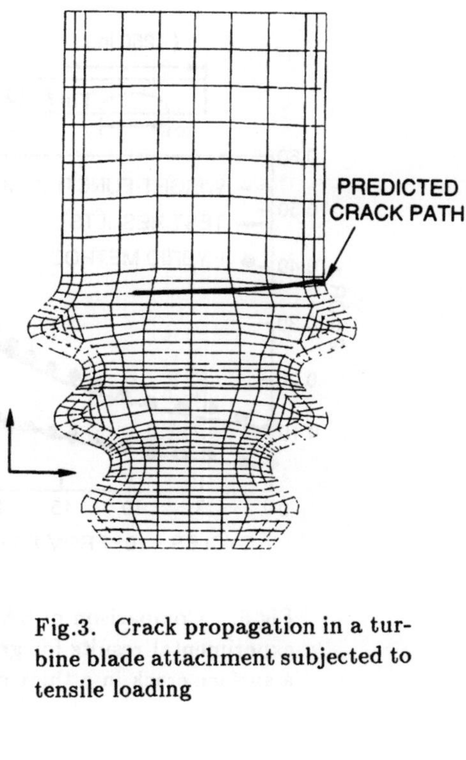

Fig.3. Crack propagation in a turbine blade attachment subjected to tensile loading

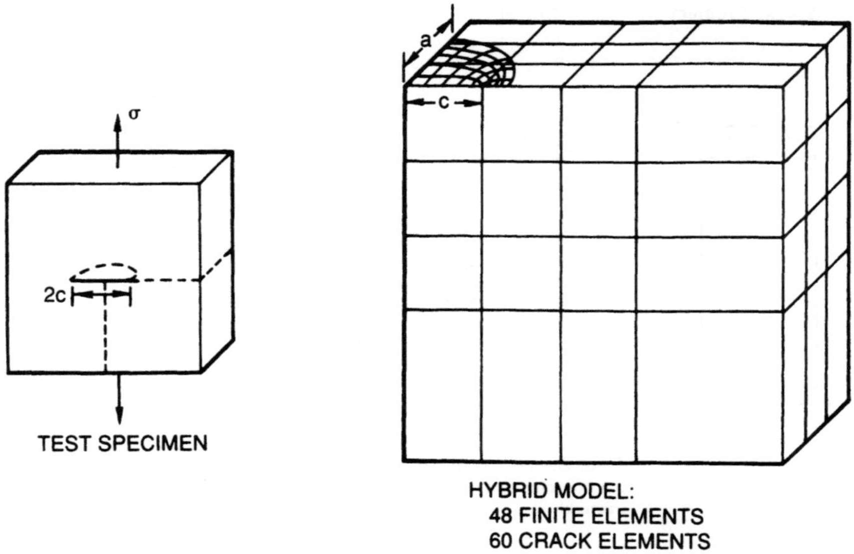

Fig.4. Specimen geometry and corresponding hybrid model used in fatigue crack growth study

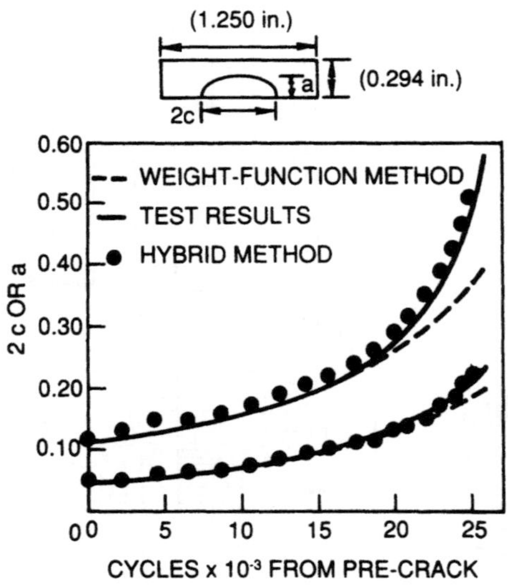

Fig.5. Comparison of SAFE and experimental results for growth of a surface crack in a thick plate

3-D Sound Generated by Moving Sources

H. ANTES and T. MEISE

Institute of Applied Mechanics
Technische Universität Carolo Wilhelmina
Braunschweig, Germany

Summary

Based on Kirchhoff's integral equation, a time-stepping boundary element procedure is formulated by which many acoustic wave propagation phenomena can be studied. Especially, exterior radiation problems are considered where sound sources can move arbitrarily in a three-dimensional surrounding, and the scattering of the acoustic waves at obstacles is treated. Some examples of environmental noise pollution problems, like noise from airplanes flying over a residential area, are presented in order to demonstrate the wide range of possible applications.

Integral / Boundary Element Equations in 3-D Acoustics with Moving Sources

A mathematical model for the propagation of small-amplitude acoustic waves through a homogeneous acoustic medium is the linear wave equation

$$\Delta p(\mathbf{x},t) - c^{-2} \frac{\partial^2}{\partial t^2} p(\mathbf{x},t) = - \rho \frac{\partial}{\partial t} \gamma(\mathbf{x},t) \qquad \text{for } \mathbf{x} \in \Omega \text{ , } t \geqq 0 \text{ .} \tag{1}$$

The scalar function $p(\mathbf{x},t)$ is the excess acoustic pressure at a point $\mathbf{x}$ at time t, c is the speed of sound waves in the medium and $\gamma(\mathbf{x},t)$ denotes the density of sound sources in the interior of the domain Ω. Assuming homogeneous initial conditions, i.e. a quiescent past of the acoustic domain, the determination of the sound distribution in the region Ω, interior or exterior to the boundary surface Γ means to find the unique solution of the equation (1) which satisfies the boundary conditions

$$p(\mathbf{x},t) \;=:\; \bar{p}(\mathbf{x},t) \qquad \text{for } \mathbf{x} \in \Gamma_1 \text{, } t \geqq 0 \text{ ,} \tag{2}$$

$$\frac{\partial}{\partial n_x} p(\mathbf{x},t) \;=:\; q(\mathbf{x},t) = \bar{q}(\mathbf{x},t) \qquad \text{for } \mathbf{x} \in \Gamma_2 \text{, } t \geqq 0 \text{ .} \tag{3}$$

The conditions (3) describe the case of acoustic scattering by perfectly hard surfaces and of the radiation by surfaces with known velocity distributions ($-\rho \frac{\partial}{\partial t} v_n = q$). Absorbing surfaces are characterized by the so-called Robin condition:

$$q(\mathbf{x},t) = -\frac{1}{c}\left[\frac{1-\alpha_r(\mathbf{x})}{1+\alpha_r(\mathbf{x})}\right]\frac{\partial}{\partial t}p(\mathbf{x},t) = -A(\mathbf{x})\frac{\partial}{\partial t}p(\mathbf{x},t) \quad , \tag{4}$$

where $\alpha_r(\mathbf{x})$ and $A(\mathbf{x})$ give the reflection factor and the admittance of the surface, respectively.

The reformulation of this differential equation problem as a boundary integral equation, the so-called Kirchhoff' integral equation can be found using the method of weighted residuals (e.g. Morse and Feshbach [2]). One obtains for three-dimensional domains Ω with the boundary Γ ($t' = t-\tau$):

$$d(\xi)\ p(\xi,t) = \int_0^t \Big\{ \oint_\Gamma \left[q(\mathbf{x},\tau)\ p^*(\mathbf{x},\xi;t') - p(\mathbf{x},\tau)\ q^*(\mathbf{x},\xi;t')\right] d\Gamma_{\mathbf{x}} + \rho \int_\Omega \frac{\partial}{\partial t}\gamma(\mathbf{x},\tau)\ p^*(\mathbf{x},\xi;t')\ d\Omega_x \Big\}\ d\tau, \tag{5}$$

where the jump coefficient $d(\xi)$ is 1 and 0.5 for points ξ in the interior of Ω and on a smooth surface boundary Γ, respectively. At 3-D corners and edges with the inner solid angle $\Delta\Omega$, it holds $d(\xi) = \frac{\Delta\Omega}{4\pi}$.

The adequate full space fundamental solution $p^*(\mathbf{x},\xi;t')$ of Equ. (1) and its normal derivative $q^*(\mathbf{x},\xi;t')$ are given by

$$p^* = \frac{1}{4\pi}\frac{1}{r}\,\delta(t'-\frac{r}{c}) \tag{6}$$

$$q^* = \frac{1}{4\pi r}\left[-\frac{1}{r}\,\delta(t'-\frac{r}{c}) + \frac{\partial}{\partial r}\delta(t'-\frac{r}{c})\right]\frac{\partial r}{\partial n} \tag{7}$$

where

$$r = |\ \mathbf{x} - \xi\ | \qquad \text{and} \qquad R = \sqrt{c^2 t'^2 - r^2} \tag{8}$$

δ is the Dirac function. If only noise propagation above horizontally oriented hard reflecting surfaces shall be studied, it is possible to use the so-called half-space fundamental solution which is obtained by introducing the mirror source points (with respect to this surface) as additional singular points [2].

In three dimensions, due to the known properties of the Dirac function and obeying the relation

$$\frac{\partial}{\partial r}\delta(t'-\frac{r}{c}) = c\,\frac{\partial}{\partial \tau}\delta(t'-\frac{r}{c}) \quad , \tag{9}$$

this integral equation (5) can easily be transformed to the integro-differential equation ("$\bullet$" marks differentiation with respect to time)

$$d(\xi)\ p(\xi,t) - \oint_{\Gamma} \frac{\partial r}{\partial n}\ \frac{1}{4\pi r}\left[\ \frac{1}{c}\ \dot{p}(\mathbf{x},t_r) + \frac{1}{r}\ p(\mathbf{x},t_r)\right] d\Gamma_{\mathbf{x}}$$
$$= \oint_{\Gamma} \frac{1}{4\pi}\left[\frac{1}{r}\ q(\mathbf{x},t_r)\right] d\Gamma_{\mathbf{x}} + \rho \int_{\Omega} \dot{\gamma}(\mathbf{x},t_r)\ \frac{1}{4\pi r}\ d\Omega_{x} \tag{10}$$

where $t_r = t - \frac{r}{c}$ denotes the so-called 'retarded' time with the property $p(\mathbf{x},t_r) = q(\mathbf{x},t_r) = \dot{\gamma}(\mathbf{x},t_r) = 0$ for $t_r \leqq 0$. As a special case, sources can be concentrated to any interior point $\mathbf{x}_l$, i.e.

$$\gamma(\mathbf{x},t) = \sum_{l=1}^{L} g_l(t)\ \delta(\mathbf{x}-\mathbf{x}_l) \ , \tag{11}$$

where $g_l(t)$ gives the time dependent strength of the l-th source. Then, the source term in Equ. (10) becomes ($t_{lr} = t - |\mathbf{x}_l - \xi|/c$)

$$\rho \int_{\Omega} \dot{\gamma}(\mathbf{x},t_r)\ \frac{1}{4\pi r}\ d\Omega_{x} = \sum_{l=1}^{L} \dot{g}_l(t_{lr})\ \frac{\rho}{4\pi}\ \frac{1}{|\mathbf{x}_l - \xi|} \ . \tag{12}$$

This expression is also correct [1,2,3], when the position $\mathbf{x}_l$ of this point source is moving with a constant speed $\mathbf{v}_l$, i.e. $\mathbf{x}_l(t_r) = \mathbf{v}_l t_r + \mathbf{x}_l(0)$.

The discrete form of this Equ. (10) is developed by the usual boundary element techniques using point collocation, where interpolation functions as simple as possible are implemented. Thus, for every plane triangular surface segment Γ^e

$$p(\mathbf{x},\tau) = \frac{1}{\Delta t}\left[\ (t_{m+1} - \tau)\ p^{em} + (\tau - t_m)\ p^{em+1}\right] \quad \text{for } \mathbf{x} \in \Gamma^e \tag{13}$$

$$q(\mathbf{x},\tau) = q^{em+1} \quad \text{for } \mathbf{x} \in \Gamma^e \tag{14}$$

is introduced at the time interval $[t_m, t_{m+1}]$ of length Δt. Finally, assuming all time steps Δt to be of the same duration, one obtains the following system of algebraic equations as a discrete analogue of equation (10):

$$(\frac{1}{2}\mathbf{I} + \mathbf{Q}^{(1)})\bullet\mathbf{p}^{(n)} + \sum_{m=1}^{n-1} \mathbf{Q}^{(n+1-m)}\bullet\mathbf{p}^{(m)} = \sum_{m=1}^{n} \mathbf{P}^{(n+1-m)}\bullet\mathbf{q}^{(m)} + \mathbf{r}^{(n)}. \quad (15)$$

The block matrices $\mathbf{Q}^{(m)}$ and $\mathbf{P}^{(m)}$, $m=1,2,\ldots,n$, and, for point sources with constant intensity change $\dot{g}_l(t) = g_{l0}$, the source term vector $r^{(n)}$ are determined to be

$$\mathbf{P}^{(m)} \triangleq \frac{1}{4\pi} \int_{\Gamma^e} \frac{1}{r_j} H(m-\frac{r_j}{c\Delta t})\, H(\frac{r_j}{c\Delta t} -m+1)\, d\Gamma_{\mathbf{x}} \quad (16)$$

$$\mathbf{Q}^{(m)} \triangleq \int_{\Gamma^e} \frac{n_i(\mathbf{x})}{4\pi r_j^2} \frac{\partial r_j}{\partial x_i} \left\{ \begin{array}{l} m\, H(m-\frac{r_j}{c\Delta t})\, H(\frac{r_j}{c\Delta t} -m+1) - \\ -(m-2)\, H(m-1-\frac{r_j}{c\Delta t})\, H(\frac{r_j}{c\Delta t} -m+2) \end{array} \right\} d\Gamma_{\mathbf{x}} \quad (17)$$

$$\mathbf{r}^{(n)} \triangleq \frac{\rho}{4\pi} \sum_{l=1}^{L} g_{l0} \frac{1}{|\mathbf{x}_l - \xi_j|} \quad . \quad (18)$$

The integrals over the "non-singular" boundary elements and segments have been evaluated approximately using 21-point triangular Gaussian quadrature, while on straight elements in the first time step containing the "singular" point ξ the integrations for matrix **P** have been carried out analytically [2,4,5] (the corresponding elements of **Q** are zero for plane elements containing both, the point ξ and the point **x**).

The Effect of Noise Source Movements on 3-D Noise Distribution

Since several years, the noise pollution due to public traffic has become an important problem in urban areas. Along highways, the noise pollution can be rectified by erecting barriers. There, as shown by Filippi [6], the continuously produced traffic noise and the mostly long, almost straight screen can be represented by a simple, two-dimensional modelization. But often in residential areas, the traffic is not such big that long noise barriers make sense. Moreover, the noise is produced by single driving cars or motor bicycles, and, fortunately only sometimes, by airplanes landing or taking-off at a nearby-located airport. For studying such noise problems, in any case a three-dimensional analysis in time domain is necessary because such noise sources are moving in a real three-dimensional surrounding.

Two samples of such moving noise source problems are considered:

(i) a car, represented by a point noise source, driving straight with a constant speed of 152 km/h through an assembly of buildings (Fig. 1)

(ii) an airplane, again modeled as a point noise source, flying with 15 grade increase of height and a constant speed of 504 km/h across one and two small houses (Figs. 6a and 6b), respectively.

Since the ground is assumed to be hard reflecting, the half-space fundamental solutions can be used. Therefore, only the boundaries of the buildings have to be discretized.

The numerical results of the first example [2] are obtained using time steps of duration Δt = 0.059 secs and only 84 boundary elements for the 4 buildings: 16 and 26 elements for modelling the small 20 m x 5 m x 10 m houses and the larger 33 m x 11 m x 20 m buildings, respectively. The car, i.e. noise source of constant intensity $1 \cdot 10^3$ Pa/m in a heigth 1 m above ground, starts 80 meters ahead of the buildings. The car arrives at a position (A-A) between the first two buildings 2 secs, i.e. $34\Delta t$ later. Then, the noise level distribution, plotted in a horizontal section view in Figure 2, shows different effects in the neighbourhood of the buildings, clearly corresponding to their different positions and sizes. Both, the big building on the right and the smaller one on the left of

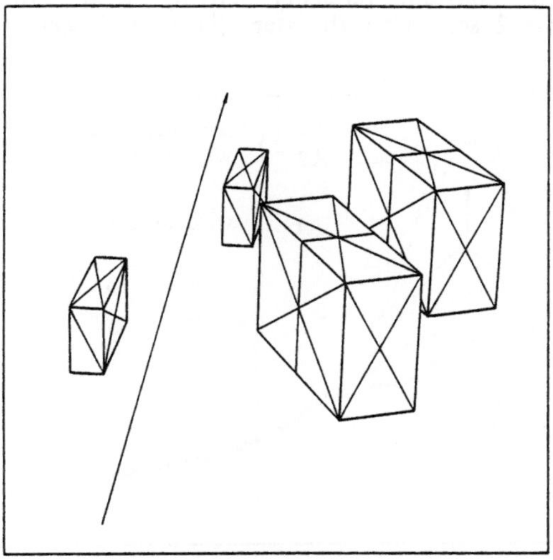

Fig. 1. Course and discretized 3-D surrounding of a moving noise source

the actual position of the noise source, act more or less as noise barriers. Moreover, as shown in Fig. 3 by the time history of the noise at points A1 and A2, located centrally at the front and the back side of the bigger building, respectively, the reflections between these two buildings even amplify the noise.

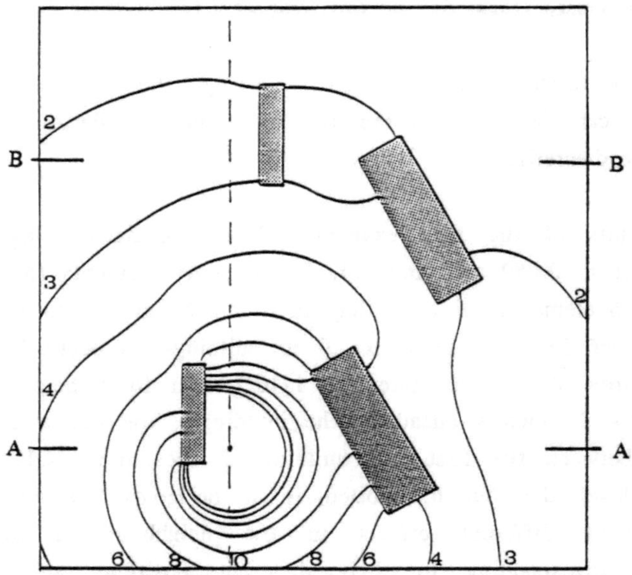

Fig.2. Noise source of constant intensity moving with 152 km/h: Noise level distribution at time 2 secs after the start (horizontal section)

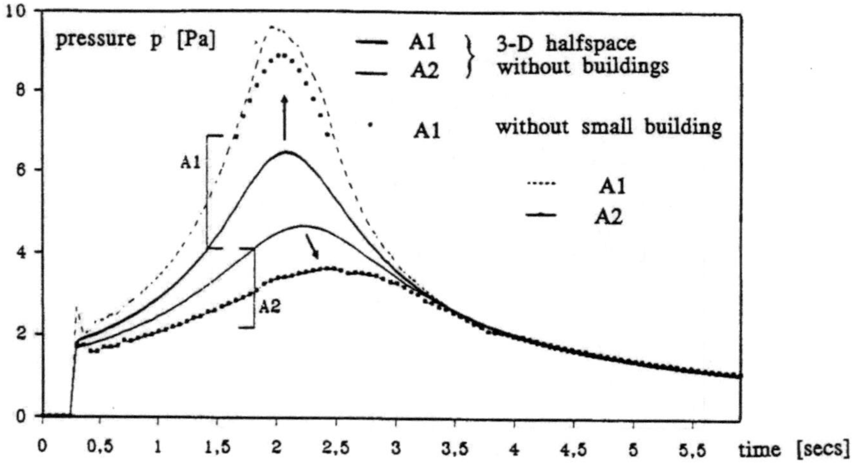

Fig.3. Time history of noise pressure at points A1 and A2: Noise amplification due to the reflections between the two buildings

On the other hand, the second small house in front of and oriented parallel to the noise source course causes almost no disturbance of the wave front. Also, when the actual position of the noise source is by the side of it, i.e. 3.42 secs after the start, this small house does not protect the area behind of it against noise very much. This is visible in Fig. 4, presenting in a vertical section view (B-B) the noise level distribution at this moment: the noise level line "8", for instance, indicates an equal noise in almost the same distance on both sides of the noise source, regardless of the existence of this buildings. Only the very loud near field noise is reduced by the barrier of the building.

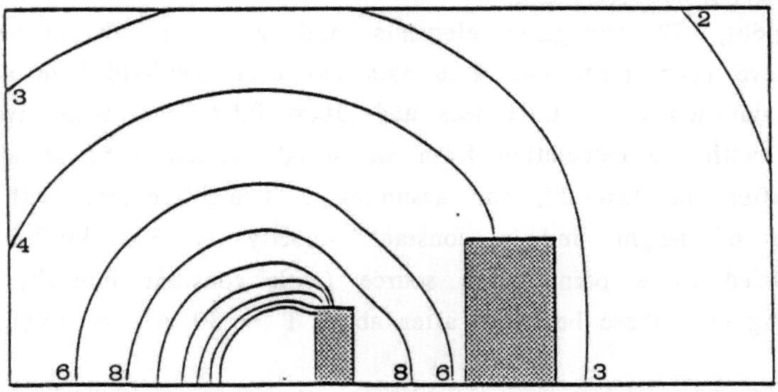

Fig.4. Noise level distribution (vertical section at position B-B) at time 3.42 secs after the start of a point noise source moving with 152 km/h

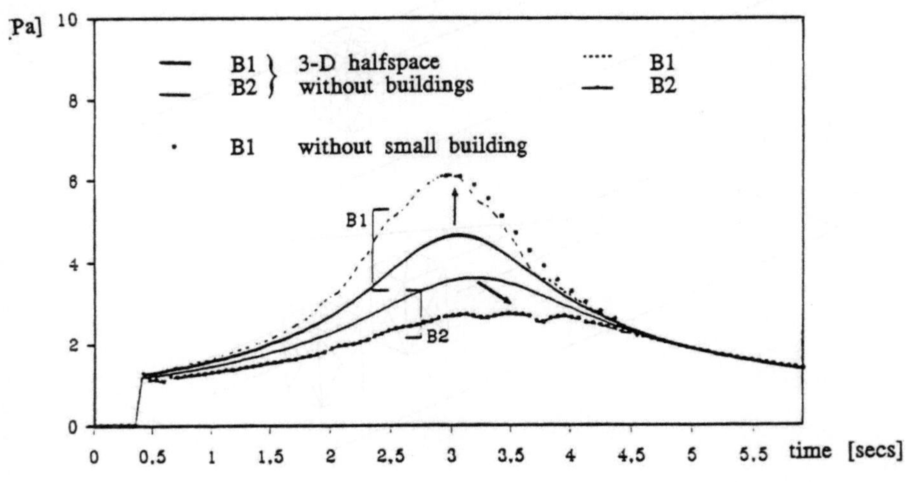

Fig.5. Time history of noise pressure at points B1 and B2: Effect of the small building's existence on the noise due to a passing 'car' (152 km/h)

In Fig.5 , the time history of the the noise pressure at the points B1 and B2, located at the front and the back center of the second big building, respectively, makes this fact evident: at both points, the noise level changes while the 'car' is passing, but all the time, there is no noise reduction due to the existence of this small building parallel to the car's course.

In the second example, again, a boundary element discretization of only the buildings has to be accomplished to study the noise distribution due to an airplane take-off. For these buildings of size (4 m x 4 m x 6 m), in the one-building example (Fig.6a) and in the configuration with two buildings (Fig.6b), 72 triangular elements and 2 times 18 elements, respectively, have been used. The time axis has been subdivided in equal time steps of duration $\Delta t_1 = 0.01$ secs and $\Delta t_2 = 0.02$ secs, respectivelly. Then, starting with the calculation from an initial position 0.87 m above ground, i.e. after the take-off, and assuming a straight course with 15 grade increase of height and a constant velocity of 504 km/h, the air-plane, modeled as a point noise source (with constant intensity 10^3 Pa/m^3), is flying over these buildings after about $T = 30\,\Delta t_1 = 15\Delta t_2 = 0.30$ secs.

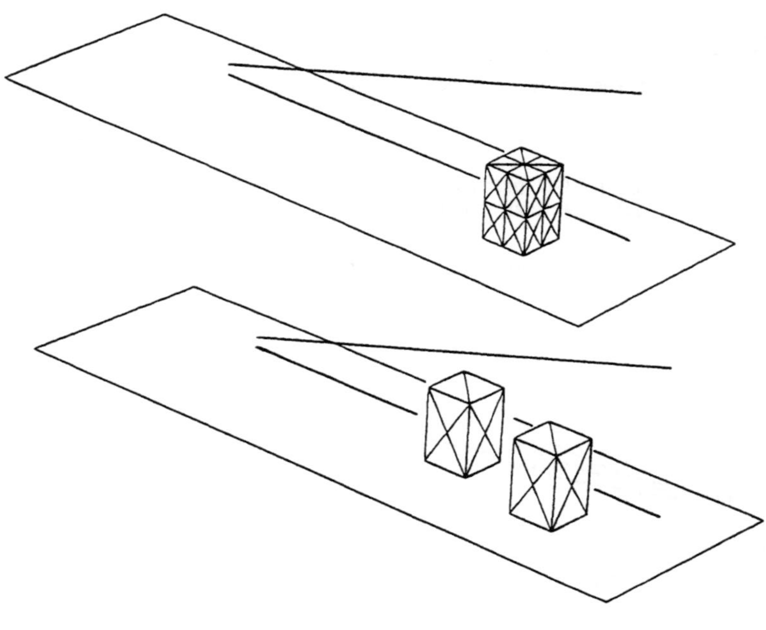

Fig.6. Course and discretized 3-D surrounding of a point noise source 'flying' (a) over one building (b) over two buildings

For such problems, the most important task is not to find the detailed time history of the noise distribution, but to know about the noise maxima ever reached at every observation point during the whole considered period T: $p_{max}(x) = \max|p(x,t), 0 \leqq t \leqq T|$.

In Figs. 7a and 7b, the distributions of these noise maxima are plotted. As expected, there is almost no noise reduction behind the buildings: the reduction is the less, the closer the airplane is flying over the buildings.

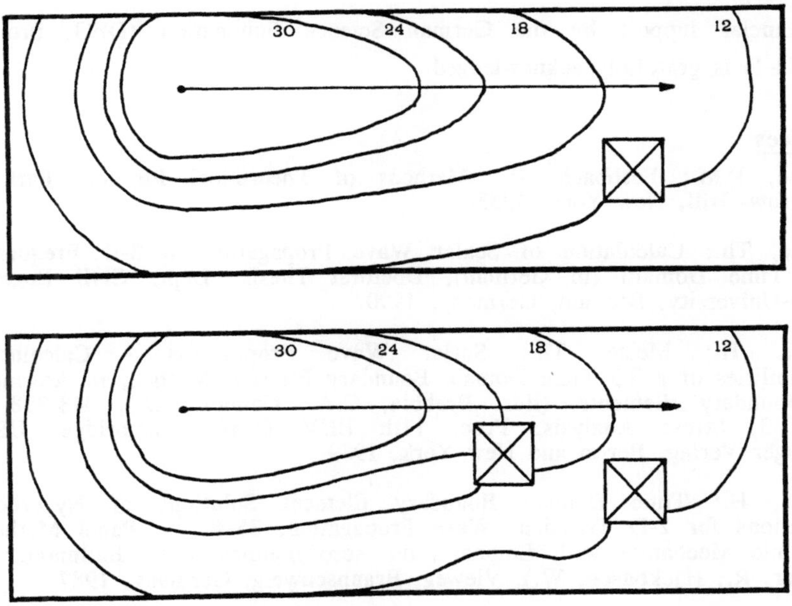

Fig.7. Noise maxima distribution of a point noise source: 'airplane' flying with 504 km/h over (a) one building (b) two buildings

More examples, e.g. parametric studies concerning 3-D noise barriers may be found in [7].

Conclusions

A 3-D time domain boundary element procedure (code name NOISE-3D) has been successfully applied to solve environmental noise problems with single noise sources moving in a three-dimensional surrounding. The range of possible and interesting applications is very wide and not so much restricted by the methodology, but by the computer facilities. Here, HP 9000/300 workstations and the IBM 3090-600J supercomputer have been used.

Thus, there is a demand for improving this computer code with regard to an optimal data storage management, to introduce parallel processing, a.s.o.

Since an extension of Kirchhoff's integral equation to sound radiating from moving closed surfaces has already been given by Farassat et al. [8,9], the incorporation of this additional possibility into the above prescribed time-stepping boundary element procedure is under work.

Acknowledgement
The financial support by the German Science Foundation (DFG, project An 140/2-1) is gratefully acknowledged.

References

1. Morse, P.M.; Feshbach, H.: Methods of Theoretical Physics, Part I, McGraw- Hill, New York, 1953.

2. Meise, Th.: Calculation of Scalar Wave Propagation in 3-D Frequency and Time Domain (in German), Doctoral Thesis, Dept. Civil Engng., Ruhr-University, Bochum, Germany, 1990.

3. Antes, H.; Meise, Th.: Scalar Wave Propagation - Calculation Capabilities of a 3D Time Domain Boundary Element Method, in Advances in Boundary Elements, (Eds. Brebbia, C.A.; Connor, J.J.), 343-358 in Vol. 3, Stress Analysis, Proc. 11th BEM Conf., Cambridge, USA, Springer Verlag, Berlin and New York, 1989.

4. Antes, H.: Time Domain Boundary Element Solutions of Hyperbolic Equations for 2-D Transient Wave Propagation, 35-42 in: Panel Methods in Fluid Mechanics with Emphasis on Aerodynamics (eds.: Ballmann, J.; Eppler, R.; Hackbusch, W.), Vieweg, Braunschweig, Germany, 1987.

5. Antes, H.: Applications of the Boundary Element Method in Elastodynamics and Fluiddynamics (in German), Mathematische Methoden in der Technik, Vol. 9, Teubner, Stuttgart, Germany, 1988.

6. Filippi, P.J.T.: Integral Equations in Acoustics, 1-49 in: Theoretical Acoustics and Numerical Techniques (ed.: Filippi, P.J.T.), CISM Courses and Lectures No. 277, Springer, Wien-New York, 1983.

7. Antes, H.: Applications in Environmental Noise, Chapt. 11 in: Advances in BEM in Acoustics (eds.: Ciskowski, R.D.; Brebbia, C.A.), Comp. Mech. Publ., Southampton (in print).

8. Farassat, F.; Myers, M.K.: Extension of Kirchhoff's Formula to Radiation from Moving Surfaces, J. Sound Vibrations 123 (1988) 451-460.

9. Farassat, F; Myers, M.K.: The Moving Boundary Problem for the Wave Equation: Theory and Application, 21-44 in: Computational Acoustics: Algorithms and Applications (eds.: Lee, D.; Sternberg, R.L.; Schultz, M.H.), Elsevier, Amsterdam, 1988.

A Boundary Element Procedure for the Analysis of Two-Dimensional Elastic Structures

M. Aristodemo and E. Turco

Dipartimento Strutture, Università della Calabria
87030 Arcavacata di Rende, Cosenza, Italy

Summary

The paper proposes a boundary element discretization for the analysis of 2-D elastic problems. The procedure is designed paying attention to the efficiency of the interpolation used to describe the mechanical quantities on the boundary and to the precision and the cost of the evaluation of boundary integrals. In particular, a non- traditional quadratic interpolation and an analytical integration are used. The features of the interpolation and the common structure of the kernels allow the use of the same discretization model for the analysis of plane stress and plate bending problems. Some applications relative to both problems are presented.

Introduction

The efficiency of a numerical model based on boundary formulation is largely dependent on the computational strategies followed to derive the algebraic system from that of boundary equation and to evaluate the solution inside the domain from the boundary one. The aspects which appear to significantly influence the performance of the model are the interpolation used for modeling the boundary quantities and the precision achieved in the evaluation of the boundary integrals. We note that both these aspects are relevant when a formulation based on a singular fundamental solution is used. In this case the saving of variables produced by an efficient interpolation is more appreciable because of the non-symmetrical structure of the system matrix, and the presence of singular and quasi-singular kernels poses serious problems of accuracy to the integral evaluation.

The paper proposes a discretization procedure designed with a particular attention to the aspects outlined above, aiming to optimize the use of the computational resources. The main feature of the procedure is the capability of giving accurate solutions making use of a very small number of unknowns.

Two-dimensional problems

The procedure is devoted to the analysis of plane elasticity and plate bending problems. This connection follows from the common features of the kernels arising in these problems which can be handled with similar procedures. Other 2-D problems, as the axisymmetric states, give rise to a quite different kernel structure.

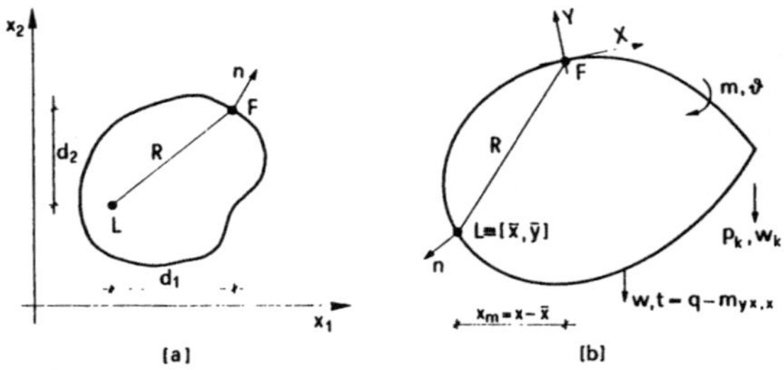

Figure 1: Plane [a] and plate bending [b] problems.

The distinction between plane problems and plate bending is restricted to the maximum order of singularity which happens in the boundary integrals and some minor formulation differences, such as the introduction of the corner reaction required in plate bending problems.

We recall here the fundamental solutions for both these problems with the aim of discussing later the problems of boundary integrals evaluation.

For plane elasticity the variables are indicated in Fig.1-a. The fundamental solution for the displacement component is expressed, in a global reference system, by the relations

$$u^*_{ij} = -\frac{1}{8\pi G(1-\nu)}\left[(3-4\nu)\delta_{ij}\ln R - \frac{d_i d_j}{R^2}\right] \tag{1}$$

while the traction components have the form

$$t^*_{ij} = -\frac{1}{4\pi(1-\nu)R^2}\left[(1-2\nu)(n_j d_i - n_i d_j) + \left((1-2\nu)\delta_{ij} + \frac{2d_i d_j}{R^2}\right) d_k n_k\right] \tag{2}$$

For plate bending using a force and a couple as sources, with the notation of Fig. 1-b the fundamental solution associated to the force can be expressed as

$$\begin{aligned}
w^*_F &= \frac{1}{8\pi D}\left[R^2\left(\ln R - \frac{1}{2}\right)\right] && (3)\\
\theta^*_F &= -\frac{\bar{y}}{4\pi D}\ln R && (4)\\
m^*_F &= -\frac{1}{4\pi}\left[k_1 \ln R + \frac{\nu x_m^2 + \bar{y}^2}{R^2}\right] && (5)\\
t^*_F &= \frac{1}{4\pi}\frac{\bar{y}\,[k_1 x_m^2 - k_4 \bar{y}^2]}{R^4} && (6)
\end{aligned}$$

while the terms associated to the couple are

$$\begin{aligned}
w^*_C &= \frac{1}{4\pi D}\ln R\,[x_m n_1 - \bar{y} n_2] && (7)\\
\theta^*_C &= \frac{1}{4\pi D}\left[-\frac{\bar{y} x_m}{r^2} n_1 + \left(\frac{\bar{y}^2}{R^2} + \ln R\right) n_2\right] && (8)
\end{aligned}$$

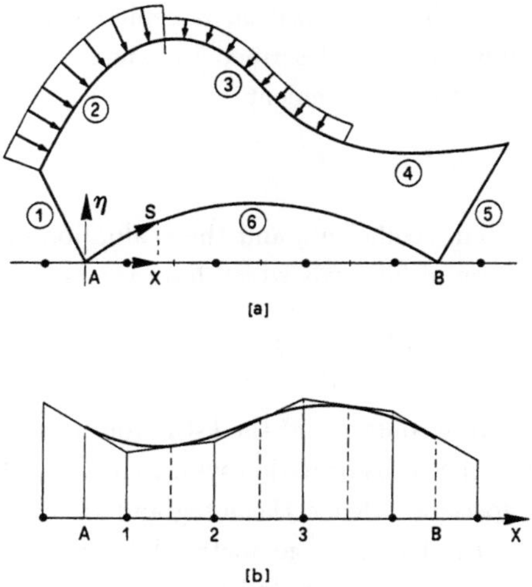

Figure 2: Boundary discretization

$$m_C^* = -\frac{1}{4\pi R^2}\left[\left(k_1 x_m^3 + k_3 x_m \bar{y}^2\right) n_1 + \left(k_4 x_m^2 \bar{y} - k_1 \bar{y}^3\right) n_2\right] \quad (9)$$

$$t_C^* = -\frac{1}{4\pi R^6}\left[2\left(k_1 x_m^3 \bar{y} + k_5 x_m \bar{y}^3\right) n_1 + \left(k_1 x_m^4 + k_4 \bar{y}^4 + 6 k_2 x_m^2 \bar{y}^2\right) n_2\right] \quad (10)$$

where D is the flexural rigidity an the constant k depend on the Poisson ration by the relations: $k_1 = 1 + \nu$, $k_2 = 1 - \nu$, $k_3 = 3\nu - 1$, $k_4 = \nu - 3$ and $k_5 = 5 - 3\nu$.

Boundary discretization

The choices concern the assumption of a shape for the data and the unknowns on the boundary, and the selection of a level of fulfilment of the continuity requirements. In order to include the singular situations which can take place both in the geometry and in the mechanical data, the presence of corners and points in which the imposed tractions or displacements are discontinuous must be considered.

In the proposed description the boundary is regarded as an assembly of macro-elements which correspond to pieces of the boundary without any geometrical or mechanical singularity. An independent interpolation is assumed within each macro-element and then the discontinuities of the data are represented while the compatibility of the solution is not imposed a priori at the points of connection between contiguous macro-elements. The element subdivision within each macro-element is associated to a C_1 continuous representation of the mechanical quantities. This interpolation uses the fulfilment of the inter-element continuity conditions to eliminate the degrees of freedom which correspond to such conditions. A continuous interpolation is then obtained everywhere on the macroelement, constituted by quadratic functions on each element. By referring to [1] for the details, we stress that the number of parameters required by this discretization is equal to the number of elements plus 2. This then brings the advantage of having a high continuity interpolation with a number of variables near to that required by a piece-wise constant interpolation.

A schematic representation of the discretization is depicted in Fig.2. We note that the curvilinear pieces of the boundary are described by using a reference straight abscissa connecting the end points of the macroelement.

Evaluation of integrals

The construction of the system coefficients and the evaluation of the solution inside the domain requires the evaluation of integrals which have the typical form

$$F^k = \int_l f[R] x^k J[x] dx \tag{11}$$

where f denotes the fundamental solution, x^k the k-th order term of a polynomial interpolation and J the jacobian of the transformation which connects the curvilinear geometry to a reference rectilinear abscissa. Hence the integrand is constituted by: the smooth term x^k, the irrational term J (which for a geometrical description isoparametric with the assumed interpolation is a square root of a 2nd order polynomial), and the term f which tends to be singular when the distance R between the field and the source points tends to zero. It is worth noting that besides the need to handle singular terms required by the construction of the system, the evaluation of the domain solution poses the problem of higher-order quasi-singular kernels.

The more popular strategies for the evaluation of integrals involved in BEM formulations are based on a numerical approach. A standard gaussian quadrature is used when the source point is far enough from the element, while specialized techniques have been proposed to handle self-effect terms. Finite-parts integration [2], subtraction technique [3], coordinate transformations [4] are methods which are able to furnish accurate results for some kinds of singularity. In the case of quasi-singular terms only the procedure proposed by Telles [4] is available.

The numerical integration method is easy to implement. The efficiency of the resulting algorithms, however, should be compared with the alternative method based on the analytical approach to the integration. The idea of developing analytical formulae for the integrals arising in boundary equations has been supported by some researchers. In 1973 Riccardella [5] presented the analytical results relative to plane problems using a linear interpolation. More recently some works on analytical integration for BEM equations have been presented. Vable [6] for a class of fundamental solutions and Abdel-Akher and Hartley [7] for plate bending problems have developed formulae in the case of rectilinear elements. Katz [8] has proposed an approach to potential problems defined on curved boundaries.

With the aim of setting up a computationally efficient procedure we have turned our attention to the analytical integration by developing it for all the quantities involved in the BEM analysis of plane and plate bending problems defined on rectilinear boundaries. To give an idea of the handiness of the final results we report here the expressions of the integrals when the plate bending fundamental solution (3-10) is used. The complete results, relative to both the construction of the boundary solution and the evaluation of the domain solution, are available in the report [10].

By denoting with capital letters the integrals of the products between the fundamental solution and the j-th order term of the interpolation function, the final results for the force take the form

$$W_F^j = \frac{1}{8\pi D}\left[(\bar{x}^2+\bar{y}^2)f_j - 2\bar{x}f_{j+1} + f_{j+2} + \alpha_j\right] \tag{12}$$

$$\Theta_F^j = -\frac{\bar{y}}{4\pi D}f_j \tag{13}$$

$$M_F^j = -\frac{1}{4\pi}\left[k_2\bar{y}^2 D_{j,1} + k_1 f_j + \beta_j\right] \tag{14}$$

$$T_F^j = \frac{\bar{y}}{4\pi}\left[\left(k_1\bar{x}^2 - k_4\bar{y}^2\right)D_{j,2} - 2k_1\bar{x}D_{j+1,2} + k_1 D_{j+2,2}\right] \tag{15}$$

The integrals resulting from the couple are

$$W_C^j = -\frac{1}{4\pi D}\left[(\bar{x}f_j - f_{j+1})\,n_1 + \bar{y}f_j n_2\right] \tag{16}$$

$$\Theta_C^j = \frac{1}{4\pi D}\left[\bar{y}\,(\bar{x}D_{j,1} - D_{j+1,1})\,n_1 + \left(\bar{y}^2 D_{j,1} + f_j\right)n_2\right] \tag{17}$$

$$\begin{aligned} M_C^j &= \frac{1}{4\pi}\Big[\Big((k_1\bar{x}^3 + k_3\bar{x}\bar{y}^2)D_{j,2} \\ &- (3k_1\bar{x}^2 + k_3\bar{y}^2)D_{j+1,2} + k_1(3\bar{x}D_{j+2,2} - D_{j+3,2})\Big)\,n_1 \\ &- \bar{y}\left(k_4\bar{x}^2 - k_1\bar{y}^2)D_{j,2} - 2k_4\bar{x}D_{j+1,2} + k_4 D_{j+2,2}\right)n_2\Big] \end{aligned} \tag{18}$$

$$\begin{aligned} T_C^j &= \frac{1}{4\pi}\Big[2\bar{y}\left((k_1\bar{x}^3 + k_5\bar{x}\bar{y}^2)D_{j,3} - (3k_1\bar{x}^2 + k_5\bar{y}^2)D_{j+1,3}\right. \\ &+ k_1(3\bar{x}D_{j+2,3} - D_{j+3,3}))\,n_1 - \left((k_1\bar{x}^4 + k_4\bar{y}^4 + 6k_2\bar{x}^2\bar{y}^2)D_{j,3}\right. \\ &- 4(k_1\bar{x}^3 + 3k_2\bar{x}\bar{y}^2)D_{j+1,3} + 6(k_1\bar{x}^2 + k_2\bar{y}^2)D_{j+2,3} \\ &- 4k_1\bar{x}D_{j+3,3} + k_1 D_{j+4,3})\,n_2\Big] \end{aligned} \tag{19}$$

We note that the above expressions are expressed in terms of the following functions

$$f_j = \frac{1}{j+1}\left[\frac{l^{j+1}}{2}cL_j - D_{j+2,1} + \bar{x}D_{j+1,2}\right] \tag{20}$$

$$cL_j = \ln[(\bar{x}-l)^2 + \bar{y}^2] + (-1)^j \ln[(\bar{x}+l)^2 + \bar{y}^2] \tag{21}$$

$$\alpha_0 = -(\bar{x}^2+\bar{y}^2)l - \frac{l^3}{3} \tag{22}$$

$$\alpha_1 = \frac{2l^3\bar{x}}{3} \tag{23}$$

$$\alpha_2 = -(\bar{x}^2+\bar{y}^2)\frac{l^3}{3} - \frac{l^5}{5} \tag{24}$$

$$\beta_0 = 2\nu l \tag{25}$$

$$\beta_1 = 0 \tag{26}$$

$$\beta_2 = 2\nu\frac{l^3}{3} \tag{27}$$

	$R/l = 0.10$			$R/l = 0.05$		
n.g.p.	$\Theta_C^{(0)}[lnR]$	$T_F^{(0)}[1/R]$	$T_C^{(0)}[1/R^2]$	$\Theta_C^{(0)}[lnR]$	$T_F^{(0)}[1/R]$	$T_C^{(0)}[1/R^2]$
4	-5.7120	3.2111	-11.271	-7.5119	2.0674	- 35.292
6	-5.7266	4.3313	-6.4078	-7.9450	3.1735	- 46.790
8	-5.6782	4.4909	9.3931	-7.7049	4.7496	- 12.812
10	-5.6073	4.2758	-2.3312	-7.8363	3.4097	- 41.991
12	-5.6686	4.5410	1.1574	-7.6267	5.8197	50.545
14	-5.6948	4.3460	1.0031	-7.8051	3.7109	- 38.618
16	-5.6778	4.4787	8.7432	-7.6470	5.6237	57.577
analityc	-5.6842	4.4274	5.6042	-7.7208	4.7124	5.7514

Table 1: Integration of quasi-singular contributions.

and of the basic integrals

$$D_{n,m} = \int_{-l}^{l} \frac{x^n}{[(x-\bar{x})^2+\bar{y}^2]^m} dx \tag{28}$$

These can be conveniently coded taking advantage of the hierarchical structure of the analytic integration formulae [11].

Here we intend highlight some aspects of the analytical integration. Due to the closeness of the kernels, the same basic results are useful in deriving the final formulae for both plane and plate bending problems. The results can be written for a generic order of the interpolation function and can be arranged in a hierarchical organization which proves to be computationally efficient: higher order terms can be computed as function of the lower order terms. The evaluation of the contributions from singular terms are easily obtained from the generic expression of the integrals by means of a specialization of the position of the source and a limiting operation to locate the source within the element. In this case the results take a more compact form.

Several advantages result from coding the results of the analytical integration within a BEM procedure. First of all, for the computation of non-singular contributions the CPU time is considerably less than the time required by a numerical integration which uses enough Gauss points to ensure the same precision in the final results. For instance, the time required to integrate with 10 points a series of element contributions in plate bending analysis is about seven times more than the time required by the procedure which uses analytical results.

The precision of the analytical computation of singular contributions is independent of the degree of singularity, while the results obtainable from numerical schemes or ad hoc strategies usually work well only for some particular singularities. We note that the stronger singularities, such as the fundamental solution (10), pose serious problems to the non-analytical methods. We point out also that the analytical integration allows high accuracy in evaluation of the solution in the domain band near to the boundary, whereas

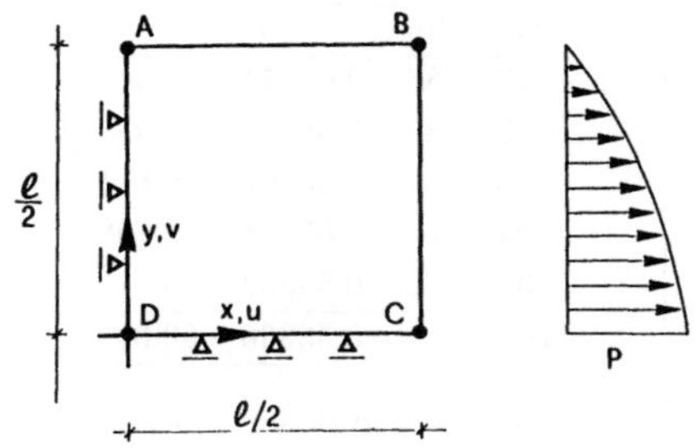

Figure 3: Square plate stretched by a parabolic load

Elements	u_C	u_B	v_B	v_A	σ_{yD}	σ_{xD}	σ_{xA}
8	0.70406	0.13449	0.02326	-0.20075	-1.2226	8.6647	4.1250
12	0.69559	0.15712	0.02363	-0.20611	-1.3553	8.6008	4.1294
16	0.69349	0.16471	0.02507	-0.20751	-1.3923	8.5807	4.1464
20	0.69302	0.16864	0.02575	-0.20763	-1.4020	8.5819	4.1499
24	0.69295	0.17090	0.02572	-0.20748	-1.4033	8.5877	4.1445
28	0.69292	0.17220	0.02544	-0.20737	-1.4032	8.5923	4.1364
40	0.69280	0.17353	0.02464	-0.20733	-1.4041	8.5955	4.1194
Exact	0.69253	0.17435	0.02435	-0.20747	-1.4095	8.5904	4.1067

Table 2: Displacements and stresses for the stretched plate.

the precision of the numerical integration tends to deteriorate. Such behaviour is shown in Table 1 where the results obtained by Telles numerical scheme [4] in the integration of three terms corresponding to different orders of singularity (indicated in square brackets) are compared with the values obtained by the analytical integration for two values of the ratio between the distance and the length of the elements.

It is evident that a general use of analytical integration should include the case of curvilinear boundaries. Here the irrational term is present in the integrand. Although it can be easily rationalized by means of standard substitutions, a high number of terms come out, and careful management is required.

Numerical results

Some test problems have been analysed to verify the effectiveness of the proposed procedure. The first problem is the square plate stretched by a parabolic load depicted in Fig.3. The data are: $l = 15m$, $E = 100N/m^2$, $\nu = 0.3$, $p_{max} = 10N/m$. The displacement and stress solution is shown in Table 2 where the comparison is made with a very refined analytical solution obtained by trigonometric and hyperbolic series expansions [12].

Table 3 refers to the problem of a cantilever ($l = 48m$, $h = 12m$, $E = 20000N/m^2$, $\nu = 0.25$) bent by an edge parabolic load having a total weight of 40 N. w represents the

Elements	dof	w	σ_x
10	36	0.68254	75.418
12	40	0.56121	61.358
18	52	0.53407	59.841
24	64	0.53321	59.798
30	76	0.53363	59.913
Series		0.53360	60.000

Table 3: Cantilever beam: Tip deflection and normal stress.

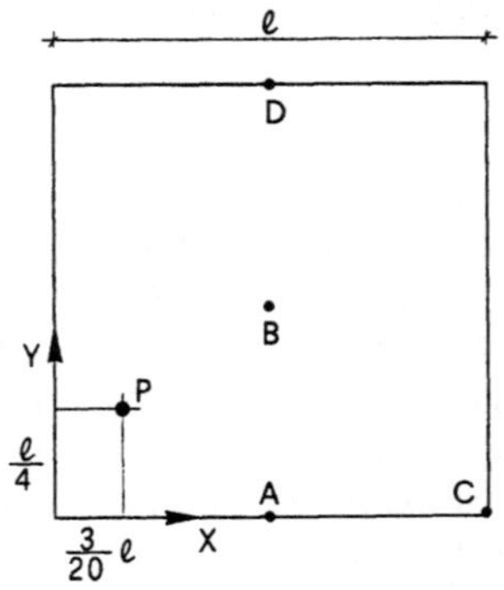

Figure 4: Simply supported square plate

tip deflection and σ_x the maximum normal stress component in the section located at a quarter of the length.

With regard to plate bending problems a simply supported square plate subjected to a uniform load has been analysed. The adimensional results shown in Table 4 are compared with another Bem solution based on cubic and linear interpolation [13] and the values of the series expansion analysis [14].

The same square plate has been analysed also in the case of an eccentric loading represented by a force acting at the point P of Fig.4. Table 5 shows some values of the adimensional solution, and Figure 5 plots the bending moment along a median line of the plate.

	Elements	dof	T_A/ql	M_B/ql^2	$w_B D/ql^4$	R_C/ql^2
[13]	8	20	0.594		0.00457	0.0470
	20	44	0.450		0.00413	0.0609
	40	84	0.423		0.00408	0.0639
Present analysis	4	28	0.4294	0.04850	0.004132	0.0665
	8	36	0.4329	0.04832	0.004112	0.0641
	12	44	0.4205	0.04810	0.004088	0.0604
	20	60	0.4210	0.04794	0.004068	0.0622
	28	76	0.4203	0.04790	0.004065	0.0635
	36	92	0.4205	0.04789	0.004064	0.0643
series			0.420	0.0479	0.00406	0.065

Table 4: Simply supported square plate under uniform load

Elements	dof	M_{xB}/ql^2	M_{yB}/ql^2	M_{xyA}/ql^2	w_B/ql^4
4	28	0.04303	0.05356	-0.03564	0.005326
12	44	0.06183	0.07465	-0.05604	0.007613
20	60	0.06133	0.07386	-0.05846	0.007536
28	76	0.06059	0.07304	-0.05715	0.007446
36	92	0.06054	0.07300	-0.05701	0.007441
series		0.06047	0.07293	-0.05709	0.007449

Table 5: Simply supported square plate under eccentric load

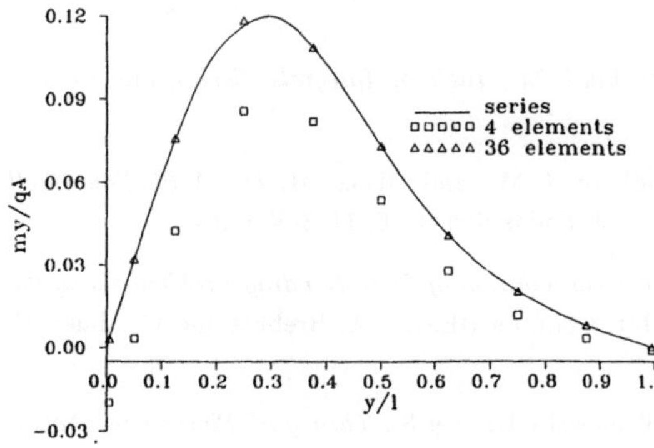

Figure 5: Simply supported square plate under eccentric load: bending moments on median line

References

[1] Aristodemo M., *A High-Continuity Finite Element Model for Two-Dimensional Elastic Problems*, Computers & Structures, vol. 21, n.5, 987-993, (1985).

[2] Kutt H. R., *The Numerical Evaluation of Principal Value Integrals by Finite-Part Integration*, Numer. Math.,vol. 24, n.3, 205-210,(1975).

[3] Guiggiani M., Casalini P., *Direct Computation of Cauchy Principal Value Integrals in Advanced Boundary Elements*, Int. j. numer. methods eng., vol.24, 1711-1720, (1987).

[4] Telles J.F.C., *A Self-Adaptive Co-ordinate Transformation for Efficient Numerical Evaluation of General Boundary Element Integrals*, Int. j. numer. methods eng., vol.24, 959-973, (1987).

[5] Riccardella, P.C., *An implementation of the boundary-integral tecnique for planar problems of elasticity and elasto-plasticity.* Ph.D. thesis, Carnegie Mellon University.

[6] Vable, M., *An Algorithm based on the Boundary Element Method for problems in Engineering Mechanics*, Int. j. numer. methods eng., vol.21, 1625-1640, (1985).

[7] Abdel-Akher, A., Hartley, G.A., *Evaluation of Boundary Integrals for Plate Bending*, Int. j. numer. methods eng., vol.28, 75-93, (1989).

[8] Katz, C., *Analytic Integration of Isoparametric 2D- Boundary Element*, BEM VII Int. Conf. on BEM, C.A. Brebbia and G. Maier eds., Springer-Verlag, Berlin, 13.115-13.130, (1985).

[9] Aristodemo M., Turco E., *A Strategy for Getting Accurate Solutions by Boundary Methods*, FEMCAD 89, Paris, 335-342, October 1989.

[10] Aristodemo, M., Turco, E., *Analytical Evaluation of Integrals arising in Boundary Element Formulations*, Dipartimento Strutture, Università della Calabria, Report n. 126, settembre 1990.

[11] Gradshteyn I. S., Ryzhik I. M., *Table of Integrals, Series, and Products*, Academic Press, New York, 1965.

[12] Cowper, G.M., Lindberg, G.M., and Olson, M. D., *A Shallow Shell Finite Element of Triangular Shape*, Int. J. Solids Struct., 6, 1133-56 (1970).

[13] Zotemantel R., *Numerical Solution of Plate Bending Problem Using the Boundary Element Method*, BEM VII Int. Conf. on BEM, C.A. Brebbia and G. Maier eds., Springer, Berlin, 4.17- 4.28, (1985).

[14] Timoshenko S.P., Woinowsky-Krieger S., *Theory of Plates and Shells*, McGraw-Hill International Book Company, 1984.

The Boundary Element Method for the Diffusion Equation: A Feasibility Study

Dorothy C. Attaway

Department of Manufacturing Engineering
Boston University
Boston, Massachusetts USA

Summary

The use of the boundary element method for solving the mathematical diffusion equation is presented. The boundary integral equation method is used to derive the integral equation. For the numerical solution, there are two basic approaches to the discretization in time, and the issues involved are discussed. The formulation has been extended to handle the case of diffusion problems with moving boundaries. One possible application of this extension, the modeling of the growth of a dendritic crystal, is described.

Introduction

The Boundary Element Method (BEM) consists of transforming a partial differential equation which holds over a given domain into an integral equation over the boundary of the domain using the boundary integral equation method (BIEM), and then solving by using a finite element representation of the unknown function on the boundary. In this paper the BEM will be described for the diffusion equation

$$\mathbf{L}u = \alpha \nabla^2 u - \frac{\partial u}{\partial t} = f \tag{1}$$

which holds over a domain A with boundary C for time $0 < t < T$. For simplicity, we will set $\alpha = 1$ although this does not affect the outcome of the solution methodology. Boundary conditions of either the Dirichlet type in which u is prescribed or the Neumann type in which $\partial u/\partial n$ is prescribed or mixed boundary conditions, and an initial condition to prescribe u at time $t = 0$ must be stated to specify the problem. For the numerical solution, there are two basic approaches to the discretization in time, the "step-by-step" approach, and the "initial value" approach. In the step by step approach, the values of the unknown at the preceding time step are used as pseudo-initial conditions in order to obtain the solution for the current time step. In the initial value approach, the previous values are taken into account by summing the boundary and domain integrals over all previous time steps in obtaining the solution at each time step.

In early works on the diffusion equation ([1,2]), the time dependence of the problem was removed by using Laplace or Fourier transforms to obtain an elliptic PDE, resulting in an integral formulation in the transform space. Others have used the indirect method of obtaining an integral formulation (see Butterfield [3]). Brebbia and Walker [4] employed a coupled boundary element - finite difference method in which the finite difference method was used to approximate the time derivatives. The boundary integral equation method as described in this paper was first used by Chang [5] in which results were obtained for two-dimensional heat conduction problems using the step-by-step approach to the time discretization. This method was also described by Brebbia and Wrobel [6] and extended to three-dimensional problems by Pina and Fernandes [7], also using the step-by-step approach. Muzzio and Solaini [8] used a boundary integral equation formulation in which the time integration was performed analytically, and a finite-element representation was used for the discretization in space. Other approaches have included substituting the fundamental solution corresponding to the steady state into the diffusion equation and approximating the temporal derivative using the finite element technique. Then, by choosing certain interpolation functions, it is possible to obtain a boundary-only formulation by applying reciprocity a second time. This approach was discussed by Nardini and Brebbia [9] and extended to nonlinear problems by Brebbia and Wrobel [10], in which it was termed the "dual reciprocity BEM".

BIEM for the Diffusion Equation

One approach for obtaining an integral formulation for this problem is to start by forming the inner product of the operator on u and an unknown function G:

$$(G,\mathbf{L}u) = \int_0^T \iint_A G\mathbf{L}u\,dA\,dt = \int_0^T \iint_A G\left(\nabla^2 u - \frac{\partial u}{\partial t}\right) dA\,dt \tag{2}$$

This is then integrated by parts twice, and the Gauss divergence theorem is applied. Assuming an interior problem in which the boundary of the domain is fixed in time, we obtain a general integral formulation for the diffusion equation over a fixed surface:

$$(u,\mathbf{L}^*G) = -\int_0^T \oint_C \left(G\frac{\partial u}{\partial n} - \frac{\partial G}{\partial n}u\right) ds\,dt - \iint_A (Gu)\Big|_0^T dA + (G,f) \tag{3}$$

where $\mathbf{L}^* = \alpha\nabla^2 + \frac{\partial}{\partial t}$ is the adjoint of $\mathbf{L}$, whereas G is the fundamental solution, e.g. the solution of $\mathbf{L}^*G = \delta(\overline{x} - \overline{x}_*, t - t_*)$ subject to no specified spatial boundary conditions but subject to the "initial" condition $G(\overline{x}, T) = 0$. This results in

$$E(\overline{x}_*)u(\overline{x}_*, t_*) = -\int_0^{t_*} \oint_C \left(G\frac{\partial u}{\partial n} - u\frac{\partial G}{\partial n}\right) ds\,dt + \iint_A (Gu)\Big|_{t=0} dA + \int_0^{t_*} \iint_A Gf\,dA\,dt \tag{4}$$

where either u or $\partial u/\partial n$ is known from the boundary conditions. Also, $E(\overline{x}_*)$ has the value 0 if $\overline{x}_*$ falls outside of the domain A, the value 1 if $\overline{x}_*$ is inside A, and the value 1/2 if $\overline{x}_*$ is on the boundary C of A. The fundamental solution is given by

$$G = \frac{H(t_* - t)e^{-r^2/4(t_*-t)}}{[4\pi(t_* - t)]^{D/2}} \tag{5}$$

for a D-dimensional space (Morse and Feshbach [11]). An equation similar to Eq. 4 holds for 3-dimensional problems.

Numerical Solution

The integral equation derived above (Eq. 4) gives a solution of the problem in terms of both known and unknown values. In order to solve for the unknown, it is necessary to introduce numerical approximations both in space and in time.

To discretize the space integrals over the domain A and its boundary C, finite-element representations are used for u, f, and v (substituting v for $\partial u/\partial n$ for notational simplicity). Using the collocation form of the weighted residual method for the spatial discretization of the boundary C and similarly for the domain A, Eq. 4 becomes

$$\begin{aligned} E(\bar{x}_k)u_k(t_*) &= \sum_j \int_{t_0}^{t_*} \hat{B}_{kj} v_j(t)dt + \sum_j \int_{t_0}^{t_*} \hat{C}_{kj} u_j(t)dt \\ &+ \sum_p \hat{F}_{kp}(t_0)u_p(t_0) + \sum_p \int_{t_0}^{t_*} \hat{H}_{kp} f(t)dt \end{aligned} \tag{6}$$

where

$$\hat{B}_{kj}(t_* - t) = -\oint_C \phi_j G_k(t_* - t)ds, \quad \hat{F}_{kp} = \iint_A \phi_p G\Big|_{t_0} dA$$

$$\hat{C}_{kj}(t_* - t) = \oint_C \phi_j \frac{\partial G_k(t_* - t)}{\partial n} ds, \quad \hat{H}_{kp} = \iint_A \phi_p G(t_* - t)dA \tag{7}$$

represent matrices of known values which can be determined using numerical integration techniques such as Gaussian integration. For the subscripts, k denotes the collocation points, j denotes the indices of boundary elements C_j, and p denotes the indices of field elements A_p.

In order to discretize the time integrals in Eq. 6, the time domain is divided into time steps, Δt. For a given step in time, n, $t^{n-1} + \Delta t = t^n$. Upon examination of Eq. 6, it is seen that depending on the given boundary conditions, only u or v on the boundary is unknown for each time step, and all of the other terms are known. The solution, however, must be obtained for every collocation point, and can then be found for other points, including points on the boundary and in the field. Therefore, the solution must first be obtained for points on the boundary, and once these are known, they may be used to find the solution for points in the field. The two approaches to the time discretization that will be discussed in this section reflect the two basic ways in which this can be achieved. In the approach called the "initial value" approach, the initial conditions given at $t = 0$ are used in the calculations for each time step. In this approach, the unknown is calculated for all collocation points on the boundary for all time steps. If desired, one may then calculate it for any point within the field. By contrast, in the "step-by-step" approach, the initial conditions given at $t = 0$ are used in the calculations for the first time step, but for all succeeding time steps, "pseudo"-initial conditions are used, i.e., the values obtained at time $t = t^{n-1}$ are used to obtain the values at time $t = t^n$. Using this approach, the unknown is calculated for all collocation points (first on the boundary and then in the field) for each time step before moving on to the next time step. Both approaches are described below.

In both cases, the discretization may be achieved using an interpolation formula much like the one used for the spatial discretization, i.e.

$$v_j(t) = \sum_{m=1}^{n} v_j^m Q_m(t) \tag{8}$$

where n is the total number of time steps, v_j^m is the value of v on element C_j during the time step from t^{m-1} to t^m, and $Q_m(t)$ is an interpolation function. For a zeroth order formulation (considered here), we assume that v_j has an "average" value during this time step, and denote this by $\bar{v}_j^m$ (i.e., $v_j(t) = \bar{v}_j^m$ for the time step from t^{m-1} to t^m, and similarly for $u_j(t)$ and $f_p(t)$).

In the initial value approach to the discretization in time, t_0 is set to 0, and t_* is set to t^n, so Eq. 6 becomes

$$E_k u_k^n = \sum_{m=1}^{n} \sum_j B_{kj}^{(n-m)} \bar{v}_j^m + \sum_{m=1}^{n} \sum_j C_{kj}^{(n-m)} \bar{u}_j^m + \sum_p F_{kp}^n u_p^0 + \sum_{m=1}^{n} \sum_p H_{kp}^{(n-m)} \bar{f}_p^m \tag{9}$$

where

$$B_{kj}^{(n-m)} = \int_{t^{m-1}}^{t^m} \hat{B}_{kj}(t^n - t)dt, \quad C_{kj}^{(n-m)} = \int_{t^{m-1}}^{t^m} \hat{C}_{kj}(t^n - t)dt$$

$$F_{kp}^n = \hat{F}_{kp}(0), \quad H_{kp}^{(n-m)} = \int_{t^{m-1}}^{t^m} \hat{H}_{kp}(t^n - t)dt \tag{10}$$

For the step-by-step approach, t_0 is set to t^{n-1}, and t_* is set to t^n. Noting that the summation $\sum_m$ in Eq. 8 ranges only over one value, $m = n$, Eq. 6 becomes

$$E_k u_k^n = \sum_j B_{kj}^0 \bar{v}_j^n + \sum_j C_{kj}^0 \bar{u}_k^n + \sum_p F_{kp}^1 u_p^{n-1} + \sum_p H_{kp}^0 \bar{f}_p^n \tag{11}$$

where

$$B_{kj}^0 = \int_{t^{n-1}}^{t^n} \hat{B}_{kj}(t^n - t)dt, \quad C_{kj}^0 = \int_{t^{n-1}}^{t^n} \hat{C}_{kj}(t^n - t)dt$$

$$F_{kp}^1 = \hat{F}_{kp}(t^{n-1}), \quad H_{kp}^0 = \int_{t^{n-1}}^{t^n} \hat{H}_{kp}(t^n - t)dt \tag{12}$$

The average values for $v_j(t)$, $u_j(t)$ and $f_p(t)$ can be found using several different methods. For example, $\bar{u}_j^m$, $\bar{v}_j^m$ and $\bar{f}_p^m$ can be taken to be the average of the values at time t^{m-1} and time t^m. However, it was found (see [12] for proof) that instabilities result from using this interpolation for $\bar{v}_j^m$ and $\bar{f}_p^m$ for a Dirichlet problem. In this case, the value at the middle of the time step can be used (i.e., $\bar{v}_j^m = v_j^{m-\frac{1}{2}}$).

Step by Step vs. Initial Value Approach

In the step by step approach, the values of u_p at the preceding time step are used as pseudo-initial conditions in order to obtain the solution for the current time step. Thus, all desired values on the boundary and in the domain must be calculated and either stored or printed before moving on to the next time step. For the initial value approach, the previous values are taken into account by summing the boundary and domain integrals over all previous time steps in obtaining the solution at each time step. Thus, only the solution at boundary nodes need be calculated at each time step and upon completion of all time steps, the solution may be obtained at any desired interior points. This approach requires the storage of all boundary values for all time steps.

While both approaches are always possible, the advantage gained of one over the other depends on the statement of the original problem. If, for example, the original diffusion equation is homogeneous and linear (i.e., $f = 0$), then the last term in Eqs. 11 and 9 becomes zero, and the summation of the domain integrals vanishes. In this case, the initial value approach gains an advantage over the step by step approach, especially if the initial condition $u_p^0 = 0$. If this is the case, then the inner grid is not needed in the formulation, so the initial value approach in which all boundary solutions are obtained for all time steps may be more efficient.

If, however, the original PDE is nonhomogeneous and/or nonlinear (i.e., $f \neq 0$), then the inner grid is needed since the domain integral is included in the formulation. In this case, since the inner grid must be utilized at each time step anyway, it may make more sense to employ the step by step approach and obtain the solution at each interior point before moving on to the next time step. Since for each time step the unknown must be found for all control points on the boundary before they can be calculated in the field, it may be convenient to separate all matrices defined in Eq. 12 into two: one for the boundary, and one for the field. Noting that the domain function $E_k = \frac{1}{2}$ on the boundary and $E_k = 1$ in the field, this implies that for each time step first the unknown is found on the boundary and then in the field.

For the initial value approach, if it is assumed that the problem is homogeneous and linear (i.e., $f_p^n = 0$), and that the initial conditions are homogeneous, (i.e., $u_p^{(0)} = 0$), and noting that $E_k = \frac{1}{2}$ on the boundary, the solution is obtained for all control points on the boundary for all time steps. If desired, the solution could then be found for any interior point, with $E_k = 1$ in A.

Numerical Validation

The validation of the formulation described here was presented in [12]. Computer programs were written to test several example problems. The programs utilized the initial value approach to the time discretization, and were exercised for cases which could be compared to exact analytical solutions. Results were obtained for two different domain geometries. Analyses were made of the convergence of the approximate solutions, and comparisons were made of the converged solutions and the exact results. The stability of the results was tested for different interpolation functions. In all cases, the numerical results were in excellent agreement with the analytical results.

One test case was the homogeneous diffusion problem with Dirichlet boundary conditions given by $\nabla^2 u - \frac{\partial u}{\partial t} = 0$ with $u_0 = 0$ and $u = 1$ on C where C is a unit circle. Figure 1 shows convergence results as Δx goes to 0, i.e. it shows solutions obtained for the unknown v for $\Delta t = .005$ for 4,8,16,32 and then 64 elements. Similar excellent convergence results were obtained as Δt goes to 0. The converged results found for 32 elements with $\Delta t = .0005$ are compared to analytical results in Figure 2. In this figure, the numerical results are plotted as horizontal lines for each time interval rather than as data points, since each represents an average value during a time interval. It can be seen that the best results are obtained when the average value is plotted at the beginning of each time interval.

Extension to Moving Boundaries

The integral formulation for an interior diffusion problem is given by

$$E(\overline{x}_*)u(\overline{x}_*,t_*) = -\int_0^T \oint_C \left(G\frac{\partial u}{\partial n} - \frac{\partial G}{\partial n}u\right) dsdt - \int_0^T \iint_A \frac{\partial}{\partial t}(Gu)dAdt + \int_0^T \iint_A GfdAdt \tag{13}$$

Assuming that the boundary is fixed in time, the second term on the right hand side of this equation is simplified and Eq. 4 is the resulting formulation. If, however, the boundary is moving, then both C and A are time dependent. To handle this, Leibnitz-type differentiation is applied, resulting in the following integral formulation for an interior problem with a moving boundary:

$$\begin{aligned} E(\overline{x}_*)u(\overline{x}_*,t_*) = & - \int_0^{t_*} \oint_{C(t)} \left[G\left(\frac{\partial u}{\partial n} + u\overline{v}\cdot\overline{n}\right) - \frac{\partial G}{\partial n}u\right] dsdt \\ & + \iint_{A(0)} (Gu)\Big|_{t=0} dA + \int_0^{t_*} \iint_{A(t)} GfdAdt \end{aligned} \tag{14}$$

where $\overline{v}$ is the velocity of the moving boundary and $\overline{n}$ is the direction normal to the surface C.

Application: Modeling the Growth of Dendritic Crystals

The results presented so far have been independent of any particular application. One application which presents some intriguing difficulties is the modeling of the growth of a dendritic crystal. As described by [13], the formation of the crystal is governed by the motion of a solidification front of a pure liquid. To predict the motion of the solidification front, the homogeneous diffusion equation

$$D\nabla^2\Theta = \frac{\partial\Theta}{\partial t} \tag{15}$$

is used, where Θ is the absolute temperature and D is the thermal diffusion constant ($D = \frac{k}{\rho c_p}$, where k is thermal conductivity, c_p is specific heat and ρ is density). We assume that the convective heat transport is negligible (see Langer [13]). This equation must hold for both the inner region, the solid, and for the outer region, the fluid. (Note: this process is unstable whereas the reverse process, fluid into solid, is stable.)

In order to have homogeneous initial conditions, we introduce the symbols $\theta_F = \Theta - \Theta_{F_0}$ and $\theta_S = \Theta - \Theta_{S_0}$ (where $\Theta_{F_0} = \Theta_\infty$ and Θ_{S_0} are the initial uniform temperatures of the fluid and of the solid regions respectively). Equation 15 yields

$$D_F\nabla^2\theta_F = \frac{\partial\theta_F}{\partial t} \qquad \text{for the fluid} \tag{16}$$

$$D_S\nabla^2\theta_S = \frac{\partial\theta_S}{\partial t} \qquad \text{for the solid} \tag{17}$$

where the thermal diffusion constant D_F for the fluid is in general different from the thermal diffusion constant D_S for the solid. The initial conditions are $\theta_{S_0} = 0$ and $\theta_{F_0} = 0$, and the condition at infinity is $\theta_F = 0$.

There are two boundary conditions on the boundary between the solid and the fluid region. The first prescribes the heat conservation at a point on the moving boundary:

$$k_S\frac{\partial\theta_S}{\partial n} - k_F\frac{\partial\theta_F}{\partial n} = Lv_n \tag{18}$$

where L is the latent heat per unit volume of solid and v_n is the normal velocity of the interface. This equates the rate at which heat is generated on the boundary with the discontinuity in the temperature gradient going from the solid to the fluid. The thermodynamic boundary condition (see [13]) at the interface is given by:

$$\Theta = \Theta_M\left(1 - \frac{\sigma\kappa}{L}\right) \tag{19}$$

where Θ_M is the melting temperature ($\Theta_M > \Theta_\infty$ where Θ_∞ is the temperature at infinity), σ is the surface tension and κ is the curvature of the interface. This boundary condition must hold in both the fluid and the solid, i.e.

$$\theta_F = \Theta_M\left(1 - \frac{\sigma\kappa}{L}\right) + \Theta_{F_0} \tag{20}$$

$$\theta_S = \Theta_M\left(1 - \frac{\sigma\kappa}{L}\right) + \Theta_{S_0} \tag{21}$$

The integral formulation for an interior problem with a moving boundary was derived above. For a homogeneous problem the solution for the temperature on the boundary for the solid portion is given by

$$\frac{1}{2}\theta_S = -D_S \int_0^T \oint_{C(t)} \left(G_S \frac{\partial \theta_S}{\partial n} - \theta_S \frac{\partial G_S}{\partial n} \right) dsdt - \int_0^T \oint_{C(t)} G_S \theta_S v_n dsdt \tag{22}$$

For the fluid, the integral formulation is

$$\frac{1}{2}\theta_F = D_F \int_0^T \oint_{C(t)} \left(G_F \frac{\partial \theta_F}{\partial n} - \theta_F \frac{\partial G_F}{\partial n} \right) dsdt + \int_0^T \oint_{C(t)} G_F \theta_F v_n dsdt \tag{23}$$

In studying the growth of a dendritic crystal, some of the properties of interest are the shapes and spacings of the tips of the dendrites and the sidebranches as they emerge, and the velocities at which the primary dendrites and sidebranches grow. To do this, a grid is placed over the solidification front of the fluid, which will change as the dendrites emerge. One of the interesting problems in this application is the choice of scheme to use for the grid. One possibility is to concentrate the collocation points on the tips of the dendrites. For any scheme, one is interested in the change in position of the solidification front, and in the curvature at each point. The change in position of the interface is given by

$$\frac{d\overline{x}}{dt} \cdot \hat{n} = v_n \tag{24}$$

and the curvature κ is given by

$$\kappa = \frac{\partial \hat{n}}{\partial s} \cdot \hat{t} = -\frac{\partial^2 \overline{x}}{\partial s^2} \cdot \hat{n} \tag{25}$$

where $\hat{n}$ is the outer normal. With this formulation, we have 7 unknowns ($\frac{\partial \theta_F}{\partial n}$, $\frac{\partial \theta_S}{\partial n}$, θ_F, θ_S, v_n, $\overline{x}$, and κ), and 7 equations, so the problem is completely specified.

The basic algorithm then consists of marching in time by repeating the following steps:

1. Given κ, find θ_F and θ_S on the boundary using Eqs. 20 and 21
2. Find $\frac{\partial \theta_S}{\partial n}$ and $\frac{\partial \theta_F}{\partial n}$ using Eqs. 22 and 23
3. Calculate the velocity v_n, using Eq. 18
4. Calculate the new position $\overline{x}$ of the collocation points, using $\overline{x} = \overline{x} + v_n \Delta t \hat{n}$
5. Calculate the new curvature

Of particular interest is the case in which $D_F = D_S$ so that $G_F = G_S = G$. In this case one obtains

$$\frac{1}{2}\left(\rho_F c_{pF} \theta_F + \rho_S c_{pS} \theta_S\right) = \int_0^T \oint_{C(t)} G w dsdt - \int_0^T \oint_{C(t)} \left(k_F \theta_F - k_S \theta_S\right) \frac{\partial G}{\partial n} dsdt \tag{26}$$

where

$$w = \left[-L + \left(\rho_F c_{pF} \theta_F - \rho_S c_{pS} \theta_S\right)\right] v_n \tag{27}$$

In the above algorithm, steps 2 and 3 would be replaced by (i) Find w using Eq. 26 and (ii) Find v_n using Eq. 27.

References

[1] Cruse, T.A. and Rizzo, F.J. 'A Direct Formulation of Numerical Solution of the General Elastodynamic Problems', I and II, Journal of Math. Anal. Appl., 22, 1968.

[2] Rizzo, F.J. and Shippy, D.J. 'A Method of Solution for Certain Problems of Transient Heat Conduction', AIAA Journal, 8, 1970.

[3] Butterfield, R. and Tomlin, G.R. 'Integral Techniques for Solving Zoned Anisotropic Continuum Problems', Proc. Intl. Conf. on Variational Methods in Engineering, Vol. 2 (Brebbia and Tottenham, eds.), Southampton University Press, Southampton, 1972.

[4] Brebbia, C.A. and Walker, S. Boundary Element Techniques in Engineering, Newnes-Butterworths, London, 1980.

[5] Chang, Y.P., Kang, C.S. and Chen, D.J. 'The Use of Fundamental Green's Functions for the Solution of Problems of Heat Conduction in Anisotropic Media', Intl. J. of Heat Mass Transfer, 16, 1973.

[6] Brebbia, C.A. and Wrobel, L.C. 'The Boundary Element Method for Steady-State and Transient Heat Conduction', in Numerical Methods in Thermal Problems (Lewis and Morgan, eds.), Pineridge Press, Swansea, 1979.

[7] Pina, H.L. and Fernandes, J.L. 'Three-Dimensional Transient Heat Conduction by the Boundary Element Method', in Boundary Elements V (Brebbia et al., eds.), Springer-Verlag, Berlin, 1983.

[8] Muzzio, A. and Solaini, G. 'Boundary Integral Equation Analysis of Three-Dimensional Transient Heat Conduction in Composite Media', J. Numerical Heat Transfer, Vol. 11, 1987.

[9] Nardini, D. and Brebbia, C.A. 'The Solution of Parabolic and Hyperbolic Problems Using an Alternative Boundary Element Solution', in Boundary Elements VII (Brebbia and Maier, eds.), Springer-Verlag, Berlin, 1985.

[10] Brebbia, C.A. and Wrobel, L.A. 'The Solution of Parabolic Problems Using the Dual Reciprocity Boundary Element', in Advanced Boundary Element Methods, (Cruse, ed.), IUTAM Symposium, San Antonio, Texas, Springer-Verlag, Berlin, 1988.

[11] Morse, P.M. and Feshbach, H. Methods of Theoretical Physics (2 vols.), McGraw-Hill, New York, 1953.

[12] Attaway, D. 'The Boundary Element Method for the Diffusion Equation,' Ph.D. Dissertation, Boston University, Boston, Massachusetts, 1988.

[13] Langer, J.S. 'Instabilities and Pattern Formation in Crystal Growth,' Reviews of Modern Physics, Vol. 52, No. 1, January 1980.

Acknowledgments

This work was partially supported by the NASA Langley Research Center through the NASA Graduate Student Researchers Program.

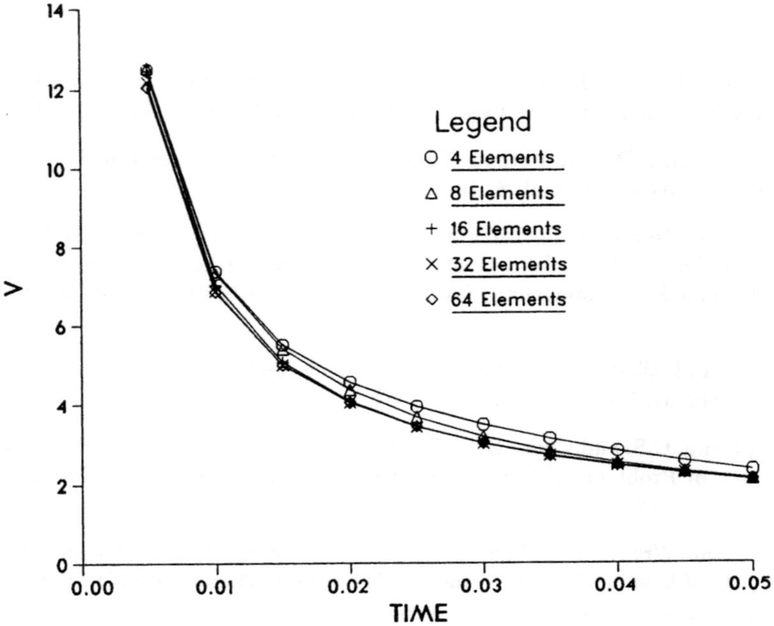

Fig. 1. Convergence results for $\Delta t = .005$

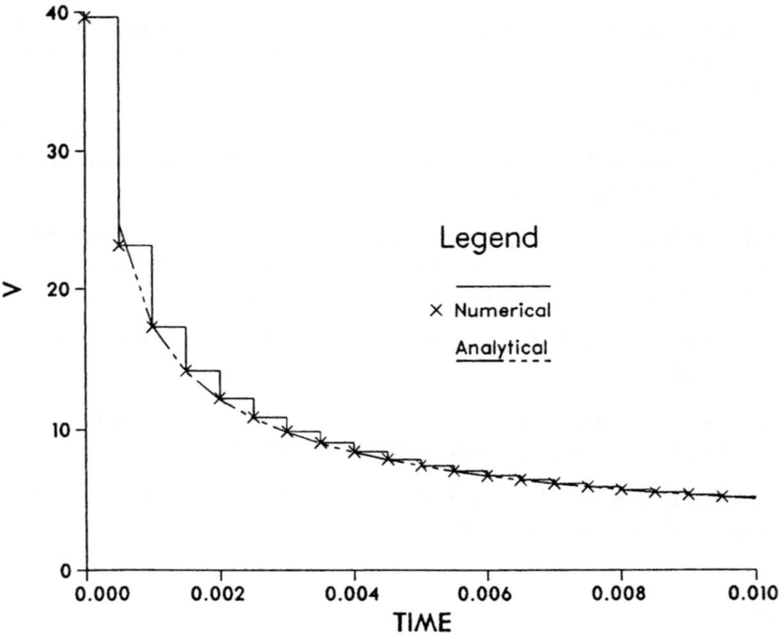

Fig. 2. Analytical versus numerical results for 32 elements with $\Delta t = .0005$

A General Integral Formulation for Rotational Flows in Aerodynamics

P. Bassanini, M.R. Lancia, R. Piva
Università di Roma La Sapienza, Roma Italy

C.M. Casciola
I.N.S.E.A.N., Roma Italy

SUMMARY

A general integral formulation based on the Poincaré representation for the velocity field is investigated in the 2D case and a few numerical results are presented. For unsteady flows the formulation is completed by a dynamical model for the evolution of the vorticity, while for steady flows a Kutta-type condition is enforced.

1 INTRODUCTION

A smooth vector field in a bounded domain can be represented in terms of its divergence and curl in the domain, and of its components on the boundary, via the Poincaré identity, which constitutes an explicit form of the well-known Clebsch-Helmholtz decomposition. The Poincaré identity involves gradients and curls of volume potentials and simple layer potentials and retains its validity on boundary points, provided the correct behavior of these potentials at the boundary is taken into account. By projection parallel and orthogonal to the normal at the boundary, two boundary integral equations are obtained for the corresponding components of the vector field. In the context of applications in Aerodynamics, these equations follow from purely kinematical considerations and do not involve any dynamical specification on the flow field.

Boundary integral equations methods in the potential framework (i. e. for inviscid flows) have been used since many years. They give rise to Fredholm equations of both first and second kind. While the latter are classical, the use of Fredholm equations of first kind is new and has been the object of recent studies [4,5,8,3]. Several important applications (e.g. Navier-Stokes equations for unsteady viscous flows [2]) lead in a natural way to such kind of equations.

A formulation based on the Poincaré identity has several advantages. First , it poses no restrictions on the vorticity and divergence of the vector field, and thus can be applied to rotational flows with wakes, sources, sinks and vorticity blobs. Further, it allows to deal in a simple and straightforward way with flows with circulation: in fact, the circulation around the body appears as a natural and intrinsic parameter in this formulation. The use of primitive variables (kinetic field rather than potential) shows obvious benefits, especially from the computational point of view. Finally, when coupled to the variational method, it leads to a well-posed problem, for which stability estimates can be proved and error estimates can be obtained.

The present paper is devoted to the 2D case. In section 2 after writing the explicit form of the weak Poincaré identity for exterior domains, we derive two scalar boundary integral equations. The existence and uniqueness analysis, carried out with several equivalent approaches, shows that the velocity field is uniquely determined provided the circulation around the body is also included among the data.

In Section 3 we describe a dynamical model for unsteady 2D attached flows past a body and we discuss its relationship with the Kutta condition for steady flows. The model is based on the replacement of the real vorticity distribution by a free vortex sheet separating at the trailing edge, as is appropriate at very high Reynolds numbers. The location and the density of the vortex layer is determined from the Euler flow equations using the concept of streakline.

A few numerical results are presented in Section 4. A first group of results concerns steady irrotational flows past airfoil profiles with a sharp trailing edge where the Kutta condition is enforced. A second group of results deals with unsteady rotational flows past airfoils with a sharp trailing edge, where the circulation and vorticity are determined from the wake model of Section 3 and from the Kelvin theorems.

2 THE POINCARÉ IDENTITY AND THE BOUNDARY INTEGRAL EQUATIONS

Consider a bounded domain $\Omega \subset \mathbb{R}^2$ with smooth connected boundary C.

Then the Poincaré identity for the exterior domain $\Omega' = \mathbb{R}^2 - \bar{\Omega}$, takes the following explicit weak form [1]:

$$\mu\, \underline{u}(x) = \text{grad}\phi(x) + J\text{grad}\psi(x) \qquad , \mu = \begin{cases} 1 & x \in \Omega' \\ 1/2 & x \in C \\ 0 & x \in \Omega \end{cases} \tag{1}$$

$$\begin{aligned} \psi(x) &= \int_D g(x,y)\omega(y)dy + V(\underline{t}\cdot\underline{u})(x) \\ \phi(x) &= -\int_D g(x,y)Q(y)dy - V(\underline{n}\cdot\underline{u})(x) \end{aligned}$$

where $J = \underline{k}\wedge, \underline{k} = \underline{n}\wedge\underline{t}$, $\underline{n}$ is inner normal to Ω', on C, $\underline{t}$ is tangent to C, ω is the scalar vorticity, $\omega = \partial u_2/\partial x_1 - \partial u_1/\partial x_2$, $Q = \partial u_1/\partial x_1 + \partial u_2/\partial x_2$, the expansion $g(x,y) = -(2\pi)^{-1}\log|x-y|$, and V is the weak extension of the simple layer operator [5]

$$V(f)(x) = \int_C g(x,y)f(y)ds_y$$

The functions ϕ and ψ (potential and stream functions, respectively) satisfy $\Delta\phi = Q$, $\Delta\psi = -\omega$. This identity holds for $\underline{u}(x)$ locally square summable in Ω', with Q and ω square summable and with compact support D in Ω', and furthermore with

$$\underline{u}(x) = \mathrm{O}(1/|\mathrm{x}|) \quad \text{for } x \to \infty \tag{2}$$

Thus the normal and the tangential traces $\underline{u} \cdot \underline{n}$, $\underline{u} \wedge \underline{t}$ belong to $H^{-1/2}(C)$ [1].

From (1) it follows the asymptotic expansion

$$\underline{u}(x) = \frac{1}{2\pi}|x|^{-2} \left\{ x \left[\int_C \underline{n} \cdot \underline{u}\, ds_y + \int_D Q dy \right] - \underline{x} \wedge \underline{k} \left[\int_C \underline{u} \cdot \underline{t}\, ds_y + \int_D \omega\, dy \right] \right\} \tag{3}$$
$$+\mathrm{O}(1/|\mathrm{x}|^2)$$

Thus the leading term depends on four quantities: the average value in Ω' of the expansion, the flux across C, the average of vorticity, and the circulation: $\Gamma = \int_C \underline{t} \cdot \underline{u}\, ds$. If $\underline{u}$ behaves at infinity according to (2), these four quantities can be assigned independently. In particular the circulation can be different from zero even in the absence of field vorticity, as is well known for two-dimensional irrotational flows in Aerodynamics. On the contrary, if we assumed $\underline{u}$ vanishing at infinity faster than (2), the flux and circulation "at infinity" would vanish and, according to Gauss' Lemma, the four quantities would be related by the equations

$$\int_{\Omega'} \mathrm{Q\, dy} + \int_{\mathrm{C}} \underline{\mathrm{n}} \cdot \underline{\mathrm{u}}\, \mathrm{ds} = 0 \ , \int_{\Omega'} \omega\, \mathrm{dy} + \int_{\mathrm{C}} \underline{\mathrm{u}} \cdot \underline{\mathrm{t}}\, \mathrm{ds} = 0 \tag{4}$$

and two only (e.g., the average expansion and vorticity) could be chosen independently. For instance, in irrotational flow the circulation would necessarily be zero.

Once Q and ω are given, $\underline{u}$ is represented (1) in terms of its traces $\underline{u} \cdot \underline{n}$, $\underline{u} \cdot \underline{t}$ on the boundary. By projecting (1) along the normal and the tangent to C and by defining [5] the boundary integral K' operator

$$K'f(x) = \int_C f(y)\underline{n}_x \cdot \mathrm{grad}_x[g(x,y)]ds_y \tag{5}$$

we find

$$\left(\frac{I}{2} + K'\right)\underline{n} \cdot \underline{u} = \frac{\partial}{\partial s}V(\underline{t} \cdot \underline{u}) + F_1(x) \qquad x \in C \tag{6}$$

$$\left(\frac{I}{2} + K'\right)\underline{t} \cdot \underline{u} = -\frac{\partial}{\partial s}V(\underline{n} \cdot \underline{u}) + F_2(x) \qquad x \in C \tag{7}$$

where $\partial/\partial s = \underline{t}_x \cdot \mathrm{grad}_x$, $\partial/\partial n = \underline{n}_x \cdot \mathrm{grad}_x$, and F_1, F_2 are known source terms.

These two equations are equivalent, in the sense that if $\underline{u}$ is a solution of one, it also satisfies the other. Thus only one of them needs to be solved. They contain four quantities, namely $\underline{t} \cdot \underline{u}, \underline{n} \cdot \underline{u}, Q$ and ω. Either $\underline{t} \cdot \underline{u}$ or $\underline{n} \cdot \underline{u}$ can be taken as unknowns, and the three remaining quantities are to be viewed as given.

From previous results [1] we have $\underline{t} \cdot \underline{u} \in H^{-1/2}(C)$, $V(\underline{t} \cdot \underline{u}) \in H^{1/2}(C)$, $K'\underline{t} \cdot \underline{u} \in H^{1/2}(C)$, and similarly for $\underline{n} \cdot \underline{u}$. Moreover $F_1, F_2 \in H^{1/2}(C)$. Now both equations (6), (7) have the same form:

$$\left(\frac{I}{2} + K'\right) p = \frac{\partial}{\partial s} V(q) + F \tag{8}$$

where either $p = \underline{n} \cdot \underline{u}$, $q = \underline{t} \cdot \underline{u}$, $F = F_1$ or $p = \underline{t} \cdot \underline{u}$, $q = -\underline{n} \cdot \underline{u}$, and $F = F_2$.

If $v = p$ is taken as unknown, eq. (8) is a Fredholm equation of the second kind in v. Its adjoint homogeneous equation has the eigenfunction $v_0 = 1$, and, from Gauss' formula, the compatibility condition

$$< F, 1 >= O \tag{9}$$

is satisfied, where $< \cdot, \cdot >$ denotes the duality pairing between $H^{1/2}(C)$, $H^{-1/2}(C)$. Trivially also $< \frac{\partial}{\partial s} V(q), 1 >= O$. Hence a solution exists and is determined up to a Robin density:

$$v = \bar{v} + a v_r \tag{10}$$

where $a \in \mathbb{R}, v_r$ is a Robin density, and $\bar{v}$ is a particular solution. Since v_r satisfies

$$< v_r, 1 > \neq O \tag{11}$$

we deduce that the constant a is determined (hence the solution is unique) if, and only if, we assign the quantity

$$< v, 1 >= A_0 \, L \tag{12}$$

which is the circulation if $v = \underline{t} \cdot \underline{u}$ or the flux if $v = \underline{n} \cdot \underline{u}$ and L is the length of the curve C.

If $v = q$ is unknown, we may integrate (8) and obtain the Fredholm equation of first kind [6]

$$V(v) = f + b \tag{13}$$

where f is a primitive of $p/2 + K'p - F$ and b is an integration constant. The system of equations (12) (13) (for given A_0) has a unique solution for (v, b). Note that equation (13) for $f = 0$ has the eigenfunction v_r, whreas the system (12) (13) (for $A_0 = 0, f = 0$) has no eigensolutions because of (11).

In the case of interest for our applications below, we have $v = \underline{t} \cdot \underline{u}$ and $p = \underline{u} \cdot \underline{n}$, so that the circulation is the quantity to be assigned.

The following variational formulation is used for the system (12) (13) :

find $(v, b) \in H^{-1/2}(C) \times \mathbb{R}$ such that, for every $\chi \in H^{-1/2}(C)$

$$\begin{aligned} < \chi, V(v) > + < \chi, b > &= < \chi, f > \qquad \forall \chi \in H^{-1/2}(C) \\ < v, 1 > &= \Gamma \end{aligned} \tag{14}$$

which is shown to be a particular case of the one proposed by Hsiao [5]. Hence his results of existence and uniqueness may be adopted in the present case. The numerical approximation of (14) implies the solution of (12) by collocation and of (13) by a Galërkin method, using e.g. a base of positive step functions.

The formulation, based on the Poincaré identity, leads to a well-posed problem for which stability and error estimates are discussed in [1].

3 WAKE MODEL AND KUTTA CONDITION

We present here a dynamical model for 2D Aerodynamics that completes the kinematical formulation obtained by representing the velocity field through the Poincaré formula.

The aim of the model, which has been elaborated and adjusted in a series of previous works (see e.g. [2], [7]) on the basis of Euler equations for an incompressible fluid, is to efficiently compute the vortical wake downstream of a body moving in a fluid at very high Reynolds numbers. We shall be interested here in incipient motions, i.e. motions starting from rest.

We shall assume that the field vorticity ω can be expressed through a vortex layer density given by the jump $[u_\tau]$ of the tangential component $u_\tau = \underline{t} \cdot \underline{u}$ of the fluid velocity through a wake separating downstream of the sharp trailing edge. The wake configuration will be represented by the parametric equation

$$W(t) : x = x_w(\sigma, t) \qquad , \qquad 0 \leq \sigma \leq t \tag{15}$$

where $t \geq 0$ is time and $\sigma, 0 \leq \sigma \leq t$ is a parameter characterizing the material point on the wake which starts from the trailing edge TE at time $t = \sigma$, $x_w(\sigma, \sigma) = x_{TE}$. Thus the total wake at time t extends from the trailing edge $x_{TE} = x_w(t,t)$ to the termination point $x_{TW} = x_w(0,t)$.

We adopt for the wake velocity components the expression

$$w_n = u_n \qquad , \qquad w_\tau = \frac{1}{2}(u_\tau^+ + u_\tau^-) \tag{16}$$

where the subscript τ denotes tangential components. Note that this choice incorporates the condition of no fluid crossing the wake. Since $\underline{w} = \partial x_W / \partial t$, this yields the differential equation for the evolution of the wake configuration

$$\begin{aligned} &\frac{\partial}{\partial t} x_W(\sigma, t) = \underline{U}(x_W(\sigma, t), t) && 0 \leq \sigma \leq t \quad , \quad t \geq 0 \\ &x_W(\sigma, \sigma) = x_{TE} && \forall \sigma \geq 0 \end{aligned} \tag{17}$$

From the Euler equations it follows that $\gamma(\sigma, t) = \gamma(x_W(\sigma, t), t)$ satisfies the conservation equation

$$\frac{\partial}{\partial t}\left[\gamma(\sigma, t) j(\sigma, t)\right] = 0 \tag{18}$$

where $j := |\partial x_W(\sigma, t)/\partial \sigma|$ $(j > O)$.

The initial value $\gamma_{TE}(\sigma)$ for the vortex layer density at the trailing edge is taken to be the limit

$$\gamma_{TE}(\sigma) = \lim_{\substack{x_+ \to TE \\ x_- \to TE}} [u_\tau(x_+,\sigma) + u_\tau(x_-,\sigma)] \tag{19}$$

where x_+, x_- are points on the upper and lower side of the profile and $u_\tau(x_\pm,\sigma)$ the corresponding components of the flow velocities tangential to the profile. These must be computed as part of the solution. The system of equations (17), (18) allows to follow the evolution by computing the fluid velocity through the Poincaré formula (1) and the boundary integral equation (8), where the term $\omega(y)dy$ is replaced by γds and the circulation is given by

$$\Gamma(t) = -\int_{W(t)} \gamma ds \tag{20}$$

In this way the dynamical problem of incipient motion is completely solved, once $\underline{n} \cdot \underline{u}$ is given and Q is taken equal to zero. From equation (20) we obtain

$$d\Gamma/dt = j(t,t)\gamma_{TE}(t) \tag{21}$$

hence, for stationary circulation around the body, the classical Kutta condition ($\gamma_{TE} = 0$) for steady flow is recovered.

4 NUMERICAL RESULTS AND DISCUSSION

In this Section we present a few numerical results obtained by the approximation methods discussed in [1] . In particular for steady flow both the collocation and the Galërkin method have been applied in connection with a formulation of the first kind for equation (8), while only the collocation method has been adopted for a formulation of the second kind for eq. (8). We shall call for brevity these three methods "first kind collocation", "first kind Galërkin" and "second kind collocation". For unsteady flows only the first kind collocation method has been used.

For steady flows we take $\omega = 0$ and we only need to solve the kinematical equations (12), (13) via one of the three methods just mentioned.

Fig. 1 shows the behaviour of tangential velocity ($v = u_\tau$) vs arc-lenght along the boundary for steady flow past an elliptic cylinder. The circulation Γ is taken equal to $-\pi$, but the analytical solution for $\Gamma = 0$ is also included for reference.

Fig. 2 shows the analogous behaviour for steady flow past a cambered Karman-Trefftz profile. The numerical results are obtained via the first kind and second kind collocation methods and are compared with the analytical solution. Here the circulation is determined by the Kutta condition in the discretized version appropriate to the collocation approach.

Next, we consider unsteady flow past a NACA0015 airfoil starting suddenly from rest with velocity U at an angle of attack of ten degrees.

The graphs of Fig. 3 report the tangential velocity along the airfoil as function of the non-dimensional time variable $\sigma^* = \sigma U/c$ (c is the chord), together with the steady velocity profile. As σ^* increases, the numerical solution evolves towards the steady solution, which according to the discussion in §3 satisfies the Kutta condition at the trailing edge.

Fig. 4 shows the graph of the function $\gamma_{TE}(\sigma^*)$ vs the non-dimensional time variable σ^*. Here γ_{TE} tends to the steady value $\gamma_{TE} = 0$ with an exponential rate faster than the corresponding rate of convergence of tangential velocities. For instance, the value of $\gamma_{TE}(\sigma^*)$ is practically zero for $\sigma^* = 4$, so that the circulation Γ is almost equal to that computed from the Kutta condition.

Finally, Figs. 5 and 6 show the time evolution for the vorticity γ_{TE} and the wake configuration when the angle of attack of the profile is suddenly reduced to zero some time after the sudden start. The main features to observe here are the sudden release of reverse vorticity as a consequence of the change in the angle of attack, and the typical configuration of the wake showing a system of two receding counterrotating vortices.

ACKNOWLEDGMENT

This research was carried out under the auspices of the GNFM of CNR, and partially supported by CNR under Contracts N. 89.01214.01 and by MURST.

References

[1] Bassanini P., Casciola C.M., Lancia M.R., Piva R., 1990, "A boundary integral formulation for the kinetic field in aerodynamics. Part I : mathematical analysis", Part II : "Applications to unsteady 2D flows" submitted to Eur. J. Mech., B/Fluids

[2] Casciola C.M., Lancia M.R., Piva R., 1989, "A general approach to unsteady flows in Aerodynamics: Classical results and perspectives". In: Proceedings ISBEM 89 , East Hartford, USA. Springer, Berlin.

[3] Costabel M., Wendland W.L., 1986, "Strong ellipticity of boundary integral operators", J.Reine Angew. Math. 372, 34-63

[4] Fichera G., 1961, "Linear elliptic equations of higher order in two independent variables and singular integral equations". In: Proc. Confer. on Partial Differential Equations and Continuum Mechanics (Madison, Wisc.), Univ. of Wisconsin Press, Madison USA

[5] Hsiao G.C., 1988, "On boundary integral equations of the first kind". In: China-US Seminar on Boundary Integral Equations and Boundary Element Methods in Physics and Engineering, Dec.27,1987- Jan. 1988, Xi'an Jiaotong Univ., Xi'an, China (PRC)

[6] Hsiao G.C., MacCamy R.C., 1973, "Solution of boundary value problems by integral equations of the first kind", SIAM Rev. 15, 4, 687-705

[7] Morino L., 1986, "Helmoltz Decomposition Revisited: Vorticity Generation and Trailing Edge Condition", Computational Mechanics 1, 65 - 90

[8] Nedelec J.C., 1977, "Approximation des équations intégrales en mécanique et en physique", Lecture Notes, Centre de Mathématiques Appliquées, Ecole Polytechnique, Palaiseaux, France

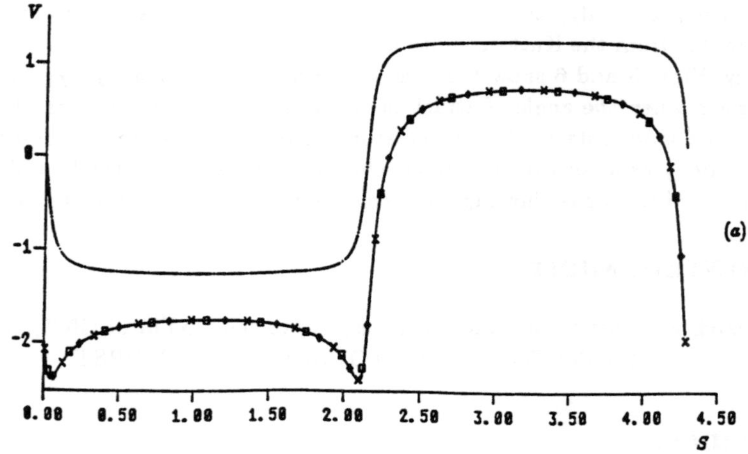

Fig. 1

Tangential velocity v vs arc length s for an elliptic cylinder. + first kind collocation, ◇ second kind collocation, □ first kind Galerkin, — exact solution ($\Gamma = -\pi$), —• exact solution ($\Gamma = 0$).

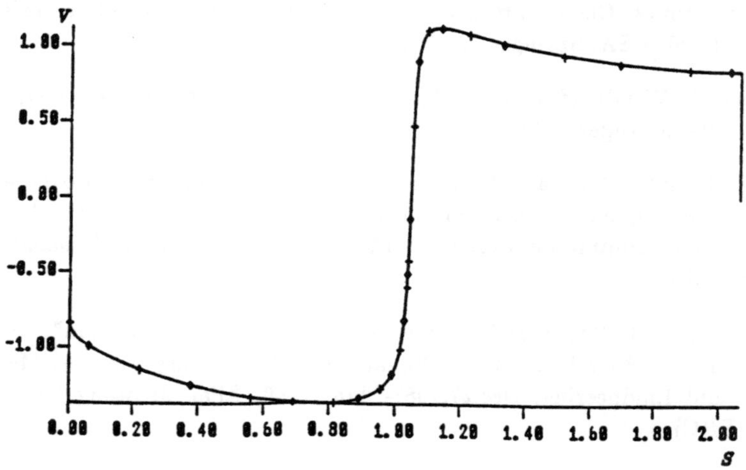

Fig. 2

Tangential velocity v vs arc length s for a cambered Karman–Trefftz airfoil. ◇ first kind collocation, + second kind collocation, — exact solution.

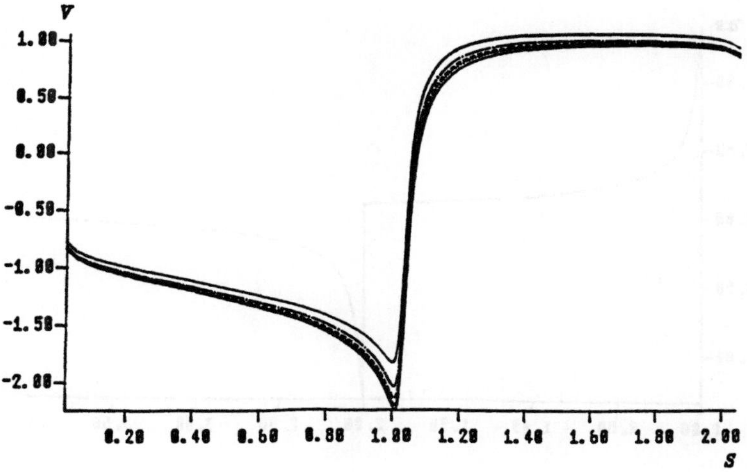

Fig. 3

Tangential velocity v vs arc length s for a $NACA0015$ airfoil. Sudden start problem: $\sigma^* = 1$, $\sigma^* = 3$, $\sigma^* = 5$, $\sigma^* = 7$, steady case.

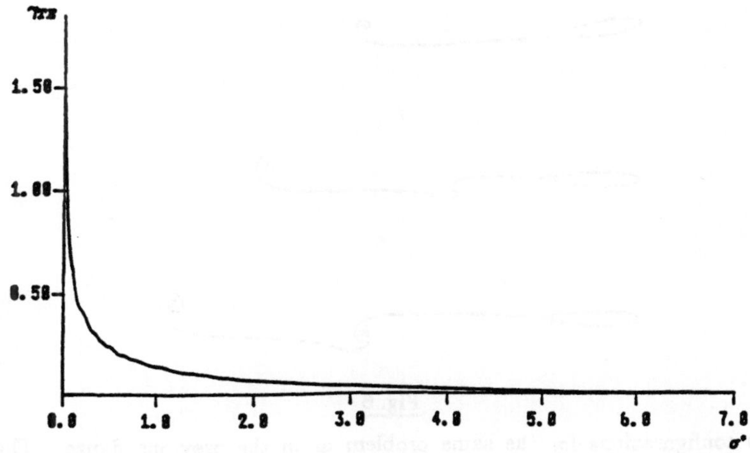

Fig. 4

Vorticity at the trailing edge γ_{TE} vs time σ^*. Same problem as in the previous figure.

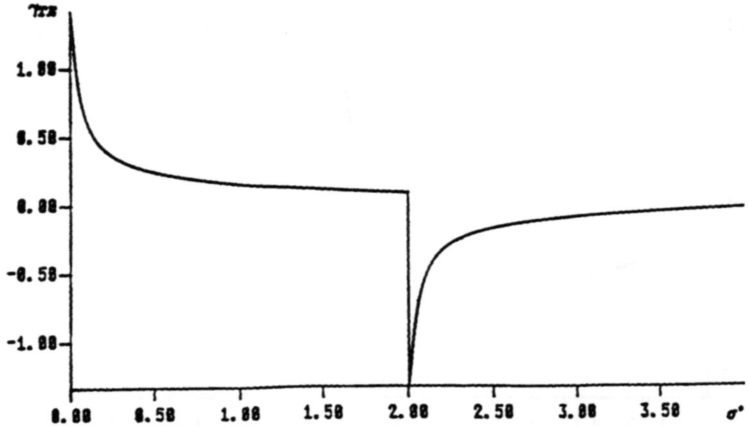

Fig. 5

Vorticity at the trailing edge γ_{TE} vs time σ^*. The angle of attack (10 degrees) of the profile, suddenly accelerated from rest at $\sigma^* = 0$, is reduced to zero at time $\sigma^* = 2$.

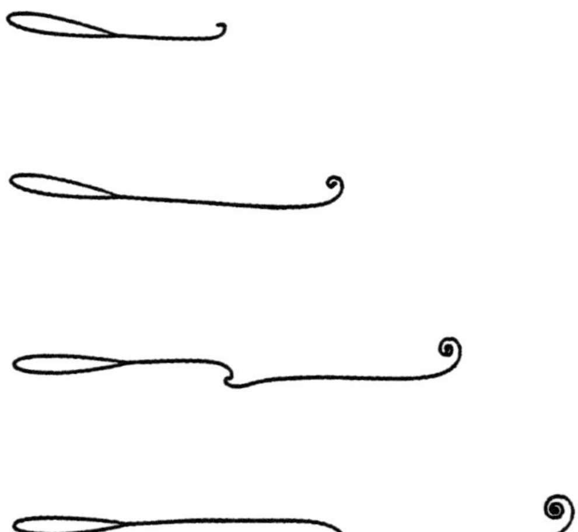

Fig. 6

Several wake configurations for the same problem as in the previous figure. The reported configurations correspond to $\sigma^* = 1, 2, 3, 4$.

Viscous Flow Analysis Using the Poincaré Decomposition

Dr. Philip Beauchamp
General Electric Co.
Lynn MA, USA

Prof. Luigi Morino
Università di Roma
Rome, ITALY

Summary

A computational technique for the analysis of two-dimensional viscous fluid flow fields is presented. The technique is based on an exact formulation of the viscous flow problem using the Poincaré decomposition for the velocity field. The numerical method developed is a hybrid technique which employs the boundary integral method to determine the velocity field while the vorticity evolution is determined from a finite-difference technique. The technique has been applied to the transient and steady-state analysis of thin airfoils in arbitrary motion. Comparison with known results have been found to be in good agreement using computational grids that would defeat most other viscous solvers.

Introduction

The ability to assume potential flow greatly simplifies the analysis of flow past streamlined bodies. Such problems are efficiently solved using the boundary-integral equation method. However, when analyzing viscous flows, the boundary-integral method loses most of its efficiency. Previously this loss has been treated by developing viscous–inviscid correction techniques which are approximate at best. In contrast, this work addresses the numerical solution of an exact formulation of the viscous problem using a method that might retain the efficiency of the boundary-integral method. The technique is based on the numerical implementation of a boundary-integral equation method using the Poincaré potential-vorticity velocity decomposition for the analysis of viscous flows past arbitrary bodies. The decomposition, which is applicable to unsteady compressible viscous flows, was initially presented in [5] for incompressible flows. Complete details of the theoretical formulation, including its relation to and advantages over the classical Helmholtz decomposition, as well as the extension to compressible flows is presented in Morino[6].

The formulation is based on decomposing the velocity field into two terms, one related to the vorticity (vortical velocity) and one that is irrotational (potential velocity). The technique solves for the vortical velocity component using the Poincaré solution, and employs a boundary-integral equation method to determine the potential velocity distribution. The vorticity in the field is determined by a finite-difference method. A key issue in the implementation is the generation of vorticity at the body surface. This vorticity generation has been examined by treating the continuous motion of the surface as the limit of a series of relaxation periods separated by discrete velocity impulses. This 'surface vorticity generation condition' has been implemented numerically in two distinct ways. In our previous work[5], the numerical methodology was based on the simultaneous solution of the surface potential and vortical velocities. This approach involves a large matrix inversion to determine the potential and vortical velocities on the surface.

The results to be presented herein focus on an alternative implementation of the surface vorticity generation condition, whereby the continuous motion of the surface is considered to be made up of a velocity impulse, during which a surface vorticity layer is generated, followed by a relaxation period during which the vorticity is immediately diffused.

Formulation

Consider the motion of a arbitrarily shaped body through an incompressible viscous fluid. The motion of the fluid about the aforementioned body is governed by the continuity equation

$$\nabla \cdot \mathbf{v} = 0 \tag{1}$$

and the Navier-Stokes equations,

$$\frac{\partial \mathbf{v}}{\partial t} + (\mathbf{v} \cdot \nabla)\mathbf{v} = \frac{-\nabla p}{\rho} + \nu \nabla^2 \mathbf{v} \tag{2}$$

where $\mathbf{v}$ is the velocity vector, ρ is the density, p is the pressure, and ν is the kinematic viscosity. The boundary conditions, assuming that the frame of reference is fixed with respect to the undisturbed air, are given by the initial condition, $\mathbf{v}(\mathbf{x}, 0) = \mathbf{v}_0(\mathbf{x})$, and by the far field conditions, $\mathbf{v} = 0$ and $p = p_\infty$. In addition, for viscous flows, the no-slip condition must be satisfied on the body, $\mathcal{S}_B$, such that $\mathbf{v} = \mathbf{v}_B$, where $\mathbf{v}_B$ is the prescribed velocity of the body.

The analysis of the flow field surrounding the body can be simplified by introducing a general potential-vorticity decomposition for the velocity,

$$\mathbf{v} = \nabla \phi + \mathbf{w} \tag{3}$$

where ϕ is called the potential, and $\mathbf{w}$, which will be referred to as the vortical velocity, is a particular solution of $\nabla \times \mathbf{w} = \boldsymbol{\zeta}$ obtained by taking the curl of Eq. 3 above and by noting that by definition the vorticity is $\boldsymbol{\zeta} = \nabla \times \mathbf{v}$.

One choice for the vortical velocity component is the well known Helmholtz decomposition. However, Morino [6] has shown that for this case the extension to compressible fields results in additional source terms in the equation for the velocity potential. These terms are non-zero even in the irrotational regions of the flow field, making it necessary to evaluate domain integrals over a portion of the domain where the vorticity is zero. This makes the Helmholtz decomposition unsuitable for the integral equation method in the compressible viscous flow problem. To overcome this problem, the solution for the vortical velocity should be chosen such that its magnitude is zero in as much of the irrotational region as possible. This may be done by using the Poincaré solution of $\nabla \times \mathbf{w} = \boldsymbol{\zeta}$. For reasons to be illustrated shortly, this solution will result in an infinite length region over which the vortical velocity component is non-zero. To minimize the extent of this region it is convenient to work in a curvilinear coordinate system. In this regard, note that the Einstein summation convention on repeated indices will be used and that lower case Greek subscripts are used for covariant components, with superscripts used for the contravariant components. The complete derivation may be found in Morino[6].

Let ξ^1, ξ^2, ξ^3 be a system of right-handed curvilinear coordinates such that a point $\mathbf{x}$ of the physical space is in a one to one correspondence with a point of the space ξ^1, ξ^2, ξ^3. Let the vorticity be expressed in terms of its contravariant components, $\boldsymbol{\zeta} = \zeta^\alpha \mathbf{g}_\alpha$, and let the vortical velocity be expressed in terms of its covariant components, $\mathbf{w} = w_\alpha \mathbf{g}^\alpha$, where the covariant base vectors are given by $\mathbf{g}_\alpha = \partial \mathbf{x}/\partial \xi^\alpha$ and the contravariant base vectors are specified such that $\mathbf{g}^\alpha \cdot \mathbf{g}_\beta = \delta^\alpha_\beta$. The Poincaré solution to $\nabla \times \mathbf{w} = \boldsymbol{\zeta}$ is then given by the system of equations,

$$\begin{aligned} w_1(\xi^1,\xi^2,\xi^3) &= 0 \\ w_2(\xi^1,\xi^2,\xi^3) &= \int_{\xi_0^1}^{\xi^1} \sqrt{g}\zeta^3(\lambda^1,\xi^2,\xi^3)d\lambda^1 \\ w_3(\xi^1,\xi^2,\xi^3) &= -\int_{\xi_0^1}^{\xi^1} \sqrt{g}\zeta^2(\lambda^1,\xi^2,\xi^3)d\lambda^1 + \int_{\xi_0^2}^{\xi^2} \sqrt{g}\zeta^1(\xi_0^1,\lambda^2,\xi^3)d\lambda^2 \end{aligned} \tag{4}$$

where g is the determinant of the metric tensor. To illustrate the nature of this solution, consider an example case in which there is an arbitrary vorticity distribution that is different from zero only in a region that is bounded by a cylinder, $\mathcal{V}_0$. Let the cylinder be oriented parallel to the x_1-axis, and defined by $x_1 \in (a_1, b_1)$ and $(x_2, x_3) \in \mathcal{D}$, where $\mathcal{D}$ is a bound domain of the plane (x_2, x_3). Then, by choosing $x_{1_0} < a_1$, the second integral in third expression for the vortical velocity in Eq. 4 is equal to zero and the solution may be rewritten as

$$\mathbf{w} = \int_{\mathcal{L}} \boldsymbol{\zeta}(\mathbf{x}') \times d\mathbf{x}' \tag{5}$$

where $\mathcal{L}$ is a line parallel to the x_1-axis and connecting the plane $x_1 = a_1$ to the point $\mathbf{x}$. It is apparent that $\mathbf{w} = 0$ except for the points in the volume, $\mathcal{V}_0$ and its "shadow" $\mathcal{V}_1$, defined by $x_1 > b_1$ and $(x_2, x_3) \in \mathcal{D}$. In the shadow, $\mathcal{V}_1$, $\mathbf{w}$ is constant along the lines parallel to the x_1-axis because the vorticity is zero outside $\mathcal{V}_0$. The magnitude is equal to the vortical velocity value of the point at the rear of $\mathcal{V}_0$ obtained by projecting back to the cylinder in the x_1 direction.

To complete the determination of the velocity, it is necessary to evaluate the contribution due to the potential velocity component. This requires that the potential be known. The governing equation for the potential ϕ is obtained by taking the divergence of Eq. 3 and combining this result with the continuity equation, Eq. 1, to obtain $\nabla^2\phi = \sigma_\mathbf{w}$, where $\sigma_\mathbf{w} = -\nabla \cdot \mathbf{w}$. Applying Green's theorem to this expression yields

$$E_*\, \phi_* = \oiint_{\mathcal{S}_B} G\, \frac{\partial \phi}{\partial n}\, d\mathcal{S} - \oiint_{\mathcal{S}_B} \phi\, \frac{\partial G}{\partial n}\, d\mathcal{S} + \iiint_{\mathcal{V}} G\, \sigma_\mathbf{w}\, d\mathcal{V} \tag{6}$$

where E_* is a function that is unity for a point in the flow field and zero for any point not in the flow field, whereas G is the fundamental solution which is given by $G = \ln|\boldsymbol{\xi} - \boldsymbol{\xi}_*|/2\pi$ in two dimensions and $G = -1/4\pi|\mathbf{x} - \mathbf{x}_*|$ in three dimensions. Equation 6 is the integral representation of ϕ_* in terms of the values of ϕ and $\partial\phi/\partial n$ on the surface and of a distribution of sources in the field whose magnitude is given by $\sigma_\mathbf{w}$. On the surface $\mathcal{S}_B$, the normal component of the boundary condition, combined with Eq. 3, yields

$$\frac{\partial \phi}{\partial n} = \chi - \mathbf{w} \cdot \mathbf{n} \tag{7}$$

where $\chi = \mathbf{v}_H \cdot \mathbf{n}$. In Eq. 6, as $\mathbf{x}_*$ approaches a surface point, Eq. 6 yields an integral equation that may be used to solve for the potential, provided that the vortical velocity, $\mathbf{w}$, is known. Equation 6 will be referred to as the distributed source formulation of the method.

One issue that must be addressed in the above expression stems from the fact that, in aerodynamic applications, the vortical region, such as the boundary-layer region around a wing, has "holes" (*e.g.*, the wing itself), which lead to discontinuities in $\mathbf{w}$. This is most easily seen for the symmetric flow about a symmetric airfoil where the vorticity has equal an opposite magnitudes on either surface. Consequently, the use of Eq. 5 along each surface yields opposite values at the trailing edge, so that the vector field $\mathbf{w}$ has a discontinuity, $\Delta\mathbf{w}$, along a surface emanating from the trailing edge and extending to infinity. This implies that $\nabla \cdot \mathbf{w}$ must be interpreted within the theory of distributions.

The problems associated with the discontinuity in the vortical velocity may be removed by integrating by parts the volume integral in Eq. 6 and using the normal boundary condition, Eq. 7, to obtain

$$E_* \, \phi_* = \oiint_{\mathcal{S}_B} G \, \chi \, d\mathcal{S} - \oiint_{\mathcal{S}_B} \phi \, \frac{\partial G}{\partial n} \, d\mathcal{S} + \iiint_{\mathcal{V}} \mathbf{w} \cdot \nabla G \, d\mathcal{V} \tag{8}$$

This representation, which will be referred to as the distributed doublet formulation, differs from that in Eq. 6 in that ϕ_* is now in terms of the values of ϕ and χ on the body surface, and in terms of a doublet distribution having intensity $|\mathbf{w}|$ oriented in the same direction as $\mathbf{w}$. Once again, as $\mathbf{x}_*$ approaches a point on the body surface $\mathcal{S}_B$, Eq. 8 yields an integral equation that may be used to solve for the potential, provided that the vortical velocity, $\mathbf{w}$, is known.

Paramount in the preceding discussion is the assumption that the vorticity in the field surrounding the body is known. This may be obtained from the vorticity evolution equation

$$\frac{D\boldsymbol{\zeta}}{Dt} = (\boldsymbol{\zeta} \cdot \nabla)\mathbf{v} + \nu \nabla^2 \boldsymbol{\zeta} \tag{9}$$

The solution of this relation for the case of an airfoil moving through an initially irrotational medium requires only that the vorticity on the surface $\mathcal{S}_B$ be known.

The only remaining issue in the formulation is the evaluation of the surface vorticity using the conceptualization introduced by Morino[4]. Consider the instantaneous start-up of a body at rest in a stationary fluid. At time $t < 0$ both the body and the fluid are at rest. Then, at time $t = 0^+$, the body is given an impulsive start whereby it moves in uniform translation with respect to the undisturbed air, such that $\boldsymbol{\zeta} = 0$ in the solid region. Also, at $t = 0^+$, the vorticity has not had time to develop in the fluid region. Thus, at time $t = 0^+$, $\boldsymbol{\zeta} = 0$ in both regions, and, as pointed out by Lighthill[3] and Batchelor[1], the flow at time $t = 0^+$ is both viscous and irrotational. However, this flow field has a discontinuity on the tangential component of the velocity, implying that there exists a zero thickness vorticity layer with intensity

$$\boldsymbol{\gamma} = \mathbf{n} \times (\nabla \phi - \mathbf{v}_B) \tag{10}$$

If the flow is viscous, $\boldsymbol{\gamma}$ immediately diffuses into the field, and the tangential boundary conditions are automatically satisfied. The process may be extended to arbitrary motion, by conceiving of a continuous motion as the limit of a sequence of 'relaxation' periods separated by instants of velocity discontinuity, when then vorticity generation occurs, (see Morino [4]).

Numerical Implementation

Based on the foregoing, a time-accurate, marching algorithm as been developed to determine the unsteady motion of an arbitrary body in a two-dimensional, incompressible, viscous flow. The algorithm is a hybrid scheme which uses classical finite-difference methods to solve for the vorticity while boundary integral methods are used to determine the potential. The solution is obtained by first discretizing the domain and then computing boundary element influence coefficients for the field and body points. The initial conditions are set and the field is then marched forward in tim using a four step procedure which is briefly outlined below.

The first step in marching the field forward in time is the determination of the vorticity in the field surrounding the body from the vorticity evolution equation, Eq. 9. This expression has been evaluated by application of a classical time-accurate finite-difference technique. The boundary conditions consist of a zero-vorticity specification at the grid edges, and the vorticity on the body computed during the previous time step.

Once the vorticity in the field is known, the vortical velocity in the field can be evaluated. For two-dimensional flows w_2 is the only non-zero component of the Poincaré solution. The numerical evaluation of Eq. 4 is done by trapezoidal rule integration starting from the upstream boundary and proceeding to the downstream boundary. The upstream boundary condition, consistent with the previous discussion, is taken to be $w_{2_0} = 0$.

Having evaluated the vortical velocity in the field, it is now possible to determine the values of the surface quantities. The unknowns that need to be found on the boundary C_B are the potential, the vorticity, and the vortical velocity. The potential on the body is found using a classical boundary element method formulation based on one of the integral representations presented earlier. The solution is obtained using M nodes on the surface of the body and N nodes in the field surrounding the body. The field nodes are the collection of points representing the intersections of the grid lines, while the surface nodes are the collection of intersection points of the grid with the body. In the discretized field ϕ, χ and w are approximated using first order boundary element shape functions. Of the two possible formulations presented, the distributed doublet formulation given by Eq. 8 has been used for the results obtained here. To determine the magnitude of the surface vorticity note that the magnitude of γ in Eq. 10 for the two-dimensional case is $\gamma = \partial\phi/\partial s - \mathbf{v}_B \cdot \mathbf{t}$, where $\mathbf{t}$ is the tangent to the surface. This expression determines the intensity of the layer of vorticity associated with the just computed potential on the surface. This layer is immediately diffused into the field, in particular to the cell adjacent to the body. Finally, the vortical velocity on the surface and along the shadow line is obtained by integration of Eq. 4.

The procedure for determining the surface quantities outlined above can be implemented in two distinct ways. In a previous paper[5] the authors demonstrated a method whereby the discretized forms of equations 4, 8, and 10 were combined into a system of equations the could be solved simultaneously. The chief drawback of this simultaneous solution method was the requirement of a square matrix for the inversion. This dictates the number of points on the surface at which the vorticity generation condition can be applied. For symmetric two-dimensional flows, a square matrix is easily obtained if the surface vorticity generation condition was evaluated at a set of staggered grid points. While this technique is good for two-dimensional symmetric flows, it is not readily applicable to non-symmetric and three-dimensional flows.

The technique used for the results presented herein starts by first applying Eq. 8 to compute the potential, without regard for satisfying the no-slip condition on the body. The no-slip condition for viscous flows, Eq. 10, is then imposed to obtain the zero thickness vortex layer adjacent to the body. This layer is immediately diffused to obtain the magnitude of the surface vorticity. From the surface vorticity, the magnitude of the vortical velocity is then found using Eq. 4. At this point, the values of the potential on the surface are corrected to obtain a potential distribution that recognizes the no-slip condition by reapplying Eq. 8 using the vortical velocities just computed. Because of the approach taken this method is referred to as the sequential solution approach to the determination of the surface quantities.

The last step of the time marching process is to determine the velocities in the field to be used by the finite-difference algorithm solving the vorticity evolution equation at the next time step. Recall that the velocity is given by $\mathbf{v} = \nabla\phi + \mathbf{w}$. Because $\mathbf{w}$ has already been found only the values of the potential velocity component, $\mathbf{v}_\phi = \nabla\phi$, need to be determined. This can be done by either finite differencing the potential solution in the field or by developing an integral expression for the obtaining the velocity directly. The preferable method for evaluating the field velocity is to develop an integral equation for the velocity thereby avoiding any grid issues. This is done by taking the gradient of either Eq. 8 or Eq. 6 with respect to an arbitrary field point.

The solution may be marched further in time by returning to the computation of the vorticity evolution equation. The size of the allowable time step in this process is determined solely by the stability and/or accuracy of the particular solution technique used for the evaluation of the vorticity evolution equation. Complete details of the procedure may be found in Beauchamp[2].

Results and Discussion

Results for the simultaneous solution method have been previously presented in Morino and Beauchamp[5]. This report focuses on results obtained from the sequential solution method. In developing this method it was found that an additional numerical step is required. If the method is implemented exactly as described the vorticity distribution along the surface exhibits a high frequency oscillation. The oscillation is initially confined to the leading and trailing edges and it propagates toward the center of the body as time continues. The results where found to be improved by using an FFT algorithm to filter out the highest frequency oscillation at the end of each time step.

To verify the sequential solution algorithm a comparative study between the two methods was performed. The problem examined was the transient-response of a flat plate at zero angle of attack at a Reynolds number of 1000. The grid consists of 35 equally spaced elements along the x-axis with 5 elements in the upstream region, 20 on the blade and 10 in the downstream region. The grid in the y direction has 5 quadratically spaced elements on either side of the flat plate extending to $y_{max} = \pm 0.1$. The time increment used for the comparison is $\Delta t = 0.001$. Figure 1 depicts the comparison in the magnitude of the velocity, $|\mathbf{v}|$, as a function of y, at $x =$ 0.25, 0.5, 0.75, and 1.0, at the time corresponding to $t = 0.1$ The numerical results obtained by the sequential solution algorithm are plotted as a solid line against the results obtained from the simultaneous solution algorithm, which are represented by the dotted line. (It should be noted that, while the plotting is done by connecting the points using straight lines, the actual computational velocity distribution between the nodes is quadratic.) The simultaneous solution algorithm results were obtained on exactly the same grid. The figures show that the two algorithms are in excellent agreement except for a very small difference in the trailing edge region. Figure 2 shows a comparison of the magnitude of the vortical velocity along the line $y = 0$ for the two algorithms at $t = 0.1$. The solid line is the sequential solution algorithm, whereas the dotted line is the simultaneous solution algorithm. The two results are in agreement to within 0.6%. The slight discrepancy can be traced to the fact that the simultaneous solution algorithm is obtaining a larger vorticity peak in the leading edge region than the sequential solution algorithm. The similarity between the two results is significant because the simultaneous solution algorithm determines the vortical velocity directly, while in the sequential solution algorithm the vortical velocity is derived from the computed vorticity.

The application of the formulation to non-symmetric flows is examined by solving for the flow past a thin airfoil which is impulsively started at an angle of attack. The problem consists of a knife-edge thin airfoil moving at a unit velocity with an angle of attack of 30 degrees in a flow with a Reynolds number of 400. The grid for this problem consists of 30 elements in the x direction and 5 elements on either side of the airfoil. In the x direction there are 5 equally spaced elements extending from the leading edge to a quarter chord upstream. The grid downstream of the trailing edge is identical. In the surface region 20 elements are distributed with equal spacing between the leading and trailing edges. The grid normal to the body is quadratically spaced and extends to $y_{max} = 0.3125$. The solution is marched forward in time using $\Delta t = .00005$.

Figures 3 and 4 respectively illustrate the velocity and the magnitude of the velocity component tangent to the surface at three distinct times. These results for the first few time steps are qualitatively quite good and clearly capture the formation of the trailing edge vortex and the associated reverse flow region. The leading edge flow acceleration is also clearly visible.

References

[1] Batchelor, G. K. : An Introduction to Fluid Dynamics, Cambridge University Press, (1967).

[2] Beauchamp, P. : " A Potential-Vorticity Decomposition for the Boundary Integral Equation Analysis of Viscous Flows," Ph. D. Thesis, Graduate School, Division of Engineering and Applied Science, Boston University, Boston, MA, USA, (1990).

[3] Lighthill, M. J. : "Introduction to Boundary Layer Theory," Part II of Laminar Boundary Layers, Ed. L. Rosenhead, Oxford University Press, pp.46-113, (1963).

[4] Morino, L. : "Helmholtz Decomposition Revisited: Vorticity Generation and Trailing Edge Condition. Part I - Incompressible Flows," Computational Mechanics, No. 1, Vol. 1, (1986).

[5] Morino, L., and Beauchamp, P. : "A Potential-Vorticity Decomposition for the Boundary-Element Analysis of Viscous Flows," Eds.: M. Tanaka and T. A. Cruse, Boundary Element Methods in Applied Mechanics, Pergamon Press, New York, NY, USA, (1988).

[6] Morino, L. : "Helmholtz and Poincaré Potential–Vorticity Decompositions for the Analysis of Unsteady Compressible Viscous Flows" Developments in Boundary Element Methods, Vol. 6: Nonlinear Problems of Fluid Dynamics, Eds. P.K. Banerjee and L. Morino, Elsevier Applied Science Publishers, Barking, UK, (1990).

Acknowledgments

This work was partially supported by NASA Langley Research Center (Grant No. NAG-1-934 to Boston University). The authors would also like to thank the General Electric Company for the support provided to Dr. Beauchamp through the Advanced Course in Engineering while working on his dissertation.

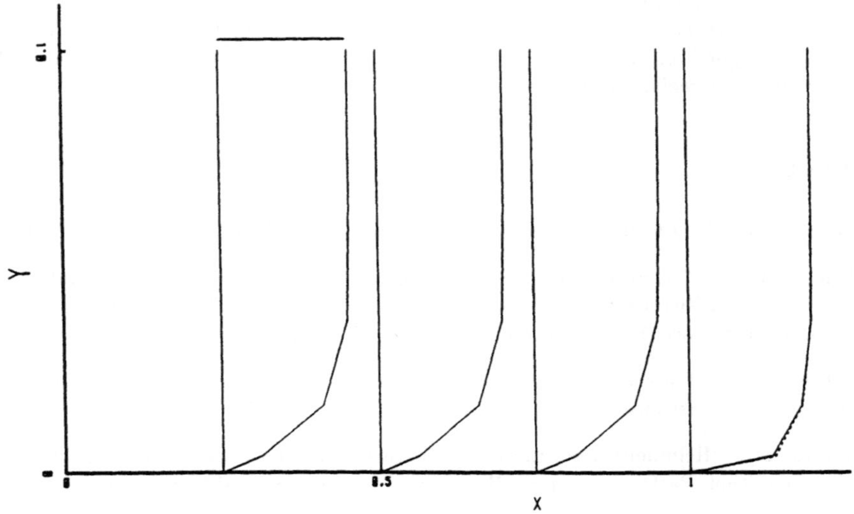

Figure 1: Velocity distribution $|\mathbf{v}|$ at $x = .25, .5, .75$, and 1.0, at $t = .1$ for no angle of attack at $\Re = 1000$; Sequential solution method (solid); Simultaneous solution method (dotted).

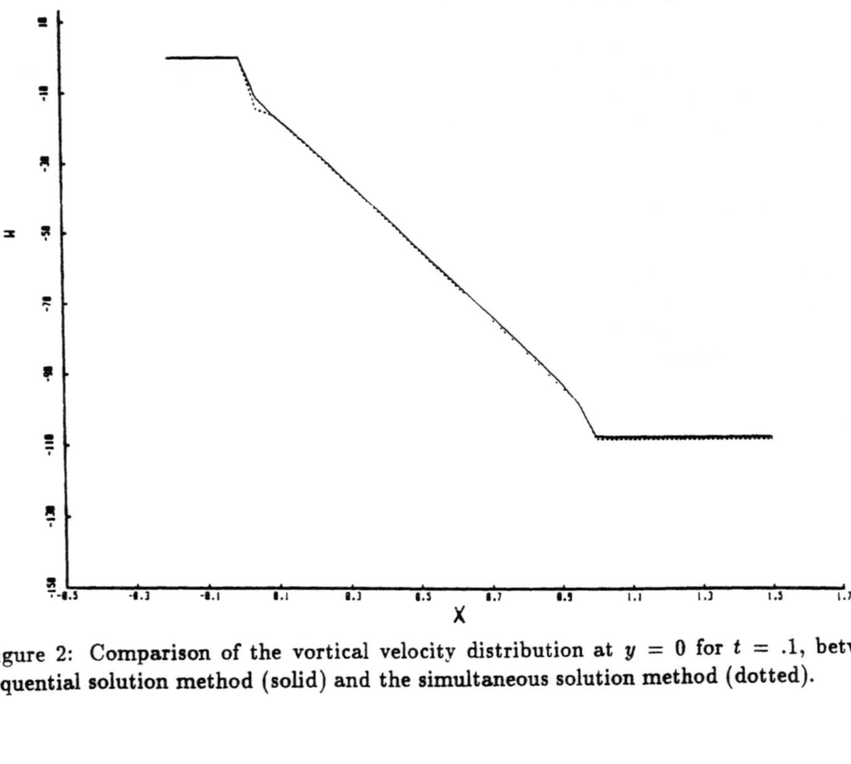

Figure 2: Comparison of the vortical velocity distribution at $y = 0$ for $t = .1$, between the sequential solution method (solid) and the simultaneous solution method (dotted).

$t = 0.002$

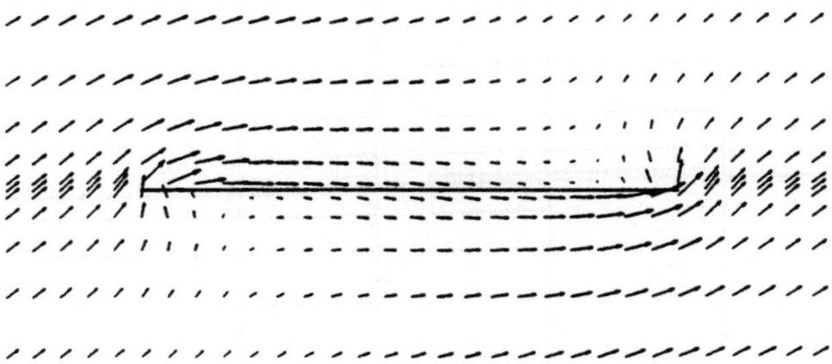

$t = 0.006$

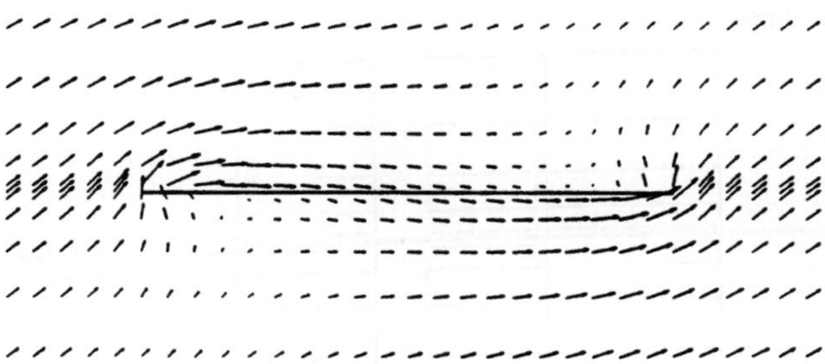

$t = 0.010$

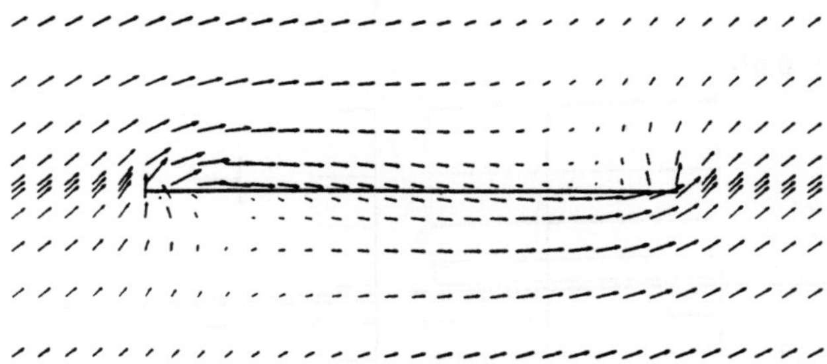

Figure 3: Velocity fields for the thin airfoil at 30 degrees angle of attack and $\Re = 400$ at $x =$ .25, .5, .75, and 0.95, at $t = .002$ .006 and .01.

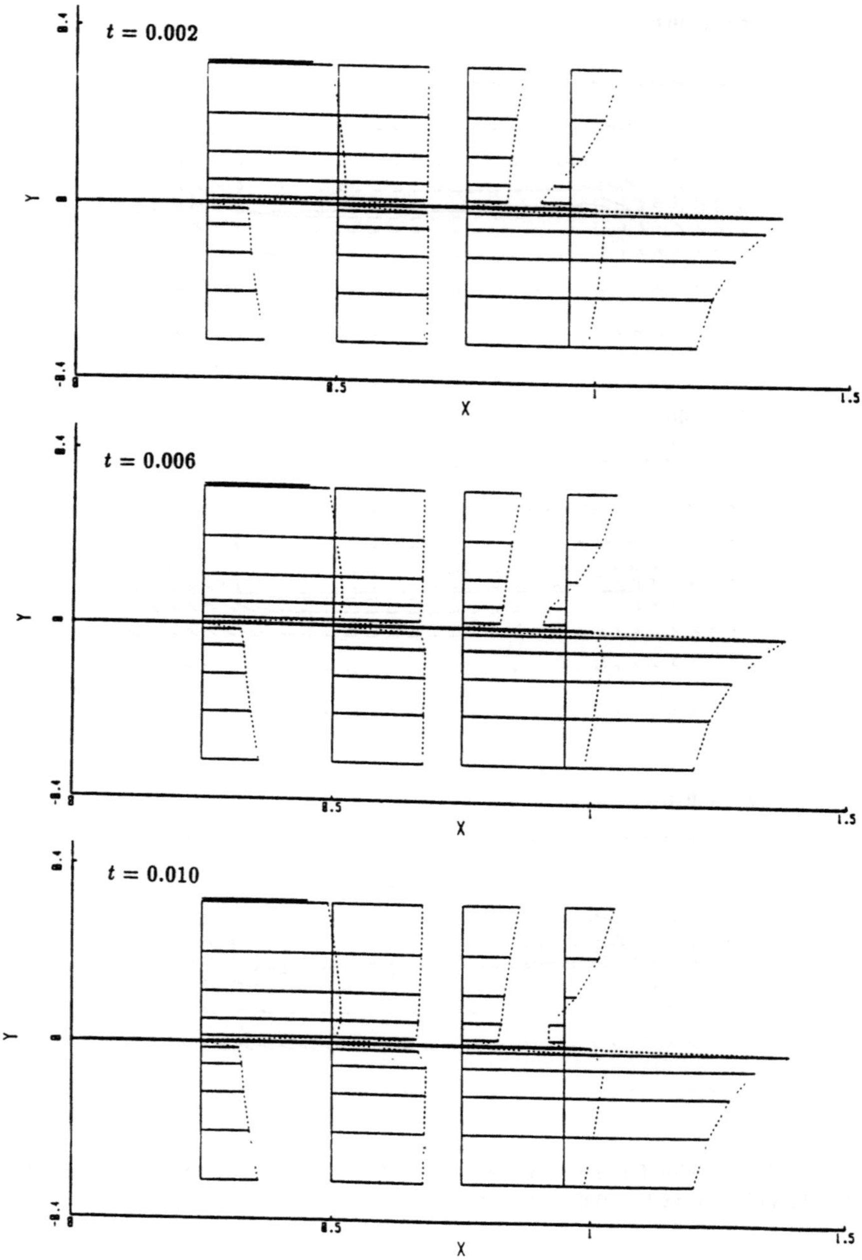

Figure 4: Velocity component u for the thin airfoil at 30 degrees angle of attack and $\Re = 400$ at $x = .25, .5, .75$, and 0.95, at $t = .002$ .006 and .01.

Application of a Low Order Panel Method to Slender Delta-Wings at High Angles of Attack

R. J. Behr and S. N. Wagner

Universität der Bundeswehr München
Institut für Luft- und Raumfahrttechnik
Werner Heisenbergweg 39, D–8014 Neubiberg

Summary

A suitable low order panel method is introduced, that allows the calculation of free vortex sheets emanating from sharp–edged wings in subsonic flow. The wing surfaces as well as the separated shear layers are discretized into panels, each carrying a constant doublet distribution. Bodies – e.g. fuselages – can be modeled by a network of source panels with elementwise constant strength.

The method makes use of a time–marching procedure, describing the development and position of the free vortex sheets step by step, starting with a solution without any shedded vorticity and ending up in a steady vortical flow around the examined configuration. Results of the numerical method concerning aerodynamic loads and flowfield characteristics are discussed and compared to other theoretical models and experimental data.

Introduction

The vortex dominated flow around slender delta–wings at moderate–to–high angles of attack plays an important role in modern aircraft design. Many of todays high performance aircrafts take advantage of the additional lift produced by leading–edge vortex separation to improve maneuvring and landing qualities.

Although the experimental investigation (see e.g. [1] ,[2] and [3]) and the development of theoretical methods (see e.g. [4], [5] or [6]) for the calculation of this type of flow started a few decades ago there are still world–wide activities in research work on this topic.

Essentially there are existing two quite different approaches towards the theoretical treatment of vortical flows. On the one hand the surface integral formulation of panel methods is used – compare e.g. [7] or [8] – and on the other hand the solution of the EULER and recently also of the NAVIER–STOKES equations is performed, see e.g. [9] and [10]. The implicit modeling of vortices and the applicability throughout the whole region of Mach numbers are remarkable benefits of these methods. However, the computational effort – both CPU–time and storage requirements – that is necessary to perform a accurate solution of the vortex flow problem by a finite difference scheme is very high (grid generation, iterative solution procedure, and post processing).

In general the panel method offers a great versatility concerning the treatment of complex geometries, is easy to handle and causes comaparatively low computational costs.

Of course, the extension of the panel method to the simulation of separated flow requires additional effort in computation and also is restricted to flow separation along aerodynamically sharp-edges. To keep the additional numerical effort resulting from the determination of the separated shear layers at a minimum the application of a low-order singularity representation of wings and vortex sheets is of advantage.

Theoretical Approach and Numerical Solution

The basic assumption of such a model is that the flow is incompressible, irrotational and homogenous over the whole region excluding the wings and free vortex sheets. Therefore a velocity potential Φ exists and the continuity equation can be written as

$$\nabla\Phi = \frac{\partial^2\phi}{\partial x^2} + \frac{\partial^2\phi}{\partial y^2} + \frac{\partial^2\phi}{\partial z^2} = 0. \tag{1}$$

The extension of the model to subsonic compressible flow is performed by applying the PRANDTL–GLAUERT transformation of the linearized potential equation. This simplification has proved to be admissable up to medium Mach numbers for single or coupled wings without leading–edge vortex separation only. The examination of wings producing strong vortices requires a more accurate analysis of compressibility effects. A combination of the present vortex–lattice method with a field-panel code for the solution of the full potential equation up to the transonic region is currently under work.

Within the present method the velocity potential Φ is represented by a set of discretely distributed doublet elements of stepwise constant strength μ, which is equivalent to a network of closed vortex rings with $\Gamma \equiv \mu$, Fig. 1. The additional presence of a fuselage is modeled by a network of source panels on its surface, each carrying a constant strength, too.

During a time dependent, quasisteady procedure, similar to the model presented in [11], starting with an impulsive motion of the examined configuration, the development of the free vortex sheets is observed until steady aerodynamic loads are reached, compare Fig. 2 and Fig. 3. The extension of the method into the time-dependent framework requires the application of Kelvin's theorem where zero net circulation is specified for all time intervals :

$$\frac{\partial\Gamma}{\partial t} = 0. \tag{2}$$

The time increment Δt is chosen as $\Delta t = a_P/U_\infty$ with a_P being the smallest distance between two neighbouring wing–bound vortices.

The kinematic boundary condition at the configuration collocation points

$$(U_\infty + \nabla\Phi) \cdot \vec{n_W} = 0. \tag{3}$$

where $\vec{n_W}$ denotes the panel normal vector of unit length, is used for determinating the a priori unknown strengths of the singularities distributed at the configuration surface

within each time step. The wake–shedding pocedure at the sharp edges of the wing and the motion of the free vortex sheets is due to the local existing velocities and all doublet elements shedded into the wake keep their circulation according equation (2) throughout the whole calculation. Thereby the Kutta–condition at the wing edges with flow separation $\Delta c_p = 0.$ is approximately fulfilled as well as the demand of vanishing normal forces over the complete free wake.
The numerical procedure starts without any free vortex sheets so that the last term on the right–hand side of equation (4) vanishes for the first solution and the wing–bound doublet strenghts μ_W at each panel i can be calculated by solving a system of linear equations.

$$[\vec{\vec{A}}_{W(i,j)} \cdot \vec{n}_{W(i)}] \cdot \mu_{W(i)}(t) = -\vec{n}_{W(i)} \cdot \vec{U}_\infty(t) - [\vec{\vec{C}}_{VS(i,k)}(t) \cdot \vec{n}_{W(i)}] \cdot \mu_{VS(k)} \qquad (4)$$

where $\vec{\vec{A}}_{W(i,j)}$ is the aerodynamic influence coefficient matrix of the wing and $\vec{\vec{C}}_{VS(i,j)}$ is the influence matrix of the wake onto the wing and μ_{VS} is the doublet strenght of a wake panel.
The induced velocities $\vec{v}_{ind}$ at all network points k of the free vortex sheets are calculated by

$$\vec{v}_{ind,k} = \vec{\vec{B}}_{W(i,k)}(t) \cdot \mu_{W(i)}(t) + \vec{\vec{C}}_{VS(j,k)}(t) \cdot \mu_{VS(j)} \qquad (5)$$

where $\vec{\vec{B}}_{W(i,k)}$ is the influence coefficient matrix of the wing onto the wake and $\vec{\vec{C}}_{VS(j,k)}$ is the influence matrix of the wake on itself.
Within each time step the length of the wake increases, see Fig. 2 and the flowfield around the wing approaches steady state conditions, see Fig. 3. During the whole calculation the induced velocities everywhere in the flowfield have to be available. Especially in the vicinity of the discrete vortex filaments a continous velocity distribution must be guaranteed so that the velocity induced by the Biot–Savart law has to be replaced by a special nearfield solution provided by the sub–vortex–technique, see [12] and compare Fig. 4, or the interpolation scheme shown in [13]. For the update of the geometry of the free wake each vortex–induced velocity is transformed into a circular motion around the center of the vortex, see Fig. 5. To make sure that no penetration of other panels occurs a local sub–time step option is applied. A point P moving through a panel within the normal time step Δt is treated separately, see also Fig. 5. Within a first part of the time step Δt_1 it moves to position P^* and with the induced velocity there, v^*_{ind} times the remaining part of the time step $\Delta t_2 = \Delta t - \Delta t_1$ it is transported to its new position P'_{new}.

Results

The comparison of the present method with two other Vortex–Lattice Methods given in Fig. 6 shows clearly the enhanced roll–up process of the leading–edge vortices emanating from the canard and the main wing that was achieved by the modifications of the model described above. Although the angle of attack is the same in the case of the solution of [15] and even 5 degrees higher in the other case both methods predict vortices that spread almost over the complete wingspan. From experimental results [16] on close coupled canard configurations it is known that there is a strong rolling up of the free vortex sheets.

The result for the AR=1 delta wing shown Fig. 7 points out that the roll–up process calculated with the present method comes very close to the experimental results of Hummel [3]. Diameter and position of the leading–edge vortex sheet in the displayed cross–flow plane are in good agreement. The spanwise pressure distribution on the upper and lower side of the wing is also predicted quite good. The aerodynamic total coefficients displayed in Fig. 8 also compared to the experimental data of [3] show some small deviations due to the fact that the wing from the experiment does not have a symmetric airfoil.The reference point for the C_M coefficient is the geometrical neutral point of the wing, so that the values itself are relatively small. The slope of the C_L over C_M diagram is again in good agreement.

A typical result of the present metod applied to another planform is presented in Fig. 9. The discretization used here is a 8x8-lattice (in Fig. 7 it is 10x10–lattice) and the wing of the experiment [2] was a thin plate so that the predicted lift and drag coefficients match the experimental data very good. The deviations in the behaviour of the C_M coefficient can be reduced by using a more fine discretization within the numerical simulation.

Fig. 10 shows some details of the solution for a delta wing. The discrete vortex lines carrying the separated vortices within the model on the one side and the well predicted geometry of rolled–up vortex sheet including the double–branched vortex emanating from the wing's trailing edge on the other side express the capabilities of the present numerical model if a fine discretization is used. The additional information given in Fig. 10 concerns the position of the vortex core in relation to the vortex–sheet geometry. It is remarkable that the vortex core and the innermost vortex filament emanating from the apex of the wing are really at the same location. The application of the method to a close–coupled canard wing configuration is shown in Fig. 11. The results obtained here also match the experimental data in a satisfying way so that further examinations on this topic are planned.

Conclusions

The presented method allows to get information about basic wing–vortex interaction problems at relatively low computational effort. The numerical procedure was successfully tested at different geometries and flow cases and the actual work on the method deals with the compatibility of the results to higher–order panel methods as they are used within the design process of modern aircrafts. A further extension of the model to thick wings is planned in the next time as well as an application to unsteady flow conditions or maneuvers.

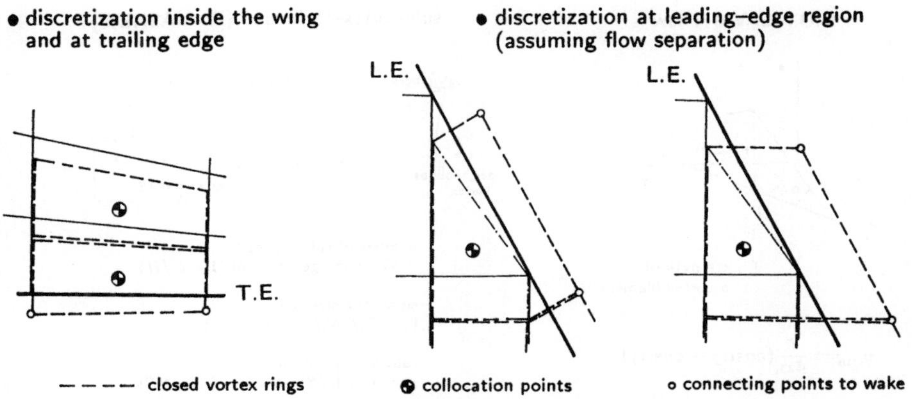

Fig. 1. Arrangement of doublet elements (vortex rings) at wing surface

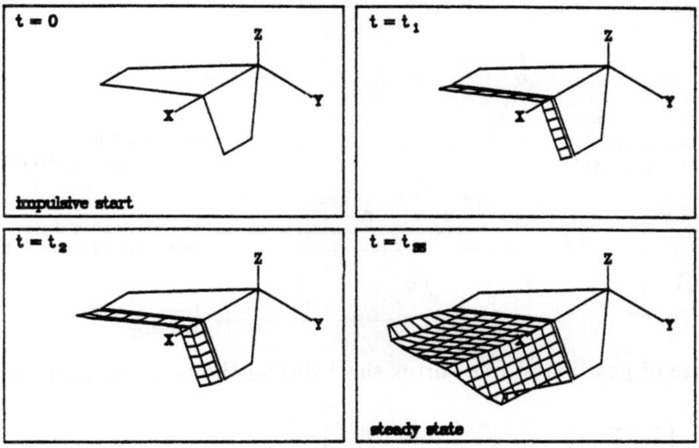

Fig. 2. Time dependent development of free vortex sheet emanating from trailing edge

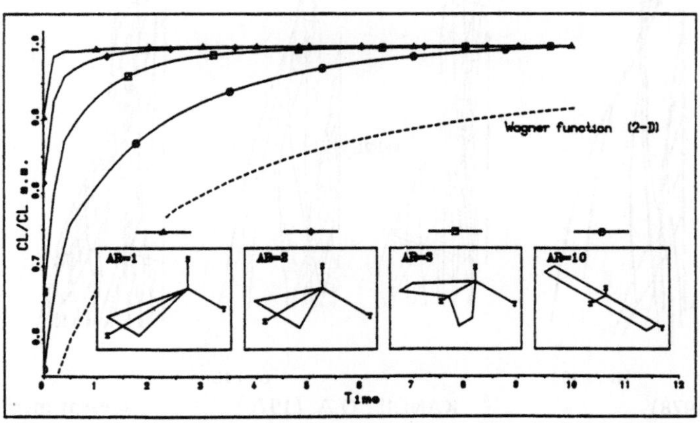

Fig. 3. Time history of wing lift from impulsive start to steady state for different aspect ratios (no leading–edge separation)

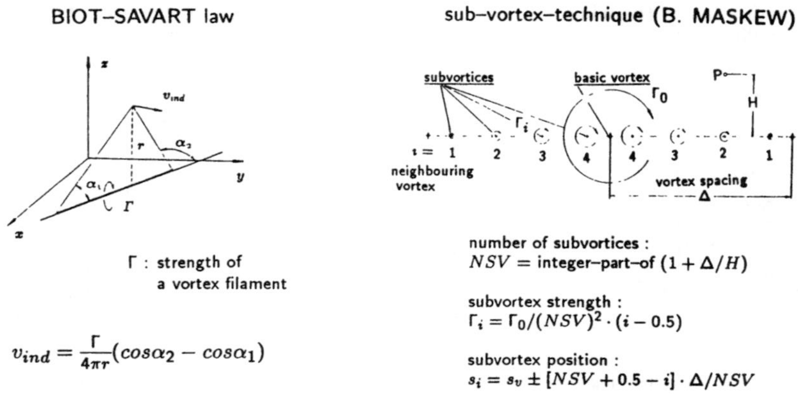

Fig. 4. Calculation of induced velocities (farfield and nearfield)

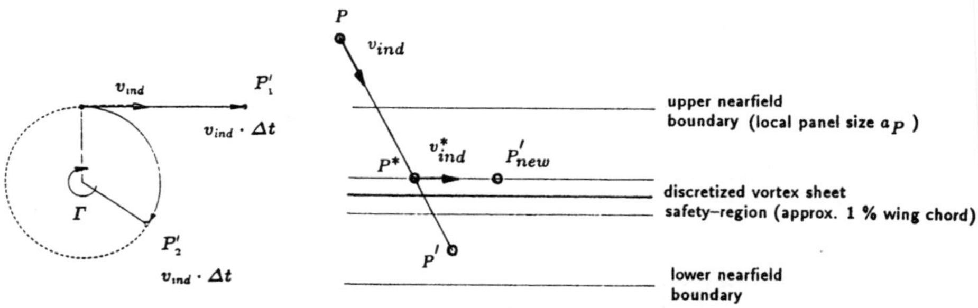

Fig. 5. Update of position of free vortex sheet during the time-marching process

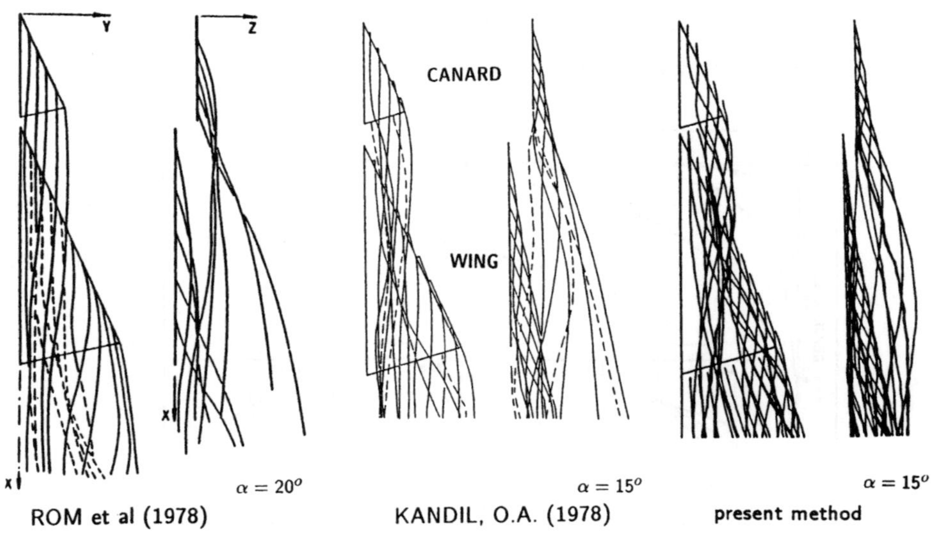

Fig. 6. Comparison of different Vortex–Lattice Methods (results taken from [14] and [15])

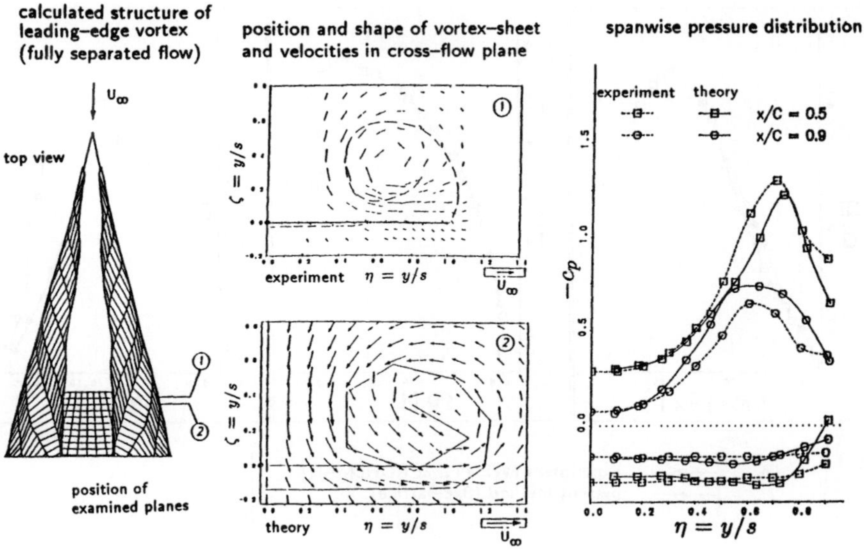

Fig. 7. Numerical and experimental (see [3]) results for the AR=1 delta wing at an angle of attack of 20.5 degrees

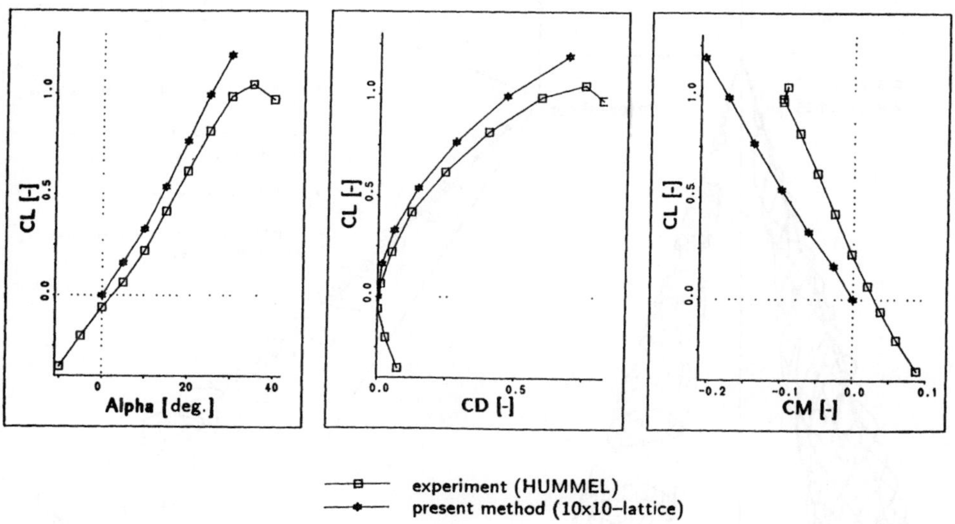

Fig. 8. Calculated and measured (see [3]) total aerodynamic coefficients for the AR=1 delta wing

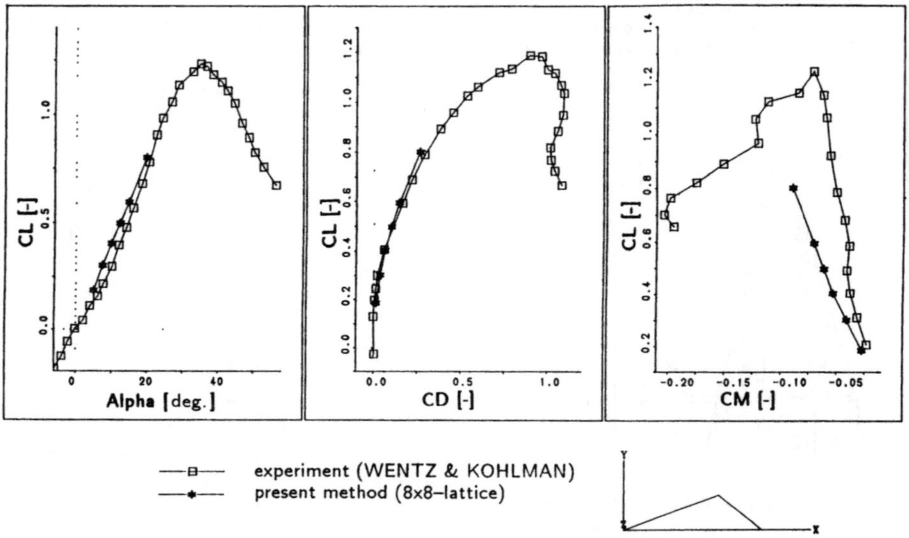

Fig. 9. Calculated and measured (see [2]) total aerodynamic coefficients for a a diamond wing with a leading–edge sweep of 70 degrees

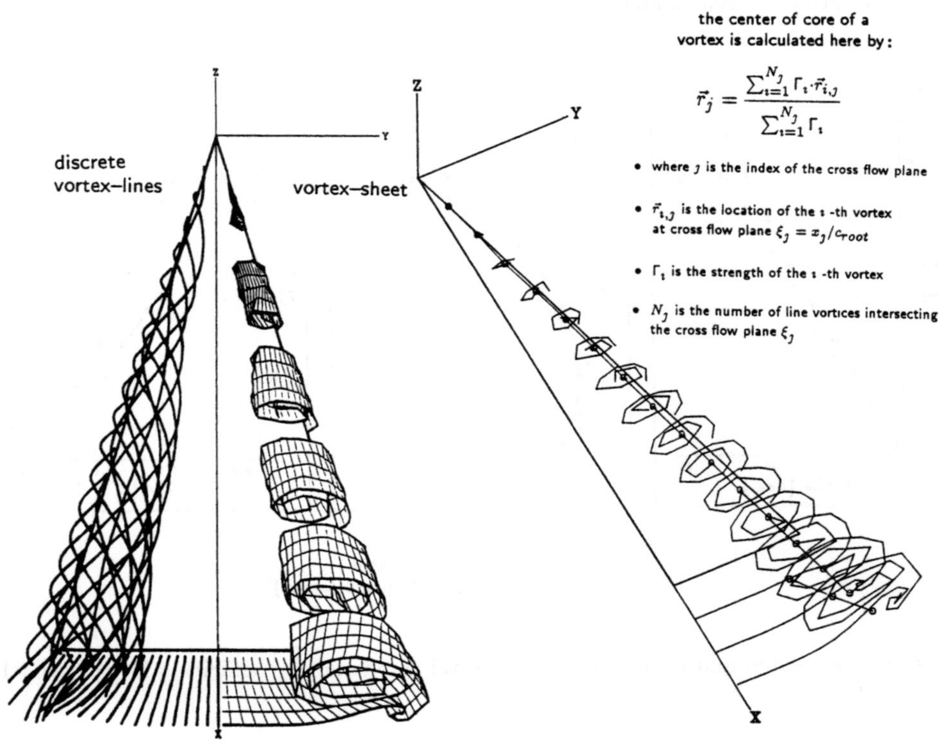

Fig. 10. Detailed view at the numerical solution for the flow around the AR=1 delta wing at an angle of attack of 20.5 degrees. (A very fine discretization of 210 panels on half wing is used)

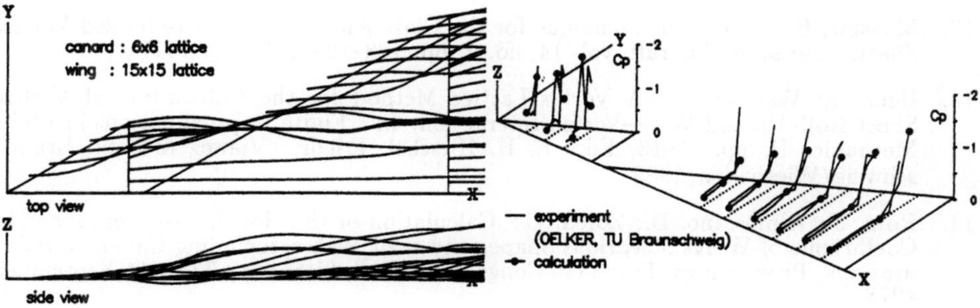

Fig. 11. Application of the present method to a close–coupled canard wing configuration (experimental results [16])

References

1. Behrbohm, H.: Basic low speed aerodynamics of the short–coupled canard configuration of small aspect ratio. SAAB TN 60, July 1965

2. Wentz, W.H.; Kohlman, D.L.: Wind Tunnel Investigations of Vortex Breakdown on Slender Sharp–edged Wings. University of Kansas Center for Research, Inc. Engineering Sciences Division, Report FRL 68–013, Nov. 1968

3. Hummel, D.: On the Vortex Formation over a Slender Wing at at Large Angles of Incidence. AGARD–CP–247, pp. 15–1 – 15–17, Oct. 1978

4. Polhamus, E.C.: Predictions of Vortex–Lift Characteristics by a Leading–Edge Suction Analogy. Journal of Aircraft, vol. 8, no. 4, pp. 193–199, April 1971

5. Lamar, J.E.; Gloss B.B.: Subsonic Aerodynamic Characteristics of Interacting Lifting Surfaces with Separated Flow around Sharp Edges predicted by a Vortex–Lattice Method. NASA TN D–7921, Sept. 1975

6. Kandil, O.A.; Mook D.T.;Nayfeh, A.H.: Nonlinear Prediction of Aerodynamic Loads on Lifting Surfaces. Journal of Aircraft, vol. 13, no. 1, January 1976

7. Hoeijmakers, H.W.M.: Computational Aerodynamics of Ordered Vortex Flows. Technical University of Delft, Thesis, May 1989

8. Gordon, R.: Numerical Simulation of Vortical Flow over a Strake–Delta Wing and a Close Coupled Delta–Canard Configuration. Proceedings of the AIAA 8th Applied Aerodynamics Conference, Portland, Oregon, pp 68–78, August 20–22, 1990

9. Hitzel, S.M.: Low and High Speed, High Angle–of–Attack Flow around a Delta–Wing by an Euler Simulation. Royal Aeronautical Society, London, April 1989

10. Longo, J.M.A.;Das, A.: Numerical Simulation of Vortical Flow over Close Coupled Canard–Wing Configuration. Proceedings of the AIAA 8th Applied Aerodynamics Conference, Portland, Oregon, pp 79–88, August 20–22, 1990

11. Levin, D.; Katz, J.: Vortex–Lattice Method for the Calculation of the Nonsteady Separated Flow over Delta Wings. Journal of Aircraft, vol. 18, no. 12, pp. 1032–1037, December 1981

12. Maskew, B.: Subvortex Technique for the Close Approach to a Discretized Vortex Sheet. Journal of Aircraft, vol. 14, no. 2, pp. 188–193, February 1977

13. Behr, R; Wagner, S.: A Vortex-Lattice Method for the Calculation of Vortex Sheet Roll–Up and Wing–Vortex Interaction. In : Finite Approximations in Fluid Mechanics II, pp. 1–13, Ed. E. H. Hirschel, Friedr. Vieweg & Sohn Braunschweig/Wiesbaden, 1989

14. Rom, J.; Almosnino, D.; Zorea, C.: Calculation of the Non Linear Aerodynamic Coefficients of Wings of Various Shapes and their Wakes, including Canard Configurations. Proceedings of the 11th Congress of ICAS, Lisbon, pp 333–344, September 1978

15. Kandil, O.A.: State of the Art of Nonlinear, Discrete-Vortex Methods For Steady and Unsteady High Angle of Attack Aerodynamics. AGARD–CP–247, pp 5–1 – 5–4, Oct. 1978

16. Hummel, D.; Oelker, H.–Chr.: Effects of Canard Position on the Aerodynamic Characteristics of a Close–Coupled Canard Configuration at Low Speed. AGARD Fluid Dynamics Panel Symposium on Aerodynamics of Combat Aircraft Controls and of Ground Effects, Madrid, Spain, pp 7–1 – 7–17, October 2–5, 1989

Acknowledgement

This paper is based on research work funded by the Deutsche Forschungsgemeinschaft DFG (Contract No. Wa 424/7).

Boundary Element Sensitivity Analysis and Optimal Design of Vibrating and Built-Up Structures

Tadeusz BURCZYNSKI and Piotr FEDELINSKI

Institute of Mechanics and Fundamentals of Machine Design, Silesian Technical University, Poland

Summary

Applications of the boundary element method (BEM) to shape sensitivity analysis and optimal design of vibrating structures with unspecified external boundaries and built-up structures with altered interfaces are discussed.

Introduction

The general variational approach to problems of sensitivity analysis and optimal design of structures with shape transformation using boundary elements was presented in Ref. [1]. In this paper the case when the shape of an external boundary of a vibrating structure is not specified in advance is considered in detail more than in previous works [2,3,9,10].

Besides the external boundary, the variation of shape of interfaces between of different materials is also studied for static built-up systems.

1. Shape sensitivity analysis and optimal design of vibrating structures

The governing differential equation for free-vibration of an isotropic homogeneous elastic body, which occupies a domain Ω with a boundary Γ, can be written as

$$\mathrm{div}\sigma(\mathbf{x}) + \rho\omega^2\mathbf{u}(\mathbf{x}) = 0 \;, \quad \mathbf{x}\in\Omega, \tag{1}$$

where σ is a stress tensor, ω a natural circular frequency, ρ a mass density and $\mathbf{u}(\mathbf{x})$ a displacement amplitude.

On the boundary Γ, there are prescribed homogeneous boundary conditions in the form of displacement amplitude $\mathbf{u}(\mathbf{x})=0$, $\mathbf{x}\in\Gamma_u$ and tractions $\mathbf{p}(\mathbf{x})=0$, $\mathbf{x}\in\Gamma_p$, where $\Gamma=\Gamma_u\cup\Gamma_p$.

It is obvious that the natural frequency ω depends on the shape of Γ. The objective is to determine the dependence of the natural frequency with respect to shape variation.

In order to solve this problem one considers an infinitesimal variation of configuration of the body by prescribing a continuous and differentiable vector field $\mathbf{g}(\mathbf{x})=(g_k(\mathbf{x}))$ (k=1,2,3 for 3-D or k=1,2 for 2-D), so that:

$$\mathbf{x}^* = \mathbf{x} + \delta\mathbf{g}(\mathbf{x}), \tag{2}$$

The transformation field $\mathbf{g}(\mathbf{x})=\mathbf{g}(\mathbf{x};\mathbf{a})$ modifies the shape of the external boundary Γ, where $\mathbf{a}=(a_r)$, (r=1,2,..R), a_r is a

shape design parameter, which specify the actual shape of the structure. The variable $\mathbf{x}$ is defined in the untransformed domain Ω with the boundary Γ and variable $\mathbf{x}^*$ is defined in the transformed domain $\Omega^*=\Omega(\mathbf{a})$ with the boundary $\Gamma^*=\Gamma(\mathbf{a})$.
The variation of the transformation field $\delta\mathbf{g}$ is expressed as

$$\delta g_k = \frac{\partial g_k}{\partial a_r}\ \delta a_r = v_k^r\ \delta a_r, \tag{3}$$

where $v_k^r=\partial g_k/\partial a_r$ can be considered as a transformation velocity field which is associated with a shape design parameter a_r.

It is convenient to treat Ω as a continuous medium and utilize the material derivative idea from continuum mechanics. Then the mapping (2) may be viewed as a dynamic process with δa_r playing the role of the time and v_k^r playing the role of the velocity field. Note that for $\delta a_r=0$ there is $\Omega^*=\Omega^o$ and $\Gamma^*=\Gamma^o$, where Ω^o and Γ^o are the initial domain and boundary, respectively.

The eigenvalue problem, described by (1), can be considered in terms of a variational equation in the form (cf.[3]):

$$\int_{\Omega(\mathbf{a})} \sigma(\mathbf{u})\cdot\varepsilon(\mathbf{u})\,d\Omega = \omega^2 \int_{\Omega(\mathbf{a})} \rho\mathbf{u}\cdot\mathbf{u}\,d\Omega, \tag{4}$$

where $\sigma(\mathbf{u})$ and $\varepsilon(\mathbf{u})$ are stress and strain tensors associated with the eigenfunction $\mathbf{u}(\mathbf{x})$.

The total material derivatives of σ, ε, $\mathbf{u}$ and a domain element $d\Omega$ and a unit vector $\mathbf{n}=(n_k)$ normal to Γ with respect to a_r can be expressed as follows (cf.[5,6]):

$$\frac{D\mathbf{q}}{Da_r} = \frac{\partial\mathbf{q}}{\partial a_r} + \mathbf{q},_k v_k^r, \qquad \mathbf{q}=\sigma,\varepsilon,\mathbf{u} \tag{5a}$$

$$\frac{D(d\Omega)}{Da_r} = v_{k,k}^r d\Omega, \qquad \frac{Dn_i}{Da_r} = (n_i n_l - \delta_{il}) n_k v_{k,l}^r\ . \tag{5b}$$

The variation of simple natural circular frequencies ω is given by:

$$\delta\omega = S_r\ \delta a_r, \tag{6}$$

where sensitivities $S_r=D\omega/Da_r$ can be evaluated by taking the total material derivative of both sides of (4) and taking into account (5) and the normalizing condition $\int_\Omega \rho\mathbf{u}\cdot\mathbf{u}\,d\Omega=1$ (cf.[2,3]):

$$S_r = \frac{1}{2\omega} \int_\Gamma (\sigma(\mathbf{u})\cdot\varepsilon(\mathbf{u}) \ -\omega^2\rho\mathbf{u}\cdot\mathbf{u}) n_k v_k^r d\Gamma. \tag{7}$$

It is interesting to notice that the relationship between the variation of the shape of the body and the variation of the natural frequency is expressed by the boundary integral and sensitivity of the frequency depends on modes determined on the boundary. This fact is of great importance in numerical calculations by means of the boundary element method.

In order to formulate the problem in terms of the boundary

unknowns only, the displacement amplitude $\mathbf{u}$ within the domain is approximated by using a set of unknown coefficients $\mathbf{s}^v$ and a set of coordinate functions $f^v(\mathbf{x})$ (cf.[12]):

$$\mathbf{u}(\mathbf{x}) = \mathbf{s}^v f^v(\mathbf{x}) , \quad v= 1,2,\ldots,V, \tag{8}$$

Using this assumption, the equation of free vibration may be written in the boundary integral form

$$\begin{aligned} &\mathbf{c}(\mathbf{x})\mathbf{u}(\mathbf{x}) - \int_\Gamma \left[\mathbf{u}^*(\mathbf{x},\mathbf{y})\mathbf{p}(\mathbf{y}) - \mathbf{p}^*(\mathbf{x},\mathbf{y})\mathbf{u}(\mathbf{y})\right] d\Gamma(\mathbf{y}) + \\ &\omega^2\rho \left\{\mathbf{c}(\mathbf{x})\hat{\mathbf{u}}^v(\mathbf{x}) - \int_\Gamma \left[\mathbf{u}^*(\mathbf{x},\mathbf{y})\hat{\mathbf{p}}^v - \mathbf{p}^*(\mathbf{x},\mathbf{y})\hat{\mathbf{u}}^v(\mathbf{y}) d\Gamma(\mathbf{y})\right\}\mathbf{s}^v = 0, \end{aligned} \tag{9}$$

where $\mathbf{u}^*$ and $\mathbf{p}^*$ are fundamental solutions of elastostatics, $\hat{\mathbf{u}}^v$ and $\hat{\mathbf{p}}^v$ are pseudo-fields of displacements and tractions, respectively, resulting from the body force $\mathbf{I}f^v$ ($\mathbf{I}$ - unit matrix).

The boundary is divided into boundary elements including a finite number of nodal points. The displacements and tractions within each element are approximated in terms of nodal values.

The boundary integral equation is applied to every node. If the same interpolation functions are used for $\mathbf{u}$, $\mathbf{p}$ and for $\hat{\mathbf{u}}^v$, $\hat{\mathbf{p}}^v$ then the equation of free vibration is obtained:

$$\mathbf{H}\ \mathbf{U} = \omega^2\ \mathbf{M}\ \mathbf{U}, \tag{10}$$

where a mass matrix $\mathbf{M}$ is given by

$$\mathbf{M} = -\rho\ [\mathbf{H}\ \hat{\mathbf{U}} - \mathbf{G}\ \hat{\mathbf{P}}]\ \mathbf{F}\ , \tag{11}$$

where $\mathbf{H}$ and $\mathbf{G}$ are the same matrices as for static problems, while $\hat{\mathbf{U}}$ and $\hat{\mathbf{P}}$ contain the node values of functions $\hat{\mathbf{u}}^v$ and $\hat{\mathbf{p}}^v$, respectively, and the matrix $\mathbf{F}$ depends on the node values of functions $f^v(\mathbf{x})$.

The matrices $\mathbf{H}$ and $\mathbf{M}$ are nonsymmetric but they can be transformed into symmetric and positive matrices (cf.[11]).

The resulting natural frequencies and mode shapes are more accurate than for the nonsymmetric method and some efficient methods can be used for an eigensolution.

In the most shape optimization problems of vibrating structures one should maximize the fundamental circular frequency. Thus, the problem is to find shape parameters $\mathbf{a}_{op}=(a_r)$ that minimize

$$J_o = -\ \omega \Rightarrow \min_{\mathbf{a}}, \tag{12}$$

subject to the volume and geometrical constraints.
This problem can be solved using gradient methods, e.g. gradient projection method (cf.[9]).

Numerical implementation

The rectangular rigid supported plate is studied to demonstrate the application of the presented method. The plate, which is shown in Fig.1.1a is divided into 50 linear boundary elements and considered as a two-dimensional elastic

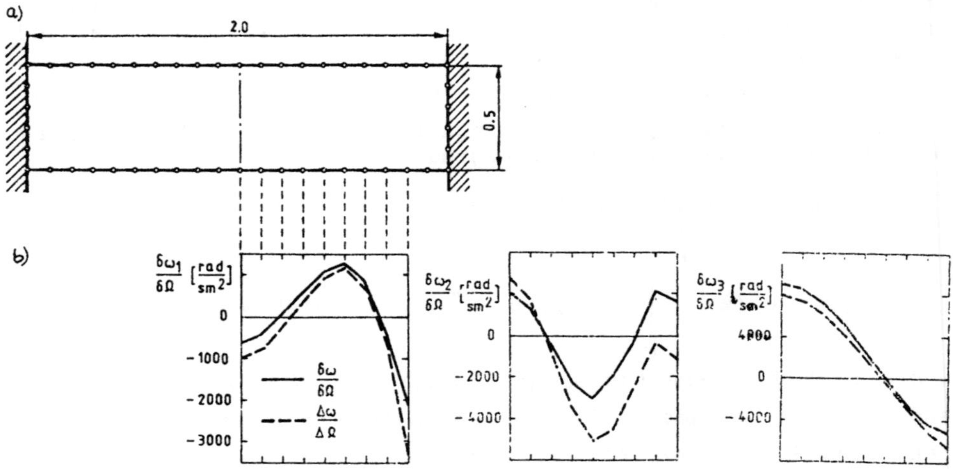
a)
2.0
0.5
b)
0
- 1000
- 2000
- 3000
0
- 2000
- 4000
0
- 4000

Fig.1.1

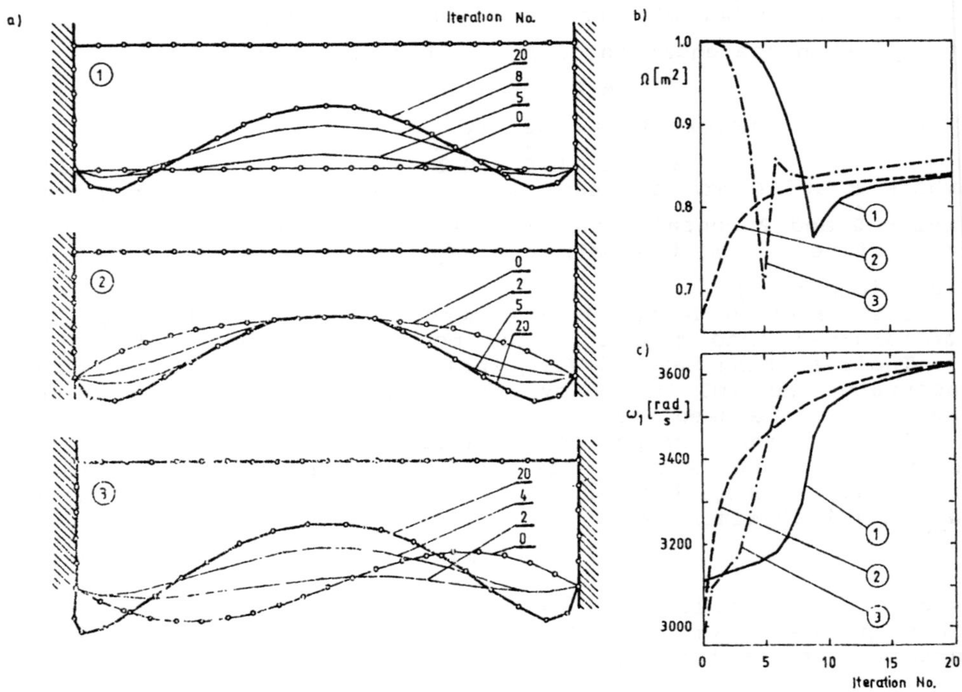
a)
Iteration No.
1
20
8
5
0
2
0
2
5
20
3
20
4
2
0
b)
1.0
0.9
0.8
0.7
c)
3600
3400
3200
3000
0
5
10
15
20
Iteration No.

Fig.1.2

body (plane strain) with the following material constants: Young's modulus $E=0.2\cdot10^{12}$[Pa], mass density $\rho=8000$[kg/m^3] and Poisson's ratio $\nu=0.3$.
The three lowest natural frequencies are:
ω_1 = 3116 [rad/s], ω_2 = 7028 [rad/s], ω_3 = 8284 [rad/s].

The sensitivities of natural frequencies caused by the modification of the lower boundary are calculated. In order to check the sensitivity accuracy, the variations of three lowest circular frequencies $\delta\omega$ caused by the normal interior modification equal to $\delta a=0.1$[m] are compared against the finite differences $\Delta\omega$. The variations are divided by the variations of the plate area $\delta\Omega$. Due to symmetry the results for one-half of the plate are considered. It can be seen from Fig.1.1b that the comparisons of the predicted variations and differences are in good agreement. The modification of the boundary in the neighborhood of the support reduces natural frequencies.

Afterwards the problem of shape optimal design of the lower boundary is considered. The objective function is to maximize the first natural frequency subject to the following geometrical constraints:
- the area of the plate should be not greater than A_o = 1[m^2],
- the height should be not less than h = 0.25[m].

The modified portion of the boundary has 19 nodes and is represented by the Bezier curve with 6 control points. Three different initial shapes, shown in Fig.1.2a are assumed. As the result of the optimization the first natural frequency increases to 3620 [rad/s] and the final area is equal to 0.840 [m^2]. In Fig.1.2b the evaluation of the area and in Fig.1.2c the objective function are presented. Practically identical final shapes are obtained. It is seen that the material tends to 'expand' near the support. The constraint imposed on the height is active at optimum.

2. Shape sensitivity analysis of built-up structures

Built-up structures are made up of combination of a variety of structural components that are interconnected by kinematic constraints at their interfaces. Our consideration is restricted to an elastic body composed of two materials occupying homogeneous subregions Ω_1 and Ω_2 ($\Omega_1 \cup \Omega_2=\Omega$) and bounded by the external boundary Γ ($\Gamma_u \cup \Gamma_p=\Gamma$) (Fig.2.1).

An interface Γ_I separates subdomains Ω_1 and Ω_2. The interface displacements ${}_I\mathbf{u}$ and tractions ${}_I\mathbf{p}$ are continuous, but their gradients and stress components exhibit discontinuities because the stiffness moduli vary discontinuously on Γ_I.

In the plane structures Γ_I can be identified not only with the interface between different materials but also with shapes and positions of stiffering ribs or thickness discontinuities.

One assumes that the external boundary Γ is unchanged but the interface Γ_I undergoes a shape transformation described by the similar mapping like (2), where the vector transformation field $\mathbf{g}(\mathbf{x})=(g_k(\mathbf{x};\mathbf{a}))$ modifies the shape of the interface Γ_I.

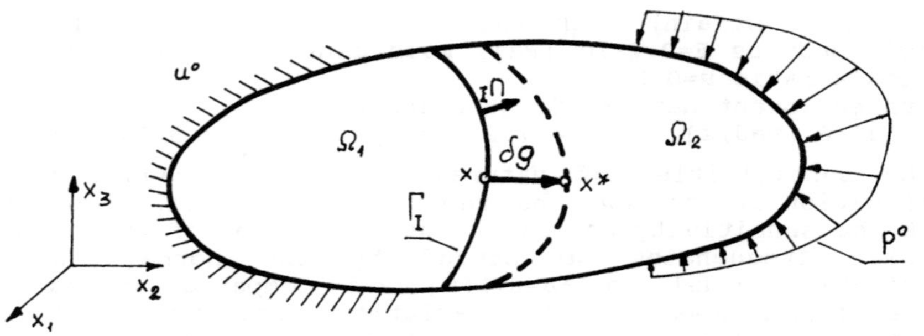

Fig.2.1

The problem of shape sensitivity analysis is considered for an arbitrary functional J of the form:

$$J = \sum_{\gamma=1}^{2} \int_{\Omega_\gamma(\mathbf{a})} \Psi^\gamma(\sigma^\gamma,\varepsilon^\gamma,\mathbf{u}^\gamma)\,d\Omega_\gamma + \int_{\Gamma} \varphi(\mathbf{u},\mathbf{p})\,d\Gamma, \qquad (13)$$

where Ψ^γ (γ=1,2) are continuous and differentiable functions of stresses σ^γ, strains ε^γ and displacements $\mathbf{u}^\gamma$ within the domains Ω_γ (γ=1,2) and φ is a continuous and differentiable function of boundary displacements $\mathbf{u}$ and tractions $\mathbf{p}$.
The first variation of the functional J can be expressed as:

$$\delta J = S_r\ \delta a_r, \qquad r=1,2,..R, \qquad (14)$$

where elements of sensitivity matrix S_r=DJ/Da_r take the form (cf.[6,7]):

$$S_r = \int_{\Gamma_I} \Big[[\![\Psi]\!] - [\![{}_I\sigma_{jl}]\!]\cdot{}_I\varepsilon^a_{jl} + [\![\mathbf{b}]\!]\cdot{}_I\mathbf{u}^a + [\![{}_I\mathbf{u},_n]\!]\cdot{}_I\mathbf{p}^a \Big]\, {}_In_k\ v^r_k\, d\Gamma_I. \qquad (15)$$

The sign $[\![\ ]\!]$ denotes a jump in a quantity across Γ_I, ${}_I\mathbf{n}$ is the unit normal vector to Γ_I, directed into the exterior of the region Ω_1, ${}_I\sigma_{jl}$ and ${}_I\varepsilon^a_{jl}$ are stress and strain components referred to the coordinate axes lying in the plane tangential to Γ_I in the primary (PS) and adjoint (AS) (denoted by superscript "a") systems, respectively, $\mathbf{b}$ are body forces.
Numerical calculations of shape sensitivity information in terms of the *adjoint variational interface boundary* (AVIB) approach require interface fields of stresses, strains and normal derivatives of displacements in the primary and the adjoint systems. Accurate evaluation of this information on the interface is the crucial problem.
When the BEM is used to solve this problem one has to check the accuracy of boundary element results for state and adjoint variables on the interface boundary.
It is well known that results of finite element analysis on the boundary are not satisfactory for interface problems. Application of finite elements to solve this problem gives

some numerical difficulties (cf.[4,8]).
Boundary integral equations for (PS) and (AS) have the form:

$$\mathbf{c}(\mathbf{x})\mathbf{u}^{\gamma w}(\mathbf{x})=\int_{\Gamma}\left[\mathbf{U}^{\gamma}(\mathbf{x},\mathbf{y})\mathbf{p}^{\gamma w}(\mathbf{y})-\mathbf{P}^{\gamma}(\mathbf{x},\mathbf{y})\mathbf{u}^{\gamma w}(\mathbf{y})\right]d\Gamma(\mathbf{y})+\mathbf{B}^{\gamma w}, \qquad \gamma=1,2;\ w=(PS),(AS) \tag{16}$$

where $\mathbf{B}^{\gamma w}$ depends on body forces for (PS) and on initial strains $\varepsilon^{\gamma ai}=\partial\Psi^{\gamma}/\partial\sigma^{\gamma}$, initial stresses $\sigma^{\gamma i}=\partial\Psi^{\gamma}/\partial\varepsilon^{\gamma}$ and pseudo-body forces $\mathbf{b}^{\gamma a}=\partial\Psi^{\gamma}/\partial\mathbf{u}^{\gamma}$ within Ω_{γ} for (AS).
Boundary conditions on the external boundary Γ of (AS) are given by:

$$\mathbf{u}^{ao}=-\partial\varphi(\mathbf{u},\mathbf{p})/\partial\mathbf{p} \text{ on } \Gamma_1 \quad \text{and} \quad \mathbf{p}^{ao}=\partial\varphi(\mathbf{u},\mathbf{p})/\partial\mathbf{u} \text{ on } \Gamma_2. \tag{17}$$

Discretizing Γ_u, Γ_p and Γ_I into boundary elements and taking into account compatibility and equilibrium conditions on the interface: ${}_I\mathbf{u}^1={}_I\mathbf{u}^2={}_I\mathbf{u}$ and ${}_I\mathbf{p}^1=-{}_I\mathbf{p}^2={}_I\mathbf{p}$ eq. (16) takes forms:

$$\begin{bmatrix} \mathbf{H}_1 & {}_I\mathbf{H}_1 & -{}_I\mathbf{G}_1 & \mathbf{0} \\ \mathbf{0} & {}_I\mathbf{H}_2 & {}_I\mathbf{G}_2 & \mathbf{H}_2 \end{bmatrix} \begin{Bmatrix} \mathbf{U}_1 \\ {}_I\mathbf{U} \\ {}_I\mathbf{P} \\ \mathbf{U}_2 \end{Bmatrix}^{w} = \begin{bmatrix} \mathbf{G}_1 & \mathbf{0} \\ \mathbf{0} & \mathbf{G}_2 \end{bmatrix} \begin{Bmatrix} \mathbf{P}_1 \\ \mathbf{P}_2 \end{Bmatrix}^{w} + \begin{Bmatrix} \mathbf{B}_1 \\ \mathbf{B}_2 \end{Bmatrix}^{w}, \qquad w=(PS),(AS) \tag{18}$$

where $\mathbf{U}_{\gamma}$ and $\mathbf{P}_{\gamma}$ (γ=1,2) are nodal displacements and tractions on the external boundary of Ω_{γ}, respectively, ${}_I\mathbf{U}$ and ${}_I\mathbf{P}$ are nodal displacements and tractions on the interface Γ_I, respectively. The discretized version of (15) can be expressed as follows:

$$S_{mp-h}=\sum_{e=1}^{E_p}\int_{\Gamma^e_{Ip}} W^e(\zeta)\,{}_I n_{m-h}\,M^e_p(\zeta)\,d\Gamma^e_I(\zeta) \qquad \begin{matrix} p=1,2,..P \\ h=2,1,0 \text{ for } 3-D \\ h=1,0 \text{ for } 2-D \end{matrix} \tag{19}$$

where $M^e_p(\zeta)$ denotes the interpolation function of the transformation field $\mathbf{g}$, $\zeta=(\zeta_1)$ is a local coordinate system placed on the boundary element Γ^e_I, E_p denotes a number of boundary elements which join themselves in boundary node *p*.

If one considers the variation of the functional (13) due to infinitesimal translation of the interface Γ_I given by a vector $\delta g_r(\mathbf{x})=\delta a_r$ (*r*=1,2,3 for 3-D or *r*=1,2 for 2-D) then the variation δJ can be expressed by (14), where $S_r=(DJ/Da_r)$ has the form (cf. [7]):

$$S_r = \int_C \left[\Psi\delta_{rj} - \sigma_{ij,r}\varepsilon^a_{il}\delta_{rj} + \sigma_{ij}u^a_{i,r} + \sigma^a_{ij}u_{i,r}\right]n_j\,dC, \tag{20}$$

where C is an arbitrary closed surface (contour) enclosing the interface Γ_I and n_j denotes the components of the normal to C. For the *adjoint variational path-independent integral* (AVPI) approach it is possible to select various integration contours. In the particular cases the contour C can be identified with the external boundary Γ or the interface Γ_I. In the last case it can be shown that (14) and (20) are the same so that the AVIB and AVPI approaches should yield the same results.

The path-independent integral (20) can be applied in sensitivity analysis with respect to translation not only of interfaces but also of internal defects such as cracks, cavities or inclusions.

The well-posed optimal design problem of the shape of the interface stated as: *find a set of parameters* $\mathbf{a}_{op}$, *so that*:

$$J \Rightarrow \min_{\mathbf{a}} \quad \textit{subject to} \quad A=\int_{\Omega_1} c_1 d\Omega_1 + \int_{\Omega_2} c_2 d\Omega_2 \le A_o \tag{21}$$

where c_1 and c_2 are specific cost of the materials, A_o is the upper bound on the material cost.

The problem described by (21) can be replaced by the problem of finding the stationarity point of the Lagrangian functional L

$$\delta L = \delta J + \lambda\ \delta A = 0, \tag{22}$$

where λ is a Lagrange multiplier and the optimality condition takes the form:

$$[\![\Psi]\!]-[\![{}_I\sigma_{jl}]\!]\cdot{}_I\varepsilon^a_{jl}+[\![\mathbf{b}]\!]\cdot{}_I\mathbf{u}^a+[\![{}_I\mathbf{u},_n]\!]\cdot{}_I\mathbf{p}^a=-\lambda(c_1- c_2)= \text{const on } \Gamma_I.$$

The above optimality condition can be directly applied in an iterative generation process of the optimal shape of the interface by the normal transformation of each boundary point on Γ_I.

Numerical experiments

Consider a thin elastic solid that is composed of two different materials (domain Ω_1=aluminium - $G_1=0.27\cdot10^{11}$[Pa], $\nu_1=0.34$ and domain Ω_2=steel - $G_2=0.85\cdot10^{11}$[Pa], $\nu_2=0.3$) subject to a simple tension (*Problem A*) and a simple bending (*Problem B*) (Fig.2.2).

The derivatives of the complementary energy (*Problem A*) and the tip displacement of the beam (*Problem B*) are calculated with respect to variation in the position of the interface Γ_I.

Problem A. Assume that functions $\Psi^\gamma=0.5\sigma^\gamma\cdot\varepsilon^\gamma$, ($\gamma$=1,2) denote the specific stress energies per unit volume and $\varphi=-\mathbf{p}\cdot\mathbf{u}^o$ on Γ_u, then the functional J (13) expresses the complementary energy Π of the structure. In the linear elastic structures the complementary energy Π is identified with the mean compliance.

Problem B. Assume that functions $\Psi^\gamma=0$ and introduce the integrand in the form

$$\varphi(\mathbf{u},\mathbf{p}) = \varphi(\mathbf{u}) = \delta(\mathbf{x} - \mathbf{x}_o)\delta_{qk}u_k. \tag{23}$$

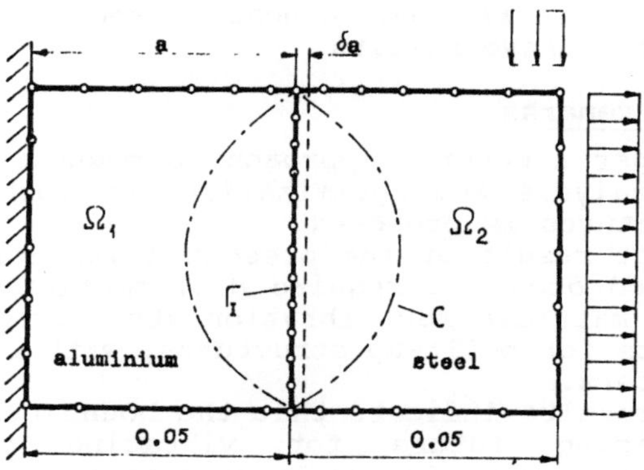

Fig.2.2

Then the functional J (13) is a displacement functional that defines the boundary displacement $\mathbf{u}=(u_q)$ in the point $\mathbf{x}_o \in \Gamma_p$

$$u_q(\mathbf{x}_o) \equiv J = \int_\Gamma \varphi(\mathbf{u})\,d\Gamma = \int_\Gamma \delta(\mathbf{x}-\mathbf{x}_o)\,\delta_{qk}\,u_k\,d\Gamma. \tag{24}$$

The adjoint structure with vanishing boundary displacement on Γ_u and vanishing fields of initial strains and stresses within Ω_γ, $\gamma=1,2$, is loaded by a simple unit load $\mathbf{p}^{ao}=(p_k^{ao})$ at point $\mathbf{x}_o$ in the positive direction of $\mathbf{u}(\mathbf{x}_o)$

$$p_k^{ao} = \frac{\partial\varphi}{\partial u_k} = \delta(\mathbf{x}-\mathbf{x}_o)\,\delta_{qk} \quad \text{on } \Gamma_p. \tag{25}$$

The boundary element mesh for each problem is the same. Numerical results of derivatives using three methods of shape sensitivity analysis, namely, the AVIB and AVPI approaches and the overall finite-difference approximation (OFD) are presented in Table 1.

Table 1

Type of problem	AVIB	AVPI	OFD
Problem A: $(D\Pi/Da)\cdot 10^{-3}$[N]	0.02617	0.02589	0.02510
Problem B: $(Du_2/Da)\cdot 10^{-2}$	0.01735	0.01707	0.01640

The step size Δa for the OFD approximation is chosen small and equals $\Delta a=0.0002a$. For the AVPI approach many different contours were selected, including also the external boundary, but this last choice gave poor results. The contour C can pass anywhere around the interface, however, one should be careful in placing the contour near the boundary due to the poor accuracy of the solution near Γ and Γ_I.

It can be seen from Table 1 that comparisons of predicted derivatives of the mean compliance (complementary energy) and the tip displacement of the beam with respect to the position of the interface show satisfactory agreement. It is possible to expect that more refined boundary element grids and higher order elements improve results.

3. Concluding remarks

In this paper a general approach to boundary element shape sensitivity analysis and optimization for both vibrating and built-up structures is proposed.

An important result of the presented variational method is that the method does not require differentiation of the mass and stiffness matrices for vibration structures, and matrices of coefficients for built-up structures, with respect to shape design parameters.

Numerical results indicate that the boundary element method has a promising future for vibrating structures and requires futher studies for built-up problems.

References

1. Burczynski, T.: *The Boundary Element Method for Selected Analysis and Optimization Problems of Deformable Bodies.* s. Mechanics, No 97, Silesian Technical Publications, Gliwice 1989 (in Polish).
2. Burczynski, T. and Fedelinski, T.:Boundary elements in shape design sensitivity analysis and optimal design of vibrating structures. *Eng. Analysis with Boundary Elements* (in print).
3. Burczynski, T. and Fedelinski, P.: Shape sensitivity analysis of natural frequencies using boundary elements. *Structural Optimization*, **2** (1990), 47-54.
4. Choi, K.K. and Seong, H.G.: A domain method for shape design sensitivity analysis of built-up structures. *Computer Methods in Applied Mechanics and Engineering*, **57** (1986), 1-15.
5. Dems, K. and Haftka, R.T.: Two approaches to sensitivity analysis for shape variation of structures. *Mechanics of Structures and Machines*, **22** (1989), 737-758.
6. Dems, K. and Mroz, Z.: Variational approach by means of adjoint systems to structural optimization and sensitivity analysis-II. *Int.J.Solids Structures* **20** (1984), 527-552.
7. Dems, K. and Mroz, Z.: On a class of conservation rules associated with sensitivity analysis in linear elasticity. *Int.J.Solids Structures* **22** (1986), 737-758.
8. Haftka, R.T. and Barthelemy, B.: On the accuracy of shape sensitivity. *Structural Optimization* (in press).
9. Fedelinski, P. and Burczynski, T.: The boundary element method for shape design sensitivity analysis and optimal design of vibrating structural elements. Proc. IX Conf. *Comp.Meth.in Mechanics*, Krakow, **1**(1989), 227-234.
10. Fedelinski, P. and Burczynski, T.: Shape optimal design of vibrating structures using boundary elements, ZAMM, **71** (in press).
11. Haisheng, R.:The symmetric dynamic boundary element method for transient elastodynamic analysis. *Boundary Elements X* (Ed.C.A.Brebbia), Springer-Verlag, Berlin (1988),375-386.
12. Nardini, D. and Brebbia, C.A.: A new approach to free vibration analysis using boundary elements. In: *Boundary Elements Methods in Engineering* (Ed.C.A.Brebbia), Springer-Verlag, Berlin (1982), 312-326.

Analysis of the Interaction Between Lifting Surfaces by Means of a Non-Linear Panel Method

G. Buresti, G. Lombardi, L. Polito

Dipartimento di Ingegneria Aerospaziale
Via Diotisalvi, 2 - 56126 PISA - Italy

Abstract

A non-linear vortex lattice method is used to study the interaction between lifting surfaces in incompressible potential flow. The model allows the wake of the upstream surface to roll up until convergence of the loads on the downstream surface is achieved. The use of a constraint to avoid spurious intersections between the upstream wake and the aft lifting surface is shown to lead to results in good agreement with available experimental data for canard-wing configurations, even in conditions of strong interference.

Introduction

The problem of the prediction of the aerodynamic loads acting on interfering lifting surfaces has a great importance in many aeronautical problems, as, for instance, the design of the horizontal tail in traditional aircraft, or of the main wing in canard configurations. In such cases it is usually essential to have a sufficiently exact geometrical description of the wake emanating from the upstream surface in order to obtain reliable estimates of the loads acting on the downstream surface. Consequently, it is necessary to use non-linear computational models, in which the geometry of the upstream wake is evaluated iteratively by imposing that it be a unloaded stream surface. This is particularly true for canard configurations, when the tip vortices of the canard strongly interact with the downstream wing, and greatly influence both the value and the distribution of the wing loads.

At the Department of Aerospace Engineering of the University of Pisa a research has been in progress for a few years with the aim of developing methods of evaluation of the aerodynamic loads acting on interfering lifting surfaces. The research also comprises experimental investigations, not only to obtain sufficient data on the influence of the variation of the various geometrical parameters involved in the problem, but also to reach a deeper understanding of the fundamental physical aspects of the phenomenon. A first stage of the experimental work is described in [1].

Together with the experimental work, a computational activity also started [2,3] with the final objective of developing codes having different sophistication and operational cost, to be used at different stages of the design process. In [2] a brief description is given of some of the several computational schemes which may be used to describe the canard-wing interaction, [4-15]. The first numerical schemes analysed during the research and described in the present paper are based on a first-order model of the physical problem. Indeed, the flow is assumed to be incompressible and potential, the wakes are represented by unloaded sheets of streamwise vorticity, and thickness effects are neglected, so that the wings are substituted by piecewise constant singularities distributed on their mean surfaces. The geometry of the rolled-up wakes is determined iteratively by relaxing initially plane configurations so that in a sufficient number of control points the wake vortices are aligned to the local velocity vectors. In spite of the simplicity of the model, a very good agreement with experimental results and a low sensitivity to the geometrical discretization was found, [2,3], at least whenever the evaluated upstream wake did not cross the downstream surface. In the present paper a procedure is described to avoid this occurrence (which is due only to the numerical approximation) by imposing a constraint to the wake displacement during the relaxation process.

Numerical approach

The code developed for the numerical approach is based on a non linear vortex lattice method, which seems to satisfy the opposite goals of adequate accuracy and low computational effort.

As previously mentioned, thickness effects are neglected, but mean surface curvature and twist can be considered. As usual, the lifting surfaces are represented by means of quadrilateral panels; a horse-shoe vortex is then placed on each panel, with its transversal portion lying at 1/4 of the panel chord, and its longitudinal sides following those of the wing panels and extending downstream to form the wake. Each wake vortex line intensity is then the algebraical sum of the intensities of the superimposed vortices emanating from the panels of two adjacent longitudinal strips. In order to permit the relaxation procedure, each wake vortex line is divided into a sufficient number of segments. The vorticity unknown values are obtained by imposing the boundary condition of tangential flow in control points, placed in the middle of the segments at 3/4 of each panel chord.

The objective of the iterative procedure is to obtain a wake which is a unloaded stream surface, aligning the vortex lines with the local velocity; several methods can be utilized, and a brief discussion on this matter is given in [2]. In the procedure used in the code described in the present paper, the wake vortex segments of the first spanwise array starting from the lifting surface are first aligned with the local velocities, evaluated at the middle point of each segment. The next step is the evaluation, making use of the new position of the first array, of the velocities on the second array of segments; then the upstream vertices of these segments are moved to coincide with the end of the corresponding segments of the first array, and their downstream vertices are moved to align the segments with the local velocities. With the new configuration the code computes the next array, and so on.

To avoid instability problems due to excessive wake roll-up (see [4]), a cut-off distance, h, is introduced in the evaluation of the velocity induced by the vortex segments. For distances from the vortex segment lower than h the induced velocity is assumed to decrease linearly to zero, as in a Rankine vortex. The cut-off distance is not a fixed value in the code, but is proportional to the square root of vortex intensity times segment length, as proposed by Almosnino, [16]. Tests showed that the variation of the cut-off distance may modify the convergence of the relaxation process, which normally improves with increasing cut-off distance.

The convergence criterion is usually based on the largest acceptable displacement between two following iterations for all wake points. However, this criterion is valid if the main interest is in the wake position and shape, and is generally applicable if isolated wings are considered. But if canard-wing interference effects are to be studied, some problems may arise. In fact, due to the close interaction between the upstream wake and the downstream surface, after several iterations the wake can have a seemingly chaotic development, i.e. its position does not seem to converge. However, as will be shown later, the vorticity in the wake of the canard remains confined to restricted zones, so that its effects on the wing loads tend to become invariant after a few iterations.

Therefore, considering also that the main objective of the code is to give reliable estimates of the loads acting on the interfering surfaces, a different convergence criterion was introduced, based of the achievement of the constancy of the local loads acting on the downstream surface.

Anyway, the problem of the convergence in interfering configurations is still being studied; at the moment we can assert that it essentially depends on the number of free vortex lines and on their distribution, as well as on the cut-off law. The best way to improve the convergence seems to be the introduction of a mechanism of amalgamation of the vortex lines at a certain stage of their evolution, by means of criteria limiting their mutual approach or their respective roll-up. The latter method is often used in pseudo-bidimensional methods (e.g. [17]), but its use in fully tridimensional problems is not immediate and does not seem to have been introduced so far.

As already pointed out, problems were found in those conditions in which some vortex lines penetrated through the wing surface, [2]. This contrasts with the condition of tangential flow to the solid surface, and arises in the numerical model because the boundary condition is imposed only in one point for each panel. Obviously, this problem should disappear by increasing the number of panels; however, to restore the physical congruence without adding too many panels, a control was inserted into the code, in order to prevent the crossing of the downstream surface by the vortex segments. Considering each wing panel, when the upstream extreme of a canard wake vortex segment is situated on the opposite side of the downstream one and the panel is crossed, the latter extreme is moved towards the same side of the former. The distance between the moved point and the panel is set by the code; the value of this distance has been varied in the range between 0.01 and 0.1 mean chords, without any significant effect on the final solution.

Some tests showed that the position of the wake of the downstream surface has negligible influence on the loads, so that it was decided to keep this wake fixed and plane, and to carry out the relaxation process only on the wake of the upstream surface.

Results

The code was applied to isolated wings in order to compare its results with those of other numerical methods and with available experimental data, [2]. The comparison showed that the code gives a good prediction for both loads and wake development.

The first canard-wing configuration analysed is the same for which experimental data had been obtained in [1], viz. a canard of aspect ratio 4 and rectangular planform placed upstream a wing with no sweep, taper ratio 0.4, aspect ratio 5.7 and the span equal to 2.11 that of the canard. Both surfaces are not twisted, with NACA 0012 wing sections, and are placed in the same plane; the longitudinal distance L of the two surfaces (between the points at 30% of the chord) is equal to 2.26 mean aerodynamic chords of the wing. Calculations were first carried out for an angle of attack of 3.8°.

Starting from a wing mesh with 26 strips along the span, with a refinement in the areas influenced by the canard tip vortex, the influence of the canard mesh on the wing lift was analysed, and the results are shown in fig 1a).

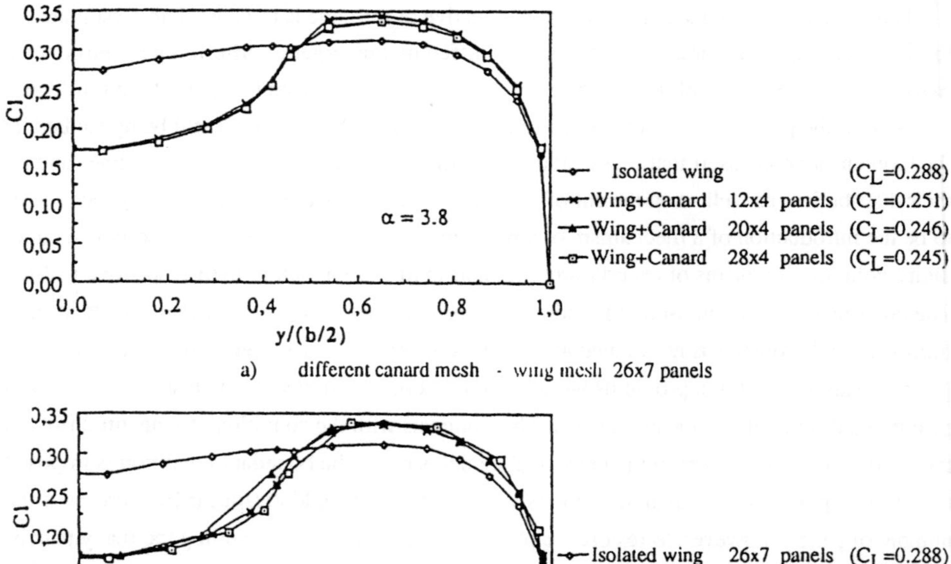

a) different canard mesh - wing mesh 26x7 panels

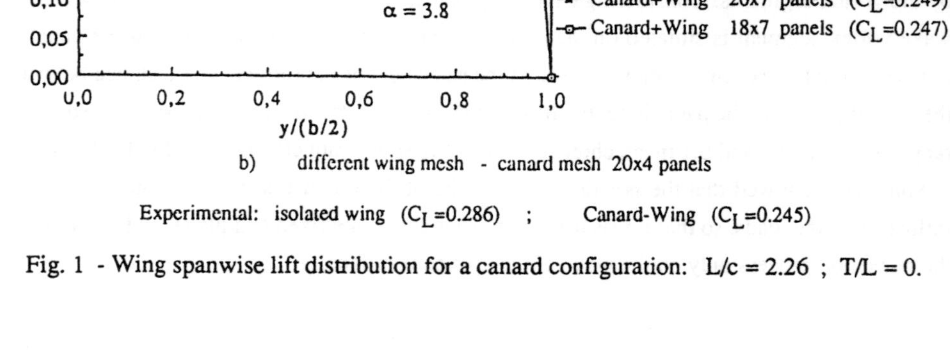

b) different wing mesh - canard mesh 20x4 panels

Experimental: isolated wing (C_L=0.286) ; Canard-Wing (C_L=0.245)

Fig. 1 - Wing spanwise lift distribution for a canard configuration: L/c = 2.26 ; T/L = 0.

The comparison with experiment is possible only with the global lift coefficient and, as can be seen, the results are in good agreement with the experimental ones, with no significant differences between the canard meshes with 20 and 28 strips. The canard mesh with 20 strips was then used to analyse the influence of reducing the number of wing panels (fig. 1b). The wing mesh with 18 strips has the aforesaid refinement in strong vortex areas, while the mesh with 20 strips has no refinement.

Once again the stability of the results is evident, and it is interesting to note that it is possible to achieve acceptable results even with a limited number of panels, provided they are appropriately distributed on the wing surface. Fig. 2, reporting graphically the result of one of the above cases, shows that, at this incidence, the whole canard wake passes above the wing (in this and in the following figures, the cross-flow lines in the wake are only for visual aid).

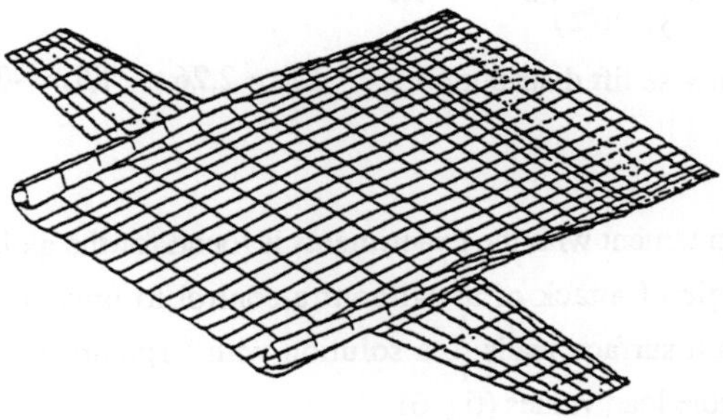

Fig. 2 - Computed wake for a canard configuration (wing wake not shown)
$L/c = 2.26$; $T/L = 0$; $\alpha = 3.8°$

To analyse the behavior of the numerical model under conditions of stronger interference, a new configuration was considered, with the canard positioned below the wing plane at a distance, T, of 0.083xL (corresponding to one of those tested in [1]). Fig. 3 shows that with an angle of attack of 3.8° the computed wake remains below the wing except for two vortex lines on each side which pass above the wing.

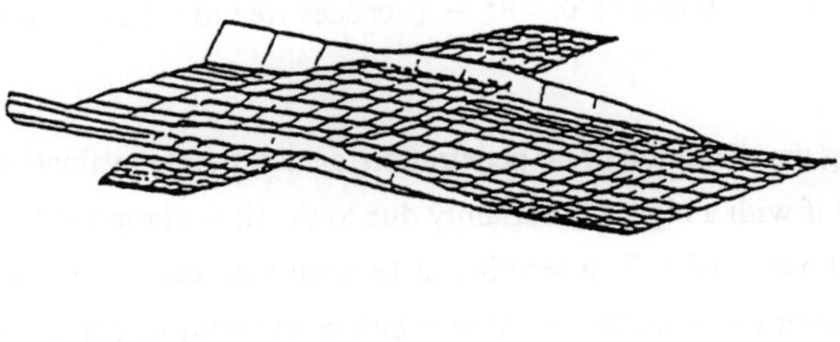

Fig. 3 - Computed wake (wing wake not shown) : $L/c = 2.26$; $T/L = -0.083$; $\alpha = 3.8°$

Figure 4 shows that in a such situation the curve of the lift coefficient along the span is similar to that obtained with coplanar surfaces (T=0), but with more accentuated gradients, as can be expected considering the stronger interference.

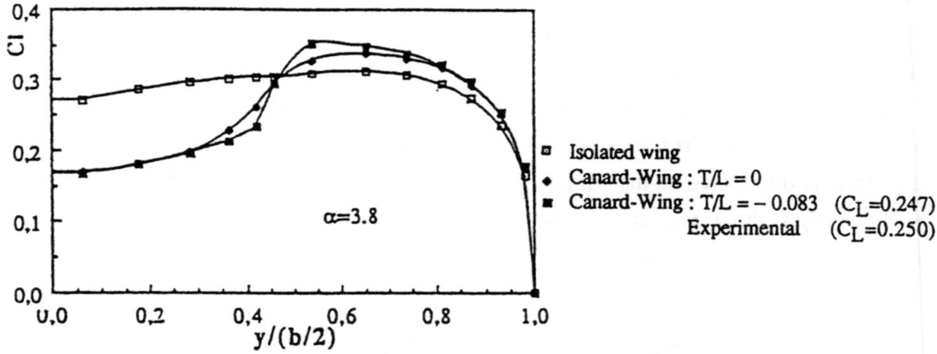

Fig. 4 - Wing spanwise lift distribution : L/c = 2.26 ; T/L = -0.083 ; α = 3.8°

Once more a good agreement with the experiments is found for the global wing lift. The same configuration at an angle of attack of 9° without a control to prevent the vortex lines from crossing the downstream surface leads to a solution with large areas of intersection (fig. 5), with completely unreliable load values (fig. 6).

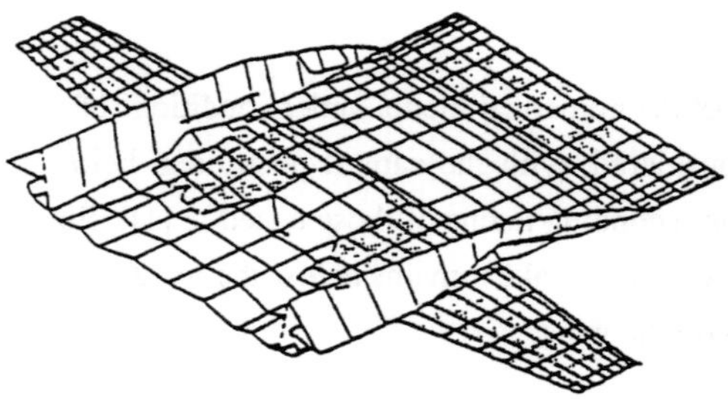

Fig. 5 - Computed wake (wing wake not shown)
L/c = 2.26 ; T/L = -0.083 ; α = 9° – (vortices free to cross the wing)

However, by using the condition of no penetration, an acceptable evaluation of the loads was again obtained, even if with a higher uncertainty due to the flow complexity. An example of the computed wake is shown in fig. 7, where it can be seen that also in this case there are some vortex lines passing over the wing and other ones below, but without compenetration and with a qualitatively plausible wake. The corresponding load distribution (fig. 8) confirms the improvement, even if the global lift value does not reach the same agreement with the experimental data as was obtained for less critical interference conditions.

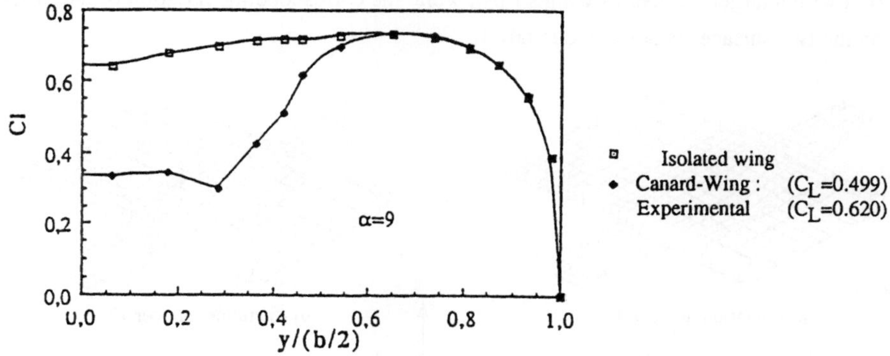

Fig. 6 - Wing spanwise lift distribution
L/c = 2.26 ; T/L = -0.083 ; α = 9° – (vortices free to cross the wing)

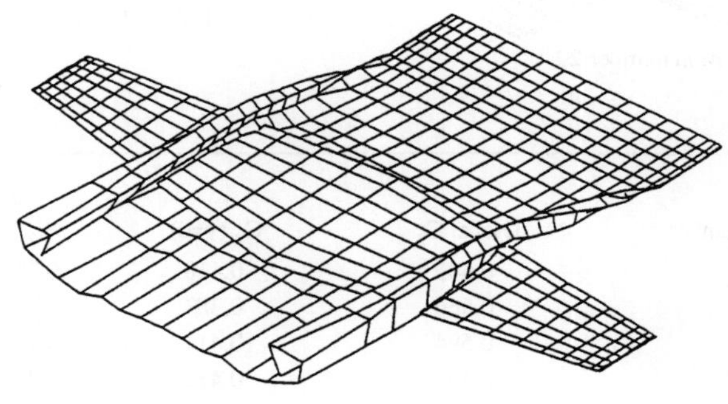

Fig. 7 - As fig. 5 with constrain of no intersection

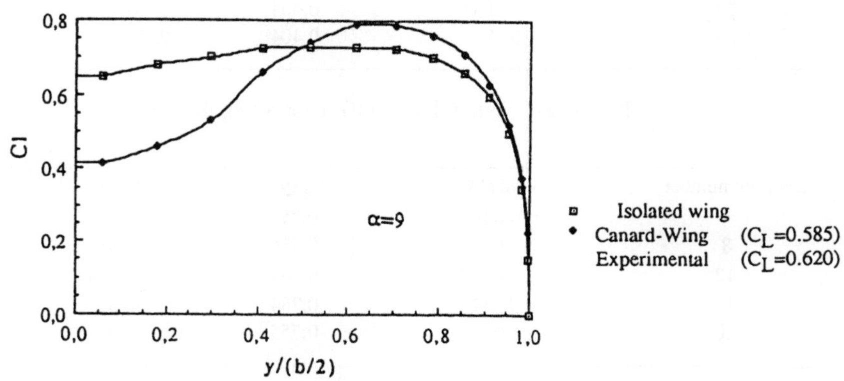

Fig. 8 - As fig. 6 with constrain of no intersection

In order to justify the use of a convergence criterion based on the constancy of the wing loads, fig. 9 shows the wake evolution at various iterations for the same configuration of fig. 7 (but with a larger number of wing strips), while the corresponding variation of the loads acting on the two surfaces is described in tab. 1.

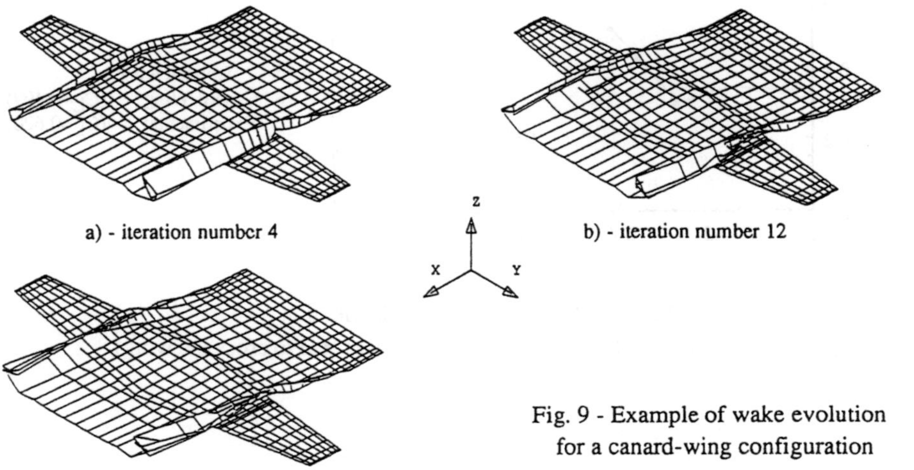

a) - iteration number 4

b) - iteration number 12

c) - iteration number 20

Fig. 9 - Example of wake evolution for a canard-wing configuration

			Maximum value	
Iteration number	Global C_z	C_z (y=0)	C_z	position (y/(b/2))
4	0.605	0.418	0.798	0.619
6	0.566	0.383	0.784	"
8	0.594	0.411	0.794	"
10	0.597	0.412	0.795	"
12	0.595	0.404	0.801	"
16	0.585	0.404	0.791	"
20	0.584	0.403	0.791	"
24	0.585	0.403	0.791	"
28	0.585	0.404	0.791	"

Tab. 1 (a) Wing loads (40 strips along the span)

Iteration number	Global C_z	C_z (y=0)	C_z(y/(b/2)=0.52)
4	0.638	0.758	0.687
8	0.636	0.756	0.686
12	0.635	0.755	0.684
16	0.634	0.754	0.683
20	0.635	0.755	0.684

Tab. 1 (b) Canard loads (20 strips along the span)

As can be seen, the shape of the rolled-up canard wake shows a seemingly chaotic development, but the vorticity remains confined in a restricted zone, so that the loads rapidly achieve almost constant values. From tab. 1b) it can also be seen that the upstream surface loads are practically not dependent on the wake position, as could be expected.

Finally, it should be pointed out that tests on several configurations showed that the same canard wake is reached irrespective of its initial position (i.e. either over or below the wing). Obviously, this important result is obtained only provided the vortex lines are divided in a sufficient number of segments; indeed, the obtainment of the same solution for any initial wake position may even be considered a test that the wake has been properly discretized.

Conclusions

In this paper the main characteristics of a potential code to evaluate the aerodynamic loads on interfering lifting surfaces are described. The model is a non linear vortex lattice, in which the geometry of the rolled-up wakes is determined iteratively by relaxing initially plane configurations so that in a sufficient number of control points the vortices are aligned to the local velocity vectors, and by using a convergence criterion based on the local loads acting on the downstream surface. This criterion allows to stop the iterative procedure at given precision levels having an immediate design interest.

In spite of the simplicity of the model, a comparison with available experimental data showed that the code is able to predict quite carefully the flow behavior around canard configurations; indeed, the results show a very good agreement with experimental data when interference effects are not too strong, and, if a suitable forcing for the condition of no penetration is used, are largely acceptable even when the canard wake directly impinges on the wing surface.

The solution is not very dependent on surface meshes and on the position of the wake of the downstream surface, even if a bad choice of the mesh for the upstream surface wake may negatively influence the roll-up process and the code convergence. Therefore, the first future developments will be aimed at avoiding the occurrence of any seemingly chaotic wake roll-up by means of techniques of amalgamation of the vortex lines.

Furthermore, as the loads on the upstream surface were confirmed to be practically constant during the relaxation process, their reevaluation after the initial step might be avoided. In this way a good saving in computational time would be possible, and more sophisticated codes might be used to obtain a better description of the upstream surface load distribution and, consequently, of the free vorticity in the field.

However, to carry out a sounder validation of the numerical model, it is essential that a deeper understanding of the physical aspects of the interference between lifting surfaces be achieved. Therefore, further experimental investigations are planned within the present research; in particular, the wing load distributions will be measured and detailed flow analyses will be carried out to obtain a sufficiently accurate description of the canard wake development in conditions of close interference.

Acknowledgment

The present research was supported by the Italian Ministry of University and of Scientific and Technological Research, M.U.R.S.T., and by the National Research Council, C.N.R.

References

1. Buresti, G.; Lombardi, G. : Indagine sperimentale sull'interferenza ala-canard. L'Aerotecnica, Missili e Spazio, Vol. 67, N. 1-4, (1988) pp. 47-57.
2. Buresti, G.; Lombardi, G.; Petagna, P. : Analisi dell'interazione fra superfici portanti mediante un modello potenziale. Atti del X Congresso Nazionale A.I.D.A.A., Pisa, (1989).
3. Vicini, A. : Calcolo degli effetti aerodinamici delle scie vorticose generate da superfici portanti subsoniche. Tesi di Laurea, Dipartimento di Ingegneria Aerospaziale, Pisa, (1989).
4. Hoeijmakers, H. W. M. : Computational vortex flow aerodynamics. AGARD-CP-342, Paper 18, (1983).
5. Smith, J. H. B. : Theoretical modelling of three-dimensional vortex flows in aerodynamics. AGARD-CP-342, Paper 17, (1983).
6. Smith, J. H. B. : Modelling three-dimensional vortex flows in aerodynamics. VKI Lect. Ser. "Introduction to vortex flow aerodynamics", (1986).
7. Suciu, E. O.; Morino, L. : A nonlinear finite element analysis of wings in steady incompressible flows with wake rollup. AIAA Paper 76-64, (1976).
8. Kandil ,O. A.; Mook, D. T.; Nayfeh, A. H. : Nonlinear prediction of the aerodynamic loads on lifting surfaces. J. of Aircraft, Vol. 13, (1976), pp. 22-28.
9. Rajeswari, B.; Dutt, H. N. V. : Nonplanar vortex-lattice method for analysis of complex multiple lifting surfaces. N.A.L. Tech. Mem. T.M. AE8606, (1986).
10. Rusak, Z.; Wasserstrom, E.; Seginer, A. : Numerical calculation of nonlinear aerodynamics of wing-body configurations. AIAA J., Vol. 21, (1983), pp. 929-936.
11. Yeh, D. T.; Plotkin, A. : Vortex panel calculation of wake roll-up behind a large aspect ratio wing. AIAA J., Vol. 24, (1986), pp. 1417-1423.
12. Smith, B. E.; Ross, J. C. : Application of a panel method to wake vortex-wing interaction and comparison with experimental data. NASA TM 88337, (1987).
13. Maskew, B. : Predicting aerodynamic characteristics of vortical flows on three-dimensional configurations using a surface-singularity panel method. AGARD CP-342, Paper 13, (1983).
14. Wagner, S.; Urban, Ch.; Behr, R. : A vortex-lattice method for the calculation of wing-vortex interaction in subsonic flow. In "Panel Methods in Fluid Mechanics with Emphasis on Aerodynamics" (Ballmann J.; Eppler R.; Hackbusch W. eds.), Notes on Numerical Fluid Mechanics Vol. 21, Vieweg, (1987), pp. 243-251.
15. Mattei, A.; Santoro, E. : Numerical computations of wake vortices behind lifting surfaces. ICAS Paper 74-28, (1974).
16. Almosnino, D.: High angle of attack calculations of subsonic vortex flow on slender bodies. AIAA Jnl., Vol 23, (1985), pp.1150-1156.
17. Moore, D. W. : A numerical study of the roll-up of a finite vortex sheet. Jnl. of Fuid Mechanics, Vol. 63, (1974), pp. 225-235.

Efficient Analysis of Complex Solids Using Adaptive Trimmed Patch Boundary Elements

M.S. Casale
PDA Engineering
2975 Redhill Ave.
Costa Mesa, CA, USA
92626

J.E. Bobrow
Department of Mechanical Engineering
University of California
Irvine, CA, USA
92717

Summary
In this paper, it is shown that boundary integral equations can be applied directly to trimmed patches, and that the resulting element, the trimmed patch boundary element, is ideally suited to adaptive analysis. The trimmed patch boundary element is derived, and the differences between this element and traditional elements are presented. Then, a local adaptive strategy is presented which is based on a hybrid error norm that uses both weighted residual and stress projector components. The paper concludes with some examples.

1.0 Introduction

Traditionally, a boundary element mesh is made by subdividing the surface of a solid into three or four sided elements. This is far easier than discretizing the entire volume, and it would seem that this by itself would be enough to make the boundary element method as popular as the finite element method. However, there are several reasons why boundary elements have yet to catch on in industry. Three of the most important of these are (1) the boundary element method has been, until recently, limited to relatively few problems, (2) the boundary element method has never been truly integrated with geometric modelers and (3) it still takes a long time to prepare a satisfactory boundary element mesh, especially for complex solids which need to be analyzed several times to insure convergence. Recent developments have helped to alleviate the first of these problems [1] [2]. In this paper, we present some developments that address the second two.

If a surface is curved or if it has several small features, subdividing it into three and four sided elements tends to produce lots of elements. This produces too many degrees of freedom for typical boundary element codes and is even more of a problem for adaptive codes which perform best when they begin with relatively coarse meshes. Solid modelers, on the other hand, routinely partition the boundary of complex solids into relatively few regions. They are able to do this because they represent surfaces using trimmed patches which have the property that they can have any number of sides [3].

In this paper, it is shown that the trimmed patch can also be treated as an n-sided boundary element [4], and that this element is ideally suited to adaptive analysis. The result is an analysis method that is integrated with the modeler. It operates on the geometry directly, and it is easy to use since the user does not have to spend a lot of time preparing the mesh.

The organization of the paper is as follows. In the next section, it is shown how the boundary integral equations of linear elastic analysis are applied to a trimmed patch. The result is a new element called the *trimmed patch boundary element*. In section 3, the adaptive strategy is outlined and section 4 concludes with some examples.

2.0 The Trimmed Patch Boundary Element

In the early days of solid modeling, models were generated by combining simple primitives [5]. More recently, model generation methods have been developed that are simultaneously more powerful and more intuitive — i.e., form feature modeling and parametric (or variational) geometry. Nevertheless, the underlying operations are still essentially combinatorial, which means that no matter how the surfaces of the primitives are initially represented, before long many will be trimmed.

It is possible to apply the boundary element method to solid models in two steps: (1) mesh the trimmed patches with triangles and quads and (2) solve the boundary integral equations on the mesh. The point of this section is that it is also possible to apply the boundary integral equations to the trimmed patches directly. The result is a new element which we call the trimmed patch boundary element. The differences between this element and traditional elements is best explained by studying the fundamental boundary integral equation term by term. For linear elastic analysis, this well known equation is

$$[c_i]\{u(\mathbf{x}_i)\}+\sum_j\int_{D_j}[P^*][M^*]|J|dD\,\{U_j\}=\sum_j\int_{D_j}[U^*][M^*]|J|dD\,\{T_j\}$$

where $[c_i]$ is the discontinuity matrix, $u(\mathbf{x}_i)$ is the displacement at point $\mathbf{x}_i$, D_j is the domain of integration for element j, $[P^*]$ and $[U^*]$ are the matrices of fundamental solutions for tractions and displacements $[M^*]$ is the matrix of shape functions, $|J|$ is the Jacobian of the element mapping and $\{U_j\}$ and $\{T_j\}$ are the displacements and tractions at the nodes of element j. For a full description of each of these terms, see for example [1]. Equation (1) is written for each node on the boundary of the solid resulting in a linear system of the form

$$[H]\{U\} = [G]\{T\} \tag{2}$$

which is solved for the unknown displacements and tractions after the boundary conditions have been added.

Traditional boundary elements model both the geometry and the field variables with relatively simple functions. The most common elements are isoparametric, which means that the geometry and the field variables are represented using the same low order polynomials. With trimmed patch boundary elements, however, the representation of the field variables and the geometry are totally independent. The geometry and the topology, that is, the number of edges of the element, are taken directly from the trimmed patch. There is no approximation. The field variables, on the other hand, are approximated as before with low order polynomials. The separation of geometry and field variables is naturally reflected in the terms of equation (1). It is useful to divide the terms into three categories: (1) those that depend on the trimmed patch geometry, (2) those that approximate the displacements and tractions (the field variables in this case) and (3) the term D_j which represents the domain of integration. This categorization will be used in what follows.

2.1 Terms that Depend on the Geometry

Three terms in (1) depend on the geometry — [P*], [U*] and |J|. The matrices [P*] and [U*] are

$$p_{ij}^* = \frac{1}{8\pi(1-\nu)r^2}\left[\frac{\partial r}{\partial n}\left\{(1-2\nu)\delta_{ij} + 3\frac{\partial r}{\partial x_i}\frac{\partial r}{\partial x_j}\right\} - (1-2\nu)\left(\frac{\partial r}{\partial x_i}n_j - \frac{\partial r}{\partial x_j}n_i\right)\right] \tag{3}$$

and

$$u_{ij}^* = \frac{1}{16\pi G(1-\nu)r}\left[(3-4\nu)\,\delta_{ij} + \frac{\partial r}{\partial x_i}\frac{\partial r}{\partial x_j}\right] \tag{4}$$

where r is the distance from the surface to the node $\mathbf{x}_i$

$$r = \left\| z(s,t) - \mathbf{x}_i \right\| \tag{5}$$

where $\mathbf{z}(s,t)$ is the local parametrization of the surface (that is, the trimmed patch mapping) ν is Poisson's ratio and G is the shear modulus, $\mathbf{n}(s,t)$ is the outward pointing surface normal, and the Jacobian is

$$|J| = \left\| \mathbf{z}_{,s}(s,t) \times \mathbf{z}_{,t}(s,t) \right\| \tag{6}$$

Consequently, to evaluate the terms [P*], [U*] and |J|, it is only necessary to compute points on the surface as a function of parametric coordinates and surface normals.

In traditional boundary element methods, the surface is parametrized locally by triangles and/or quads. Here, it is assumed that the modeler has parametrized the surface by a set of trimmed patches.

Mathematically, a trimmed patch is a one-to-one mapping **z** of a general two-dimensional region D into Euclidean 3 space:

$$\mathbf{z}(s,t)\; D \subset \Re^2 \rightarrow \Re^3 \tag{7}$$

An example of a trimmed patch is shown in Figure 1. Methods for determining the domain are presented in [5], and choosing parametric coordinates to evaluate is the job of the integration scheme, which is presented later. For now, note that the evaluation of points and normals has been made quite a bit more complex, since the mapping $\mathbf{z}(s,t)$ comes directly from the modeler and may be a bicubic, a Bezier or a Non-Uniform Rational B-spline (NURB).

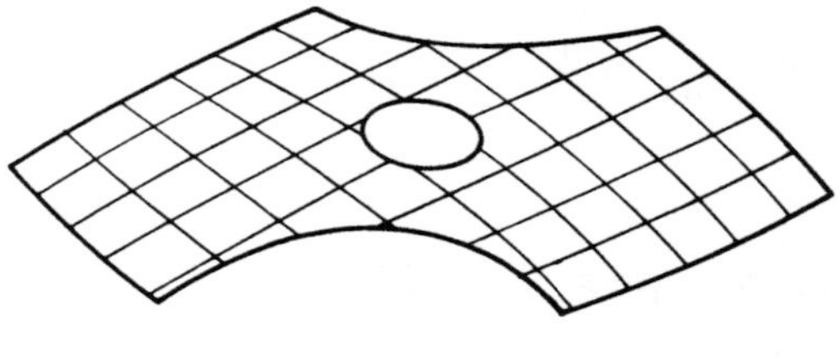

Figure 1. A trimmed patch, which is the basis of modern geometric modeling, can have any number of edges.

There are a number of methods for reducing the effect of high order mappings on the performance of the system. One of these is presented in the section on the integration scheme. For more information on evaluating common surface types, see [6] and [7].

2.2 Terms that Approximate Displacements and Tractions

The terms that approximate the displacements and tractions are [M*], $\{U_j\}$ and $\{T_j\}$. The terms in the matrix [M*] are interpolation or shape functions and the $\{U_j\}$ and $\{T_j\}$ are nodal values of displacement and traction at the nodes of element j. If a component of displacement u(s,t) is approximated by a bicubic, then u can be written as

$$u(s,t) = \sum_{i=1}^{16} N_i(s,t) P_i = [N]\{P\} \tag{8}$$

In this case, the matrix [M*] would be

$$[M^*] = \begin{bmatrix} N_1 & 0 & 0 & N_2 & 0 & 0 & & N_{16} & 0 & 0 \\ 0 & N_1 & 0 & 0 & N_2 & 0 & \cdots & 0 & N_{16} & 0 \\ 0 & 0 & N_1 & 0 & 0 & N_2 & & 0 & 0 & N_{16} \end{bmatrix} \tag{9}$$

The major difference between trimmed patch boundary elements and traditional elements, as far as the approximation of the displacements and tractions are concerned, is that the locations of the nodes cannot be fixed. The node locations have to be optimized for each patch to fit the range of the patch mapping. Two node placement strategies are shown in Figures 2 and 3.

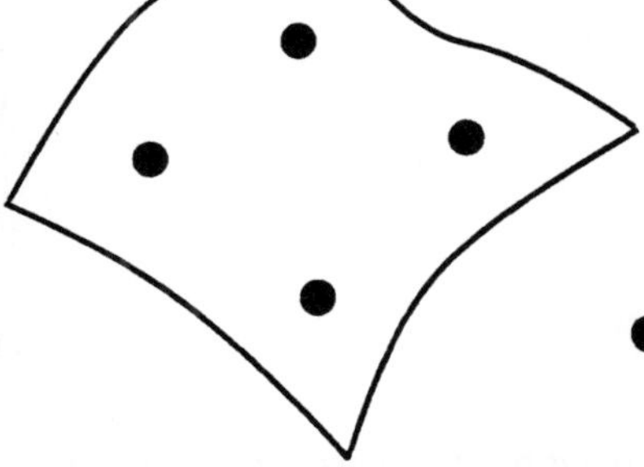

Figure 2. A bilinear discontinuous element.

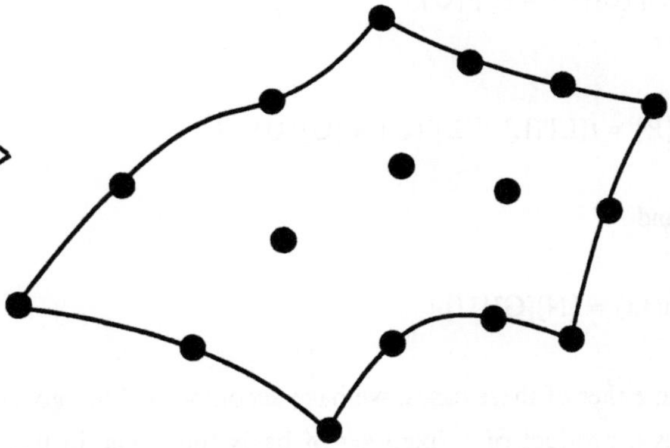

Figure 3. A bicubic element.

Note that the element in Figure 3 has five sides. Such elements arise naturally when dealing with trimmed patches. Since the nodes are not fixed, the matrix [N] is not either. However, it is possible to write [N] as the product of a matrix of fixed functions time a matrix of constants, which has the advantage that the matrix of constants can be factored out of the integral. This can be done as follows. Let u be given as in (8) where it is assumed that [N] is a matrix of fixed bicubic interpolation functions and {P} is the matrix of values of u at the fixed third points [4]. Then, suppose that values of u(s,t) are given at J or more other points. If u(s,t) interpolates these values, then

$$u(s_j,t_j) = \sum_{i=1}^{16} N_i(s_j,t_j)P_i = U_j \quad j=1,\ldots,J \tag{10}$$

where (s_j,t_j) are the parametric coordinates that correspond to the values U_j. Equation (10) can be written in matrix form as

$$[L]\{P\} = \{U\} \tag{11}$$

If L is square, that is, if J is 16 in (10), then P can be solved for directly:

$$\{P\} = [L]^{-1}\{U\} = [G]\{U\} \tag{12}$$

If J is greater than 16, then {P} can be solved for in the least squares sense:

$$[L]^t[L]\{P\} = [L]^t\{U\} \tag{13}$$

or

$$\{P\} = ([L]^t[L])^{-1}[L]^t\{U\} = [G]\{U\} \tag{14}$$

and

$$u(s,t) = [N][G]\{U\} \tag{15}$$

In either of these cases, we have accomplished our goal of writing the shape function matrix as a product of a fixed set of basis functions, in this case the fixed format Lagrangian interpolation functions, times a matrix of constants. Note also that we have a least squares shape function — one that allows more nodes to be specified than coefficients in the interpolation function. The significance of this is that we now have much greater flexibility in the node placement algorithm, and the resulting shape function is smoother than with typical interpolation methods. Furthermore, as will be seen later, the least squares shape function makes it relatively easy to compute an error norm.

2.3 The Domain of Integration

If the node $\mathbf{x}_i$ is considered to be fixed, then the integrals in (1) are of the form

$$\int_D f(s,t)dD \tag{16}$$

where D, the domain of integration, is the trimmed domain of a trimmed patch boundary element. This integral can be evaluated by summing over the triangles

$$\int_D f(s,t)dD = \sum_{k=1}^{K} \int_{T_k} f(s,t)dD \tag{17}$$

where T_k $k = 1, \ldots ,K$ is a triangulation of the domain. Note that the triangulation is used just to keep track of the domain. The triangles are not elements, since they do not contribute any degrees of freedom to the analysis, and they do not approximate the geometry, just the trimmed domain. Whenever any geometric operation is done (see section 2.1), the parametric coordinates are mapped to the exact geometry through the patch mapping.

Performing all geometric operations directly on high order surfaces is more expensive than using an approximation, especially considering the complexity of modern representations like NURBS. Nevertheless, the extra effort is often needed to achieve the required accuracy. Furthermore, the expense can be minimized by using a concept called Reusable Intrinsic Sample Points (RISP) [8]. With this method, the geometric and shape function information at the integration points are evaluated once for each triangle. This information is stored and used repeatedly for all source points. Of course, a variety of integration schemes will be needed for each triangle depending on the distance to the source point. The idea in RISP is not to eliminate the need for multiple rules but to minimize them so that as few geometric and shape function evaluations are done as is possible. For example, three non-singular rules might be chosen, one with three integration points, one with 13 and one with 52. No matter where a particular node is located, as long as the node is not on the element, the geometric and shape function information remains the same. Consequently, this information can be pre-computed and reused. A different strategy is used for singular and near singular nodes.

3.0 The Adaptive Strategy

It is well known that unlike finite element methods, boundary element methods do not lend themselves to iterative linear system solvers [9]. This fact has led us to develop a different strategy than the traditional ones that work well for finite elements, which are usually based, in one way or another, on a pre-conditioning strategy in which the results of coarse analyses are used as starting points for more refined analyses.

The idea behind our adaptive strategy is that if an intermediate solution is reasonably close to the true solution, then it is possible to iteratively increase the accuracy locally a few elements at a time. This method is called local reanalysis [4]. Suppose, for example, that after an analysis, it is determined that a single element, say element k, needs to be refined. Then, the order of the shape function on element k can be increased and a new solution computed from (1) where it is now assumed that both displacements and tractions are known over the other elements. Equation (1) becomes

$$[c_i]\{u(x_i)\}+\int_{D_k}[P^*][M^*]|J|dD\,\{U_k\}+\sum_{\substack{j\\ j\neq k}}\int_{D_j}[P^*][M^*]|J|dD\,\{U_j\}=$$

$$\int_{D_k}[U^*][M^*]|J|dD\,\{T_k\}+\sum_{\substack{j\\ j\neq k}}\int_{D_j}[U^*][M^*]|J|dD\,\{T_j\} \tag{18}$$

where this time $\{U_j\}$ and $\{T_j\}$ are known. Also, i varies from 1 to N where N is the new number of nodes over element k. Note that this number has increased due to the increase in order of the shape function of element k. Equation (18) can now be written as

$$[H]\{U\} + \{A\} = [G]\{T\} + \{B\} \tag{19}$$

where $\{A\}$ and $\{B\}$ are the products of integrals with known values of displacements and tractions, and [H] and [G] correspond to the unknowns on element k only. Hence [H] and [G] are far smaller than their counterparts in (1). For example, if element k was refined by raising the polynomial order of its shape function matrix [M] from quadratic to cubic (note that the geometry mapping **z** never changes), then there would be 48 equations.

The above description was given to outline the idea of local reanalysis. In practice, there will usually be more than one element being refined. Equation (19) would therefore be larger. Consequently, there needs to be some mechanism for choosing the elements to be refined.

Furthermore, there must be some provision for insuring that the process is truly converging. Both of these issues, choosing the elements to be refined and insuring convergence, depend upon a robust error norm.

The error norm chosen for this work is a hybrid. The error e_j in each element is

$$e_j = e_\sigma + e_w \quad (20)$$

where e_σ is the least squares residual error in stress and e_w is the weighted residual error. These quantities are best explained by an example. A typical set of nodes for a quadratic least squares element are shown in Figure 4. The 9 nodes represented by darkened circles can be considered to be the standard interpolation nodes and the 4 represented by empty circles are auxiliary nodes. The least squares shape function is computed from all 13 nodes using equation (14).

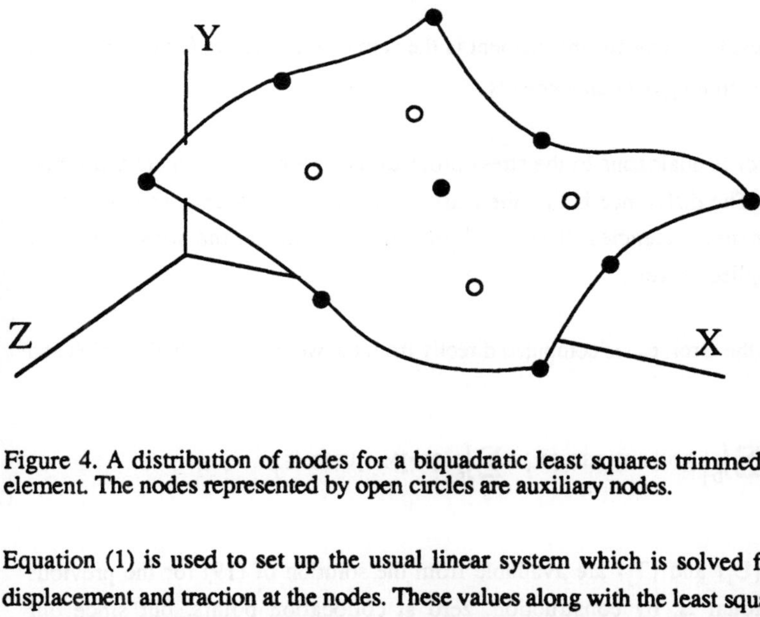

Figure 4. A distribution of nodes for a biquadratic least squares trimmed patch boundary element. The nodes represented by open circles are auxiliary nodes.

Equation (1) is used to set up the usual linear system which is solved for the values of displacement and traction at the nodes. These values along with the least squares operator and the matrix [N] define quadratic surfaces. Suppose, for example, that $\{U^x_j\}$ is the vector of the x components of displacement on element j. Then,

$$u_x(s,t) = [N]([L]^t[L])^{-1}[L]^t\{U^x_j\} \quad (21)$$

is a biquadratic polynomial that approximates the values at the nodes in the least squares sense. These surfaces are computed for all three components of displacement and traction and can be used to compute the values of the six components of stress [4].

Suppose that this procedure has been used to compute the stress at the 13 nodes. Then, these 13 values can also be fit with a quadratic surface using the least squares operator, i.e., for a component, σ, of stress

$$\sigma(s,t) = [N]([L]^t[L])^{-1}[L]^t\{\sigma_j\} \qquad (2$$

where $\{\sigma_j\}$ is the vector of stress values at all 13 nodes. If σ_k is the value at node k which has parametric value (s_k,t_k), the least squares residual for node k is then

$$(\sigma_k - \sigma(s_k,t_k))^2 \qquad (2$$

The least squares residual error for the element is the sum of these errors for all 13 nodes, and e_σ is found by summing over all elements.

This part of the error is analogous to the stress projector error which has recently become popular [10] [11] . The difference is that the least squares residual error does not require knowledge of element adjacencies. If such information is available, the stress projector method could be applied as well.

The second term in the error, e_w, is computed directly from the weighted residual r which at a point τ is

$$r(\tau) = -[c_i]\{u(\tau)\} - \sum_j \int_{D_j} [P^*][M^*]|J|dD\,\{U_j\} + \sum_j \int_{D_j} [U^*][M^*]|J|dD\,\{T_j\} \qquad (2$$

where the values $\{U_j\}$ and $\{T_j\}$ are available from the solution of (19) for the previous iteration. The residual is, by construction, zero at collocation points, but since our formulation was based on a least squares shape function the residual will not, in general, be zero anywhere. To compute an error norm, the residual is computed at the auxiliary nodes of each element. Then, e_w is the sum of this error over all the elements.

The two parts of the error norm presented above complement one another. The weighted residual error insures that the local reanalysis strategy converges, since if r is zero over the entire surface, then the problem has converged. However, determining a reasonable stopping tolerance based on the residual is difficult since the user is really interested in stress which is related to displacement by a differential operator. Consequently, in areas of large stress gradients, looking at the residual error alone may suggest that the solution has converged when in fact it hasn't.

A sketch of the adaptive strategy is therefore as follows.

(1) Perform a coarse analysis.

(2) Compute the error norm for all elements using (x).

(3) If the error norm for all elements is less than a specified tolerance, stop.

(3) Choose a set of elements to refine.

(4) Increase the order of the elements and/or subdivide.

(5) Go back to step 1.

For details on when to sub-divide and when to increase the order of the element, see [12]. Note also that if the analysis does not include all elements, then the error norm e_σ does not change for those elements that were not refined. The error e_w does, however, and this needs to be recomputed. The new error norm can be used to detect if the local reanalysis is causing the analysis to diverge.

4.0 Examples

The first example is a cantilever beam with a hole, Figure 5. The dimensions of the beam are 20 inches in x, 4 inches in y and 4 inches in z. The hole is 1 inch in radius. The beam is clamped on the right, and a 100 PSI load is applied in the -y direction at the right end. The Young's modulus is 24×10^7 and the Poisson's ratio is 0. The computed stress σ_x is shown in the figure. The theoretical maximum for this problem is 3000 psi. This loading condition is interesting because of the stress concentration around the hole. Similar results were obtained using a fine finite element mesh, but with far greater computational effort.

The second example is a stepped shaft. The model is 6 units in x , the radius of the larger part of the shaft is 1, and the radius of the smaller part of the shaft is 1/2. The model is pulled to

the right with 100 psi. Given the same material properties as in the previous example, the theoretical maximum stress σ_x is 133.2. The computed values are shown in Figure 6.
A more complicated model representing a crank shaft is shown in Figure 7. The geometry for this model was generated using 66 bicubic trimmed patches. The loading condition for this model were simulated by fixing the lower circular face and pulling in the z direction on the upper circular face. The displacement fringes are shown in the figure. A stress plot showing a stress concentration in the fillet is shown in Figure 8. It took a very fine finite element model to approximate the stress concentration with similar accuracy. The displayed results were computed in 6 hours on a workstation using the methods given in the paper, while the finite element analysis required a weekend on the same machine. This comparison in time becomes even more dramatic when added to the savings in the user's time since no mesh was required to do the analysis.

Conclusion

In this paper, the boundary integral equations of linear elastic analysis have been applied to the trimmed patch which is the primary surface representation used by modern geometric modelers. The result is a new element which we called the trimmed patch boundary element. Some of the similarities and differences between trimmed patch boundary elements and other elements were presented, and a simple but effective adaptive analysis strategy was outlined.

The appeal of the boundary element method has always been its ability to solve solid problems from just surface information. This goal is shared by modern modelers. Nevertheless, much research, and even more software, is needed before structural analysis becomes as integrated with CAD as other design applications. But the methods presented here demonstrate, at least for relatively simple problems, that this goal is achievable.

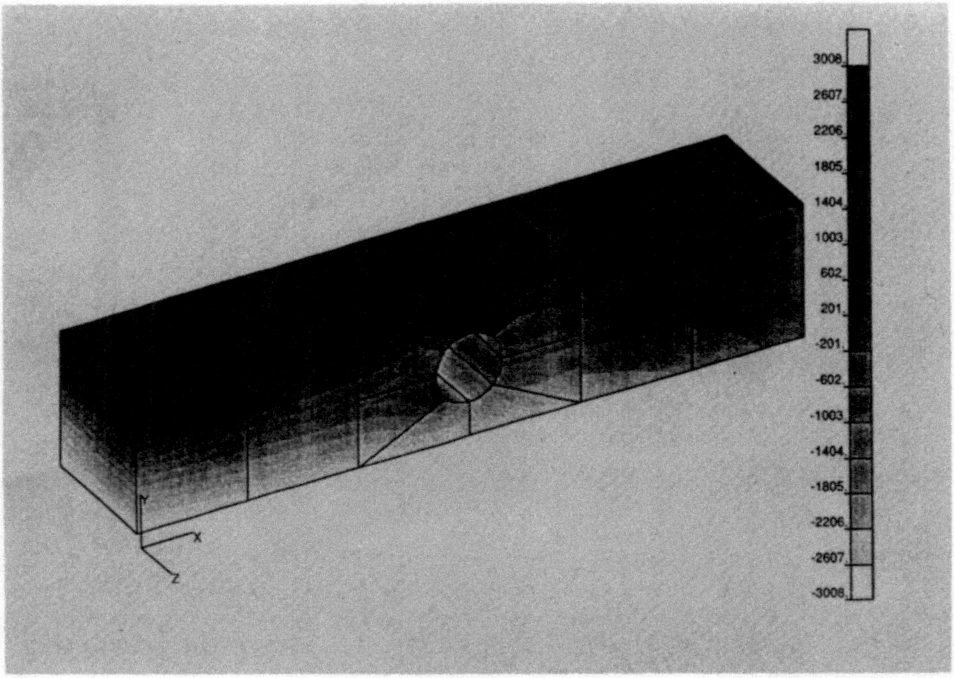

Figure 5.

A cantilever beam with a hole. The fringe plot shows the value of σ_x. The theoretical value is 3000. The stress concentration near the top of the hole was accurately predicted using the methods in the paper.

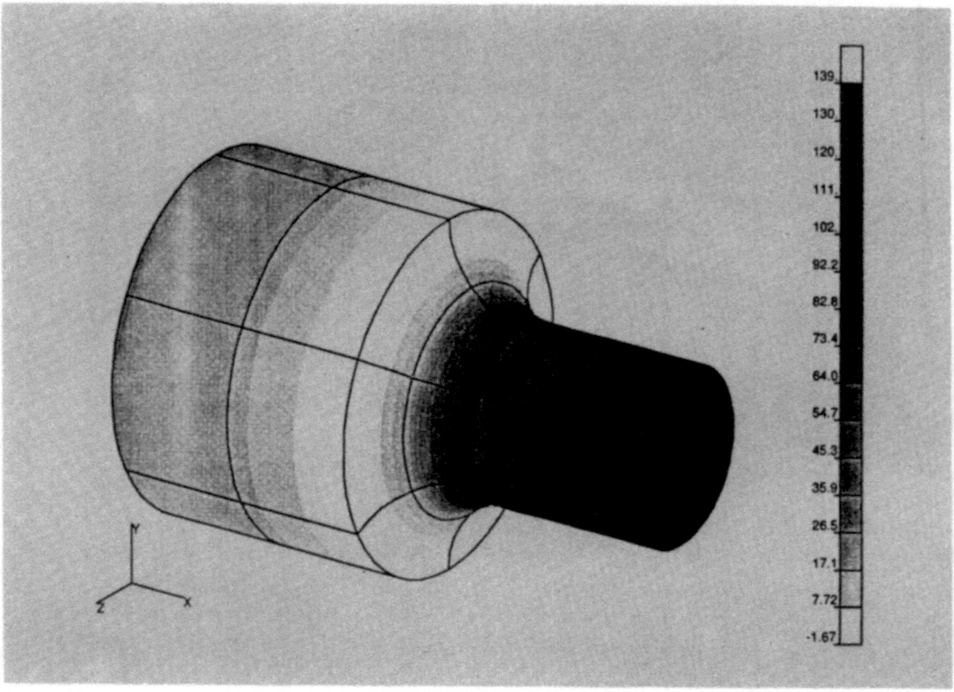

Figure 6.

A stepped shaft model. The theoretical value of σ_x in the fillet was computed to be 139 compared with 132.2 theoretical using a relatively coarse model.

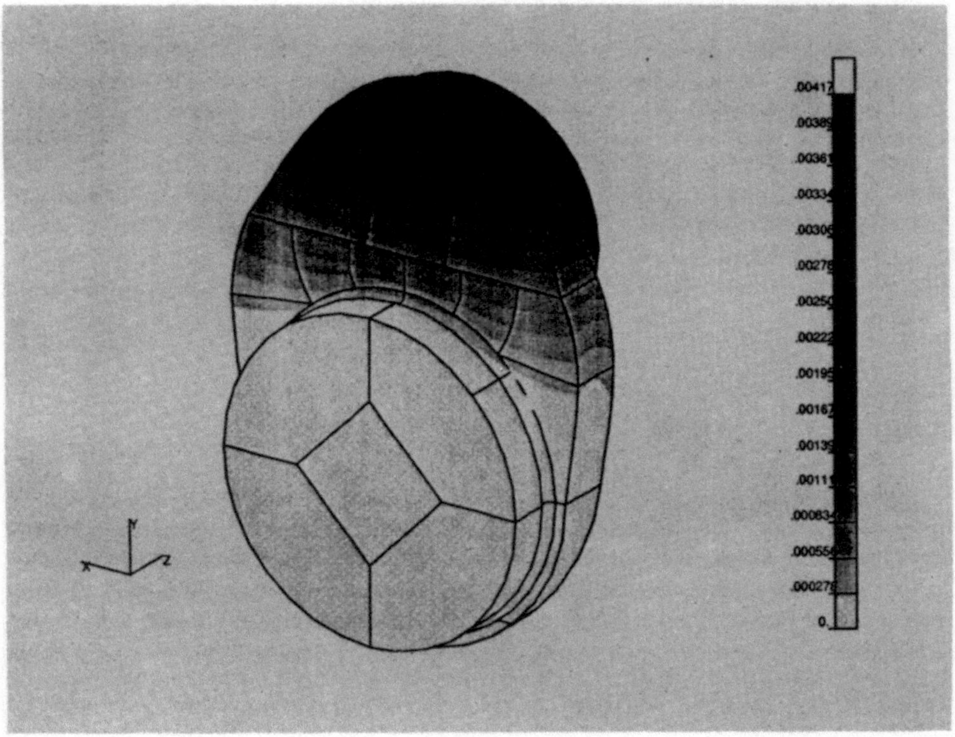

Figure 7.

The displacement fringe plot of a crankshaft model.

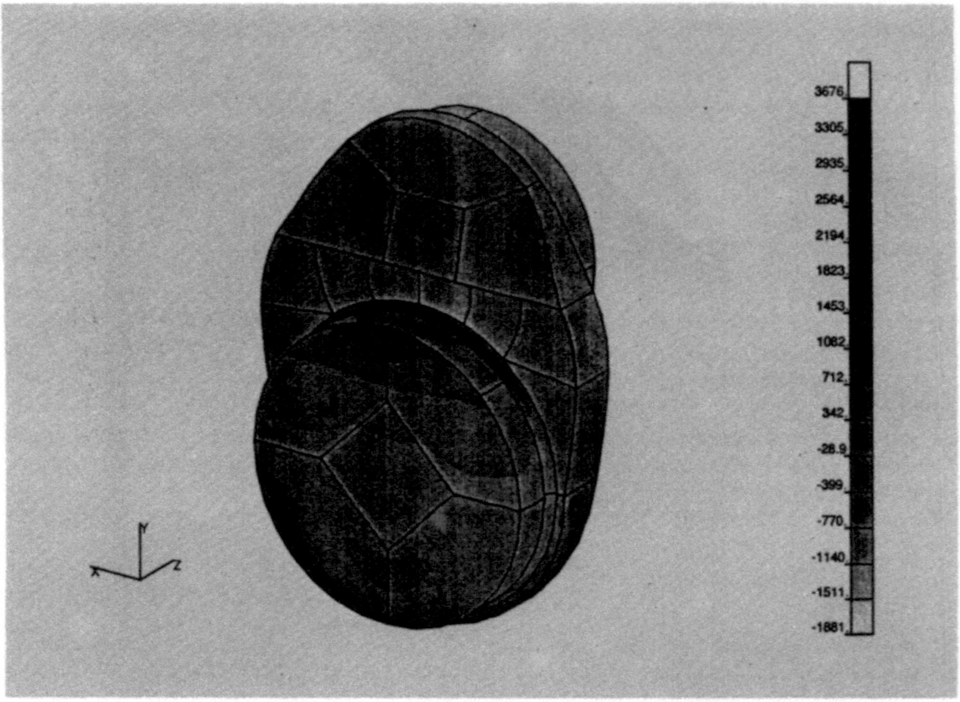

Figure 8.

A fringe plot of σ_z on the crankshaft. The stress concentration was accurately predicted in much less time than a finite element analysis and the analysis was done directly on the boundary representation produced by the modeler.

Acknowledgements

Thanks gò to the staff of PDA Engineering for their support. Special thanks go to Shan Nageswaren, Pat Sankar and Randy Underwood who are software engineers that make it all happen and to Karen Shirley for assistance with the manuscripting.

References

1. Brebbia, C., J. C. Telles and L. C. Wrobel, *Boundary Element Techniques*, Springer Verlag, New York,1984.

2. Cruse, T. A., *Boundary Element Analysis in Computational Fracture Mechanics*, Kluwer Academic Publishers, Boston, 1988.

3. Casale, M. S., Freeform Solid Modeling with Trimmed Patches, IEEE Computer Graphics and Applications, Jan., 1987, 33-43.

4. Casale, M. S., The Integration of Geometric Modeling and Structural Analysis Using Trimmed Patches, Ph.D. dissertation, University of California, Irvine, 1989.

5. Casale, M.S., A Set Operation Algorithm for Sculptured Solids Modeled with Trimmed Patches, Computer Aided Geometric Design (6), 1989, 235-247.

6. Farin, G., *Curves and Surfaces for Computer Aided Geometric Design*, Academic Press, 1988.

7. Faux, I. D. and Pratt, M. J., *Computational Geometry for Design and Manufacture*, John Wiley and Sons, New York, 1979.

8. Kane, J. H., Gupta A., Saigal, S., Reusable Intrinsic Sample Point (RISP) Algorithm for the Efficient Numerical Integration of Three Dimensional Curved Boundary Elements, International Journal for Numerical Methods in Engineering, (28), 11661-1676, 1989.

9. Mullen, R. L., Rencis, J. J., Iterative Methods for Solving Boundary Element Equations, Computers and Structures, (25), 713-723, 1987.

10. Zienkiewicz, O. C., Zhu, J. Z., A Simple Error Estimator and Adaptive Procedure for Practical Engineering Analysis, International Journal of Numerical Methods in Engineering, (24), 337-357, 1987.

11. Ainsworth, M., Zhu, J. Z., Craig, A. W., Zienkiewicz, O. C., Analysis of the Zienkiewicz~Zhu A-posterior Error Estimator in the Finite Element Method, International Journal for Numerical Methods in Engineering, (28), 2161-2174, 1990.

12. Rank, E., Adaptive h-,p- and hp-Versions for Boundary Integral Element Methods, International Journal for Numerical Methods in Engineering, (28), 1335-1349, 1989.

Stochastic Boundary Elements for Groundwater Flow with Random Hydraulic Conductivity

A. H-D. Cheng
DEPARTMENT OF CIVIL ENGINEERING, UNIVERSITY OF DELAWARE, NEWARK, DELAWARE, USA.
O. E. Lafe
OLTECH CORPORATION, MAYFIELD VILLAGE, OHIO, USA.

Abstract

The boundary element method has been successfully applied to groundwater flow problems with stochastic boundary conditions and forcing functions. The solution system is now extended to cases with random hydraulic conductivity using a perturbation technique.

Introduction

The traditional approach of modeling groundwater flow is by deterministic solution which assumes that the boundary conditions, such as the piezometric head and precipitation recharge, and the material coefficients, such as the hydraulic conductivity, are known with certainty. In real life, however, those quantities are uncertain, either due to the lack of information, or due to the intrinsic randomness of hydrologic processes in the nature. Take for example the aquifer hydraulic conductivity. Geological formations are highly heterogeneous at both large and small scales. For a problem with geometry of a given size, the correlation length scale of hydraulic conductivity is always much smaller than the domain size such that the homogeneous assumption is inadequate.[1] In an actual problem, the hydraulic conductivities are typically available from borehole cores or pumping test at intervals much greater than the correlation length scale. The values in between measurements, which are required in deterministic solution, are more or less arbitrarily fitted. The reliability of the deterministic solution is therefore doubtful. The random solution acknowledges these facts. It takes in the uncertain data in the form of statistical quantities such as mean, covariance, etc., and provide a solution in similar manner.

The current work is aimed at developing a solution capability for such stochastic boundary value problems. In particular, the boundary elements technique will be formulated. Earlier we have successfully derived the stochastic integral equations and boundary element solutions for the special case of problems with deterministic hydraulic conductivity, but under random boundary conditions and forcing functions.[2,3] We shall demonstrate in this paper that by utilizing the perturbation technique, the random hydraulic conductivity problems can be decomposed into a series of problems which can be solved using the same boundary element technique presented earlier. It is of interest to remark that although the focus of the current paper is on groundwater, the stochastic boundary element technique can be applied to many other governing equations as well.

Perturbation Equations

The governing equation for steady state groundwater flow is

$$\nabla \cdot [K(\boldsymbol{x},\gamma)\nabla\phi(\boldsymbol{x},\gamma)] = f(\boldsymbol{x},\beta) \tag{1}$$

in which ϕ is the piezometric head, K the hydraulic conductivity, $\boldsymbol{x}$ the spatial coordinates, f the forcing function which, under the vertically integrated flow assumption, corresponds to the vertical recharge caused by leakage from adjacent aquifers, or infiltration from precipitation. We have used in the above γ and β to denote ensemble spaces over which the average is taken. Two distinct parameters are used to emphasize that the randomness of the geological process (hydraulic conductivity) and the hydrologic process (forcing function and boundary condition) are totally uncorrelated. The above equation is subject to the boundary condition

$$\begin{aligned} \phi(\boldsymbol{x},\beta) \quad &\text{prescribed on} \quad \boldsymbol{x} \in \Gamma_\phi \\ q(\boldsymbol{x},\beta) = \frac{\partial \phi(\boldsymbol{x},\beta)}{\partial n(\boldsymbol{x})} \quad &\text{prescribed on} \quad \boldsymbol{x} \in \Gamma_q \end{aligned} \tag{2}$$

in which n is the unit outward normal of the boundary Γ, which consists of a Dirichlet part, Γ_ϕ, and a Neumann part, Γ_q.

Equation (1) can also be written as

$$\nabla^2 \phi(\boldsymbol{x},\gamma,\beta) + \nabla Y(\boldsymbol{x},\gamma) \cdot \nabla \phi(\boldsymbol{x},\gamma,\beta) = e^{-Y(\boldsymbol{x},\gamma)} f(\boldsymbol{x},\beta) \tag{3}$$

where Y is the logarithmic of hydraulic conductivity

$$Y(\boldsymbol{x},\gamma) = \ln K(\boldsymbol{x},\gamma) \tag{4}$$

Following a popular approach, Y can be expressed as a mean and a perturbation[4,5]

$$Y(\boldsymbol{x},\gamma) = \overline{Y}(\boldsymbol{x}) + Y'(\boldsymbol{x},\gamma) \tag{5}$$

It is assumed that the perturbation part is of $O(\varepsilon)$, where ε is a small number. This assumption limits the current solution to cases with small fluctuation in Y. Since Y is the logarithmic of K, the variation of hydraulic conductivity is larger, and can lie within one order of magnitude.

We also perturb the piezometric head into a series of descending magnitude

$$\phi(\boldsymbol{x},\gamma,\beta) = \phi_0(\boldsymbol{x},\beta) + \phi_1(\boldsymbol{x},\gamma,\beta) + \phi_2(\boldsymbol{x},\gamma,\beta) + \cdots \tag{6}$$

Substituting (5) and (6) into (1) and separating terms of different perturbation orders we obtain the following system of equations:

zeroth order

$$\nabla^2 \phi_0(\boldsymbol{x},\beta) + \nabla \overline{Y}(\boldsymbol{x}) \cdot \nabla \phi_0(\boldsymbol{x},\beta) = e^{-\overline{Y}(\boldsymbol{x})} f(\boldsymbol{x},\beta) \tag{7}$$

subject to B.C.

$$\begin{aligned} \phi_0(\boldsymbol{x},\beta) &= \phi(\boldsymbol{x},\beta), \quad \boldsymbol{x} \in \Gamma_\phi \\ q_0(\boldsymbol{x},\beta) &= q(\boldsymbol{x},\beta), \quad \boldsymbol{x} \in \Gamma_q \end{aligned} \tag{8}$$

first order

$$\begin{aligned} &\nabla^2 \phi_1(\boldsymbol{x},\gamma,\beta) + \nabla \overline{Y}(\boldsymbol{x}) \cdot \nabla \phi_1(\boldsymbol{x},\gamma,\beta) \\ &\quad = e^{-\overline{Y}(\boldsymbol{x})} Y'(\boldsymbol{x},\gamma) f(\boldsymbol{x},\beta) - \nabla Y'(\boldsymbol{x},\gamma) \cdot \nabla \phi_0(\boldsymbol{x},\beta) \end{aligned} \tag{9}$$

B.C.

$$\begin{aligned} \phi_1(\boldsymbol{x},\gamma,\beta) &= 0, \quad \boldsymbol{x} \in \Gamma_\phi \\ q_1(\boldsymbol{x},\gamma,\beta) &= 0, \quad \boldsymbol{x} \in \Gamma_q \end{aligned} \tag{10}$$

higher orders

$$\nabla^2\phi_i(\boldsymbol{x},\gamma,\beta) + \nabla\overline{Y}(\boldsymbol{x})\cdot\nabla\phi_i(\boldsymbol{x},\gamma,\beta)$$
$$= \frac{1}{i!}e^{-\overline{Y}(\boldsymbol{x})}Y'^i(\boldsymbol{x},\gamma)f(\boldsymbol{x},\beta) - \nabla Y'(\boldsymbol{x},\gamma)\cdot\nabla\phi_{i-1}(\boldsymbol{x},\gamma,\beta) \tag{11}$$

B.C.

$$\begin{aligned}\phi_i(\boldsymbol{x},\gamma,\beta) &= 0, \quad \boldsymbol{x}\in\Gamma_\phi\\ q_i(\boldsymbol{x},\gamma,\beta) &= 0, \quad \boldsymbol{x}\in\Gamma_q\end{aligned} \tag{12}$$

The above equations can be solved in succession.

For simplicity, we shall examine in this paper the special case in which the hydraulic conductivity is stationary in space. In other word, the mean of hydraulic conductivity, $\overline{Y}$, is a constant. The nonstationary case, $\overline{Y} = \overline{Y}(\boldsymbol{x})$, will require special techniques such as those used in deterministic heterogeneous groundwater flow,[6–8] which will be investigated later. With the stationarity assumption, the perturbation equations simplifies to the following

zeroth order

$$\nabla^2\phi_0(\boldsymbol{x},\beta) = e^{-\overline{Y}}f(\boldsymbol{x},\beta) \tag{13}$$

first order

$$\nabla^2\phi_1(\boldsymbol{x},\gamma,\beta) = e^{-\overline{Y}}Y'(\boldsymbol{x},\gamma)f(\boldsymbol{x},\beta) - \nabla Y'(\boldsymbol{x},\gamma)\cdot\nabla\phi_0(\boldsymbol{x},\beta) \tag{14}$$

higher orders

$$\nabla^2\phi_i(\boldsymbol{x},\gamma,\beta) = \frac{1}{i!}e^{-\overline{Y}}Y'^i(\boldsymbol{x},\gamma)f(\boldsymbol{x},\beta) - \nabla Y'(\boldsymbol{x},\gamma)\cdot\nabla\phi_{i-1}(\boldsymbol{x},\gamma,\beta) \tag{15}$$

The boundary conditions are the same as those in the nonstationary case.

Furthermore, if the boundary condition and forcing function are deterministic, the perturbation equations become

zeroth order

$$\nabla^2\phi_0(\boldsymbol{x}) = e^{-\overline{Y}}f(\boldsymbol{x}) \tag{16}$$

first order

$$\nabla^2\phi_1(\boldsymbol{x},\gamma) = e^{-\overline{Y}}Y'(\boldsymbol{x},\gamma)f(\boldsymbol{x}) - \nabla Y'(\boldsymbol{x},\gamma)\cdot\nabla\phi_0(\boldsymbol{x}) \tag{17}$$

higher orders

$$\nabla^2\phi_i(\boldsymbol{x},\gamma) = \frac{1}{i!}e^{-\overline{Y}}Y'^i(\boldsymbol{x},\gamma)f(\boldsymbol{x}) - \nabla Y'(\boldsymbol{x},\gamma)\cdot\nabla\phi_{i-1}(\boldsymbol{x},\gamma) \tag{18}$$

It is of interest to point out that ϕ_0 in this case becomes deterministic.

Statistical Properties

We actually are not interested in directly solving the governing equations in the preceding section on case by case basis. We are rather interested in finding the solution in terms of statistical properties, such as the mean, covariance, etc. The mean for ϕ and q are

$$\begin{aligned}\overline{\phi} &= \overline{\phi_0} + \overline{\phi_2} + \overline{\phi_3} + \cdots\\ \overline{q} &= \overline{q_0} + \overline{q_2} + \overline{q_3} + \cdots\end{aligned} \tag{19}$$

where we have used the overbar to denote the expectation, i.e., the ensemble mean, of a random parameter. It is of interest to point out that we have used in the above the fact $\overline{\phi_1} = 0$ and $\overline{q_1} = 0$, which is a direct consequence of (9) and (10). The correlation function is defined as

$$R_{ab}(\boldsymbol{x}, \boldsymbol{y}) = \overline{a(\boldsymbol{x}, \gamma, \beta) b(\boldsymbol{y}, \gamma, \beta)} \tag{20}$$

where a and b denote any random variable. A fluctuation of a random variable from its mean is denoted by a prime

$$a'(\boldsymbol{x}, \gamma, \beta) = a(\boldsymbol{x}, \gamma, \beta) - \overline{a}(\boldsymbol{x}) \tag{21}$$

The covariance is defined as

$$c_{ab}(\boldsymbol{x}, \boldsymbol{y}) = \overline{a'(\boldsymbol{x}, \gamma, \beta) b'(\boldsymbol{y}, \gamma, \beta)} \tag{22}$$

According to the perturbation equation (6), we have

$$c_{ab}(\boldsymbol{x}, \boldsymbol{y}) = c_{a_0 b_0} + c_{a_0 b_1} + c_{a_1 b_0} + c_{a_1 b_1} + c_{a_0 b_2} + c_{a_2 b_0} + c_{a_1 b_2} + \cdots \tag{23}$$

where a and b are ϕ or q. If we include only terms up to the first perturbation order, known as the Born approximation in stochastic wave propagation,[9] we then have

$$\begin{aligned} \overline{a} &\approx \overline{a_0} \\ c_{ab}(\boldsymbol{x}, \boldsymbol{y}) &\approx c_{a_0 b_0} + c_{a_0 b_1} + c_{a_1 b_0} + c_{a_1 b_1} \end{aligned} \tag{24}$$

Further, for deterministic boundary condition and forcing function, the covariance becomes

$$c_{ab}(\boldsymbol{x}, \boldsymbol{y}) \approx c_{a_1 b_1} \tag{25}$$

Stochastic Integral Equations

We begin with the integral equation representation of (13). Based on the simple layer potential method,[10] we have

$$\phi_0(\boldsymbol{x}, \beta) = \int_\Gamma \mu_0(\boldsymbol{x}', \beta) g(\boldsymbol{x}, \boldsymbol{x}')\, d\boldsymbol{x}' + e^{-\overline{Y}} \int_\Omega f(\boldsymbol{x}', \beta) g(\boldsymbol{x}, \boldsymbol{x}')\, d\boldsymbol{x}' \tag{26}$$

In the above, Γ is the solution boundary, Ω the solution domain, μ_0 the unknown source density function, and

$$\begin{aligned} g &= \frac{\ln r}{2\pi} \quad \text{(2-D)} \\ &= -\frac{1}{4\pi r} \quad \text{(3-D)} \end{aligned} \tag{27}$$

is the free space Green function, in which $r = |\boldsymbol{x} - \boldsymbol{x}'|$ is the distance between the base point $\boldsymbol{x}$ and a field point $\boldsymbol{x}'$. For $\boldsymbol{x} \in \Gamma$, (26) can be differentiated with respect to the boundary outward normal n to obtain

$$q_0(\boldsymbol{x}, \beta) = \int_\Gamma \mu_0(\boldsymbol{x}', \beta) g_n(\boldsymbol{x}, \boldsymbol{x}')\, d\boldsymbol{x}' + e^{-\overline{Y}} \int_\Omega f(\boldsymbol{x}', \beta) g_n(\boldsymbol{x}, \boldsymbol{x}')\, d\boldsymbol{x}' \tag{28}$$

where $g_n(\boldsymbol{x}, \boldsymbol{x}') = \partial g(\boldsymbol{x}, \boldsymbol{x}')/\partial n(\boldsymbol{x})$. If we take expectation of (26) and (28), we obtain equations for the mean

$$\overline{\phi_0}(\boldsymbol{x}) = \int_\Gamma \overline{\mu_0}(\boldsymbol{x}') g(\boldsymbol{x}, \boldsymbol{x}')\, d\boldsymbol{x}' + e^{-\overline{Y}} \int_\Omega \overline{f}(\boldsymbol{x}') g(\boldsymbol{x}, \boldsymbol{x}')\, d\boldsymbol{x}' \tag{29}$$

$$\overline{q_0}(\boldsymbol{x}) = \int_\Gamma \overline{\mu_0}(\boldsymbol{x}') g_n(\boldsymbol{x}, \boldsymbol{x}')\, d\boldsymbol{x}' + e^{-\overline{Y}} \int_\Omega \overline{f}(\boldsymbol{x}') g_n(\boldsymbol{x}, \boldsymbol{x}')\, d\boldsymbol{x}' \tag{30}$$

The above pair of equations, subject to the boundary condition

$$\begin{aligned}\overline{\phi_0}(\boldsymbol{x},\beta) &= \overline{\phi}(\boldsymbol{x},\beta), \quad \boldsymbol{x}\in\Gamma_\phi\\ \overline{q_0}(\boldsymbol{x},\beta) &= \overline{q}(\boldsymbol{x},\beta), \quad \boldsymbol{x}\in\Gamma_q\end{aligned} \tag{31}$$

have the same form as a deterministic integral equation solution system. The boundary element technique, which involves discretization of boundary into elements, interpolation of functions, numerical integration, etc. can be utilized for the solution of the mean quantities $\overline{\phi_0}$ and $\overline{q_0}$.

Next, we examine the solution of covariance. Subtracting (29) and (30) from (26) and (28), we obtain equations for fluctuation quantities

$$\phi_0'(\boldsymbol{x},\beta) = \int_\Gamma \mu_0'(\boldsymbol{x}',\beta)g(\boldsymbol{x},\boldsymbol{x}')\,d\boldsymbol{x}' + e^{-\overline{Y}}\int_\Omega f'(\boldsymbol{x}',\beta)g(\boldsymbol{x},\boldsymbol{x}')\,d\boldsymbol{x}' \tag{32}$$

$$q_0'(\boldsymbol{x},\beta) = \int_\Gamma \mu_0'(\boldsymbol{x}',\beta)g_n(\boldsymbol{x},\boldsymbol{x}')\,d\boldsymbol{x}' + e^{-\overline{Y}}\int_\Omega f'(\boldsymbol{x}',\beta)g_n(\boldsymbol{x},\boldsymbol{x}')\,d\boldsymbol{x}' \tag{33}$$

Taking self and cross products of the above equations and performing ensemble average over the result, we obtain integral equations for variance

$$\begin{aligned}c_{\phi_0\phi_0}(\boldsymbol{x},\boldsymbol{y}) = {} & \int_\Gamma\int_\Gamma c_{\mu_0\mu_0}(\boldsymbol{x}',\boldsymbol{y}')g(\boldsymbol{x},\boldsymbol{x}')g(\boldsymbol{y},\boldsymbol{y}')\,d\boldsymbol{x}'\,d\boldsymbol{y}'\\ & -e^{-2\overline{Y}}\int_\Omega\int_\Omega c_{ff}(\boldsymbol{x}',\boldsymbol{y}')g(\boldsymbol{x},\boldsymbol{x}')g(\boldsymbol{y},\boldsymbol{y}')\,d\boldsymbol{x}'\,d\boldsymbol{y}'\\ & +e^{-\overline{Y}}\int_\Omega c_{\phi_0 f}(\boldsymbol{x},\boldsymbol{x}')g(\boldsymbol{y},\boldsymbol{x}')\,d\boldsymbol{x}' + e^{-\overline{Y}}\int_\Omega c_{\phi_0 f}(\boldsymbol{y},\boldsymbol{x}')g(\boldsymbol{x},\boldsymbol{x}')\,d\boldsymbol{x}'\end{aligned} \tag{34}$$

$$\begin{aligned}c_{q_0q_0}(\boldsymbol{x},\boldsymbol{y}) = {} & \int_\Gamma\int_\Gamma c_{\mu_0\mu_0}(\boldsymbol{x}',\boldsymbol{y}')g_n(\boldsymbol{x},\boldsymbol{x}')g_n(\boldsymbol{y},\boldsymbol{y}')\,d\boldsymbol{x}'\,d\boldsymbol{y}'\\ & -e^{-2\overline{Y}}\int_\Omega\int_\Omega c_{ff}(\boldsymbol{x}',\boldsymbol{y}')g_n(\boldsymbol{x},\boldsymbol{x}')g_n(\boldsymbol{y},\boldsymbol{y}')\,d\boldsymbol{x}'\,d\boldsymbol{y}'\\ & +e^{-\overline{Y}}\int_\Omega c_{q_0 f}(\boldsymbol{x},\boldsymbol{x}')g_n(\boldsymbol{y},\boldsymbol{x}')\,d\boldsymbol{x}' + e^{-\overline{Y}}\int_\Omega c_{q_0 f}(\boldsymbol{y},\boldsymbol{x}')g_n(\boldsymbol{x},\boldsymbol{x}')\,d\boldsymbol{x}'\end{aligned} \tag{35}$$

$$\begin{aligned}c_{\phi_0q_0}(\boldsymbol{x},\boldsymbol{y}) = {} & \int_\Gamma\int_\Gamma c_{\mu_0\mu_0}(\boldsymbol{x}',\boldsymbol{y}')g(\boldsymbol{x},\boldsymbol{x}')g_n(\boldsymbol{y},\boldsymbol{y}')\,d\boldsymbol{x}'\,d\boldsymbol{y}'\\ & -e^{-2\overline{Y}}\int_\Omega\int_\Omega c_{ff}(\boldsymbol{x}',\boldsymbol{y}')g(\boldsymbol{x},\boldsymbol{x}')g_n(\boldsymbol{y},\boldsymbol{y}')\,d\boldsymbol{x}'\,d\boldsymbol{y}'\\ & +e^{-\overline{Y}}\int_\Omega c_{\phi_0 f}(\boldsymbol{x},\boldsymbol{x}')g_n(\boldsymbol{y},\boldsymbol{x}')\,d\boldsymbol{x}' + e^{-\overline{Y}}\int_\Omega c_{q_0 f}(\boldsymbol{y},\boldsymbol{x}')g(\boldsymbol{x},\boldsymbol{x}')\,d\boldsymbol{x}'\end{aligned} \tag{36}$$

$$\begin{aligned}c_{q_0\phi_0}(\boldsymbol{x},\boldsymbol{y}) = {} & \int_\Gamma\int_\Gamma c_{\mu_0\mu_0}(\boldsymbol{x}',\boldsymbol{y}')g_n(\boldsymbol{x},\boldsymbol{x}')g(\boldsymbol{y},\boldsymbol{y}')\,d\boldsymbol{x}'\,d\boldsymbol{y}'\\ & -e^{-2\overline{Y}}\int_\Omega\int_\Omega c_{ff}(\boldsymbol{x}',\boldsymbol{y}')g_n(\boldsymbol{x},\boldsymbol{x}')g(\boldsymbol{y},\boldsymbol{y}')\,d\boldsymbol{x}'\,d\boldsymbol{y}'\\ & +e^{-\overline{Y}}\int_\Omega c_{q_0 f}(\boldsymbol{x},\boldsymbol{x}')g(\boldsymbol{y},\boldsymbol{x}')\,d\boldsymbol{x}' + e^{-\overline{Y}}\int_\Omega c_{\phi_0 f}(\boldsymbol{y},\boldsymbol{x}')g_n(\boldsymbol{x},\boldsymbol{x}')\,d\boldsymbol{x}'\end{aligned} \tag{37}$$

We can also form the following auxiliary equations

$$c_{\phi_0 f}(\boldsymbol{x},\boldsymbol{y}) = e^{-\overline{Y}}\int_\Gamma c_{\mu_0 f}(\boldsymbol{x}',\boldsymbol{y})g(\boldsymbol{x},\boldsymbol{x}')\,d\boldsymbol{x}' + e^{-2\overline{Y}}\int_\Omega c_{ff}(\boldsymbol{x}',\boldsymbol{y})g(\boldsymbol{x},\boldsymbol{x}')\,d\boldsymbol{x}' \tag{38}$$

$$c_{q_0 f}(\boldsymbol{x},\boldsymbol{y}) = e^{-\overline{Y}}\int_\Gamma c_{\mu_0 f}(\boldsymbol{x}',\boldsymbol{y})g_n(\boldsymbol{x},\boldsymbol{x}')\,d\boldsymbol{x}' + e^{-2\overline{Y}}\int_\Omega c_{ff}(\boldsymbol{x}',\boldsymbol{y})g_n(\boldsymbol{x},\boldsymbol{x}')\,d\boldsymbol{x}' \tag{39}$$

$$c_{\phi_0\phi_0}(\boldsymbol{x},\boldsymbol{y}) = \int_\Gamma c_{\mu_0\phi_0}(\boldsymbol{x}',\boldsymbol{y})g(\boldsymbol{x},\boldsymbol{x}')\,d\boldsymbol{x}' + e^{-\overline{Y}}\int_\Omega c_{\phi_0 f}(\boldsymbol{y},\boldsymbol{x}')g(\boldsymbol{x},\boldsymbol{x}')\,d\boldsymbol{x}' \tag{40}$$

$$c_{q_0\phi_0}(\boldsymbol{x},\boldsymbol{y}) = \int_\Gamma c_{\mu_0\phi_0}(\boldsymbol{x}',\boldsymbol{y})g_n(\boldsymbol{x},\boldsymbol{x}')\,d\boldsymbol{x}' + e^{-\overline{Y}}\int_\Omega c_{\phi_0 f}(\boldsymbol{y},\boldsymbol{x}')g_n(\boldsymbol{x},\boldsymbol{x}')\,d\boldsymbol{x}' \tag{41}$$

$$c_{q_0q_0}(\boldsymbol{x},\boldsymbol{y}) = \int_\Gamma c_{\mu_0q_0}(\boldsymbol{x}',\boldsymbol{y})g_n(\boldsymbol{x},\boldsymbol{x}')\,d\boldsymbol{x}' + e^{-\overline{Y}}\int_\Omega c_{q_0f}(\boldsymbol{y},\boldsymbol{x}')g_n(\boldsymbol{x},\boldsymbol{x}')\,d\boldsymbol{x}' \quad (42)$$

$$c_{\phi_0q_0}(\boldsymbol{x},\boldsymbol{y}) = \int_\Gamma c_{\mu_0q_0}(\boldsymbol{x}',\boldsymbol{y})g(\boldsymbol{x},\boldsymbol{x}')\,d\boldsymbol{x}' + e^{-\overline{Y}}\int_\Omega c_{q_0f}(\boldsymbol{y},\boldsymbol{x}')g(\boldsymbol{x},\boldsymbol{x}')\,d\boldsymbol{x}' \quad (43)$$

$$c_{\mu_0\phi_0}(\boldsymbol{x},\boldsymbol{y}) = \int_\Gamma c_{\mu_0\mu_0}(\boldsymbol{x},\boldsymbol{x}')g(\boldsymbol{y},\boldsymbol{x}')\,d\boldsymbol{x}' + e^{-\overline{Y}}\int_\Omega c_{\mu_0f}(\boldsymbol{x},\boldsymbol{x}')g(\boldsymbol{y},\boldsymbol{x}')\,d\boldsymbol{x}' \quad (44)$$

$$c_{\mu_0q_0}(\boldsymbol{x},\boldsymbol{y}) = \int_\Gamma c_{\mu_0\mu_0}(\boldsymbol{x},\boldsymbol{x}')g_n(\boldsymbol{y},\boldsymbol{x}')\,d\boldsymbol{x}' + e^{-\overline{Y}}\int_\Omega c_{\mu_0f}(\boldsymbol{x},\boldsymbol{x}')g_n(\boldsymbol{y},\boldsymbol{x}')\,d\boldsymbol{x}' \quad (45)$$

$$\begin{aligned}
c_{\phi_0\phi_0}(\boldsymbol{x},\boldsymbol{y}) = & \int_\Gamma\int_\Gamma c_{\mu_0\mu_0}(\boldsymbol{x}',\boldsymbol{y}')g(\boldsymbol{x},\boldsymbol{x}')g(\boldsymbol{y},\boldsymbol{y}')\,d\boldsymbol{x}'\,d\boldsymbol{y}' \\
& + e^{-2\overline{Y}}\int_\Omega\int_\Omega c_{ff}(\boldsymbol{x}',\boldsymbol{y}')g(\boldsymbol{x},\boldsymbol{x}')g(\boldsymbol{y},\boldsymbol{y}')\,d\boldsymbol{x}'\,d\boldsymbol{y}' \\
& + e^{-\overline{Y}}\int_\Omega\int_\Gamma c_{\mu_0f}(\boldsymbol{x}',\boldsymbol{y}')g(\boldsymbol{x},\boldsymbol{x}')g(\boldsymbol{y},\boldsymbol{y}')\,d\boldsymbol{x}'\,d\boldsymbol{y}' \\
& + e^{-\overline{Y}}\int_\Omega\int_\Gamma c_{\mu_0f}(\boldsymbol{x}',\boldsymbol{y}')g(\boldsymbol{y},\boldsymbol{x}')g(\boldsymbol{x},\boldsymbol{y}')\,d\boldsymbol{x}'\,d\boldsymbol{y}' \quad (46)
\end{aligned}$$

$$\begin{aligned}
c_{q_0q_0}(\boldsymbol{x},\boldsymbol{y}) = & \int_\Gamma\int_\Gamma c_{\mu_0\mu_0}(\boldsymbol{x}',\boldsymbol{y}')g_n(\boldsymbol{x},\boldsymbol{x}')g_n(\boldsymbol{y},\boldsymbol{y}')\,d\boldsymbol{x}'\,d\boldsymbol{y}' \\
& + e^{-2\overline{Y}}\int_\Omega\int_\Omega c_{ff}(\boldsymbol{x}',\boldsymbol{y}')g_n(\boldsymbol{x},\boldsymbol{x}')g_n(\boldsymbol{y},\boldsymbol{y}')\,d\boldsymbol{x}'\,d\boldsymbol{y}' \\
& + e^{-\overline{Y}}\int_\Omega\int_\Gamma c_{\mu_0f}(\boldsymbol{x}',\boldsymbol{y}')g_n(\boldsymbol{x},\boldsymbol{x}')g_n(\boldsymbol{y},\boldsymbol{y}')\,d\boldsymbol{x}'\,d\boldsymbol{y}' \\
& + e^{-\overline{Y}}\int_\Omega\int_\Gamma c_{\mu_0f}(\boldsymbol{x}',\boldsymbol{y}')g_n(\boldsymbol{y},\boldsymbol{x}')g_n(\boldsymbol{x},\boldsymbol{y}')\,d\boldsymbol{x}'\,d\boldsymbol{y}' \quad (47)
\end{aligned}$$

$$\begin{aligned}
c_{\phi_0q_0}(\boldsymbol{x},\boldsymbol{y}) = & \int_\Gamma\int_\Gamma c_{\mu_0\mu_0}(\boldsymbol{x}',\boldsymbol{y}')g(\boldsymbol{x},\boldsymbol{x}')g_n(\boldsymbol{y},\boldsymbol{y}')\,d\boldsymbol{x}'\,d\boldsymbol{y}' \\
& + e^{-2\overline{Y}}\int_\Omega\int_\Omega c_{ff}(\boldsymbol{x}',\boldsymbol{y}')g(\boldsymbol{x},\boldsymbol{x}')g_n(\boldsymbol{y},\boldsymbol{y}')\,d\boldsymbol{x}'\,d\boldsymbol{y}' \\
& + e^{-\overline{Y}}\int_\Omega\int_\Gamma c_{\mu_0f}(\boldsymbol{x}',\boldsymbol{y}')g(\boldsymbol{x},\boldsymbol{x}')g_n(\boldsymbol{y},\boldsymbol{y}')\,d\boldsymbol{x}'\,d\boldsymbol{y}' \\
& + e^{-\overline{Y}}\int_\Omega\int_\Gamma c_{\mu_0f}(\boldsymbol{x}',\boldsymbol{y}')g_n(\boldsymbol{y},\boldsymbol{x}')g(\boldsymbol{x},\boldsymbol{y}')\,d\boldsymbol{x}'\,d\boldsymbol{y}' \quad (48)
\end{aligned}$$

$$\begin{aligned}
c_{q_0\phi_0}(\boldsymbol{x},\boldsymbol{y}) = & \int_\Gamma\int_\Gamma c_{\mu_0\mu_0}(\boldsymbol{x}',\boldsymbol{y}')g(\boldsymbol{y},\boldsymbol{y}')g_n(\boldsymbol{x},\boldsymbol{x}')\,d\boldsymbol{x}'\,d\boldsymbol{y}' \\
& + e^{-2\overline{Y}}\int_\Omega\int_\Omega c_{ff}(\boldsymbol{x}',\boldsymbol{y}')g(\boldsymbol{y},\boldsymbol{y}')g_n(\boldsymbol{x},\boldsymbol{x}')\,d\boldsymbol{x}'\,d\boldsymbol{y}' \\
& + e^{-\overline{Y}}\int_\Omega\int_\Gamma c_{\mu_0f}(\boldsymbol{x}',\boldsymbol{y}')g_n(\boldsymbol{x},\boldsymbol{x}')g(\boldsymbol{y},\boldsymbol{y}')\,d\boldsymbol{x}'\,d\boldsymbol{y}' \\
& + e^{-\overline{Y}}\int_\Omega\int_\Gamma c_{\mu_0f}(\boldsymbol{x}',\boldsymbol{y}')g(\boldsymbol{y},\boldsymbol{x}')g_n(\boldsymbol{x},\boldsymbol{y}')\,d\boldsymbol{x}'\,d\boldsymbol{y}' \quad (49)
\end{aligned}$$

Various combinations of the above integral equations can be used in conjunction with boundary element procedures to solve for the covariance quantities $c_{\phi_0\phi_0}$, $c_{\phi_0q_0}$, $c_{q_0\phi_0}$, and $c_{q_0q_0}$. In fact, there are two procedures suggested.[2] In the first, an $N \times N$ unknown system is directly solved, where N is the number of nodes used in the boundary discretization. Due to the size of the linear system (an $N^2 \times N^2$ matrix), an iterative boundary element technique[11] is needed for an efficient solution. In the second procedure, we solve each time N unknowns ($N \times N$ matrix), but for N times. The second procedure is found to be much more efficient. See Cheng & Lafe[2] for details.

The above solution involves the zeroth order perturbation quantities. Before moving on to the first order quantities, it is of interest to first point out a few things. The zeroth order solution of covariances is necessary only for cases with random boundary condition and forcing function. If those conditions are deterministic (but the hydraulic conductivity is random), then

ϕ_0 is deterministic, as it is clear from (16). As a direct consequence, $c_{\phi_0\phi_0} = c_{\phi_0 q_0} = c_{q_0\phi_0} = c_{q_0 q_0} = 0$.

The first order perturbation quantity ϕ_1 is governed by (14). The corresponding integral equations are

$$\begin{aligned}\phi_1(\boldsymbol{x},\gamma,\beta) &= \int_\Gamma \mu_1(\boldsymbol{x}',\gamma,\beta)g(\boldsymbol{x},\boldsymbol{x}')\,d\boldsymbol{x}' + e^{-\overline{Y}}\int_\Omega Y'(\boldsymbol{x}',\gamma)f(\boldsymbol{x}',\beta)g(\boldsymbol{x},\boldsymbol{x}')\,d\boldsymbol{x}' \\ &\quad - \int_\Omega \nabla Y'(\boldsymbol{x}',\gamma)\cdot\nabla\phi_0(\boldsymbol{x}',\beta)g(\boldsymbol{x},\boldsymbol{x}')\,d\boldsymbol{x}' \qquad (50)\\ q_1(\boldsymbol{x},\gamma,\beta) &= \int_\Gamma \mu_1(\boldsymbol{x}',\gamma,\beta)g_n(\boldsymbol{x},\boldsymbol{x}')\,d\boldsymbol{x}' + e^{-\overline{Y}}\int_\Omega Y'(\boldsymbol{x}',\gamma)f(\boldsymbol{x}',\beta)g_n(\boldsymbol{x},\boldsymbol{x}')\,d\boldsymbol{x}' \\ &\quad - \int_\Omega \nabla Y'(\boldsymbol{x}',\gamma)\cdot\nabla\phi_0(\boldsymbol{x}',\beta)g_n(\boldsymbol{x},\boldsymbol{x}')\,d\boldsymbol{x}' \qquad (51)\end{aligned}$$

The mean of the above is

$$\overline{\phi_1}(\boldsymbol{x}) = \int_\Gamma \overline{\mu_1}(\boldsymbol{x}')g(\boldsymbol{x},\boldsymbol{x}')\,d\boldsymbol{x}' \qquad (52)$$

$$\overline{q_1}(\boldsymbol{x}) = \int_\Gamma \overline{\mu_1}(\boldsymbol{x}')g_n(\boldsymbol{x},\boldsymbol{x}')\,d\boldsymbol{x}' \qquad (53)$$

The last two integrals in (50) and (51) drop out because events in the sample space γ and β are uncorrelated, and Y' has a zero mean. With the trivial boundary condition (10), it is clear that $\overline{\phi_1} = 0$ and $\overline{q_1} = 0$.

For the covariances, taking the self and cross product of the fluctuation part of (50) and taking expectation, we obtain equations similar to (34) to (37)

$$\begin{aligned}0 &= \int_\Gamma\int_\Gamma c_{\mu_1\mu_1}(\boldsymbol{x}',\boldsymbol{y}')g(\boldsymbol{x},\boldsymbol{x}')g(\boldsymbol{y},\boldsymbol{y}')\,d\boldsymbol{x}'\,d\boldsymbol{y}' \\ &\quad - \int_\Omega\int_\Omega \sum_{j=1}^{3}\sum_{k=1}^{3} \frac{\partial^2 c_{YY}(\boldsymbol{x}',\boldsymbol{y}')}{\partial x_j'\partial y_k'}\frac{\partial^2 c_{\phi_0\phi_0}(\boldsymbol{x}',\boldsymbol{y}')}{\partial x_j'\partial y_k'} g(\boldsymbol{x},\boldsymbol{x}')g(\boldsymbol{y},\boldsymbol{y}')\,d\boldsymbol{x}'\,d\boldsymbol{y}' \\ &\quad - e^{-2\overline{Y}}\int_\Omega\int_\Omega c_{YY}(\boldsymbol{x}',\boldsymbol{y}')c_{ff}(\boldsymbol{x}',\boldsymbol{y}')g(\boldsymbol{x},\boldsymbol{x}')g(\boldsymbol{y},\boldsymbol{y}')\,d\boldsymbol{x}'\,d\boldsymbol{y}' \\ &\quad + e^{-\overline{Y}}\int_\Omega\int_\Omega \sum_{j=1}^{3} \frac{\partial c_{YY}(\boldsymbol{x}',\boldsymbol{y}')}{\partial x_j'}\frac{\partial c_{\phi_0 f}(\boldsymbol{x}',\boldsymbol{y}')}{\partial x_j'} g(\boldsymbol{x},\boldsymbol{x}')g(\boldsymbol{y},\boldsymbol{y}')\,d\boldsymbol{x}'\,d\boldsymbol{y}' \\ &\quad + e^{-\overline{Y}}\int_\Omega\int_\Omega \sum_{j=1}^{3} \frac{\partial c_{YY}(\boldsymbol{x}',\boldsymbol{y}')}{\partial y_j'}\frac{\partial c_{\phi_0 f}(\boldsymbol{y}',\boldsymbol{x}')}{\partial y_j'} g(\boldsymbol{x},\boldsymbol{x}')g(\boldsymbol{y},\boldsymbol{y}')\,d\boldsymbol{x}'\,d\boldsymbol{y}'; \quad \boldsymbol{x},\boldsymbol{y}\in\Gamma_\phi \qquad (54)\end{aligned}$$

$$\begin{aligned}0 &= \int_\Gamma\int_\Gamma c_{\mu_1\mu_1}(\boldsymbol{x}',\boldsymbol{y}')g_n(\boldsymbol{x},\boldsymbol{x}')g_n(\boldsymbol{y},\boldsymbol{y}')\,d\boldsymbol{x}'\,d\boldsymbol{y}' \\ &\quad - \int_\Omega\int_\Omega \sum_{j=1}^{3}\sum_{k=1}^{3} \frac{\partial^2 c_{YY}(\boldsymbol{x}',\boldsymbol{y}')}{\partial x_j'\partial y_k'}\frac{\partial^2 c_{\phi_0\phi_0}(\boldsymbol{x}',\boldsymbol{y}')}{\partial x_j'\partial y_k'} g_n(\boldsymbol{x},\boldsymbol{x}')g_n(\boldsymbol{y},\boldsymbol{y}')\,d\boldsymbol{x}'\,d\boldsymbol{y}' \\ &\quad - e^{-2\overline{Y}}\int_\Omega\int_\Omega c_{YY}(\boldsymbol{x}',\boldsymbol{y}')c_{ff}(\boldsymbol{x}',\boldsymbol{y}')g_n(\boldsymbol{x},\boldsymbol{x}')g_n(\boldsymbol{y},\boldsymbol{y}')\,d\boldsymbol{x}'\,d\boldsymbol{y}' \\ &\quad + e^{-\overline{Y}}\int_\Omega\int_\Omega \sum_{j=1}^{3} \frac{\partial c_{YY}(\boldsymbol{x}',\boldsymbol{y}')}{\partial x_j'}\frac{\partial c_{\phi_0 f}(\boldsymbol{x}',\boldsymbol{y}')}{\partial x_j'} g_n(\boldsymbol{x},\boldsymbol{x}')g_n(\boldsymbol{y},\boldsymbol{y}')\,d\boldsymbol{x}'\,d\boldsymbol{y}' \\ &\quad + e^{-\overline{Y}}\int_\Omega\int_\Omega \sum_{j=1}^{3} \frac{\partial c_{YY}(\boldsymbol{x}',\boldsymbol{y}')}{\partial y_j'}\frac{\partial c_{\phi_0 f}(\boldsymbol{y}',\boldsymbol{x}')}{\partial y_j'} g_n(\boldsymbol{x},\boldsymbol{x}')g_n(\boldsymbol{y},\boldsymbol{y}')\,d\boldsymbol{x}'\,d\boldsymbol{y}'; \quad \boldsymbol{x},\boldsymbol{y}\in\Gamma_q \qquad (55)\end{aligned}$$

$$0 = \int_\Gamma\int_\Gamma c_{\mu_1\mu_1}(\boldsymbol{x}',\boldsymbol{y}')g(\boldsymbol{x},\boldsymbol{x}')g_n(\boldsymbol{y},\boldsymbol{y}')\,d\boldsymbol{x}'\,d\boldsymbol{y}'$$

$$
\begin{aligned}
&-\int_\Omega\int_\Omega \sum_{j=1}^{3}\sum_{k=1}^{3} \frac{\partial^2 c_{YY}(x',y')}{\partial x'_j \partial y'_k}\frac{\partial^2 c_{\phi_0\phi_0}(x',y')}{\partial x'_j \partial y'_k} g(x,x')g_n(y,y')\,dx'\,dy' \\
&-e^{-2\overline{Y}}\int_\Omega\int_\Omega c_{YY}(x',y')c_{ff}(x',y')g(x,x')g_n(y,y')\,dx'\,dy' \\
&+e^{-\overline{Y}}\int_\Omega\int_\Omega \sum_{j=1}^{3} \frac{\partial c_{YY}(x',y')}{\partial x'_j}\frac{\partial c_{\phi_0 f}(x',y')}{\partial x'_j} g(x,x')g_n(y,y')\,dx'\,dy' \\
&+e^{-\overline{Y}}\int_\Omega\int_\Omega \sum_{j=1}^{3} \frac{\partial c_{YY}(x',y')}{\partial y'_j}\frac{\partial c_{\phi_0 f}(y',x')}{\partial y'_j} g(x,x')g_n(y,y')\,dx'dy';\ x\in\Gamma_\phi, y\in\Gamma_q \qquad (56)
\end{aligned}
$$

$$
\begin{aligned}
0 = &\int_\Gamma\int_\Gamma c_{\mu_1\mu_1}(x',y')g_n(x,x')g(y,y')\,dx'\,dy' \\
&-\int_\Omega\int_\Omega \sum_{j=1}^{3}\sum_{k=1}^{3} \frac{\partial^2 c_{YY}(x',y')}{\partial x'_j \partial y'_k}\frac{\partial^2 c_{\phi_0\phi_0}(x',y')}{\partial x'_j \partial y'_k} g_n(x,x')g(y,y')\,dx'\,dy' \\
&-e^{-2\overline{Y}}\int_\Omega\int_\Omega c_{YY}(x',y')c_{ff}(x',y')g_n(x,x')g(y,y')\,dx'\,dy' \\
&+e^{-\overline{Y}}\int_\Omega\int_\Omega \sum_{j=1}^{3} \frac{\partial c_{YY}(x',y')}{\partial x'_j}\frac{\partial c_{\phi_0 f}(x',y')}{\partial x'_j} g_n(x,x')g(y,y')\,dx'\,dy' \\
&+e^{-\overline{Y}}\int_\Omega\int_\Omega \sum_{j=1}^{3} \frac{\partial c_{YY}(x',y')}{\partial y'_j}\frac{\partial c_{\phi_0 f}(y',x')}{\partial y'_j} g_n(x,x')g(y,y')\,dx'dy';\ x\in\Gamma_q, y\in\Gamma_\phi \qquad (57)
\end{aligned}
$$

The above system allows us to solve for the source density covariance $c_{\mu_1\mu_1}$. We also need the following equations to solve for two more fictitious densities c^* and c^{**}

$$
\begin{aligned}
0 \;=\; &\int_\Gamma c^*(x',y)g(x,x')\,dx' + e^{-\overline{Y}}\int_\Omega c_{YY}(x',y)c_{ff}(x',y)g(x,x')\,dx' \\
&-\int_\Omega \sum_{j=1}^{3} \frac{\partial c_{YY}(x',y)}{\partial x'_j}\frac{\partial c_{\phi_0 f}(x',y)}{\partial x'_j} g(x,x')\,dx';\quad y\in\Gamma_\phi \qquad (58) \\
0 \;=\; &\int_\Gamma c^*(x',y)g_n(x,x')\,dx' + e^{-\overline{Y}}\int_\Omega c_{YY}(x',y)c_{ff}(x',y)g_n(x,x')\,dx' \\
&-\int_\Omega \sum_{j=1}^{3} \frac{\partial c_{YY}(x',y)}{\partial x'_j}\frac{\partial c_{\phi_0 f}(x',y)}{\partial x'_j} g_n(x,x')\,dx';\quad y\in\Gamma_q \qquad (59) \\
0 \;=\; &\int_\Gamma c^{**}(x',y)g(x,x')\,dx' + e^{-\overline{Y}}\int_\Omega \sum_{j=1}^{3} \frac{\partial c_{YY}(x',y)}{\partial y_j}\frac{\partial c_{\phi_0 f}(y',x')}{\partial y_j} g(x,x')\,dx' \\
&-\int_\Omega \sum_{j=1}^{3}\sum_{k=1}^{3} \frac{\partial^2 c_{YY}(x',y)}{\partial x'_j \partial y_k}\frac{\partial^2 c_{\phi_0\phi_0}(x',y)}{\partial x'_j \partial y_k} g(x,x')\,dx';\quad y\in\Gamma_\phi \qquad (60) \\
0 \;=\; &\int_\Gamma c^{**}(x',y)g_n(x,x')\,dx' + e^{-\overline{Y}}\int_\Omega \sum_{j=1}^{3} \frac{\partial c_{YY}(x',y)}{\partial y_j}\frac{\partial c_{\phi_0 f}(y',x')}{\partial y_j} g_n(x,x')\,dx' \\
&-\int_\Omega \sum_{j=1}^{3}\sum_{k=1}^{3} \frac{\partial^2 c_{YY}(x',y)}{\partial x'_j \partial y_k}\frac{\partial^2 c_{\phi_0\phi_0}(x',y)}{\partial x'_j \partial y_k} g_n(x,x')\,dx';\quad y\in\Gamma_q \qquad (61)
\end{aligned}
$$

Finally, the covariances anywhere on the boundary or in the domain can be evaluated as

$$
\begin{aligned}
c_{\phi_1\phi_1}(x,y) = &\int_\Gamma\int_\Gamma c_{\mu_1\mu_1}(x',y')g(x,x')g(y,y')\,dx'\,dy' \\
&+\int_\Omega\int_\Omega \sum_{j=1}^{3}\sum_{k=1}^{3} \frac{\partial^2 c_{YY}(x',y')}{\partial x'_j \partial y'_k}\frac{\partial^2 c_{\phi_0\phi_0}(x',y')}{\partial x'_j \partial y'_k} g(x,x')g(y,y')\,dx'\,dy'
\end{aligned}
$$

$$+e^{-2\overline{Y}}\int_\Omega\int_\Omega c_{YY}(\boldsymbol{x}',\boldsymbol{y}')c_{ff}(\boldsymbol{x}',\boldsymbol{y}')g(\boldsymbol{x},\boldsymbol{x}')g(\boldsymbol{y},\boldsymbol{y}')\,d\boldsymbol{x}'\,d\boldsymbol{y}'$$
$$-e^{-\overline{Y}}\int_\Omega\int_\Omega\sum_{j=1}^{3}\frac{\partial c_{YY}(\boldsymbol{x}',\boldsymbol{y}')}{\partial x_j'}\frac{\partial c_{\phi_0 f}(\boldsymbol{x}',\boldsymbol{y}')}{\partial x_j'}g(\boldsymbol{x},\boldsymbol{x}')g(\boldsymbol{y},\boldsymbol{y}')\,d\boldsymbol{x}'\,d\boldsymbol{y}'$$
$$-e^{-\overline{Y}}\int_\Omega\int_\Omega\sum_{j=1}^{3}\frac{\partial c_{YY}(\boldsymbol{x}',\boldsymbol{y}')}{\partial y_j'}\frac{\partial c_{\phi_0 f}(\boldsymbol{y}',\boldsymbol{x}')}{\partial y_j'}g(\boldsymbol{x},\boldsymbol{x}')g(\boldsymbol{y},\boldsymbol{y}')\,d\boldsymbol{x}'\,d\boldsymbol{y}'$$
$$+e^{-\overline{Y}}\int_\Omega\int_\Gamma c^*(\boldsymbol{x}',\boldsymbol{y}')g(\boldsymbol{x},\boldsymbol{x}')g(\boldsymbol{y},\boldsymbol{y}')\,d\boldsymbol{x}'\,d\boldsymbol{y}'+e^{-\overline{Y}}\int_\Omega\int_\Gamma c^*(\boldsymbol{x}',\boldsymbol{y}')g(\boldsymbol{y},\boldsymbol{x}')g(\boldsymbol{x},\boldsymbol{y}')\,d\boldsymbol{x}'\,d\boldsymbol{y}'$$
$$-\int_\Omega\int_\Gamma c^{**}(\boldsymbol{x}',\boldsymbol{y}')g(\boldsymbol{x},\boldsymbol{x}')g(\boldsymbol{y},\boldsymbol{y}')\,d\boldsymbol{x}'\,d\boldsymbol{y}'-\int_\Omega\int_\Gamma c^{**}(\boldsymbol{x}',\boldsymbol{y}')g(\boldsymbol{y},\boldsymbol{x}')g(\boldsymbol{x},\boldsymbol{y}')\,d\boldsymbol{x}'\,d\boldsymbol{y}' \tag{62}$$

$$c_{q_1q_1}(\boldsymbol{x},\boldsymbol{y})=\int_\Gamma\int_\Gamma c_{\mu_1\mu_1}(\boldsymbol{x}',\boldsymbol{y}')g_n(\boldsymbol{x},\boldsymbol{x}')g_n(\boldsymbol{y},\boldsymbol{y}')\,d\boldsymbol{x}'\,d\boldsymbol{y}'$$
$$+\int_\Omega\int_\Omega\sum_{j=1}^{3}\sum_{k=1}^{3}\frac{\partial^2 c_{YY}(\boldsymbol{x}',\boldsymbol{y}')}{\partial x_j'\partial y_k'}\frac{\partial^2 c_{\phi_0\phi_0}(\boldsymbol{x}',\boldsymbol{y}')}{\partial x_j'\partial y_k'}g_n(\boldsymbol{x},\boldsymbol{x}')g_n(\boldsymbol{y},\boldsymbol{y}')\,d\boldsymbol{x}'\,d\boldsymbol{y}'$$
$$+e^{-2\overline{Y}}\int_\Omega\int_\Omega c_{YY}(\boldsymbol{x}',\boldsymbol{y}')c_{ff}(\boldsymbol{x}',\boldsymbol{y}')g_n(\boldsymbol{x},\boldsymbol{x}')g_n(\boldsymbol{y},\boldsymbol{y}')\,d\boldsymbol{x}'\,d\boldsymbol{y}'$$
$$-e^{-\overline{Y}}\int_\Omega\int_\Omega\sum_{j=1}^{3}\frac{\partial c_{YY}(\boldsymbol{x}',\boldsymbol{y}')}{\partial x_j'}\frac{\partial c_{\phi_0 f}(\boldsymbol{x}',\boldsymbol{y}')}{\partial x_j'}g_n(\boldsymbol{x},\boldsymbol{x}')g_n(\boldsymbol{y},\boldsymbol{y}')\,d\boldsymbol{x}'\,d\boldsymbol{y}'$$
$$-e^{-\overline{Y}}\int_\Omega\int_\Omega\sum_{j=1}^{3}\frac{\partial c_{YY}(\boldsymbol{x}',\boldsymbol{y}')}{\partial y_j'}\frac{\partial c_{\phi_0 f}(\boldsymbol{y}',\boldsymbol{x}')}{\partial y_j'}g_n(\boldsymbol{x},\boldsymbol{x}')g_n(\boldsymbol{y},\boldsymbol{y}')\,d\boldsymbol{x}'\,d\boldsymbol{y}'$$
$$+e^{-\overline{Y}}\int_\Omega\int_\Gamma c^*(\boldsymbol{x}',\boldsymbol{y}')g_n(\boldsymbol{x},\boldsymbol{x}')g_n(\boldsymbol{y},\boldsymbol{y}')\,d\boldsymbol{x}'d\boldsymbol{y}'+e^{-\overline{Y}}\int_\Omega\int_\Gamma c^*(\boldsymbol{x}',\boldsymbol{y}')g_n(\boldsymbol{y},\boldsymbol{x}')g_n(\boldsymbol{x},\boldsymbol{y}')\,d\boldsymbol{x}'d\boldsymbol{y}'$$
$$-\int_\Omega\int_\Gamma c^{**}(\boldsymbol{x}',\boldsymbol{y}')g_n(\boldsymbol{x},\boldsymbol{x}')g_n(\boldsymbol{y},\boldsymbol{y}')\,d\boldsymbol{x}'\,d\boldsymbol{y}'-\int_\Omega\int_\Gamma c^{**}(\boldsymbol{x}',\boldsymbol{y}')g_n(\boldsymbol{y},\boldsymbol{x}')g_n(\boldsymbol{x},\boldsymbol{y}')\,d\boldsymbol{x}'\,d\boldsymbol{y}' \tag{63}$$

$$c_{\phi_1q_1}(\boldsymbol{x},\boldsymbol{y})=\int_\Gamma\int_\Gamma c_{\mu_1\mu_1}(\boldsymbol{x}',\boldsymbol{y}')g(\boldsymbol{x},\boldsymbol{x}')g_n(\boldsymbol{y},\boldsymbol{y}')\,d\boldsymbol{x}'\,d\boldsymbol{y}'$$
$$+\int_\Omega\int_\Omega\sum_{j=1}^{3}\sum_{k=1}^{3}\frac{\partial^2 c_{YY}(\boldsymbol{x}',\boldsymbol{y}')}{\partial x_j'\partial y_k'}\frac{\partial^2 c_{\phi_0\phi_0}(\boldsymbol{x}',\boldsymbol{y}')}{\partial x_j'\partial y_k'}g(\boldsymbol{x},\boldsymbol{x}')g_n(\boldsymbol{y},\boldsymbol{y}')\,d\boldsymbol{x}'\,d\boldsymbol{y}'$$
$$+e^{-2\overline{Y}}\int_\Omega\int_\Omega c_{YY}(\boldsymbol{x}',\boldsymbol{y}')c_{ff}(\boldsymbol{x}',\boldsymbol{y}')g(\boldsymbol{x},\boldsymbol{x}')g_n(\boldsymbol{y},\boldsymbol{y}')\,d\boldsymbol{x}'\,d\boldsymbol{y}'$$
$$-e^{-\overline{Y}}\int_\Omega\int_\Omega\sum_{j=1}^{3}\frac{\partial c_{YY}(\boldsymbol{x}',\boldsymbol{y}')}{\partial x_j'}\frac{\partial c_{\phi_0 f}(\boldsymbol{x}',\boldsymbol{y}')}{\partial x_j'}g(\boldsymbol{x},\boldsymbol{x}')g_n(\boldsymbol{y},\boldsymbol{y}')\,d\boldsymbol{x}'\,d\boldsymbol{y}'$$
$$-e^{-\overline{Y}}\int_\Omega\int_\Omega\sum_{j=1}^{3}\frac{\partial c_{YY}(\boldsymbol{x}',\boldsymbol{y}')}{\partial y_j'}\frac{\partial c_{\phi_0 f}(\boldsymbol{y}',\boldsymbol{x}')}{\partial y_j'}g(\boldsymbol{x},\boldsymbol{x}')g_n(\boldsymbol{y},\boldsymbol{y}')\,d\boldsymbol{x}'\,d\boldsymbol{y}'$$
$$+e^{-\overline{Y}}\int_\Omega\int_\Gamma c^*(\boldsymbol{x}',\boldsymbol{y}')g(\boldsymbol{x},\boldsymbol{x}')g_n(\boldsymbol{y},\boldsymbol{y}')\,d\boldsymbol{x}'\,d\boldsymbol{y}'+e^{-\overline{Y}}\int_\Omega\int_\Gamma c^*(\boldsymbol{x}',\boldsymbol{y}')g_n(\boldsymbol{y},\boldsymbol{x}')g(\boldsymbol{x},\boldsymbol{y}')\,d\boldsymbol{x}'\,d\boldsymbol{y}'$$
$$-\int_\Omega\int_\Gamma c^{**}(\boldsymbol{x}',\boldsymbol{y}')g(\boldsymbol{x},\boldsymbol{x}')g_n(\boldsymbol{y},\boldsymbol{y}')\,d\boldsymbol{x}'\,d\boldsymbol{y}'-\int_\Omega\int_\Gamma c^{**}(\boldsymbol{x}',\boldsymbol{y}')g_n(\boldsymbol{y},\boldsymbol{x}')g(\boldsymbol{x},\boldsymbol{y}')\,d\boldsymbol{x}'\,d\boldsymbol{y}' \tag{64}$$

$$c_{q_1\phi_1}(\boldsymbol{x},\boldsymbol{y})=\int_\Gamma\int_\Gamma c_{\mu_1\mu_1}(\boldsymbol{x}',\boldsymbol{y}')g_n(\boldsymbol{x},\boldsymbol{x}')g(\boldsymbol{y},\boldsymbol{y}')\,d\boldsymbol{x}'\,d\boldsymbol{y}'$$
$$+\int_\Omega\int_\Omega\sum_{j=1}^{3}\sum_{k=1}^{3}\frac{\partial^2 c_{YY}(\boldsymbol{x}',\boldsymbol{y}')}{\partial x_j'\partial y_k'}\frac{\partial^2 c_{\phi_0\phi_0}(\boldsymbol{x}',\boldsymbol{y}')}{\partial x_j'\partial y_k'}g_n(\boldsymbol{x},\boldsymbol{x}')g(\boldsymbol{y},\boldsymbol{y}')\,d\boldsymbol{x}'\,d\boldsymbol{y}'$$
$$+e^{-2\overline{Y}}\int_\Omega\int_\Omega c_{YY}(\boldsymbol{x}',\boldsymbol{y}')c_{ff}(\boldsymbol{x}',\boldsymbol{y}')g_n(\boldsymbol{x},\boldsymbol{x}')g(\boldsymbol{y},\boldsymbol{y}')\,d\boldsymbol{x}'\,d\boldsymbol{y}'$$

$$-e^{-\overline{Y}}\int_\Omega\int_\Omega\sum_{j=1}^{3}\frac{\partial c_{YY}(\boldsymbol{x}',\boldsymbol{y}')}{\partial x'_j}\frac{\partial c_{\phi_0 f}(\boldsymbol{x}',\boldsymbol{y}')}{\partial x'_j}g_n(\boldsymbol{x},\boldsymbol{x}')g(\boldsymbol{y},\boldsymbol{y}')\,d\boldsymbol{x}'\,d\boldsymbol{y}'$$

$$-e^{-\overline{Y}}\int_\Omega\int_\Omega\sum_{j=1}^{3}\frac{\partial c_{YY}(\boldsymbol{x}',\boldsymbol{y}')}{\partial y'_j}\frac{\partial c_{\phi_0 f}(\boldsymbol{y}',\boldsymbol{x}')}{\partial y'_j}g_n(\boldsymbol{x},\boldsymbol{x}')g(\boldsymbol{y},\boldsymbol{y}')\,d\boldsymbol{x}'\,d\boldsymbol{y}'$$

$$+e^{-\overline{Y}}\int_\Omega\int_\Gamma c^*(\boldsymbol{x}',\boldsymbol{y}')g_n(\boldsymbol{x},\boldsymbol{x}')g(\boldsymbol{y},\boldsymbol{y}')\,d\boldsymbol{x}'\,d\boldsymbol{y}'+e^{-\overline{Y}}\int_\Omega\int_\Gamma c^*(\boldsymbol{x}',\boldsymbol{y}')g(\boldsymbol{y},\boldsymbol{x}')g_n(\boldsymbol{x},\boldsymbol{y}')\,d\boldsymbol{x}'\,d\boldsymbol{y}'$$

$$-\int_\Omega\int_\Gamma c^{**}(\boldsymbol{x}',\boldsymbol{y}')g_n(\boldsymbol{x},\boldsymbol{x}')g(\boldsymbol{y},\boldsymbol{y}')\,d\boldsymbol{x}'\,d\boldsymbol{y}'-\int_\Omega\int_\Gamma c^{**}(\boldsymbol{x}',\boldsymbol{y}')g(\boldsymbol{y},\boldsymbol{x}')g_n(\boldsymbol{x},\boldsymbol{y}')\,d\boldsymbol{x}'\,d\boldsymbol{y}' \quad (65)$$

The above system can once again be solved using techniques described in Cheng & Lafe.[2] For problems with deterministic boundary condition and forcing function, the above system provides a complete solution in the first perturbation order. When random boundary conditions are involved, the mixed order covariances, $c_{\phi_0\phi_1}$, $c_{\phi_1\phi_0}$, $c_{q_0q_1}$, $c_{q_1q_0}$, $c_{\phi_0q_1}$, etc., also need to be evaluated. The integral equations can be constructed in the same way as the above.

References

[1] Gelhar, L., "Stochastic subsurface hydrology from theory to applications", Water Resour. Res., **22**, 135S-145S, 1986.

[2] Cheng, A.H-D. and Lafe, O.E., "Boundary element solution for stochastic groundwater flow: Random boundary condition and recharge," to appear in Water Resour. Res.

[3] Cheng, A.H-D., Abousleiman, Y. and Lafe, O.E., "Stochastic BEM for transient groundwater flow with stationary random boundary condition," *Computational Engineering with Boundary Elements, Vol. 1: Fluid and Potential Problems*, BETECH90, Univ. Delaware, eds. S. Grilli, C.A. Brebbia and A.H-D. Cheng, Comp. Mech. Publ., 157-165, 1990.

[4] Dagan, G., "Stochastic modeling of groundwater flow by unconditional and conditional probabilities. 1. Conditional simulation and the direct problem", Water Resour. Res., **18**, 813-833, 1982.

[5] Bakr, A., Gelhar, L.W., Gutjahr, A.L. and MacMillan, J.R., "Stochastic analysis of spatial variability in subsurface flows. 1. Comparison of one- and three-dimensional flows", Water Resour. Res., **14**, 263-271, 1978.

[6] Cheng, A.H-D., "Darcy's flow with variable permeability—a boundary integral solution," Water Resour. Res., **20**, 980-984, 1984.

[7] Cheng, A.H-D., "Heterogeneities in flows through porous media by the boundary element method", Chap. 6 in *Topics in Boundary Element Research, Vol. 4: Applications in Geomechanics*, ed. C.A. Brebbia, Springer-Verlag, 129-144, 1987.

[8] Lafe, O.E. and Cheng, A.H-D., "A perturbation boundary element code for groundwater flow in heterogeneous aquifer", Water Resour. Res., **23**, 1079-1084, 1987.

[9] Sobczyk, K., *Stochastic wave propagation*, Elsevier, 1985.

[10] Jaswon, M. and Symm, G.T., *Integral Equation Methods in Potential Theory and Elastostatics*, Academic Press, 1977.

[11] Cahan, B.D. and Lafe, O.E., "On the iterative boundary element method," *Computational Engineering with Boundary Elements, Vol. 2: Solid and Computational Problems*, BETECH90, Univ. Delaware, eds. A.H-D. Cheng, C.A. Brebbia and S. Grilli, Comp. Mech. Pub., 367-375, 1990.

A New Integration Algorithm for Nearly Singular BIE Kernels

T. A. Cruse
Vanderbilt University
Nashville, Tennessee 37235 USA

R. Aithal, Consultant
Southwest Research Institute
San Antonio, Texas 78228 USA

Overview

The boundary integral equation for the elasticity problem is written in terms of the boundary tractions t_j and boundary displacements u_j in the usual manner [1]

$$C_{ij}u_j(P) + \int\int_{<S>} T_{ij}(P,Q)u_j(Q)dS(Q) = \int\int_{<S>} U_{ij}(P,Q)t_j(Q)dS(Q) \tag{1}$$

where $<S(Q)>$ denotes the principal value of the integrals on the boundary surface. The points Q(y) and P(x) respectively denote the integration point and the source point, corresponding to the point of application of the point load influence function. The tractions and displacements for the point load solution are written as $T_{ij}(P,Q)$ and $U_{ij}(P,Q)$, respectively. The C_{ij} matrix corresponds to the value of the jump in the first integral as the interior displacement evaluation point p(x) is taken to the boundary point P(x).

Following the usual procedures [2][1] for a numerical quadrature of the boundary integral equation (BIE), we replace the actual surface by a set of boundary elements, ΔS_n, over which the boundary shape and boundary data are replaced by the usual quadratic shape functions and nodal values of the variables

$$x_i, u_i, t_i = M^{\alpha}(\xi_1,\xi_2)x_i^{\alpha}, u_i^{\alpha}, t_i^{\alpha} \tag{2}$$

The superscript α has the range of six or eight depending on whether the boundary element is a triangle or a quadrilateral.

Each of the integrals in Eq (1) are represented by sums of integrals over each boundary element, which is illustrated for the traction kernel, as follows

$$\sum \int\int_{\Delta S_n} T_{ij}(P,Q)u_j(Q)dS(Q) = \sum u_j^{\alpha} \int\int_{\Delta A_n} T_{ij}(P,Q)M^{\alpha}(\xi)\; J(\xi)dA(\xi) \tag{3}$$

---

1 The integration algorithm in this reference is adopted herein for illustration and numerical compari The comments developed herein apply to the totality of Gaussian integration strategies.

where $J(\xi)$ is the Jacobian of the transformation to the reference (unit) area, A. The principal value notation will be imposed on any boundary element implementation of Eq (3) for which the boundary element contains the singular point P(x).

Eq (3) is normally integrated using Gaussian quadrature in boundary element codes. Unfortunately, the use of Gaussian quadrature does not produce exact results, since the integrated functions are singular to varying degrees depending on the kernel function and on the location of P(x) relative to the element. As a result, various numerical integration schemes have been used over the years in order to control the error of the numerical quadratures of the boundary integrals. Element subdivision, polar coordinates, and very high-order Gaussian integrations are generally used in these schemes.

The convergence characteristics of the Gaussian integration schemes is generally very poor for problems with graded meshes or thin sections, such that the source point $P \rightarrow Q$ on the integration element. We take the distance to be close in the sense of distance of P(x) relative to the size of the boundary element. Codes with specific error control algorithms generally fail to converge within reasonable distances, while those codes without error control yield very inaccurate results. Such problems as fracture mechanics modeling and geometries with one thin dimension fall into this challenging category.

Recently, some very powerful concepts have been introduced in BEM implementations for the potential theory problem that eliminate the singular character of the kernels for Gaussian integration at $P = Q$. The approach taken by Lean and Wexler [3] is to regularize the singularity for $P = Q$ through particular coordinate mappings that produce the desired, singularity-cancelling character in the mapping Jacobian. The modified mapping is applied to a new kernel, which is used to regularize the original BIE kernel. The numerical results were very encouraging for the $P = Q$ case and indicated the important role of higher-order expansions of the mapping Jacobian.

A second approach published at the same time [4] is fundamentally different. The terms in the singular integrals for $P = Q$ are individually expanded in a Taylor series manner. The leading singular terms are integrated exactly. Gaussian integration is applied only to terms which have been fully regularized. This approach has been recently extended to the elasticity case for $P = Q$ [5]. The Taylor series expansion approach is taken in the current work for $P \neq Q$.

Integration Algorithm

The proposed algorithm reduces the singular kernel functions to regular functions, for which the Gaussian integrations yield very accurate results for low orders of integration. The algorithm is based on Taylor series expansions of all terms in the integrands of Eq (3) such that explicit integrations of singular and weakly singular terms are performed analytically, and that numerical quadrature is only performed on the fully regularized terms.

The first step in the development of the new algorithm is to project the curved boundary element onto a flat plane. The flat plane is taken to be tangent to the curved boundary element (as distinct from the actual surface) at one of the boundary element nodes, taken to be Q_0, which is the node on the boundary element closest to the source point, P(**x**). The mapping from the boundary element to the flat plane is given by

$$dS(Q) = J(Q,Q')dS(Q') = J'dS' \tag{4}$$

where the prime denotes the flat projection plane.

The new algorithm applied herein is based on integrations of the singular terms in a local coordinate system, shown in Fig. 1. The ζ_1, ζ_2 coordinates denote points in the plane containing the reference field point, Q_0, and the tangent to the boundary element at that point. The normal to the plane is easily computed from the Jacobian elements, evaluated at Q_0. The ζ_1 coordinate is taken for convenience to be aligned with the isoparametric integration direction given by ξ_1. The origin of the local coordinate system is taken to be the projection of the source point P(**x**) onto the ζ_1-ζ_2 plane. Coordinates of the integration point Q(**y**) are given in this coordinate system by

$$\begin{aligned} \zeta_1 &= \rho\cos(\theta) \\ \zeta_2 &= \rho\sin(\theta) \\ \zeta_3 &= \Delta(\zeta_1,\zeta_2) \end{aligned} \tag{5}$$

This coordinate system was first used by Cruse [6] for exact integration of the BIE formulation for flat boundary elements and linear data interpolations. The earlier analytical integrations are applied in the current work.

We begin the regularization process by expanding the Jacobian of the transformation in terms of the ζ_1-ζ_2 directions, relative to the value of the Jacobian at the reference point Q_0. The first-order expansion terms are given by

$$J'(\xi) = J'(Q_0) + \frac{\partial J'}{\partial \zeta_1}\Big|_{Q_0}\Delta\zeta_1 + \frac{\partial J'}{\partial \zeta_2}\Big|_{Q_0}\Delta\zeta_2 + O(\Delta^2) = L_0(J') + O(\Delta^2) \tag{6}$$

The variable Δ is the distance of the integration point Q(**x**) from the flat integration surface, that is $\zeta_3(Q)$. For quadratic isoparametric elements, the value of Δ is proportional to δ^2, where δ is the projected distance of the integration point from the reference point, Q_0. Thus, we see that the explicit expansion is to terms of order δ^2.

The boundary data for the quadratic isoparametric problem is given by Eq 2. The linear part of the boundary data is given by the following form, illustrated for the boundary displacements

$$u_i(Q) - u_i(Q_0) + \frac{\partial u_i}{\partial \zeta_1}\Big|_{Q_0} \Delta\zeta_1 + \frac{\partial u_i}{\partial \zeta_2}\Big|_{Q_0} \Delta\zeta_2 + O(\delta^2) - L_0(u_i) + O(\delta^2) \qquad (7)$$

The use of the linear expansion of the displacements has been previously used to regularize the traction integral for the linear element case [7]. This earlier regularization led directly to the explicit derivation of the surface stress term in BIE analysis. The difference term for the quadratic variation is shown to be of order Δ, or δ^2.

If we now expand the traction (or displacement) kernel in Eq 1 with respect to the integration point in the mapped plane A,' rather than the mapped boundary element area, A, the following is obtained

$$T_{ij}(P,Q) - T_{ij}(P,Q') + O\left(\frac{\Delta}{r(P,Q)}\right) \qquad (8)$$

where the truncation term is of the order of Δ divided by the distance r(P,Q). Now substitute Eq 6-8 into Eq 3 for the flat integration element, A,' to obtain

$$\Delta I_n - \int_\theta \int_{\rho(\theta)} T_{ij}(P,Q') L_0(u_i)\, L_0(J')\, \rho\, d\rho\, d\theta \qquad (9)$$

The $\rho(\theta)$ integral in Eq 8 can be integrated analytically, using the approach in [7]. When the element has straight sides in the projected plane, the integral with θ may also be done in closed form.[2]

A major point to be made is that the singular nature of the kernel functions is entirely removed by the radial integration process. The remaining integral with respect to θ is totally regular. In the work reported herein, the θ integration is computed numerically.

The original integral, Eq 2, may be fully regularized by subtracting Eq 9. The difference integral is of the following form

$$\left\{\int\int_{\Delta S_n}\right\} - \int\int_{\Delta A}\{T_{ij}(P,Q)u_j(Q)J dA - T_{ij}(P,Q')L_0(u_i)L_0(J')J'dA'\} \qquad (10)$$

2 The forms of these integrals and the analytical results are available from the first author.

such that Eq 3 is given by the sum of Eq 10 and Eq 9. The fact that Eq 10 is fully regular is seen by carrying out the expansion of all the terms with the following result

$$\Delta\{\int\int_{\Delta S}\} - \int\int_{\Delta A'} [L(u_i(\xi_0))O\,(\frac{\Delta}{r})J(\xi_0) + T_{ij}(P,Q')\,O(\Delta)J(\xi) + T_{ij}(P,Q')L(u_i(\xi_0))O(\Delta) + O(\Delta^2)]\,dA' \tag{11}$$

The result in Eq 11 is of the order of Δ. As $P \rightarrow Q$, the terms are totally regular, so long as the polar form [2] of the Gaussian integration is used to cancel the r(P,Q) in the denominator of the term from the kernel function expansion. The limit for $P = Q$ also exists, and is regular. This result is consistent with that of [5]. Thus, Eq 11 may be integrated with standard, low-order Gaussian integration. No element subdivisions or other elaborate error correction system is required. As will be shown in the examples, low order Gaussian integration is sufficient to produce nearly exact results in most cases. While the above discussion is for the traction kernel, the extension to the displacement kernel results in the same conclusions.

Numerical Results

The numerical evaluation of the above algorithm has been made by computing the traction and displacement kernels for a single element with the source point P(x) approaching the element along a normal to the corner or midside node. The element is taken to be flat as well as curved. The integral results for the constant, linear and quadratic boundary data cases are also computed. The data shown are taken only from the U_{11} and T_{11} terms in the kernels, but these are representative of all other terms. A more comprehensive set of combinations of source point and integration element would add nothing to the conclusions we can draw from these numerical results.

Figure 2 and 3 show the numerical integration results for the U_{11} and T_{11} terms for a flat element ten units square, considering the boundary data to be constant over the element. The legend indicates the distance of the source point from the element. The new algorithm described above gives constant values of the integral results versus the integration order, since Eq 10 for this case gives a zero result, even with the source point within 0.3% distance from the element. These figures simply illustrate the fact that the use of the Gaussian integration algorithm requires significant integration order for accuracy as $P \rightarrow Q$, especially for the traction kernel (an expected result!).

The results in Figure 4 and 5 are obtained by letting the integration element be curved into a cylindrical shape, with the displacement of the midside node relative to the corners given by Delta. Thus, in this first case the element is covering more than a 135° arc. The source point is taken from a distance of 10% of the element size from the element, but along the normal to a corner node in Figure 4, and a midside node in Figure 5. Clearly, the new algorithm far outperforms Gaussian integration.

Figure 6 is a more realistic case of element curvature. The element curvature is roughly equivalent to four elements on a 90° arc. The distance of the source point from the element is again 10% of the element size. The standard Gaussian integration scheme is not much better than before, while the new algorithm provides very accurate answers with very low integration orders. The Gaussian integration algorithm used in these examples retains the square array of Gauss points which is likely leading to the oscillatory behavior of all of these results.

Figure 7 considers the case of linear boundary data and a flat boundary element. Again, the new algorithm gave fully converged results, independent of Gaussian integration order, as expected. The standard Gaussian system is seen to be very slowly convergent, for the source point distances selected.

Figure 8 is for the case of quadratic boundary data and compares the new algorithm directly with the standard Gaussian integration scheme. Even for points 5% of the distance from the element, the standard scheme requires nearly the full complement of Gauss points to converge. The new algorithm converges with excellent accuracy within a 4x4 integration order.

Conclusions

The results confirm that the use of a semi-analytical approach to integrating the boundary integral equation kernels eliminates the need for higher order Gaussian integration, element subdivision, and elaborate error control schemes. Continuing work places emphasis now on developing a fast implementation of the new algorithm, as well as the logic for mixing the new algorithm and the standard algorithm, for the highest possible code efficiency. Application of the new algorithm to problems with steeply graded meshes is also planned.

References

1. Cruse, T. A., **Boundary Element Analysis in Computational Fracture Mechanics**, Kluwer Academic Publishers, The Netherlands (1988).

2. F. J. Rizzo and D. J. Shippy, An Advanced Boundary Integral Equation Method for Three-Dimensional Thermoelasticity, **Int. J. Num. Meth. Eng. 11**, 1753-1768 (1977).

3. Meng H. Lean and A. Wexler, Accurate Numerical Integration of Singular Boundary Element Kernels over Boundaries with Curvature, **Int. J. Numer. Meth. Eng., 21**, 211-228 (1985).

4. M. H. Aliabadi, W. S. Hall, and T. G. Phemister, Taylor Expansions for Singular Kernels in the Boundary Element Method, **Int. J. Numer. Meth. Eng., 21**, 2221-2236 (1985).

5. M. Guiggiani and A. Gigante, A General Algorithm for Multidimensional Cauchy Principal Value Integrals in the Boundary Element Method, submitted for publication.

6. T. A. Cruse, An Improved Boundary-Integral Equation Method for Three Dimensional Elastic Stress Analysis, **Comp. & Struct., 4**, 741-754 (1974).

7. T. A. Cruse, Three-Dimensional Elastic Stress Analysis of a Fracture Specimen with an Edge Crack, **Int. J. Fract. Mech., 7**, 1-15 (1971).

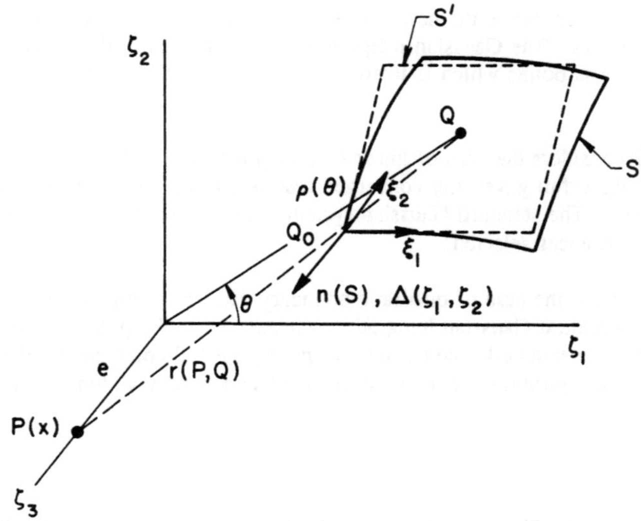

Figure 1: Transformation of Isoparametric Element Area to Flat Integration Element

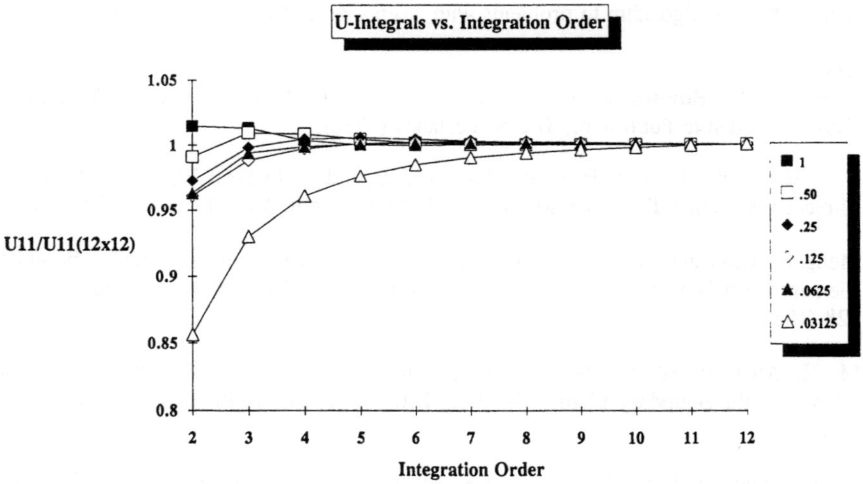

Figure 2: Gaussian Integration Results for U_{11} Kernel Function: Element Size = 10x10; Variable Distance of P(x) from Element

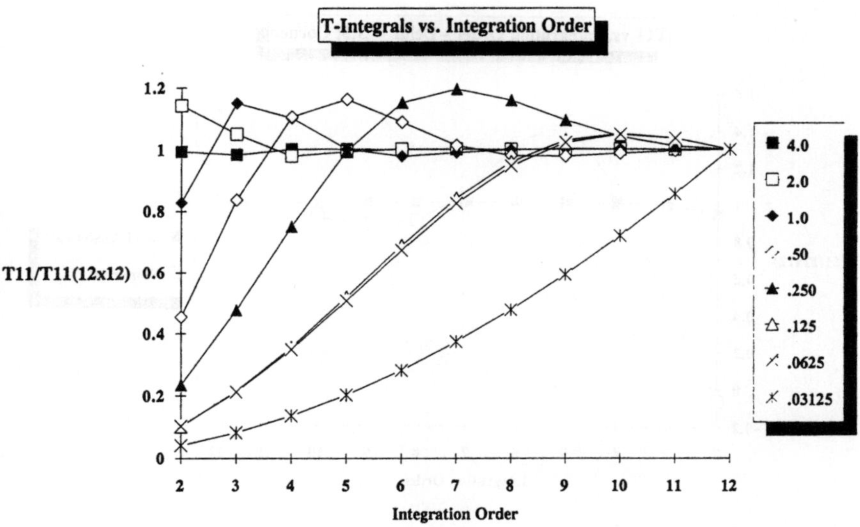

Figure 3: Gaussian Integration Results for T_{11} Kernel: Element Size = 10x10; Variable Distance of P(x) from Element

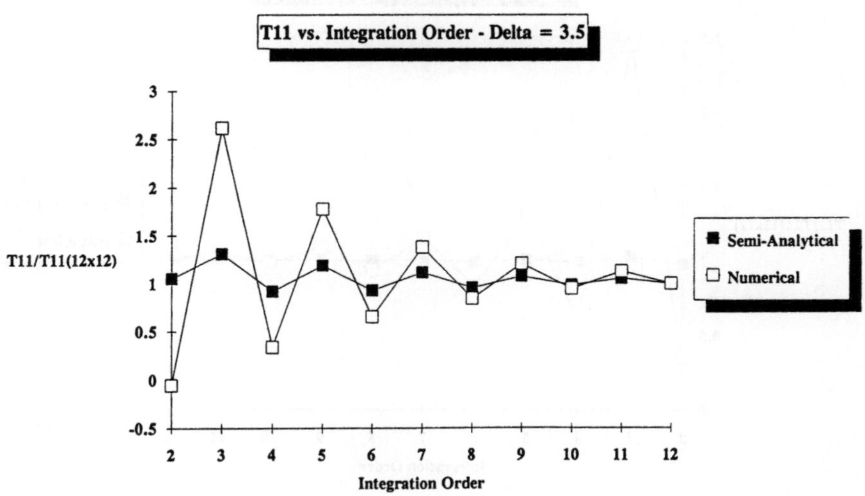

Figure 4: Integration of T_{11} for Curved Element (Delta/Length = 0.35); P(x) Distance from Mid Side Node = 1

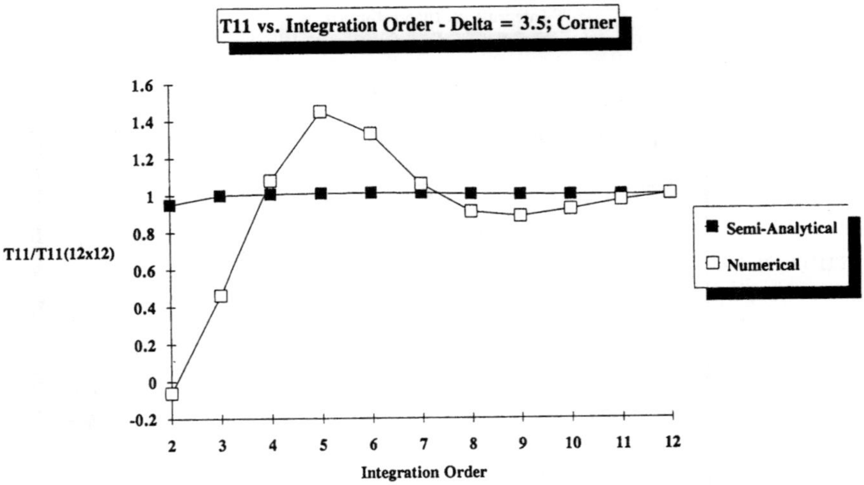

Figure 5: Integration of T_{11} Kernel for Cuved Element (Delta/Length = 0.35); P(x) Distance from Corner Node = 1

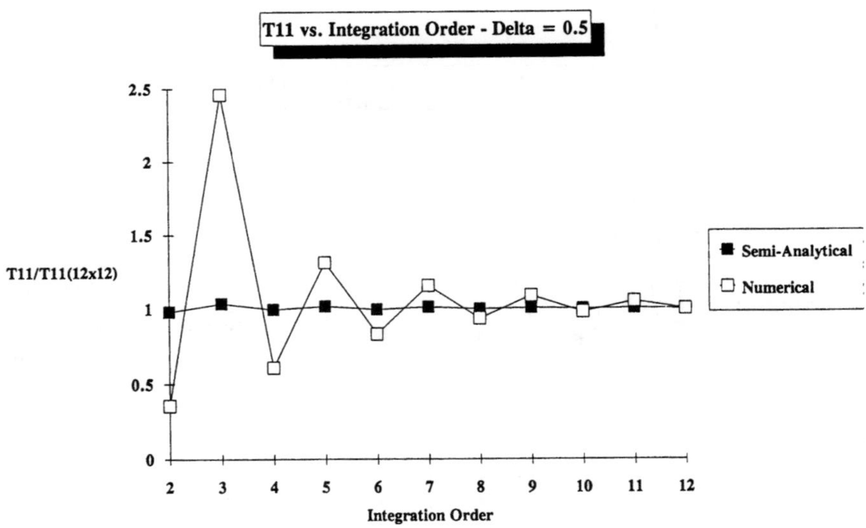

Figure 6: Integration of T_{11} for Curved Element (Delta/Length = 0.5); P(x) Distance from Element Midside Node = 1

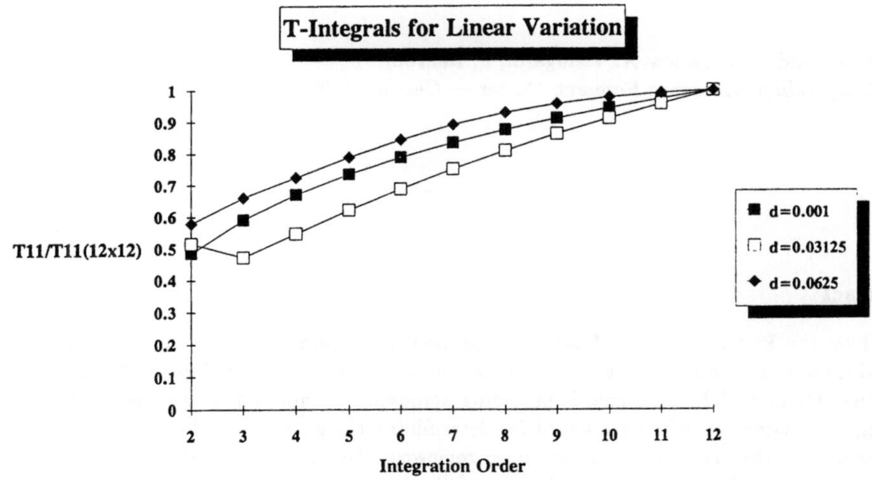

Figure 7: Gaussian Integration Results for Linear Variation Boundary Terms: 10x10 Element Size and Variable Distance of P(x) from Element

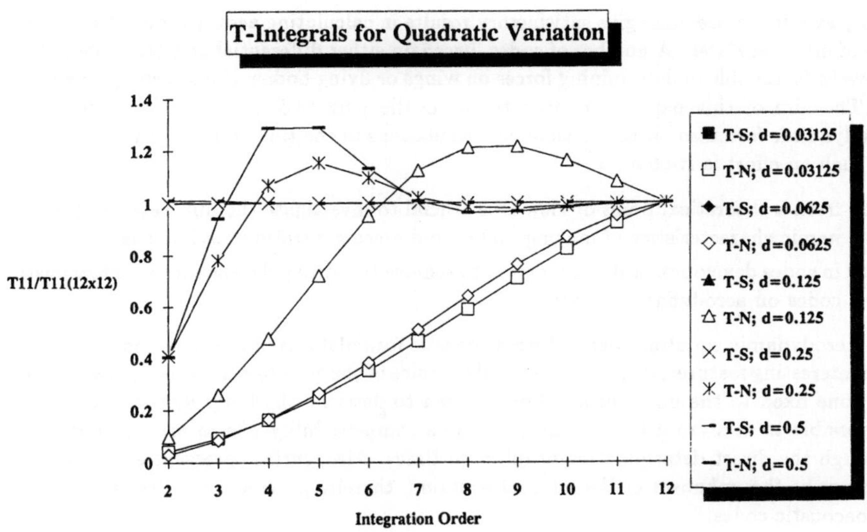

Figure 8: Integration Results for Quadratic Boundary Data Term: 10x10 Element Size and Variable Distance of P(x) from Element

A Contribution to Lifting Surfaces Aerodynamics Based on Time Domain Aeroacoustics

E. De Bernardis, D. Tarica, A. Visingardi, P. Renzoni
C.I.R.A., Italian Aerospace Research Center — Capua (Italy)

Abstract

Based on the Ffowcs Williams-Hawkings equation, a boundary integral method is presented for the calculation of the aerodynamic loading on thin lifting surfaces in linearized compressible flow. The final goal is providing aerodynamic inputs to aeroacoustic codes through the same procedure employed for determining the sound field. Several methods proposed over the last few years have been reviewed; then an aerodynamic formula has been chosen, which was derived by Milliken following an acoustic formulation proposed by Farassat. The method has been improved by adding some correcting terms, as a first step to turn the original lifting surface approach into an actual BEM formulation.

Introduction

Many existing procedures give satisfactory results in calculating aerodynamic characteristics of lifting surfaces. A number of codes, based on either differential or integral methods, prove to be reliable in determining forces on wings or flying bodies of more complex shape.

The aim of this paper is related to the challenging task of developing new linear aerodynamic formulas based on acoustic formulations in the time domain. the motivation for such an effort is twofold:

i) from a theoretical point of view it is crucial to investigate the link between aerodynamic characteristics of moving bodies and noise generating mechanisms;

ii) in codes development it is important to remove the strong dependence of aeroacoustic codes on aerodynamic inputs.

Aerodynamic equations derived from acoustic formulations in the time domain exhibit two interesting features, due to the particular choice of the variables involved. A rest frame, *i.e.* one fixed to the undisturbed fluid, is used to describe the body motion: this make it possible to analyze arbitrary motions while compressibility effects are accounted for through the direct determination of retarded times. Also, surface pressure automatically appears as the unknown in the integral equation, thereby providing the desired input to aeroacoustic codes.

In view of the above considerations several methods for aerodynamic calculations have recently been developed, based on aeroacoustic formulas. Integral expressions representing the formal solution to the Ffowcs Williams-Hawkings equation [1] provide the link between the linear acoustic pressure field generated by bodies in motion and their aeodynamic characteristics.

A great deal of theoretical work has been done by F. Farassat from 1975 to 1985. After developing several forms of solutions to the linearized version of the Ffowcs Williams-Hawkings equation, he has started a research effort aimed at highlighting the close relationship between some well-known aerodynamic formulas [2] and integral equations that might be derived from his acoustic formulations [3,4]. He also developed new regularization techniques for singular integrals, based on generalized function theory [5].

It is worth mentioning the research work conducted on a similar issue, over the same decade, by D.B. Hanson [6,7]. He integrated in the frequency domain the Goldstein [8] version of the Ffowcs Williams-Hawkings equation, and followed a more traditional approach to state the link between acoustics and unsteady aerodynamics.

Based on Farassat's early work, several papers have been published which attempt to exploit the advantages of the direct calculation of retarded times in developing numerical procedures for compressible, arbitrary motion aerodynamics. Long [9] derived a subsonic integral equation from the so-called Farassat's formulation 1-A [3]. Due to the structure of that formula, the choice of the pressure as the unknown in the resulting integral equation makes it impossible to properly account for the wake contribution.

This drawback could be overcome by employing a suitable basic principle of fluid motion in conjunction with the integral formula based on the acoustic pressure. Thus, using the linearized form of the momentum equation, Milliken [10] formulated the aerodynamic problem through an integral equation derived from Farassat's formulation 1. In Milliken's work the application of this formula was restricted to a plane wing in steady rectilinear motion. The wing surface was approximated by the mean chord surface and a numerical technique was obtained that proved to be equivalent to the Vortex Lattice Method. The procedure developed by Milliken was then refined by Farassat and Myers [11], who applied it to propellers. Finally Long and Watts [12] used a similar mathematical model in an attempt to deal with arbitrarily shaped bodies in arbitrary motion.

Two more comprehensive works have recently been presented on the subject of relating aerodynamics and acoustics. Brandão [13] has laid down a more general form of the Ffowcs Williams-Hawkings equation, thereby deriving aerodynamic integral equations for different flight speed ranges: he also developed a new regularization technique in order to treat singular integrals [14]. A different approach has been followed by Lee [15]: he proposed to relate aerodynamic and acoustic fields of moving bodies through the solution of a potential based boundary integral equation derived by an inhomogeneous wave equation. Interesting results obtained with this formulation have recently been presented by Lee and Yang [16].

On the basis of the research work outlined above we aim to develop a numerical procedure to be included within the framework of a code for the prediction of the noise generated by moving bodies. Since Farassat's formulation 1 is employed in our acoustic calculations, the methods proposed by Milliken [10] and Long and Watts [12] seem to establish a suitable ground for this task. The latter method provide an accurate modelling of the mathematical problem: this allows a more general applicability, but the size and cost of the computational work make it difficult to employ this technique as a part of an aeroacoustic code. On the contrary the Milliken formulation is based mostly on analytical integration in its application to steady body motion. Thus it has been selected as a reference method in this work, and corrections have been introduced in order to apply the procedure to more complex cases, while maintaining its computational burden within reasonable limits.

Milliken's Lifting Surface Method

A form of solution to the linearized Ffowcs Williams-Hawkings equation, known as Farassat's formulation 1 [3], is taken as the starting point in Milliken's analysis.

$$
\begin{aligned}
4\pi\tilde{p}(\mathbf{x},t) &= \frac{1}{c_0}\frac{\partial}{\partial t}\int_{S^*}\left\{\frac{\rho_0 c_0 v_n + \tilde{p}\cos\theta}{r|1-m_r|}\right\}_{\tau=\tau^*} dS \\
&+\int_{S^*}\left\{\frac{\tilde{p}\cos\theta}{r^2|1-m_r|}\right\}_{\tau=\tau^*} dS
\end{aligned} \tag{1}
$$

This describes the acoustic pressure generated at the field point $\mathbf{x}$ and time t by the motion of a body. In the above equation $\tilde{p} = p - p_0$ represents the pressure perturbation with respect to the value at rest, both at a field point and on the surface S of the moving body; S^* is the region of the body surface which contributes to $\tilde{p}(\mathbf{x},t)$ at the emission time $\tau = \tau^*$ (where integrands are to be evaluated), and it coincides with the entire surface S for fully subsonic body motion; ρ_0 and c_0 are the density and the speed of sound, respectively, for the fluid at rest. Denoting by $\mathbf{v}$ the local absolute velocity of the body surface, $v_n = \mathbf{v}\cdot\mathbf{n}$ is the local normal velocity ($\mathbf{n}$ being the local unit normal vector to S), while $m_r = \mathbf{m}\cdot\hat{\mathbf{r}}/c_0$ is the Mach number in the radiation direction, with $\mathbf{m} = \mathbf{v}/c_0$ the Mach vector and $\hat{\mathbf{r}} = \mathbf{r}/r$ the unit vector in the (instantaneous) source-observer direction; finally, θ is the angle between $\mathbf{n}$ and $\hat{\mathbf{r}}$.

Using a thin wing approximation, equation (1) is written on the mean chord surface $S_{m.c.}$; thus, referring to fully subsonic motion (resulting in a linear flow field) one has:

$$
\begin{aligned}
4\pi\tilde{p}(\mathbf{x},t) &= \frac{1}{c_0}\frac{\partial}{\partial t}\int_{S_{m.c.}}\left\{\frac{2\rho_0 c_0 \bar{v}_n - [p]\cos\theta}{r(1-m_r)}\right\}_{\tau=\tau^*} dS \\
&-\int_{S_{m.c.}}\left\{\frac{[p]\cos\theta}{r^2(1-m_r)}\right\}_{\tau=\tau^*} dS
\end{aligned} \tag{2}
$$

where $\bar{v}_n$ represents the averaged normal velocity of the lower and upper faces of the wing surface, and $[p]$ the pressure jump between them.

The first attempt of deriving an aerodynamic integral equation from an acoustic time domain formulation was carried out by Long [3] using a slightly different formula. In that case, the appearance of the pressure as the only field quantity in the acoustic formula leads to completely disregard some effects of the subsonic lifting body motion on the structure of the related flow field.

In order to obtain a more refined model, equation (2) is then combined with the linearized, inviscid momentum equation:

$$
\rho_0\phi(\mathbf{x},t) = -\int_{-\infty}^{t}\tilde{p}(\mathbf{x},t')dt'
$$

where ϕ is the fluid velocity potential function, and the following expression is thus obtained:

$$
\begin{aligned}
4\pi\rho_0\phi(\mathbf{x},t) &= -\frac{1}{c_0}\int_{S_{m.c.}}\left\{\frac{2\rho_0 c_0 \bar{v}_n - [p]\cos\theta}{r(1-m_r)}\right\}_{\tau=\tau^*} dS \\
&+\int_{-\infty}^{t}\int_{S_{m.c.}}\left\{\frac{[p]\cos\theta}{r^2(1-m_r)}\right\}_{\tau=\tau^{*\prime}} dS dt'
\end{aligned} \tag{3}
$$

In the above equation $\tau^{*'}$ represents the emission time of a signal received by the observer at t'. The t'-integration accounts for the wake produced by the body throughout the history of its motion.

According to this fully linearized analysis, it is possible to define ϕ_T and ϕ_L as the velocity potential distributions corresponding to thickness and lifting effects respectively; equation (3) can thus be split up as follows:

$$4\pi\rho_0\phi_T(\mathbf{x},t) = -\frac{2}{c_0}\int_{S_{m.c.}}\left\{\frac{\rho_0 c_0 \bar{v}_n}{r(1-m_r)}\right\}_{\tau=\tau^*} dS \tag{4}$$

$$\begin{aligned} 4\pi\rho_0\phi_L(\mathbf{x},t) &= \frac{1}{c_0}\int_{S_{m.c.}}\left\{\frac{[p]\cos\theta}{r(1-m_r)}\right\}_{\tau=\tau^*} dS \\ &+ \int_{-\infty}^{t}\int_{S_{m.c.}}\left\{\frac{[p]\cos\theta}{r^2(1-m_r)}\right\}_{\tau=\tau^{*'}} dS\,dt' \end{aligned} \tag{5}$$

Only the lifting problem is considered by Milliken. Taking the derivative of equation (5) normal to the mean chord surface, while letting the observer approach the surface itself, the following singular integro-differential equation is obtained:

$$\begin{aligned} 4\pi\rho_0 v_n(\mathbf{x},t) &= \frac{1}{c_0}\frac{\partial}{\partial n}\int_{S_{m.c.}}\left\{\frac{[p]\cos\theta}{r(1-m_r)}\right\}_{\tau=\tau^*} dS \\ &+ \frac{\partial}{\partial n}\int_{-\infty}^{t}\int_{S_{m.c.}}\left\{\frac{[p]\cos\theta}{r^2(1-m_r)}\right\}_{\tau=\tau^{*'}} dS\,dt' \end{aligned} \tag{6}$$

The numerical solution of the problem is carried out by collocation method. The mean chord surface is discretized into a set of N zeroth order rectangular boundary elements, so that equation (6) is turned into the following algebraic system:

$$4\pi\rho_0 v_n(\mathbf{x}_i,t) = \sum_{j=1}^{N}\left\{[p]_j\frac{\partial}{\partial n}(IB_{ij}+IW_{ij})\right\} \qquad i=1,2,\ldots,N \tag{7}$$

Here IB_{ij} and IW_{ij} are expressed by:

$$IB_{ij} = \frac{1}{c_0}\int_{S_j}\left\{\left(\frac{\cos\theta}{r(1-m_r)}\right)_{ij}\right\}_{\tau=\tau^*} dS \tag{8}$$

$$IW_{ij} = \int_{-\infty}^{t}\int_{S_j}\left\{\left(\frac{\cos\theta}{r^2(1-m_r)}\right)_{ij}\right\}_{\tau=\tau^{*'}} dS\,dt' \tag{9}$$

S_j being the surface of the j-th panel, *i.e.* the j-th portion of the mean chord surface.

The tasks of integrating singular kernels on the panels and taking their normal derivative at the collocation points are simultaneously achieved by Milliken without exploiting any well established regularization technique. The above integrals are first analytically evaluated with the observer a small distance above the surface; then the limit of those (convergent) integrals is taken as the observer-surface distance approaches zero. Then, for the j-th panel the expression on the right-hand side of equation (7) one gets:

$$\frac{\partial}{\partial n}(IB_{ij}+IW_{ij}) = \lim_{\varepsilon\to 0}\frac{\partial}{\partial\varepsilon}(IB_{ij}+IW_{ij}) \tag{10}$$

In the above equation ε is the parameter representing the fictitious distance between the observer and the surface.

Corrections to Milliken's formula

The procedure outlined above has been applied by Milliken to a finite thin wing, and it proved to be equivalent to the Vortex Lattice Method.

However, assuming the mean chord surface to lie on a plane leads to difficulties at the leading edge. Thus correction terms have been added to Milliken's formula in order to improve the behaviour of the solution near this critical point, regarding two aspects of Milliken's regularization procedure:

i) starting with the observer a small distance ε above the plane of the mean chord surface forces the numerator in the integrand of equations (8) and (9) to have the expression:

$$\cos\theta = \frac{\varepsilon}{r}$$

ii) taking the ε-derivative and letting ε go to zero can not even roughly approximate the normal derivative near the leading edge.

In the attempt of achieving better results in the vicinity of the leading edge, while preserving the simplicity of Milliken's formulation, the following corrections have been added in our procedure:

i) full expressions for the actual surface unit normal vector and for the radiation vector are used in equations (8) and (9), leading to the representation

$$\cos\theta = \frac{(x_i^3 - x_j^3) + (x_i^1 - x_j^1)\cot\alpha}{r}$$

where terms in brackets represent components of the radiation vector, while α is the angle between the local normal vector to the (actual) surface of the wing and the velocity of the mean flow. The first term corresponds to the one obtained by Milliken, while the second one provides further integrals IB_{ij}^c and IW_{ij}^c which are evaluated analytically to get the correcting terms;

ii) these latter are treated following a procedure similar to the one proposed by Milliken, but accounting for the actual direction of the unit normal vector. This is in fact nearly parallel to the mean chord surface in the neighbourhood of the leading edge, thereby requiring a parameter like ε to be defined in the flow direction. This is first used as a differentiation variable in taking the derivative of IB_{ij}^c and IW_{ij}^c; then, letting it approach zero, the normal derivative at the leading edge is approximated.

A corrected formula is obtained without increasing computational costs. Like the Milliken formula it can be applied to thin finite wings, achieving results of reasonable accuracy even for compressible flow.

Numerical results

This section presents results for a high aspect ratio wing moving in steady rectilinear motion. This simple case has been chosen in order to analyze the fundamental properties of Milliken's lifting surface formulation and of the corrections that have been proposed.

The initial numerical experiments found instabilities in the solution of equation (7) with and without corrections. Milliken states that the instabilities are strongly dependent on the control point location. Figure 1 shows the oscillations found in the pressure distribution at midspan of a high aspect ratio wing ($AR = 10$) with NACA 0012 airfoils. It was found that, for the case considered, a control point position of 95% chord gave the best results. This value is in accordance with Milliken's study which uses the Galerkin method to solve the linear system resulting from equation (7). However, the present calculation using the collocation method are less sensitive to the position of the control point. In fact good results are obtained even at 70% chord. It is expected that the control point location will change when considering more complex geometries in more general motions.

In this formulation compressibility is accounted for via retarded time (the time between emission and reception of a signal). The effect of compressibility on the pressure distribution is shown in Figure 2 for the corrected formulation. Figure 3 shows the effect of compressibility calculated from the full potential equation using a finite difference method. A comparison of Figures 2 and 3 shows that compressibility is properly accounted for in this formulation and is consistent with the Prandtl-Glauert rule.

The effect of the corrections made to Milliken's formulation can be seen in Figure 4 for incompressible flow and in Figures 5 and 6 for compressible flow. The corrective terms remove the singularity at the leading edge. Both the incompressible and the compressible cases show the leading edge behaviour of a typical airfoil section. The corrected formulation gives a higher ΔC_p peak than the full potential calculation but the overall behaviour is captured correctly. The discrepancies with the full potential calculation are due to the fact that in the corrected formulation the real airfoil is still represented by its mean chord surface.

More extensive numerical tests are needed in order to properly account for the effect of the corrective terms.

Concluding remarks

This paper presents a boundary integral method for the calculation of the aerodynamic loading on thin lifting surfaces in linearized compressible flow which is based on an acoustic formulation proposed by Farassat. The first application of this formulation was carried out by Milliken for a thin wing in steady rectilinear subsonic motion.

We have taken an initial step in an effort to apply his formulation to more realistic configurations. The corrections we have proposed have shown to effectively remove the leading edge singularity. More extensive tests are needed in order to determine the extent of the leading edge region where the corrective terms are important.

The major objective of developing a simple and cost-effective numerical procedure for the calculation of aerodynamic loads on thin lifting surfaces to be used in aeroacoustic codes has been achieved. The initial results obtained by correcting Milliken's formula are encouraging and can be easily integrated in our aeroacoustic code, also based on Farassat's formulation 1. We expect that this will greatly reduce the cost of developing the aeroacoustic code.

Future efforts will be devoted at developing a full boundary element formulation from Farassat's acoustic formulation. The results we have presented clearly indicate that the representation of a real airfoil through its mean chord surface is not entirely satisfactory and thus the actual surface must be considered.

It is expected that the full BEM formulation will require the numerical evaluation of the integrals although it is hoped that some analytic evaluations will be possible in order to keep the computational burden within reasonable limits.

References

[1] Ffowcs Williams, J.E., Hawkings, D.L.: Sound generation by turbulence and surfaces in arbitrary motion. Phil. Trans. of the Royal Society A264 (1969) 321–342.

[2] Ashley, H., Landahl, M.T.: Aerodynamics of wings and bodies. Reading, MA: Addison-Wesley 1965.

[3] Farassat, F.: Advanced theoretical treatment of propeller noise. Von Karman Institute Lecture Series 81–82/10 (1982), Brussels (Belgium).

[4] Farassat, F.: A new aerodynamic integral equation based on an acoustic formula in the time domain. AIAA J. 22 (1984) 1337–1340.

[5] Farassat, F.: Discontinuities in aerodynamics and aeroacoustics: the concept and applications of generalized derivatives. J. of Sound and Vibration 55 (1977) 165–193.

[6] Hanson, D.B.: Compressible helicoidal surface theory for propeller aerodynamics and noise. AIAA J. 21 (1983) 881–888.

[7] Hanson, D.B.: Compressible lifting surface theory for propeller performance calculation. J. of Aircraft 22 (1985) 609–617.

[8] Goldstein, M.E.: Aeroacoustics. New York: McGraw-Hill 1976.

[9] Long, L.N.: The compressible aerodynamics of rotating blades based on an acoustic formulation. NASA TP-2197 (1983).

[10] Milliken, R.L.: A new lifting surface method: an acoustic approach. M.Sc. Thesis, School of Engineering and Applied Sciences, George Washington University, Washington, DC, 1986.

[11] Farassat, F., Myers, M.K.: Aerodynamics via acoustics: application of acoustic formulas for aerodynamic calculation. AIAA Paper 86-1877 (1986).

[12] Long, L.N., Watts, G.A.: Arbitrary motion aerodynamics using an aeroacoustic approach. AIAA J. 25 (1987) 1442–1448.

[13] Brandão, M.P.: On the aeroacoustics, aerodynamics and aeroelasticity of lifting surfaces. Ph.D. Thesis, Stanford University, Stanford, CA, 1988.

[14] Brandão, M.P.: Improper integrals in theoretical aerodynamics: the problem revisited. AIAA J. 25 (1987) 1258–1260.

[15] Lee, Y.J.: On the integral formulation of wave equation for arbitrary moving boundary and its applications to aerodynamics and aeroacoustics. Ph.D. Thesis, National Taiwan University, 1988.

[16] Lee,Y.J., Yang, J.Y.: Panel method for arbitrary moving boundaries problems. AIAA J. 28 (1990) 432–438.

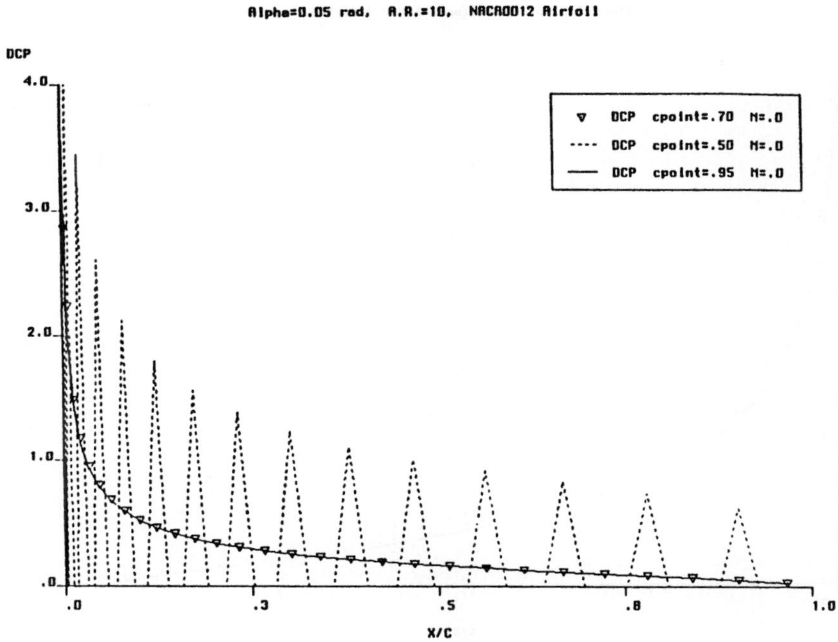

Figure 1: Effect of control point location on the pressure distribution at midspan.

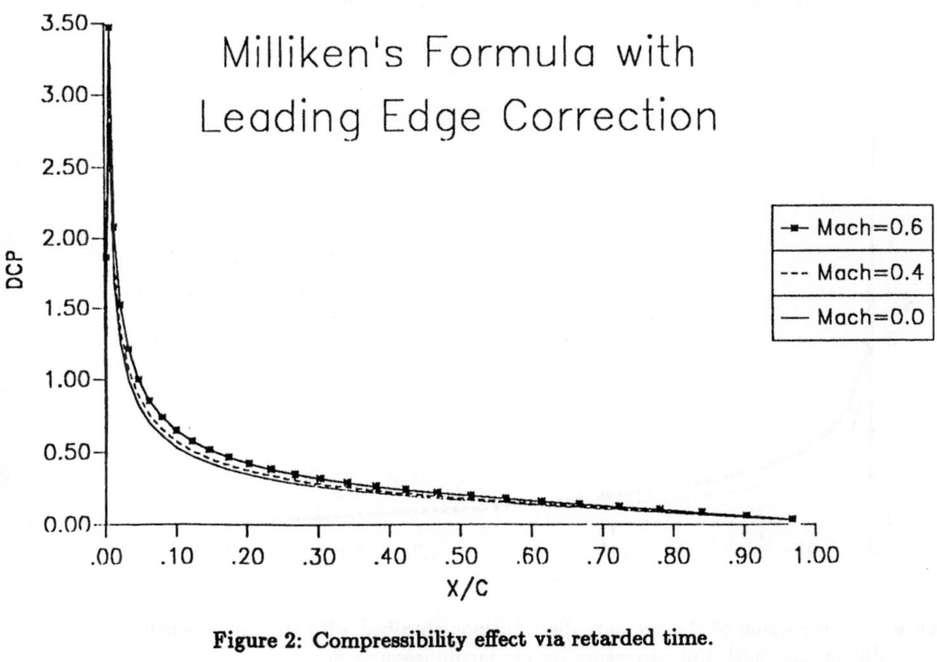

Figure 2: Compressibility effect via retarded time.

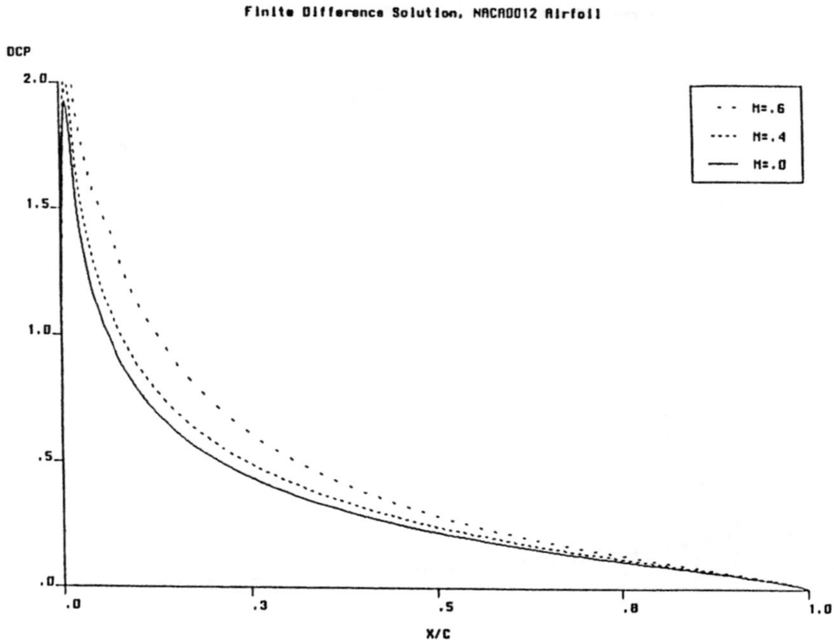

Figure 3: Compressibility effect using the full potential equation.

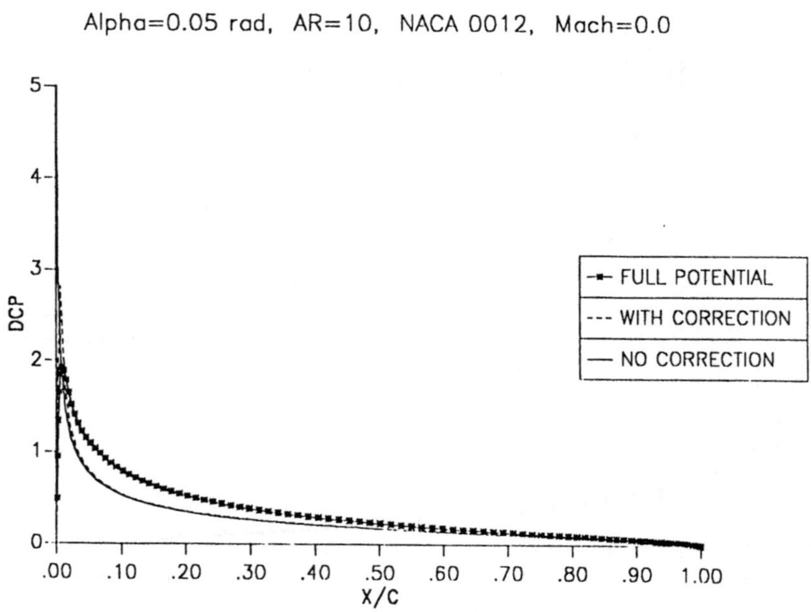

Figure 4: Comparison of the pressure distributions obtained with the full potential equation, Milliken, and with the correction terms: incompressible flow.

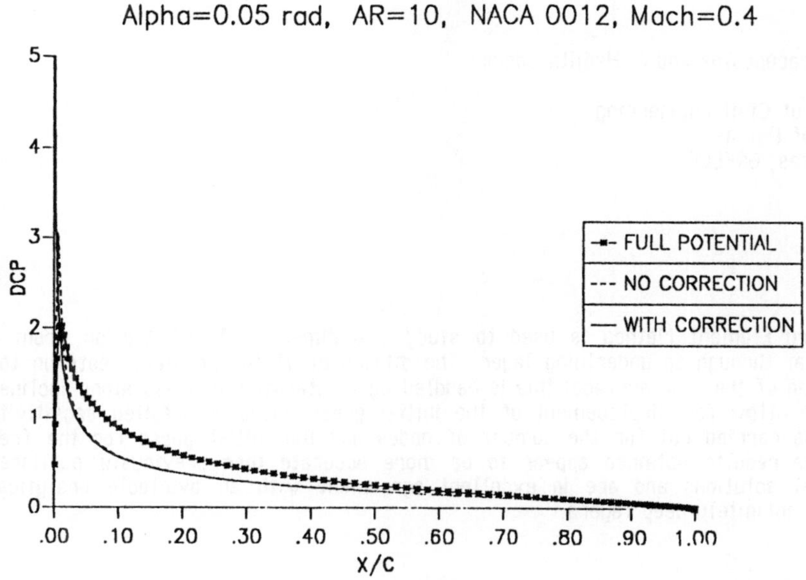

Figure 5: Comparison of the pressure distributions obtained with the full potential equation, Milliken and with the correction terms: compressible flow.

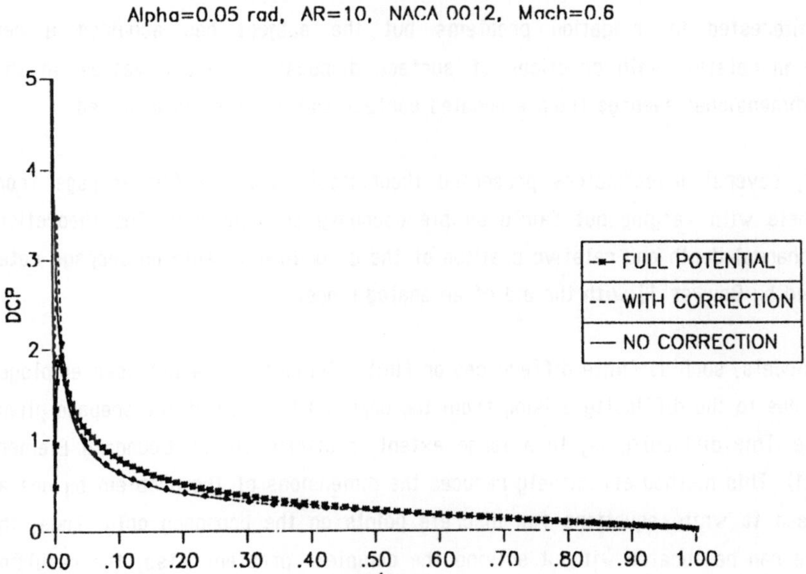

Figure 6: Comparison of the pressure distributions obtained with the full potential equation, Milliken and with the correction terms: compressible flow.

Infiltration from Surface Water Bodies

A. C. Demetracopoulos and C. Hadjitheodorou

Department of Civil Engineering
University of Patras
261 10 Patras, GREECE

Abstract
The Boundary Element Method is used to study two-dimensional infiltration from a surface canal through an underlying layer. The difficulty of the problem rests on the determination of the free surface; this is handled by an iteration process along inclined lines, which allow for displacement of the initial guess nodes. A detailed sensitivity analysis was carried out for the number of nodes and the initial guess for the free surface. The results obtained appear to be more accurate than previously published nonanalytical solutions and are in excellent agreement with an available analytical solution for infinitely deep layers.

Introduction

Seepage from surface water bodies through the underlying soil strata is a problem of great importance in many facets of engineering. Traditionally, it has been examined by engineers interested in irrigation problems but the subject has acquired a new significance in relation with practices of surface disposal of liquid wastes. In this study, two-dimensional seepage from elongated surface water bodies is analyzed.

In the past, several investigators presented theoretical solutions for seepage from single channels with varying but fairly simple boundary conditions[4]. The theoretical effects of channel depth and relative position of the groundwater table on seepage rates was examined by Bouwer[1] with the aid of an analog model.

Numerical models, such as finite differences or finite elements, have not been employed extensively due to the difficulty arising from the unknown location of the seepage plume free surface. This difficulty is, to a large extent, rectified by the Boundary Element Method (BEM). This method effectively reduces the dimensions of the problem by one as it can be used to write equations for discrete points on the boundary only. Thus, the free surface can be located without solving the complete problem. Also, the resulting algebraic equations completely define the location of the free surface and, theoretically, no iterations are necessary. In practice, however, since the equations are nonlinear, an iterative solution is obtained. A first detailed presentation regarding the location of the free surface in groundwater flow problems was given by Liggett[6] and an improved version of the same method was described in Liggett and Liu[7]. The BEM has been used in the study of seepage through and below dams by several investigators [2,6,8,9]. The seepage from a surface stream towards an underlying aquifer was tackled by Dillon and Liggett[3] but with emphasis on the variation of the phreatic surface.

It is apparent from the above that a problem which was treated theoretically in the past is amenable to solution with the BEM for more complex geometries and boundary conditions. This method also allows for the treatment of time variable problems and, in general, permits the investigation of real problems for which the existing theoretical solutions are not satisfactory.

Theoretical Background

The two-dimensional (vertical plane), saturated flow through porous media is governed by Laplace's equation:

$$\frac{\partial^2\varphi}{\partial^2 x} + \frac{\partial^2\varphi}{\partial^2 y} = 0 \tag{1}$$

where $\varphi = \varphi(x,y)$ is the piezometric head, defined as:

$$\varphi(x,y) = \frac{p(x,y)}{\rho g} + y \tag{2}$$

with $p(x,y)$ the water pressure, ρ the water density and y the elevation above an arbitrary datum. The boundary conditions, for the problem described earlier in general terms, can be classified as follows.

(a) Dirichlet condition:

$$\varphi = \varphi_b \quad \text{on } \Gamma_1 \in \Gamma \tag{3}$$

where φ_b is the prescribed piezometric head along Γ_1, which is a portion of the solution domain boundary, Γ.

(b) Neumann condition:

$$\mathbf{q}\cdot\mathbf{n} = q_b \quad \text{on } \Gamma_2 \in \Gamma \tag{4}$$

where $\mathbf{q}$ is the Darcy velocity, $\mathbf{n}$ is the unit outward vector normal to the boundary and q_b is the prescribed volumetric flux across Γ_2, which is part of the boundary Γ.

Making use of Darcy's Law

$$\mathbf{q} = -K\nabla\varphi \tag{5}$$

boundary condition (4) becomes:

$$\frac{\partial\varphi}{\partial n} = \nabla\varphi\cdot\mathbf{n} = -\frac{q_b}{K} \quad \text{on } \Gamma_2 \in \Gamma \tag{6}$$

where K the conductivity and $\partial\varphi/\partial n$ the normal derivative of the piezometric head.

(c) Free surface condition:

For steady flow the free surface is a streamline, thus the normal flux vanishes. Also, at the free surface the pressure is zero and the boundary condition becomes:

$$\frac{\partial\varphi}{\partial n} = 0 \quad \text{on } \Gamma_3 \in \Gamma \tag{7a}$$

$$\varphi = y \quad \text{on } \Gamma_3 \in \Gamma \tag{7b}$$

The solution of Eq. (1) with boundary conditions Eqs. (3), (6) and (7) is accomplished herein by the BEM. This method is based on the integral equation[7]:

$$\sigma\varphi(P) = \int_\Gamma \left[\varphi(Q)\frac{\partial}{\partial n}(\ln r) - \ln r\frac{\partial}{\partial n}\varphi(Q)\right]ds \tag{8}$$

where P is any point on the boundary Γ or inside the domain, Q is a series of points (nodes) on Γ over which the integration is performed, σ=2π if P is inside the domain, σ=π if P is on a smooth portion of Γ and σ is the interior angle of the boundary if P is at a point on Γ where an angle is formed. The quantity r is the distance between point P and point Q.

If φ and $\partial\varphi/\partial n$ are known everywhere on Γ, then Eq. (8) yields the solution for φ anywhere in the domain by a simple boundary integration. In most problems, however, either φ or $\partial\varphi/\partial n$ are given at each boundary point and Eq. (8) is used to find the "missing data" by choosing P at a succession of boundary nodes and assuming the behavior of φ and $\partial\varphi/\partial n$ between nodes. Herein the behavior is assumed to be linear.

The integration for all chosen boundary nodes can be carried out explicitly, resulting in a set of simultaneous equations of the form:

$$[H]\{\varphi\} = [G]\left\{\frac{\partial\varphi}{\partial n}\right\} \tag{9}$$

where [H] and [G] are coefficient matrices and $\{\varphi\}$, $\{\partial\varphi/\partial n\}$ column matrices containing the unknown nodal values of φ and $\partial\varphi/\partial n$. The coordinates of the nodes, which belong to a free surface, are assumed before the computation begins. Equation (7a) is taken as the boundary condition for these nodes and the values of φ are found from Eq. (9). The prior estimation of the free surface serves to fix the values of [H] and [G] in Eq. (9) and thus removes the nonlinearity. The free surface location can be adjusted and the process is repeated until the solution coincides with the estimated position.

Problem Definition

The problem solved herein is depicted in Fig. 1. Water infiltrates from a long canal of constant head through an underlying layer with free drainage conditions at its bottom. The infiltration plume will have the shape shown in Fig. 1. The exact location of the free surface and the infiltration rate are computed in the present study. Due to symmetry, only the solution in the half domain is computed (see Fig. 2), which is now defined as ABCDEA. Therein the boundary conditions are:

$$\varphi = 0 \qquad \text{on AE} \tag{10a}$$

$$\frac{\partial\varphi}{\partial n} = 0 \qquad \text{on AB} \tag{10b}$$

$$\varphi = D_W \qquad \text{on BCD} \tag{10c}$$

$$\varphi = y \text{ and } \frac{\partial\varphi}{\partial n} = 0 \qquad \text{on DE} \tag{10d}$$

Determining the location of the free surface (DE) is the main difficulty of this problem. As explained in the previous section, an initial guess DF is defined, so that the piezometric head is underestimated (line to the left of "true" free surface). Potential use of the classical approach of vertical corrections for the position of boundary nodes along line DF would, at best, produce the portion of the free surface DF' and would fail to give the full free surface profile DE.

The novelty of the present approach lies in the use of inclined lines passing through an origin, O. Such a line is shown in Fig. 2, connecting the origin with a node b. The initial guess of φ for this node is ab. The updated piezometric head, which is obtained from the solution of Eq. (9), is under-relaxed to the value ac and displaced to a new position de dictated by the straight line Obe. The process is continued until two successive iterations give practically the same result. This is considered the true piezometric head of the node, which has now taken the position h on the free surface. Under-relaxation is used to avoid divergence phenomena which are associated with the angle of inclination of lines Ob, as it will be explained in the following section. This approach is applied from node D up to a node m, which is taken close to F. The iteration process displaces this node to the position n. Since the free surface is orthogonal to the drainage surface RR, the location of E is obtained by considering that its horizontal coordinate is the same as that of node n. It is obvious from Fig. 2 that this approximation improves as the initial guess for the free surface is closer to the "true" free surface. The numerical solution of the problem is based on a computer code, FRSURF, originally developed by J.A. Liggett (personal communication) and modified by the authors.

Results

The BEM solution was compared with Bouwer's[1] and Jeppson's[5] results. The first solved the problem with an electric analog model while the second used finite differences in a transformed solution domain. In order to check the overall behavior of the present solution, a reference problem was used. The pertinent quantities (see Fig. 1) are $W_b=1$, $H_w=0.75$, $D_p=3.5$ and the channel slopes are 1:1. The above values yield $D_w=4.25$ and $W_s=1.25$. For this problem Bouwer[1] gives both the rate of infiltration and the free surface of the plume.

For different initial guesses of the free surface, an investigation was undertaken for the location of the point of origin, O, of the inclined lines. The location of this point affects the determination of the free surface, especially its lower part (see Fig. 2). When the angle θ is very small the last node, m, is diplaced too far up, thus distorting the free surface, while for large values of θ instabilities tend to occur as the lower computational nodes are displaced too far to the right, hence diverging from the "true" free surface. From the aforementioned analysis it was determined that the maximum value of θ (corresponding to node m) must be in the range $0.4 \leq \tan\theta \leq 0.5$.

Regarding the number of nodes used, it was found that when the nodes on the initial guess for the free surface are taken at locations such that $\Delta y \approx 0.2$, good accuracy is obtained for both the free surface and the rate of infiltration. Additional attention was given to the nodes on the other boundaries of the solution domain. The cases examined are summarized in Table 1 (see also Fig. 2) and the free surface results are shown in Fig. 3. The normalized infiltration rate, I/KW_s, is computed along the canal bottom, BCD. It is based on the piezometric head gradient, $\partial\Phi/\partial n$, computed at the nodes, and the distance between contiguous nodes.

Table 1. Node distribution and resulting infiltration rates.

Case No.	Number of Nodes					I/KWs
	AB	BD	DE	EA	Total	
1	3	3	22	3	27	1.828
2	3	5	22	3	29	1.802
3	5	3	22	5	31	1.818
4	5	5	22	5	33	1.791
5	8	6	21	5	36	1.782
6*	8	6	21	5	36	1.786

*different initial guess

An increase in nodes on boundaries AB and EA does not affect the results significantly(see Cases 1 and 3). On the other hand, an increase in nodes on boundary BD has a much more pronounced effect on both the location of the free surface and the infiltration rate (see Cases 1,2,4 and 5). As the nodes on BD increase, both the infiltration rate and the width of the infiltration plume decrease. A denser grid on this boundary emulates the singular behavior of node C better.

Solution of the problem for different combinations of the parameters W_b, H_w and D_p lead to some interesting observations regarding the choice of the initial guess for the free surface. While this choice did not affect the infiltration rate, it had a significant influence on the shape of the free surface. Thus, when the angle α (see Fig. 2) was very small, the solution was unable to predict the shape of the free surface correctly. Gradual increase of α led to the computation of an optimum, smooth free surface profile. These tests resulted in Fig. 4 from which it can be seen that the angle of the initial profile decreases with increasing thickness of the drainage layer. This behavior agrees with the physical mechanism governing two-dimensional infiltration. As the drainage layer thickness increases, the flow resistance of the medium increases, leading to reduced infiltration and hence to reduced width of the infiltration plume. The aforementioned physical mechanisms are also evident from the behavior of the normalized infiltration rate shown in Fig. 5.

It is noteworthy that the present results are in good agreement with those of Jeppson and, in general, underpredict the infiltration rates determined by Bouwer. However, the present results, for large values of the ratio D_w/W_b, are in excellent agreement with Vedernikov's[4] theoretical solutions for infinitely deep drainage layers (see Table 2).

Table 2. Normalized infiltration rates

H_w/W_b	Bouwer	Jeppson	Vedernikov	Present Study
0.75	1.81	——	1.78	1.782
0.50	1.70	1.689	1.67	1.700
0.25	1.57	1.534	1.53	1.535

Summary

The BEM was used for the solution of the problem of two-dimensional infiltration from a canal through a layer with free drainage conditions at the bottom. A novel iteration technique was incorporated for the location of the free surface. It was found that the solution is sensitive to the number of boundary points located at the canal bottom. The free surface initial guess appears to be crucial in the correct determination of the free surface, while it does not affect the infiltration rate. The results appear to be more accurate than previously obtained solutions as evidenced by the agreement with Vedernikov's analytical solution for infinitely deep drainage layers.

References

1. Bouwer H.: Theoretical aspects of seepage from open channels. J. Hydraulics Div., ASCE, 91(HY3), 1965, 37-59.

2. Chang C.S.: Boundary element method in seepage analysis with a free surface, in Implementations of Computer Procedures and Stress-Strain Laws in Geotechnical Engineering, Desai C.S. and Saxena S.K. (eds.), Acorn Press, Durham, 1981.

3. Dillon P.J. and Liggett J.A.: An ephemeral stream-aquifer interaction model. Water Resour. Res., 19(3), 1983, 621-626.

4. Harr M.E.: Groundwater and Seepage. McGraw-Hill, New York, 1962.

5. Jeppson R.W.: Seepage from ditches-Solution by finite differences. J. Hydraulics Div., ASCE, 94(HY1), 1968, 259-283.

6. Liggett J.A.: Location of free surface in porous media. J. Hydraulics Div., ASCE, 103(HY4), 1977, 353-365.

7. Liggett J.A. and Liu P. L-F.: The Boundary Integral Equation Method for Porous Media Flow. George Allen and Unwin, London, 1983.

8. Liu P. L-F. and Liggett J.A.: Boundary solutions to two problems in porous media. J. Hydraulics Div., ASCE, 105(HY3), 1979, 171-183.

9. Niwa et al.: An application of the integral equation method to seepage problems, in Theoretical and Applied Mechanics Volume 24, University of Tokyo Press, 1974.

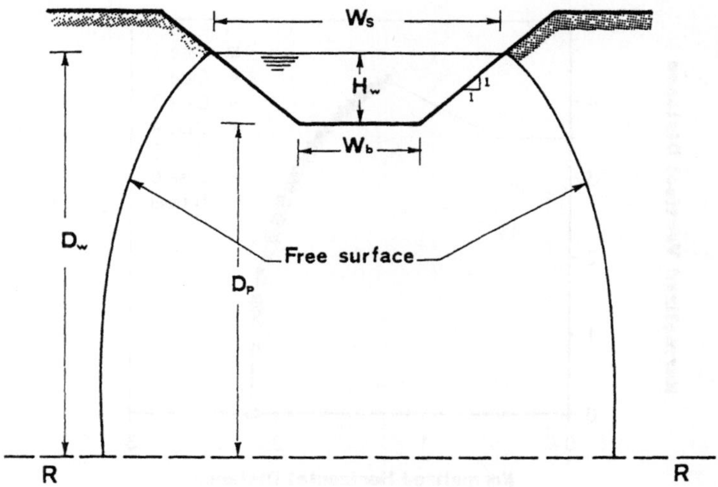

Figure 1. Definition sketch for two-dimensional infiltration from a canal

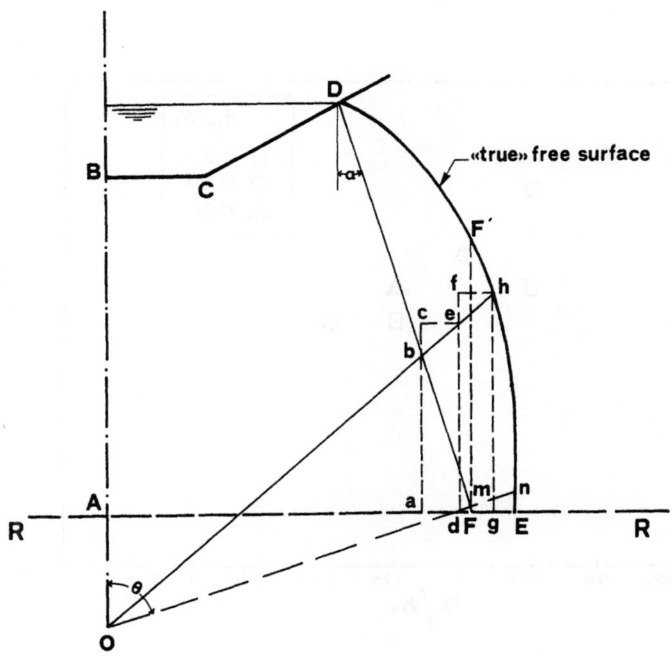

Figure 2. Iteration technique used for computation of free surface

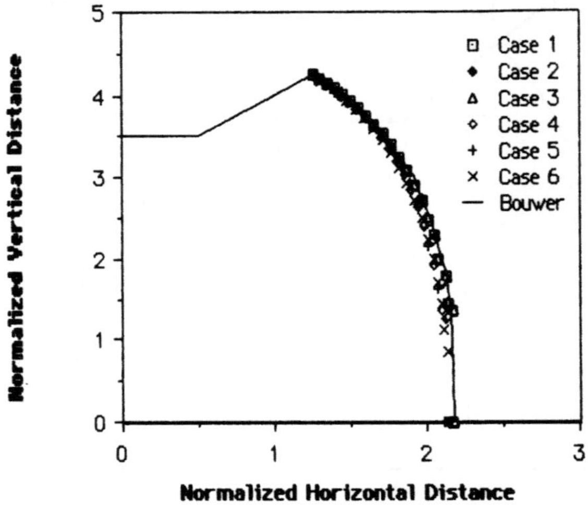

Figure 3. Effect of nodes on shape of infiltration plume

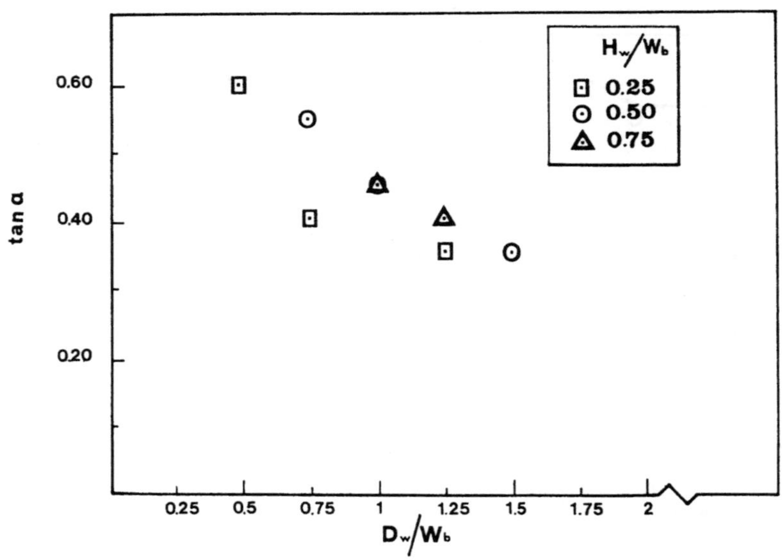

Figure 4. Angle α as a function of drainage layer thickness

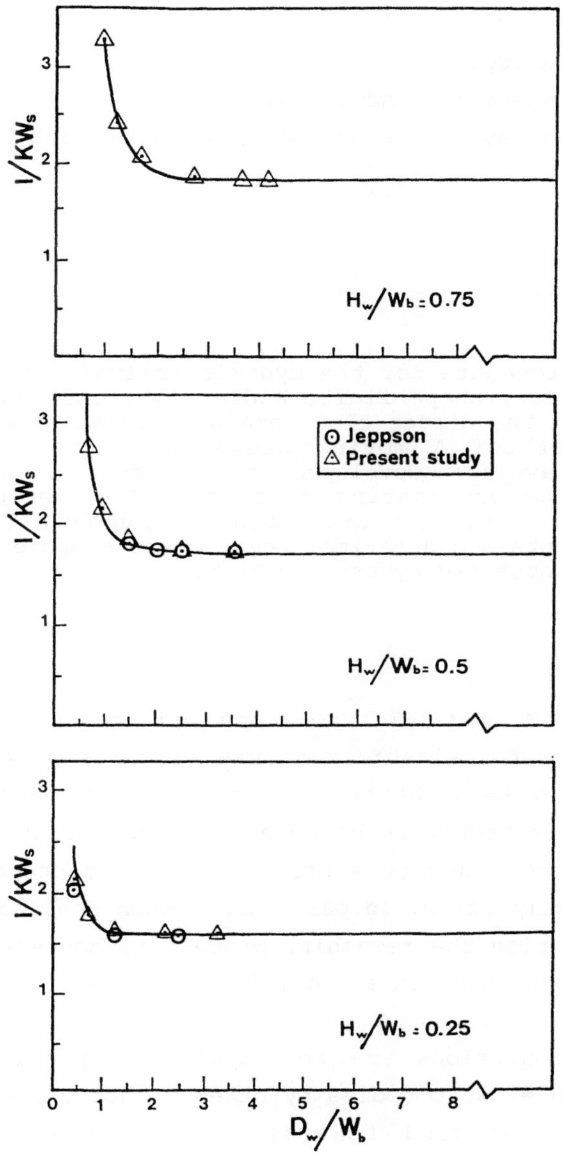

Figure 5. Normalized infiltration rates as a function of drainage layer thickness. Solid line represents Bouwer's results

Dynamic Crack Propagation Using Boundary Elements

J. Dominguez and R. Gallego

Escuela Superior de Ingenieros Industriales

Universidad de Sevilla, Av. Reina Mercedes, 41012-Sevilla, SPAIN

Summary

A boundary element procedure for the dynamic analysis of crack propagation in arbitrary shape finite bodies is presented. The procedure is based on the direct time domain formulation of the Boundary Element Method. A moving singular element and a remeshing technique have been developed to model the solution of the stresses near the propagating crack tip. The method is applied to a problem of dynamic crack propagation in a finite elastic domain. The obtained numerical results are compared with available solutions obtained by other authors.

Introduction

The Boundary Element Method (BEM) has appeared recently as a competent alternative for elastic fracture mechamic problems. This method seems to be a better choice tham the FEM for elastodynamic fracture mechamics because the discretization is restricted only to the boundary surface and the concept of singular element is simplified. In particular, when dealing with dynamic crack propagation the remeshing process is conceptually much simpler in the BEM than in any domain technique.

Integral equation formulations have been applied to problems of propagating cracks in elastic bodies by other authors. However, all these studies, directed towards the simulation of earthquake sources, are limited to infinite domains and use either the BEM in conjunction with a node release mechanism (Das and Kostrov, 1987) or are restricted to particular formulations related to the BEM (Burridge, 1969).

In the present work, the direct time domain formulation of the BEM is used in combination with a moving singular quarter-point (SQP) element to study dynamic crack propagation in elastic bodies. To do so a remeshing technique have been developed. These ideas are used in the general context of time domain BE with shape functions for space and time discretization. To start out with, the direct time domain formulation of the BEM and the SQP element are summarized. Next the remeshing technique and the boundary element formulation for moving elements are studies. In doing this, mumerical aspects of the process leading to the final system of equations are discussed. A numerical example including a finite length central crack that propagates with costant velocity in a rectangular plane body is studied. Results are compared with those obtained by other authors using analytical and numerical methods. The example illustrates the efficiency and accuracy of the present procedure.

Time Domain B.E. Formulation

The integral representation of the displacement of a point inside an elastic domain Ω or on its boundary Γ at time t can be written in terms of the boundary integral of the time convolution of the boundary tractions with the fundamental solution displacements, and the boundary integral of the time convolution of the boundary displacements and velocities, with fundamental solution displacement derivatives (see Antes, 1985).

To accomplish the numerical solution of the integral equation, displacements and tractions are approximated using interpolation functions

$$u_j = \sum_q \sum_m \varphi^q(r)\, \eta^m(\tau)\, u_j^{mq}$$

$$p_j = \sum_q \sum_m \psi^q(r)\, \mu^m(\tau)\, p_j^{mq} \qquad (1)$$

where u_j^{mq} and p_j^{mq} stand for the j displacement and traction, respectively, of the node q at the time $t_m = m\,\Delta t$. The functions $\varphi^q(r)$ and $\psi^q(r)$ are space interpolation functions and $u^m(\tau)$ and $\eta^m(\tau)$ time interpolation functions. The approximation for the velocities is taken consistently with that of the displacements.The time interpolation functions $\mu^m(\tau)$ and $\eta^m(\tau)$ are assumed to be piecewise constant and piecewise linear, respectively.

The elements used in the present work are three-node space quadratic elements except for those in contact with the crack tips which are singular quarter-point (SQP) elements. The domain is subdivided by a boundary running along the crack which leaves each side of the crack on the boundary of different subregions. If the first element from the crack tip is a SQP element and follows the directon of the crack, the SIF are given directly by the nodal values as:

$$K_I = \bar{p}_2^{1} \, (2\pi l)^{\frac{1}{2}}$$

$$K_{II} = \bar{p}_1^{1} \, (2\pi l)^{\frac{1}{2}} \tag{2}$$

Details of the SQP boundary element formulation may be seen in the work by Martinez and Dominguez (1984).

Moving singular element

Consider a mode-I dynamic crack propagation problem as shown in Figure 1. A boundary is introduced along the crack, following the known direction of propagation. Because of the simmetry only one half of the domain is analysed. Assume that the crack propagates at a certain speed C, and that at time t the position of the crack tip and the discretization of the part of the boundary that contains one side of the crack is as shown in

Figure 2.a. Two elements, one before and one after the tip are SQP. After a time increment Δt, the crack tip has moved C Δt (Figure 2.b). As the crack tip advances from left to right, the size of the elements to the left is increased and that of the elements to the right is decreased. The SQP elements are excluded of this adjustment and their size remains the same along the crack propagation process. The translation of the crack tip for each time step can take any value and is not related to the assumed discretization. After some time, the elements on the left hand side would be much bigger than those on the right hand side. To avoid this, when the ratio of the sizes is greater than 5/3 (figure 2.c) the number of elements to the left is increased by one and the number of elements to the right is decreased by one. The elements discretization of both sides is then redefined (Figure 2.d).

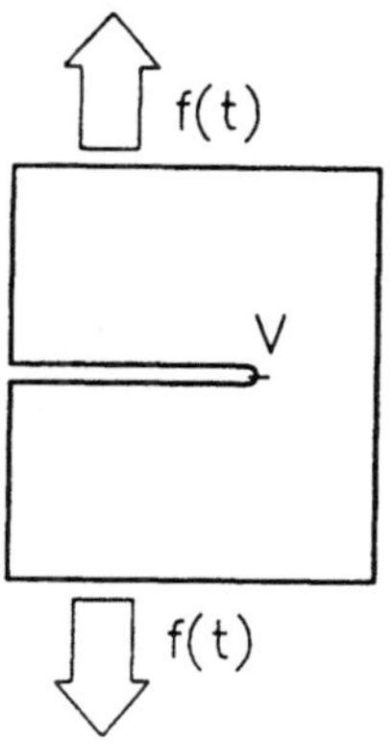

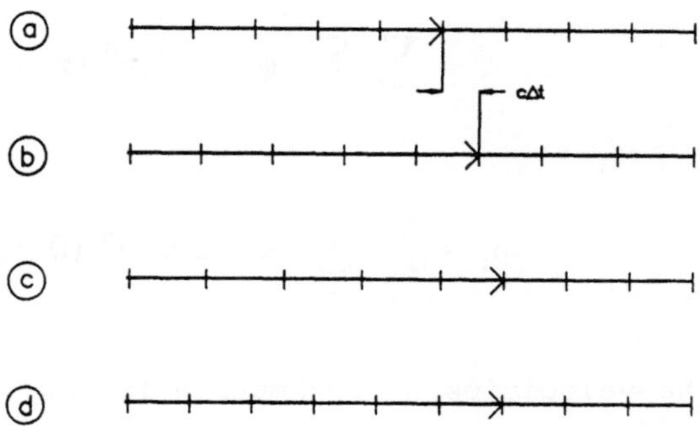

Figure 1. Mode-I dynamic crak propagation problem.

Figure 2. Movement of the elements and remeshing process.

The BE matrix equation for time step n can be written in a subdivided form as

$$\sum_{m=1}^{n} \begin{bmatrix} \mathbf{H}_{ff}^{nm} & \mathbf{H}_{fc}^{nm} \\ \mathbf{H}_{cf}^{nm} & \mathbf{H}_{cc}^{nm} \end{bmatrix} \begin{Bmatrix} \mathbf{u}_{f}^{m} \\ \mathbf{u}_{c}^{m} \end{Bmatrix} = \sum_{m=1}^{n} \begin{bmatrix} \mathbf{G}_{ff}^{nm} & \mathbf{G}_{fc}^{nm} \\ \mathbf{G}_{cf}^{nm} & \mathbf{G}_{cc}^{nm} \end{bmatrix} \begin{Bmatrix} \mathbf{p}_{f}^{m} \\ \mathbf{p}_{c}^{m} \end{Bmatrix} \tag{3}$$

where the subindex "f" stands for the nodes which possition remains fixed during the remeshing process and "c" for those that change position during that process. The translation property of the fundamental solution can not be used to compute the submatrices corresponding to changing collocation points and/or to changing integration elements. In those cases $H_{ij}{}^{nmpq}$ and $G_{ij}{}^{nmpq}$ depend on the n and m values and not only on the difference n - m.

If the general time domain BE discretization is used for moving elements, the results are poor because the space and time dependence of the variables are not uncoupled. The space interpolation functions of the variables move with the elements and therefore this functions are not only space dependent but also time dependent. Displacements and tractions are represented as follows:

$$u_j = \sum_q \sum_m \varphi^q(r,\tau)\, \eta^m(\tau)\, u_j^{mq}$$

$$p_j = \sum_q \sum_m \psi^q(r,\tau)\, \mu^m(\tau)\, u_j^{mq} \tag{4}$$

The velocities approximation is now

$$\dot{u}_j = \sum_q \sum_m [\, \varphi^q(r,\tau)\, \eta^m(\tau) + \varphi^q(r,\tau)\, \eta^m(\tau)\,]\, u_j^{mq} \tag{5}$$

In the regular formulation of the time domain BEM, the space shape functions are taken away from the time integrals which are done analytically. This is not possible when ψ^q and φ^q are time dependent. In order to do the time integration, an approximation of the space shape functions time dependence is done. Two terms of their series expanssion are taken:

$$\psi^q(r,\tau) = \psi^q(r,\tau_{m-1}) + (\tau - \tau_{m-1})\, \psi^q(r,\tau_{m-1})$$

$$\phi^q(r,\tau) = \phi^q(r,\tau_{m-1}) + (\tau - \tau_{m-1})\, \phi^q(r,\tau_{m-1}) \tag{6}$$

By substitution of equation (6) into the integral representation the time integrals can be separated from the space integrals and the integral representation can be written as the typical BE equation.

The computation of the coefficients of the system requires a time integration followed by a space integration. The time integration is done analytically. Explicit expressions for those integrals may be found in the work by Gallego (1990).

The space integration requires first the computation of the time derivatives of the shape functions φ^q and ψ^q. These derivatives are easily computed assuming that for each time step the elements move with a constant speed C.

In the present work the shape functions of the moving elements are singular of the kind $r^{-1/2}$. Their derivatives are singular of the kind $r^{-3/2}$. These singularities are integrated numerically. The integration can be done because the Heaviside function makes only necessary the computation of the finite part of the integrals (Kutt, 1975). The numerical formulas given by Kutt (1975) for finite part computations have been used. A ten points integration squeme produces accurate results for the finite parts of the integrals containing $r^{-3/2}$ singularities.

Center Crack Propagating in a Rectangular Plate

In order to evaluate the application of the present approach to crack propagation in finite bodies under dynamic loading conditions a problem of this kind, previously studied by Nishioka and Atluri (1980) using a different numerical procedure, is analysed. The problem is that of a rectangular plane domain with a central crack (Figure 3). A uniformly distributed traction, with a Heaviside-function time-dependence, is applied at the two sides parallel to the crack.The material properties are: Shear Modulus, $\mu = 2.94\ 10^{10}$ Pa; Poisson's ratio $\upsilon = 0.286$ and density, $\rho = 2450$ Kg/m^3. The crack, with an

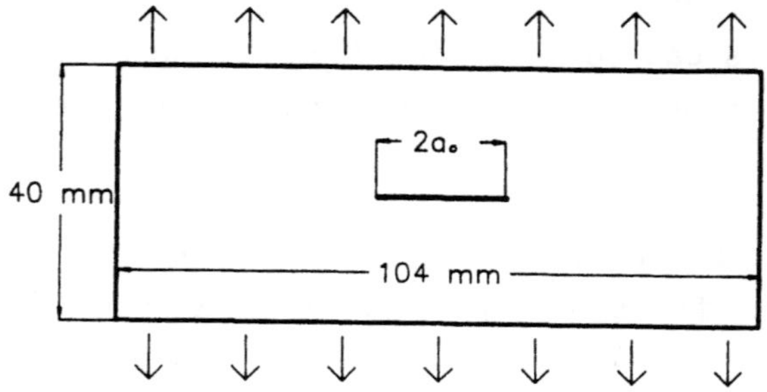

Figure 3. Center-cracked rectangular plate subject to step-function normal stress.

initial length 2a = 24 mm, remains stationary until a time t_0=4.4μs and then propagates with a constant velocity C = 1000 m/sec.

Due to simmetry, only one quater of the plate is discretized by boundary elements. The discretization at the initial crack length is shown in Figure 4. The two elements containing the crack tip are moving SQP elements and the rest are standard space quadratic elements. The boundary elements discretization is the same that results from Nishioka and Atluri's (1980) finite element discretization of the boundary. In modelling the crack propagation, the size of the two moving SQP do not change. The regular quadratic elements are readjusted in accordace with the aforementioned criterium. The time step Δt = 0.34 μs is such that the parameter β for the equal size regular elements that model most of the boundary is:

$\beta = c_1 \, \Delta t/L = 1.08.$

The computed values of the mode-I SIF normalized by $\sigma(\pi a)^{1/2}$ are shown versus time in Figure 5, along with analytical resuls by Freund (1973) and the Finite Element results by Nishioka and Atluri (1980). The present boundary element results are in

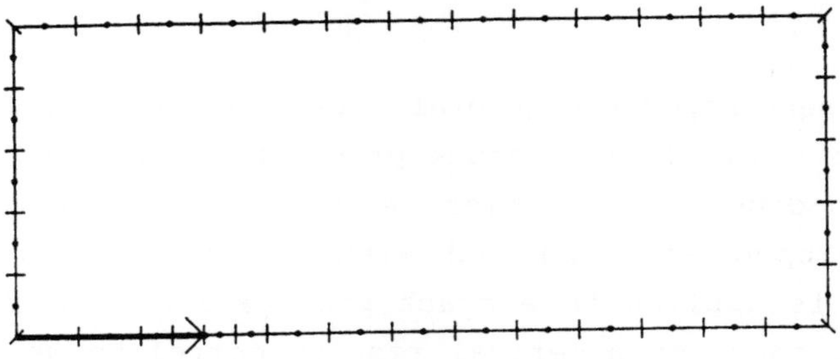

Figure 4. Boundary element mesh for center-cracked rectangular plate.

excellent agreement with the half-plane crack in infinite domain results by Freund (1973), until interaction of waves coming fron one crack tip with the other crack tip takes place. The present results are also in good agreement with those computed by finite elements by Nishioka and atluri (1980), for times after the time for which the infinite domain solution may be considered valid.

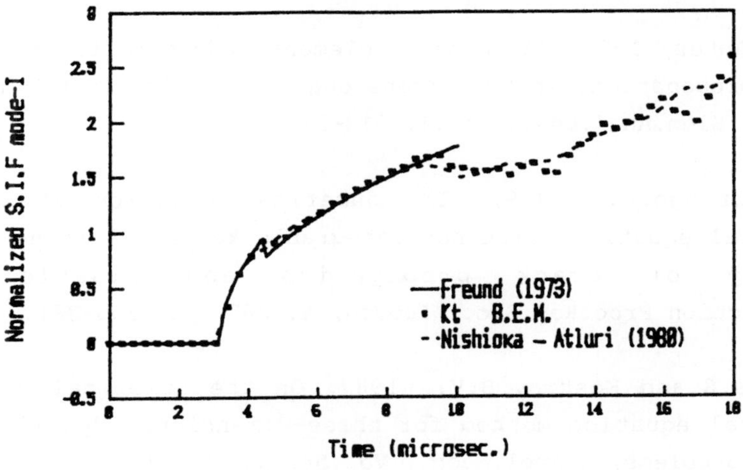

Figure 5. Time dependence of dynamic SIF for a central crack propagating with C = 1000 m/sc in a rectangular plate.

Closure

In the present paper the general direct boundary element method is used to study dynamic crack propagation in arbitrary shape finite two-dimensional bodies. A moving singular element has been developed and combined with a remeshing scheme. The procedure is applied to a crack propagating in a rectangular plate. The computed numerical results correlate well with the available analytical solution, for infinite domains, when this solution is valid; i.e., before wave interaction effects are noticeable. The computed solution is in good agreement, for all times, with existing numerical results obtained by other authors using a different method.

Acknowledgement

The autors would like to express their gratitude to the spanish "Comision Interministerial de Ciencia y Tecnologia" for supporting this work under a research grant.

References

1. H.Antes, 1985, "A boundary element procedure for transient wave propagations in two-dimensional isotropic elastic media", Finite Elem.Anal.Des., Vol.1, 313-322.

2. Burridge,R., 1969, "The numerical solution of certain integral equations with non-integrable kernels arising in the theory of crack propagation and elastic wave diffraction Proc.Roy.Soc., London, A. 265, pp. 353-381.

3. Das,S.and Kostrov,B.V., 1987, On the numerical boundary integral equation method for three-dimensional dynamic shear crack problems, J.Appl.Mech., Vol.54, pp. 99-104.

4. Freund,L.B., 1973, Crack Propagation in an elastic solid subjected to general loading-III: Stress wave loading, J.Mech.Phys.Solids, Vol.21, pp.47-61.

5. R.Gallego, 1990, "Numerical studies of elastodynamic fracture mechamics problems" (In spanish). Ph.D.Thesis, Universidad de Sevilla.

6. Kutt,H.R., 1975, "Quadrature formulae for finite-part integrals" ,CSIR Specia Report, WISK 178, National Research Institute for Mathematical Sciences, Pretoria, Replublic of South Africa.

7. J.Martinez and J.Dominguez, 1984, "On the use of quarter-point boundary elements for stress intensity factor computations", Int.J.Num.Mech.Eng. Vol. 20, pp.1941-1950.

8. Nishioka,T., Atluri,S.N., 1980, "Numerical modelling of dynamic crack propagation in finite bodies by moving singular element, Part I: Formulation; Part II: Results, J.Appl.Mech, Vol.47, pp.570-576 & pp.577-582.

The Shock Noise of High Speed Rotating Blades - The Supersonic Shock Problem

F. FARASSAT

Applied Acoustics Branch
NASA Langley Research Center
Hampton, Virginia

SUMMARY

Shock waves associated with high speed rotating blades, such as helicopter rotors, are potent noise generators. In general, these shocks are time dependent in shape and extent. The shock noise generation is described by part of the quadrupole source term of the Ffowcs Williams-Hawkings (FW-H) equation. An efficient calculation of such shock noise is an aim of aeroacousticians. Earlier a formulation for prediction of shock noise produced by subsonically moving shocks was published. The main difference between subsonically and supersonically moving shock noise generation is that in the latter case multiple emission times may exist for some regions of the shock. This introduces additional complexity into the analytic formulation and the possibility of singularities in the solution. Here we present two formulations for supersonic shock noise prediction. Both are based on the boundary element method with the boundary data supplied by nonlinear aerodynamic codes. The emphasis in this paper is on the analytical derivation of results suitable for efficient computation in applications of interest.

Introduction

The Ffowcs Williams-Hawkings (FW-H) equation indicates that there are three kinds of acoustic sources for rotating blades– thickness, loading, and quadrupole sources [1]. The thickness and loading sources are distributed on the blade surface while the quadrupole sources are in the volume around the blades. The quadrupole sources generate broadband noise by turbulence in the flow field. They also generate discrete frequency noise due to the rotating deterministic (steady or periodic) stress field around the blades. We are concerned here with the discrete frequency noise only. The quadrupole source term appearing in the governing acoustic equation, the FW-H equation [1], can be written in such a way that the contributions of regions of high gradients, such as shock surfaces, are clearly identified [2,3]. One important result of this analysis is that shocks around high speed rotating blades are potent noise generators [2,4]. Thus, the shock noise constitutes an important part of the quadrupole noise. Fortunately, the shock sources are easier to incorporate in noise prediction than the full quadrupole sources.

In an earlier paper, a formula for prediction of the noise of shocks travelling at subsonic speed was derived [4]. Here we extend that result to shocks travelling at supersonic speed. In this paper, the terms subsonic and supersonic are defined based on the speed of sound in the undisturbed medium. The specific application we have in mind is for helicopter rotors. Above some advancing tip Mach number, the shocks over the blades which normally extend up to the blade tips suddenly extend far beyond the tip while still remaining attached to the blade. In general, part of the shock structure will be travelling at supersonic speed. This phenomenon was discovered by Schmitz and Yu and was named transonic delocalization by them (see [5], p. 210). Other applications include rotating shocks around advanced propellers and in ducted fans.

The governing equation of shock noise generation which is derived in the next section is based on the FW-H equation. This equation is an inhomogeneous wave equation with sources on moving and deformable shock surfaces. In practice, the shock structure and strength are obtained from some sophisticated unsteady nonlinear aerodynamics code. This implies that the inhomogeneous source term of the governing wave equation is known. The solution of the governing equation is then obtained by the Green's function technique. For this reason our approach to the problem of shock noise prediction is based on the boundary element method. As mentioned in reference [4], the basic difficulty in the application of BEM is that the shock surface is deformable and is time dependent in extent. The subsonic shock problem was solved by mapping the shock surface to a time-independent domain and a useful analytic expression was derived for acoustic calculations [4]. For supersonic shocks, we have the additional complications of multiple emission times from some regions of shock surface and the possibilities of mathematical singularities. Thus, the problem of supersonic shock noise prediction is considerably more complicated than the subsonic case.

In this paper, we derive several equivalent formulations for supersonic shock noise prediction based on the Green's function solution of the governing equation. We use some results from generalized function theory and differential geometry. In general, many factors must be considered as to the selection of a particular formulation for coding on a computer. We will not address this aspect of the problem here.

The Governing Equation

Let T_{ij} be the Lighthill stress tensor [1]. The quadrupole term of the FW-H eq. is

$$E=\frac{\overline{\partial^2}}{\partial x_i \partial x_j}\left[T_{ij}H(f)\right] \tag{1}$$

where $f(\vec{x},t)=0$ describes the surface of the blade ($f>0$ outside the body) and $H(\cdot)$ is the Heaviside function. The bar on the partial derivatives in eq. (1) stands for generalized differentiation [6,7]. We assume that $f(\vec{x},t)$ is defined so that $\nabla f=\vec{n}$ where $\vec{n}$ is the unit outward normal to the blade. Assume that the shock surfaces around the blade are all described by $k(\vec{x},t)$ such that $\nabla k=\vec{n}'$ where $\vec{n}'$ is the unit vector pointing to the downstream side of the shock. We first identify the contribution of the shocks on the right of eq. (1).

To take generalized derivatives of the right of eq. (1), we should identify the discontinuities in the function $T_{ij}H(f)$. There is a discontinuity in T_{ij} across the blade surface and another across the shocks. We thus have [6,7]

$$\frac{\overline{\partial}}{\partial x_j}\left[T_{ij}H(f)\right]=\frac{\partial T_{ij}}{\partial x_j}H(f)+T_{ij}n_j\delta(f)+H(f)\Delta T_{ij}n'_j\delta(k) \tag{2}$$

where $\delta(\cdot)$ the Dirac delta function. Also we have defined the jump $\Delta=[\ ]_2-[\ ]_1$, where the subscripts 1 and 2 refer to upstream and downstream regions of shocks, respectively. We next define the following two vectors.

$$Q_i=T_{ij}n_j; \qquad q_i=\Delta T_{ij}n'_j \tag{3-a,b}$$

Substitute in eq. (2) and take the generalized derivative of both sides of eq. (2) with respect to x_i

$$E=\frac{\partial^2 T_{ij}}{\partial x_i \partial x_j}H(f)+\frac{\partial T_{ij}}{\partial x_j}n_i\delta(f)+\Delta\left[\frac{\partial T_{ij}}{\partial x_j}\right]n'_i\delta(k)+\frac{\partial}{\partial x_i}\left[Q_i\delta(f)\right]+\frac{\partial}{\partial x_i}\left[q_i\delta(k)\right] \tag{4}$$

From this equation, we see that the contribution of shocks to E is

$$E_S=\Delta\left[\frac{\partial T_{ij}}{\partial x_j}\right]n'_i\delta(k)+\frac{\partial}{\partial x_i}\left[q_i\delta(k)\right] \tag{5}$$

We will need this form in the next section for one of the formulations we will derive.

We write E_S in another form by working on the last term of eq. (5). Let the edge curve of the shock surfaces in terms of the local surface coordinates be given by $\tilde{k}=0$ such that $\tilde{k}>0$ on the shock surface. Assume that the surface gradient $\nabla_2\tilde{k}$ of $\tilde{k}$ is the local inward geodesic normal $\vec{v}'$ to the edge curve $k=\tilde{k}=0$, i.e. $\nabla_2\tilde{k}=\vec{v}'$ [8]. This can always be done by redefining $\tilde{k}$ [3].We include the fact that the shock surfaces are open by using the Heaviside function $H\left[\tilde{k}\right]$ in the last term of eq. (5). For reasons explained in reference [3], it is preferable to use restriction of functions multiplying a Dirac delta function to the support of the delta function. We use $(\hat{})$ for the restriction operation and write $q_i\delta(k) \equiv \hat{q}_i\delta(k)$. We are therefore interested in evaluating

$$\frac{\partial}{\partial x_i}[q_i\delta(k)] \equiv \bar{\nabla}\cdot\left[\bar{\hat{q}}H(\tilde{k})\delta(k)\right] \tag{6}$$

We now use the definition of divergence in three-dimensional space in terms of surface divergence [3] in eq. (6). We have

$$\nabla\cdot\left[\bar{\hat{q}}H(k)\delta(k)\right] = \bar{\nabla}_2\cdot\left[\bar{\hat{q}}_T H(\tilde{k})\right]\delta(k) - 2H_k\hat{q}_{n'}H(\tilde{k})\delta(k) + \hat{q}_{n'}H(\tilde{k})\delta'(k) \tag{7}$$

where H_k is the local mean curvature of the shock and we have defined

$$\bar{q}_{n'} = q_i\bar{n}_i'; \qquad \bar{q}_T = \bar{q} - q_{n'}\bar{n}' \tag{8-a,b}$$

We can write the first term of eq. (7) as follows

$$\nabla_2\cdot\left[\bar{\hat{q}}_T H(\tilde{k})\right] = H(\tilde{k})\nabla_2\cdot\bar{\hat{q}}_T + \bar{\hat{q}}_T\cdot\bar{\nu}'\delta(\tilde{k}) \tag{9}$$

By using unsteady shock relations, it can be shown that $q_{\nu'} \equiv q_T\cdot\bar{\nu}' = 0$. We substitute eq. (9) in eq. (7), and then in eq. (5). Finally, we drop $H(\tilde{k})$ in the resulting equation since it is obvious that shock sources exist on open surfaces. The resulting equation is

$$E_S = \left\{\nabla_2\cdot\bar{q}_T - 2H_k q_{n'} + \Delta\left[\frac{\partial T_{ij}}{\partial x_j}\right]n_i'\right\}\delta(k) + \hat{q}_{n'}\delta'(k) \equiv \psi_1(\bar{x},t)\delta(k) + \hat{q}_{n'}\delta'(k) \tag{10}$$

where ψ_1 refers to the expression in curly brackets. Note that we have dropped $(\hat{})$ in $\psi_1(\bar{x},t)$ but not in $\hat{q}_{n'}\delta'(k)$. The reason is that in ψ_1, all q's are multiplied by $\delta(k)$. We maintain $(\hat{})$ on $\hat{q}_{n'}$ here to remind us that $\partial\hat{q}_{n'}/\partial n' = 0$ in the Green's function solution. The governing wave equation for shock noise generation is, therefore,

$$\Box^2 p' = E_S \tag{11}$$

where E_S is given by eq. (5) or eq. (10). Here $p'(\bar{x},t)$ is the acoustic pressure.

Solution Based on Green's Function Technique

We will briefly derive two equivalent solutions of eq. (11) based on the Green's function technique. The readers should consult references [3, 6-13] for the mathematical basis and the details of the derivations. Studying the terms of E_S in eqs. (5) and (10), we note that we need to solve the wave equation with essentially two types of sources as follows

$$\Box^2 p_1' = \phi_1(\bar{x},t)\delta(k) \tag{12-a}$$

$$\Box^2 p_2' = \hat{\phi}_2(\bar{x},t)\delta'(k) \tag{12-b}$$

Here, we are again using $(\hat{})$ on ϕ_2 to signify restriction of ϕ_2 to the surface $k=0$ [3,13]. We note that the Green's function of the wave equation in the unbounded domain is $\delta(g)/4\pi r$ where $g=\tau-t+r/c$. Here $r=|\bar{x}-\bar{y}|$ and $(\bar{x},t)$ and $(\bar{y},t)$ are the observer and the source space-time coordinates. The speed of sound in the undisturbed medium is c.

The formal solution of eq. (12-a), using the above Green's function, is

$$4\pi p_1'(\bar{x},t)=\int\frac{\phi_1(\bar{y},\tau)}{r}\delta(k)\delta(g)d\bar{y}d\tau \tag{13}$$

The interpretation of the product of two delta functions is as follows. We first let $\tau\rightarrow g$ and integrate the right side of eq. (13) with respect to g. The result is

$$4\pi p_1'(\bar{x},t)=\int\frac{1}{r}\left[\phi_1(\bar{y},\tau)\right]_{ret}\delta(K)d\bar{y} \tag{14}$$

where $K(\bar{y};\bar{x},t)=k\left(\bar{y},t-\frac{r}{c}\right)=\left[k(\bar{y},\tau)\right]_{ret}$. Also, we have used the subscript ret for evaluation at the retarded time $t-r/c$. We now use the following relation to integrate $\delta(K)$ in eq. (14) [9,10,12]

$$d\bar{y}=\frac{dy_1dy_2dK}{|\partial K/\partial y_3|}=\frac{dKd\Sigma}{\Lambda} \tag{15-a}$$

$$\Lambda=\left[1+M_{n'}^2-2M_{n'}\cos\theta\right]^{1/2} \tag{15-b}$$

where $d\Sigma$ is the element of the surface area of the surface K=const., $M_{n'}$, is the local normal Mach number of the shock surface $k=0$ and θ is the angle between the local normal $\bar{n}'$ to $k=0$ and the radiation direction $\bar{r}=\bar{x}-\bar{y}$. We note that the reason for appearance of Λ in the denominator of the expression on the right side of eq. (15-a) is that $\Lambda=|\nabla K|$. Using eq. (15-a) in eq. (14) and integrating with respect to K, we obtain

$$4\pi p_1'(\bar{x},t)=\int_{k=0}\frac{1}{r}\left[\frac{\phi_1(\bar{y},\tau)}{\Lambda}\right]_{ret}d\Sigma \tag{16}$$

The construction of the surface Σ is conceptually easy. It is the surface formed by the intersection of the collapsing sphere $g=0$ with the moving and deformable shock surface $k=0$. In practice, it is constructed in steps as follows. Let us define the Γ-curve as the curve formed by the intersection of surfaces $g=0$ and $k=0$ at a given source time τ. We can then show that [13]

$$\frac{d\Sigma}{\Lambda}=\frac{cd\Gamma d\tau}{\sin\theta} \tag{17}$$

We can thus write eq. (16) as follows

$$4\pi p_1'(\bar{x},t)=\int_{-\infty}^{t}\frac{d\tau}{c(t-\tau)}\int_{\substack{k=0\\g=0}}\frac{\phi_1(\bar{y},\tau)}{\sin\theta}d\Gamma \tag{18}$$

This method of integration is known as the collapsing sphere method.

From the solution of eq. (12-a), we can immediately write the solution of the following wave equation

$$\Box^2 p_3' = \frac{\partial}{\partial x_i}\left[\phi_i(\bar{x},t)\delta(k)\right] \tag{19}$$

which has an inhomogeneous source term like the second term of eq. (5). The formal solution using Green's function technique is

$$\begin{aligned} 4\pi p_3'(\bar{x},t) &= \frac{\partial}{\partial x_i}\int \frac{\phi_i(\bar{y},\tau)}{r}\delta(k)\delta(g)d\bar{y}d\tau \\ &= \int \phi_i(\bar{y},\tau)\delta(k)\frac{\partial}{\partial x_i}\left[\frac{\delta(g)}{r}\right]d\bar{y}d\tau \end{aligned} \tag{20}$$

We next use the following relation in the above equation

$$\frac{\partial}{\partial x_i}\left[\frac{\delta(g)}{r}\right] = -\frac{1}{c}\frac{\partial}{\partial t}\left[\frac{\hat{r}_i\delta(g)}{r}\right] - \frac{\hat{r}_i\delta(g)}{r^2} \tag{21}$$

where $\hat{r}_i = (x_i - y_i)/r$ is the unit radiation vector. We bring $\partial/\partial t$ out of the integral and interpret the products of delta functions as in the case of solution of eq. (12-a). We can then write the solution of eq. (19) as follows

$$4\pi p_3'(\bar{x},t) = -\frac{1}{c}\frac{\partial}{\partial t}\int_{K=0}\frac{1}{r}\left[\frac{\phi_r(\bar{y},\tau)}{\Lambda}\right]_{ret} d\Sigma - \int_{K=0}\frac{1}{r^2}\left[\frac{\phi_r(\bar{y},\tau)}{\Lambda}\right]_{ret} d\Sigma \tag{22}$$

where $\phi_r = \phi_i\hat{r}_i$.

The solution of eq. (11) with the source term given by eq. (5) can now be written based on the solutions of eqs. (12-a) and (19) as follows

$$4\pi p'(\bar{x},t) = -\frac{1}{c}\frac{\partial}{\partial t}\int_{K=0}\frac{1}{r}\left[\frac{q_r(\bar{y},\tau)}{\Lambda}\right]_{ret} d\Sigma + \int_{K=0}\left\{\frac{1}{r}\left[\frac{n_i^{\prime}\Delta\left(\partial T_{ij}/\partial x_j\right)}{\Lambda}\right]_{ret} - \frac{1}{r^2}\left[\frac{q_r(\bar{y},\tau)}{\Lambda}\right]_{ret}\right\}d\Sigma \tag{23}$$

where $q_r = q_i\hat{r}_i$. This is our first formulation for the prediction of the supersonic shock noise. We propose that the observer time differentiation be taken numerically in eq. (23).

We now concentrate on the solution of eq. (12-b) by Green's function technique. The formal solution after using the transformation $\tau \to g$ and integrating with respect to is

$$4\pi p_2'(\bar{x},t) = \int \frac{\hat{\phi}_2(\bar{y},\tau)}{r}\delta'(K)d\bar{y} \tag{24}$$

This can now be easily integrated by using eq. (15-a) and the operational property of $\delta'(\cdot)$ [6,7,10]. The algebraic manipulations are as follows

$$\begin{aligned} 4\pi p_2'(\bar{x},t) &= \int \frac{\hat{\phi}_2(\bar{y},\tau)}{r\Lambda}\delta'(K)dKd\Sigma \\ &= \int \frac{\hat{\phi}_2\sqrt{g_{(2)}}}{r\Lambda}\delta'(K)dKdu^1du^2 \end{aligned} \tag{25}$$

where in the last integral, we have introduced surface coordinates $\left(u^1,u^2\right)$ on the surface K=constant. The determinant of the coefficients of the first fundamental form [8] is denoted as $g_{(2)}$. We now integrate eq. (25) with respect to K. We get

$$4\pi p_2'(\vec{x},t) = -\int_{K=0} \frac{\partial}{\partial K}\left[\frac{\hat{\phi}_2\sqrt{g_{(2)}}}{r\Lambda}\right] du^1 du^2$$
$$= -\int_{K=0} \frac{1}{\Lambda}\frac{\partial}{\partial N'}\left[\frac{\hat{\phi}_2\sqrt{g_{(2)}}}{r\Lambda}\right] du^1 du^2 \qquad (26)$$

where $\partial/\partial N'$ is directional derivative in the direction of $\vec{N}'$, the unit normal to the surface $K=0$. This unit vector is given by the relation

$$\vec{N}' = \frac{\vec{n}' - M_{n'}\vec{\hat{r}}}{\Lambda}$$
$$= \frac{1}{\Lambda}\left[(1 - M_{n'}\cos\theta)\vec{n}' - M_{n'}\sin\theta\,\vec{\hat{t}}\right] \qquad (27)$$

where $\vec{\hat{t}}$ is the unit vector tangent to the shock surface $k=0$ in the direction of projection of $\vec{\hat{r}}$ on the local tangent plane of $k=0$.

We can show that [14]

$$\frac{\partial}{\partial N'}\sqrt{g_{(2)}} = -2H_K\sqrt{g_{(2)}} \qquad (28)$$

where H_K is the local mean curvature of the surface Σ: $K=0$. This curvature can be related to the local curvature of the shock surface $k=0$ and other kinematic parameters [14]. Using this relation in eq. (26), we obtain

$$4\pi p_2'(\vec{x},t) = -\int_{K=0}\left\{\frac{1}{\Lambda}\frac{\partial}{\partial N'}\left[\frac{\hat{\phi}_2}{r\Lambda}\right]\right\}_{ret} d\Sigma + \int_{K=0}\left[\frac{2\hat{\phi}_2 H_K}{r\Lambda^2}\right]_{ret} d\Sigma \qquad (29)$$

Since $\partial\hat{\phi}_2/\partial n' = 0$, we have from eq. (27)

$$\frac{\partial\hat{\phi}_2}{\partial N'} = -\frac{M_{n'}\sin\theta}{\Lambda}\frac{\partial\hat{\phi}_2}{\partial\hat{t}} \qquad (30)$$

where $\partial/\partial\hat{t}$ denotes the directional derivative of $\hat{\phi}_2$ along the tangent vector $\vec{\hat{t}}$ to the shock surface $k=0$ (not the surface Σ: $K=0$). We, thus, write the solution of eq. (12-b) in the form

$$4\pi p_2'(\vec{x},t) = \int_{K=0}\frac{1}{r}\left[\frac{M_{n'}\sin\theta}{\Lambda^2}\frac{\partial\phi_2}{\partial\hat{t}}\right]_{ret} d\Sigma + \int_{K=0}\left\{\frac{\phi_2}{\Lambda}\left[\frac{2H_K}{r\Lambda} - \frac{\partial}{\partial N'}\left(\frac{1}{r\Lambda}\right)\right]\right\}_{ret} d\Sigma \qquad (31)$$

Note that now we can drop the $(^\wedge)$ on ϕ_2 since in the first integral, $\partial/\partial\hat{t}$ only uses the restriction of ϕ_2 to the surface $k=0$ and in the second integral $K=0$ implies that $k=0$.

We note that the inhomogeneous source terms of eq. (11) given by eq. (10) are of the types of eqs. (12-a) and (12-b). Using the solutions of these latter equations derived above, we get the second formulation for supersonic shock noise prediction:

$$4\pi p'(\vec{x},t)=\int_{K=0}\frac{1}{r}\left[\frac{\psi_1}{\Lambda}+\frac{M_{n'}\sin\theta}{\Lambda^2}\frac{\partial q_{n'}}{\partial\hat{t}}\right]_{ret}d\Sigma+\int_{K=0}\left\{\frac{q_{n'}}{\Lambda}\left[\frac{2H_K}{r\Lambda}-\frac{\partial}{\partial N'}\left(\frac{1}{r\Lambda}\right)\right]\right\}_{ret}d\Sigma \qquad (32)$$

This closed form expression can be integrated on a computer by the collapsing sphere method mentioned above.

Concluding Remarks

In this paper, we have derived the solution of wave equation with sources on deformable boundaries moving at supersonic speed. The solution was given in two equivalent forms for use in the boundary element method for shock noise prediction. Existing noise prediction codes can be modified to incorporate the results of this paper. However, it must be emphasized that the results of this paper have applications in other areas involving wave propagation, such as aerodynamics and electromagnetism.

The fact that Green's function technique is used in finding closed form solutions incorporating the outgoing boundary condition eliminates difficulties associated with other methods, such as computational fluid dynamics. Problems with reflections from far-field boundaries in CFD do not appear in the boundary element methods for which our results were developed. It can be shown that solution of the wave equation with moving boundaries can develop integrable singularities in the field. Readers are referred to the references [15] and [16] for information on the singularities.

References

1. Ffowcs Williams, J. E.; Hawkings, D. L.: Sound Generated by Turbulence and Surfaces in Arbitrary Motion. Phil. Trans. Roy. Soc. A264 (1969) 321-342.

2. Farassat, F.; Brentner, Kenneth S.: The Uses and Abuses of the Acoustic Analogy in Helicopter Rotor Noise Prediction. AHS Jour. 33(1) (1988) 29-36.

3. Farassat, F.; Myers, M. K.: An Analysis of the Quadrupole Noise Source of High Speed Rotating Blades. Computational Acoustics (volume 2) Scattering, Gaussian Beams and Aeroacoustics (Ding Lee, Ahmet Cakmak and Robert Vichnevetsky, eds.), North Holland. (1990) 227-240.

4. Farassat, F.; Tadghighi, H.: Can Shock Waves on Helicopter Rotors Generate Noise?– A Study of Quadrupole Source. Presented at the AHS 46th Annual Forum & Technology Display, May 21-23, 1990, Washington, D.C.

5. Schmitz, F. H.; Yu, Y. H.: Helicotper Impulsive Noise: Theoretical and Experimental Status. Recent Advances in Aeroacoustics (A. Krothapalli and C. A. Smith, eds.), Springer-Verlag. (1986) 149-243.

6. Gel'fand, I. M.; Shilov, G. E.: Generalized Functions. (volume 1) Properties and Operations. Academic Press (1964).

7. Kanwal, R. P.: Generalized Functions– Theory and Technique. Academic Press (1983).

8. McConnell, A. J.: Applications of Tensor Analysis. Dover Publications (1957).

9. Farassat, F.: Linear Acoustic Formulas for Calculation of Rotating Blade Noise. AIAA Jour. 19(8) (1981) 1122-1130.

10. Farassat, F.: Discontinuities in Aerodyanmics and Aeroacoustics: The Concept and Applications of Generalized Derivatives. Jour. Sound and Vib. 55(2) (1977) 165-193.

11. Farassat, F.: Theoretical Analysis of Linearized Acoustics and Aerodynamics of Advanced Supersonic Propellers. AGARD-CP-366 (10) (1985) 1-15.

12. Farassat, F.: Lectures on the Aeroacoustics of Rotating Blades in time Domain. Lectures Delivered in the Department of Mechanics and Aeronautics, University of Rome, "La Sapienza," July 1989 (Unpublished).

13. Farassat, F.: Theory of Noise Generation from Moving Bodies with an Application to Helicotper Rotors. NASA Tech. Rep. R-451 (1975).

14. Farassat, F.; Myers, M. K.: The Moving Boundary Problem for the Wave Equation: Theory and Application. Computational Acoustics– (volume 2) Algorithms and applications. (D. Lee, R. L. Sternberg, M. H. Schultz, eds.), North Holland (1988).

15. De Bernardis, E.: On a New Formulation for the Aeroacoustics of Rotating Blades (in Italian). Ph.D. Thesis, University of Rome, "La Sapienza." (1989).

16. De Bernardis, E.; Farassat, F.: On the Possibility of Singularities in the Acoustic Field of Supersonic Sources When BEM is Applied to a Wave Equation. Presented at the International symposium on Boundary Element Methods. East Hartford, Conn., Oct. 2-4, 1989.

Hypersingular Boundary Integral Equations: A New Approach to Their Numerical Treatment

M. GUIGGIANI
Dipartimento di Costruzioni Meccaniche e Nucleari
Università degli Studi di Pisa, 56126 Pisa, Italy

G. KRISHNASAMY, F. J. RIZZO
Department of Theoretical and Applied Mechanics
University of Illinois at Urbana-Champaign, Urbana, IL 61801, USA

T. J. RUDOLPHI
Department of Mechanical Engineering and Engineering Mechanics
Iowa State University, Ames, IA 50011, USA

Summary

In the first part, hypersingular boundary integral equations are obtained through proper consideration of the limiting process. It is proved that no divergent terms actually arise, and that interpretations of the integrals are not required. In the second part, a general algorithm for the direct numerical treatment of hypersingular integrals in the BEM is developed. The proposed approach operates in terms of intrinsic coordinates and shows any hypersingular integral in the BEM to be equivalent to a sum of two regular integrals. Numerical results on *curved* elements are presented.

1 Introduction

Hypersingular kernels arise whenever the gradient of a standard integral equation is taken. The well-known integral equation for the stress tensor at internal points in elastic problems is a typical example of integral equation with hypersingular (and strongly singular) kernels.

In some important BEM analyses, such as crack problems, elastoplasticity, viscoplasticity, shape optimization, plate bending, resolution of fictitious eigenfrequencies, symmetric formulations, etc., it is of vital importance to employ *boundary* integral equations with *hypersingular* kernels, that is to take the source point on the boundary. Moreover, in many other fields a reliable application of hypersingular boundary integral equations (HBIE) would be beneficial. For instance, it would be possible to compute the whole stress tensor directly on the boundary by using the aforementioned integral equation for the stress tensor.

So far, the usual approach has been for avoiding the direct use of HBIE's by first employing various regularization techniques. Integration by parts (in various forms) [12, 1, 10, 9, 2], Stokes' theorem [7], simple analytical solutions [11] were all employed to transform hypersingular integrals to less singular ones. In [3] a direct analytical integration is carried out, along with a limiting process, although only on flat elements.

In this paper, a fresh approach is developed. In the first part, the limiting process leading to hypersingular boundary integral equations is thourougly discussed. It is shown that theoretical difficulties in dealing with HBIE's are only apparent since no unbounded quantities actually arise.

As a result of this analysis, a new unambiguous way of writing HBIE's in terms of the original variables is presented. It can be regarded as the extension to the hypersingular case of the usual way of writing boundary integral equations.

Careful analysis of the limiting process has also strong relevance for the development of an appropriate numerical algorithm.

In the second part, a new approach to the direct evaluation of integrals with hypersingular kernels is presented. Since the suggested method operates in terms of intrinsic coordinates, special care is exercised to preserve the features of the limiting process when the discretization of the geometry is introduced. This is a key point for the rigorous treatment of singular integrals in intrinsic coordinates. The use of intrinsic coordinates gives full generality to the proposed method.

Numerical results confirm that a *direct numerical* treatment of HBIE's is very effective (this is somehow contrary to general belief). The actual computation only requires standard Gaussian quadrature fomulae of low order.

2 Limiting Process for Hypersingular BIE's

Analysis is presented for elastic problems since they are a paradigma for any other elliptic problem.

Let us first consider the well-known standard boundary integral equation for a three-dimensional domain Ω, bounded by the surface S with unit outward normal $\vec{n}(\mathbf{x}) = \{n_i\}$

$$\lim_{\varepsilon\to 0}\left\{\int_{(S-e_\varepsilon)+s_\varepsilon}[T_{ij}(\mathbf{y},\mathbf{x})u_j(\mathbf{x}) - U_{ij}(\mathbf{y},\mathbf{x})t_j(\mathbf{x})]dS_x\right\} = 0 \tag{1}$$

where u_j and t_j denote the displacement and traction components, respectively.

If $r = [(x_j - y_j)(x_j - y_j)]^{1/2}$ denotes the distance between the source point $\mathbf{y} = \{y_i\}$ and integration point $\mathbf{x} = \{x_i\}$, the fundamental solution U_{ij} has a *weak* singularity of order r^{-1}, when $r \to 0$, while the other kernel function T_{ij} has a *strong* singularity of order r^{-2}.

In eq. (1) the source point $\mathbf{y} \in S$. Since eq. (1) stems from Green's second identity, it may be only formulated on a domain *not* including the singular point $\mathbf{y}$. Therefore, a (vanishing) neighbourhood v_ε of $\mathbf{y}$ has been removed from the original domain Ω (Figure 1). The integration is thus performed on the boundary $(S - e_\varepsilon) + s_\varepsilon$ of the new domain. Of course, the integration must be done before taking the limit.

Notice that equation (1) already states that the value of the overall limit is zero. Thus, all divergent parts (if any) will be cancelled out in the end.

It is not necessary to take a sphere (a circle in 2D) to exclude the point $\mathbf{y}$. Indeed, equation (1) (i.e., the Green's second identity) holds *whatever* the shape of the chosen neighbourhood v_ε. A sphere is merely the most convenient shape.

However, the shape of e_ε must be *consistent* with the shape of s_ε throughout the process.

Since all functions in equation (1) are regular, we can differentiate it with respect to any coordinate y_k of the source point, thus obtaining

$$\lim_{\varepsilon\to 0}\left\{\int_{(S-e_\varepsilon)+s_\varepsilon}[V_{ikj}(\mathbf{y},\mathbf{x})u_j(\mathbf{x}) - W_{ikj}(\mathbf{y},\mathbf{x})t_j(\mathbf{x})]dS_x\right\} = 0 \tag{2}$$

where, for 3D elastostatic problems, we have

$$V_{ikj} = \frac{\partial T_{ij}}{\partial y_k} = -\frac{A}{r^3}\Big\{ (1-2\nu)\left[3r_{,k}\left(n_i r_{,j} - n_j r_{,i}\right) - n_i\delta_{jk} + n_j\delta_{ik} - n_k\delta_{ij}\right]$$

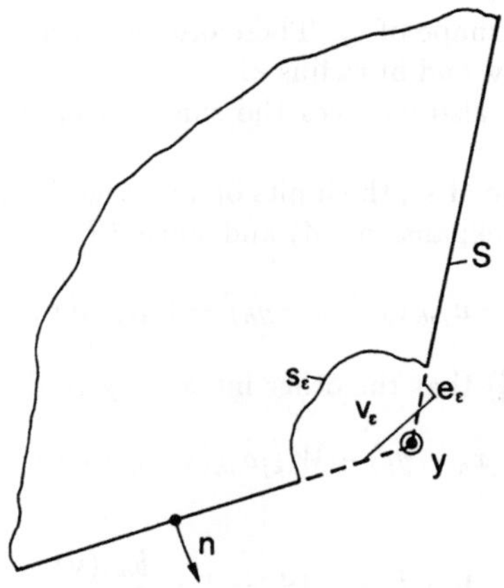

Figure 1: Exclusion of the singular point **y** by a vanishing volume.

$$- 3n_k r_{,i} r_{,j} + 3\left[5r_{,i} r_{,j} r_{,k} + (1-2\nu)\delta_{ij} r_{,k} - \delta_{ik} r_{,j} - \delta_{jk} r_{,i}\right] \frac{\partial r}{\partial n}\bigg\} ;$$

$$W_{ikj} = \frac{\partial U_{ij}}{\partial y_k} = \frac{A}{2G\,r^2}\left[3r_{,i} r_{,j} r_{,k} + (3-4\nu)\delta_{ij} r_{,k} - \delta_{ik} r_{,j} - \delta_{jk} r_{,i}\right] \qquad (3)$$

where $r_{,i} = \partial r/\partial x_i = -\partial r/\partial y_i$. As expected, the kernel W_{ikj} shows a *strong* singularity of order r^{-2} while the kernel V_{ikj} is *hypersingular* of order r^{-3}, as $r \to 0$. The above expressions also represent the asymptotic behaviour of the corresponding kernels for time-harmonic elastodynamics (e.g. [1, 7]).

Now, we assume that the displacement $u_j \in C^{1,\alpha}$, at **y** (i.e., u_j is differentiable at **y**, with its derivatives satisfying a Hölder condition). Accordingly, we have the following expansions

$$\begin{aligned} u_j(\mathbf{x}) &= u_j(\mathbf{y}) + u_{j,h}(\mathbf{y})(x_h - y_h) + O(r^{1+\alpha}); \\ t_j(\mathbf{x}) = \sigma_{jh}(\mathbf{x})n_h(\mathbf{x}) &= \sigma_{jh}(\mathbf{y})n_h(\mathbf{x}) + O(r^\alpha) \end{aligned} \qquad (4)$$

where $\alpha > 0$ (usually, $\alpha = 1$). Essentially, expansions (4) mean that u_j *and* its gradient $u_{j,h}$ are continuous at **y**. This fact has also relevance in the selection of the discretization scheme, as it has to satisfy the same continuity requirements ($u_j \in C^{1,\alpha}$, and $t_j \in C^{0,\alpha}$) at each collocation point. As also stated in [8, 7, 3, 2], these continuity requirements are demanded by the nature of the hypersingularity, and hence need to be satisfied no matter what method is used.

By adding and subtracting in (2) the relevant terms of expansions (4), a more convenient form of the HBIE (2) is obtained as

$$\begin{aligned} \lim_{\varepsilon\to 0} \bigg\{ &\int_{(S-e_\varepsilon)} [V_{ikj}(\mathbf{y},\mathbf{x})u_j(\mathbf{x}) - W_{ikj}(\mathbf{y},\mathbf{x})t_j(\mathbf{x})]dS_x \\ &+ \int_{s_\varepsilon} \Big(V_{ikj}[u_j(\mathbf{x}) - u_j(\mathbf{y}) - u_{j,h}(\mathbf{y})(x_h - y_h)] - W_{ikj}[t_j(\mathbf{x}) - \sigma_{jh}(\mathbf{y})n_h(\mathbf{x})] \Big) dS_x \\ &+ u_j(\mathbf{y}) \int_{s_\varepsilon} V_{ikj}\, dS_x + \int_{s_\varepsilon} [V_{ikj}u_{j,h}(\mathbf{y})(x_h - y_h) - W_{ikj}\sigma_{jh}(\mathbf{y})n_h(\mathbf{x})]dS_x \bigg\} = 0 \end{aligned} \qquad (5)$$

The value of the limit taken as a whole in either equations (2) or (5) is completely independent on the selected shape of v_ε. Therefore, we select the most convenient shape, that is a sphere centered at $\mathbf{y}$ and of radius ε.

The selected shape of s_ε also enforces the shape of e_ε, which becomes a *symmetric* neighbourhood (Figure 1).

Owing to the simple shape of s_ε, the limits of all integrals on s_ε in (5) can be evaluated analytically. Because of the expansions (4) and since $dS_x = O(\varepsilon^2)$ on s_ε, it follows that

$$\lim_{\varepsilon\to 0} \int_{s_\varepsilon} \Big(V_{ikj}[u_j(\mathbf{x}) - u_j(\mathbf{y}) - u_{j,h}(\mathbf{y})(x_h - y_h)] - W_{ikj}[t_j(\mathbf{x}) - \sigma_{jh}(\mathbf{y})n_h(\mathbf{x})] \Big) dS_x = 0 \quad (6)$$

It is easy to show (see [6]) that the other integrals give rise to the following terms

$$\lim_{\varepsilon\to 0} \int_{s_\varepsilon} [V_{ikj}u_{j,h}(\mathbf{y})(x_h - y_h) - W_{ikj}\sigma_{jh}(\mathbf{y})n_h(\mathbf{x})]dS_x = c_{ikjh}(\mathbf{y})u_{j,h}(\mathbf{y}) \quad (7)$$

and

$$\lim_{\varepsilon\to 0} \int_{s_\varepsilon} V_{ijk}dS_x = \lim_{\varepsilon\to 0} \frac{b_{ikj}(\mathbf{y})}{\varepsilon} \quad (8)$$

where c_{ikjh} and b_{ikj} are (bounded) coefficients that only depend upon the local geometry of S at $\mathbf{y}$.

The coefficients $c_{ikjh}(\mathbf{y})$ are the *free-term coefficients* of the hypersingular boundary integral equation for displacement derivatives. Notice that, in general, both kernels W_{ikj} and V_{ikj} in (7) contribute to them. At smooth boundary points the free-term simply become $c_{ikjh}u_{j,h} = 0.5\,u_{i,k}$.

Expression (8) shows that the limit on s_ε of the integral of V_{ikj} is either zero or unbounded, depending on the value of $b_{ikj}(\mathbf{y})$. $b_{ikj} = 0$ if $\mathbf{y}$ is an internal point for Ω. If $\mathbf{y}$ is a boundary point, then $b_{ikj} \neq 0$ (in general), and the limit in (8) is unbounded of order ε^{-1}. However, this problem is only apparent. As a matter of fact, it arose only because we artificially separated the integrals on s_ε from the integral on $(S - e_\varepsilon)$. If they are considered *together* as they are in the original equation (5) (or (2)), no unbounded quantities arise at all.

The separation into integrals on s_ε and on the remaining surface $(S - e_\varepsilon)$ is allowed only when each single term remains bounded by itself, which is not always the case in HBIE's.

According to the analysis above, the *hypersingular boundary integral equation* for vector problems can be written in the following form

$$c_{ikjh}(\mathbf{y})u_{j,h}(\mathbf{y}) + \lim_{\varepsilon\to 0}\left\{ \int_{(S-e_\varepsilon)} [V_{ikj}(\mathbf{y},\mathbf{x})u_j(\mathbf{x}) - W_{ikj}(\mathbf{y},\mathbf{x})t_j(\mathbf{x})]\, dS_x + u_j(\mathbf{y})\frac{b_{ikj}(\mathbf{y})}{\varepsilon} \right\} = 0 \quad (9)$$

This is the first fundamental result of this paper. Contrary to common practice, in equation (9) the limiting process is still indicated explicitely. Hypersingular boundary integral equations in this form are not only rigorous, but also unambiguous. All terms have a clear meaning, and well defined mathematical concepts have been used in all steps.

A few comments are in order here. Equation (9) has the same formal structure of the classical Somigliana identity, for $\mathbf{y} \in S$. Indeed, the same basic steps have been used in its derivation. The same kind of limiting process have been employed. We still have a free-term and a singular integral. Moreover, the problem is still formulated in the original (physical) variables. We can truly speak of hypersingular Somigliana identity when referring to eq. (9).

The higher order of singularity required stronger continuity requirements on u_j and t_j, not on the geometry. $\mathbf{y}$ may well be at a non-smooth boundary point. This fact clearly stems from the derivation of eq. (9), and, apparently, it is reported here for the first time.

Equation (9) also show that the concepts of Cauchy principal value and Hadamard finite part are not necessary. The explicit consideration of the limiting process solves all theoretical problems. The balance between unbounded quantities is made explicit. The terms $u_j(\mathbf{y})\, b_{ikj}/\varepsilon$ cancel the singular terms arising from the integral on $(S - e_\varepsilon)$.

Equations like (9) can be used to compute all displacement derivatives $u_{i,j}\,(\mathbf{y})$ at any boundary point, which, e.g., allows for the evaluation of the whole stress tensor $\sigma_{ij}(\mathbf{y})$ directly on the boundary.

The numerical treatment proceeds *directly* from equations with hypersingular kernels in the form of (9).

It is important noting that the shape of the vanishing neighbourhood e_ε must be consistent with the already evaluated coefficients c_{ikjh} and b_{ikj}, even after the geometry is represented by boundary elements.

3 Numerical Evaluation of Hypersingular Integrals

Since W_{ikj} is only strongly singular, the evaluation of the limit (which, by virtue of the symmetric shape of e_ε, coincides with the Cauchy principal value of the surface integral)

$$\lim_{\varepsilon \to 0} \left\{ - \int_{(S-e_\varepsilon)} W_{ikj}(\mathbf{y},\mathbf{x}) t_j(\mathbf{x}) dS_x \right\} \tag{10}$$

can be achieved, in full generality, by the direct numerical method presented by Guiggiani and Gigante [5, 4].

On the other hand, a new method is necessary for the evaluation of the quantity

$$\lim_{\varepsilon \to 0} \left\{ \int_{(S-e_\varepsilon)} V_{ikj}(\mathbf{y},\mathbf{x}) u_j(\mathbf{x}) dS_x + u_j(\mathbf{y}) \frac{b_{ikj}(\mathbf{y})}{\varepsilon} \right\} \tag{11}$$

where V_{ikj} is hypersingular, as $r \to 0$.

Notice that, *consistently* with the already obtained $c_{ikjh}(\mathbf{y})$, and $b_{ikj}(\mathbf{y})$, the neighbourhood e_ε on S is given by $e_\varepsilon = \{\mathbf{x} \in S |\, |\mathbf{x} - \mathbf{y}| \le \varepsilon\}$.

Limits (10) and (11) can be considered separately. They are both bounded. In fact, the singular kernels are multiplied by different functions, and reciprocal cancelling effects are therefore not possible, in general.

We denote that portion of S containing the singular point $\mathbf{y}$ by S_s. If discontinuous elements with collocation at element interiors are used, then S_s consists of just one element; whereas if $C^{1,\alpha}$-continuous element are used to represent u_i, S_s consists of all adjacent elements connected to the singular point $\mathbf{y}$.

For simplicity, the case of $\mathbf{y}$ belonging to just one element is considered here. However, the analysis is in no way restricted to this case.

As usual, on each boundary element, the displacement is represented in terms of shape functions and nodal values, $u_i(\mathbf{x}) = N^c(\boldsymbol{\xi}(\mathbf{x}))u_i^c$, where $\boldsymbol{\xi} = (\xi_1, \xi_2)$ are the intrinsic coordinates.

From a computational standpoint, we need the evaluation of the quantity I defined on the element S_s

$$I = \lim_{\varepsilon \to 0} \left\{ \int_{(S_s-e_\varepsilon)} V_{ikj}(\mathbf{y},\mathbf{x}) N^a(\boldsymbol{\xi}(\mathbf{x})) dS_x + N^a(\boldsymbol{\eta}) \frac{b_{ikj}(\mathbf{y})}{\varepsilon} \right\} \tag{12}$$

where N^a represents those shape functions (usually just one) that are not zero at $\boldsymbol{\eta} = \boldsymbol{\xi}(\mathbf{y})$, the *image* in the local plane of the collocation point $\mathbf{y}$ (quite often, $N^a(\boldsymbol{\eta}) = 1$).

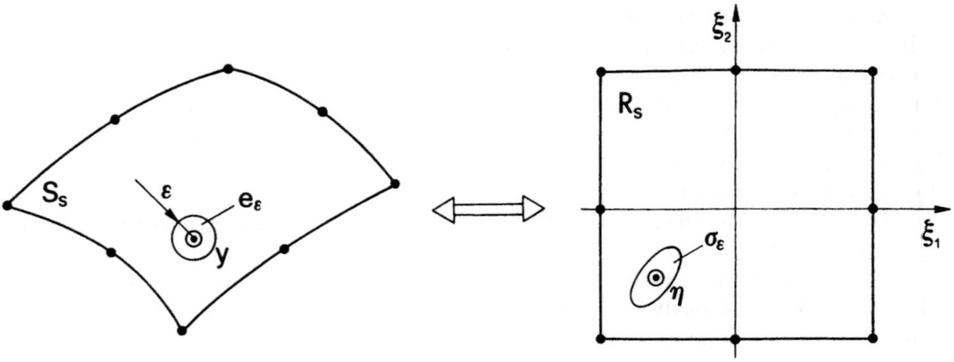

Figure 2: Image in the space of intrinsic coordinates of the boundary element and of the vanishing neighbourhood.

By means of the usual representation for the geometry in terms of shape functions and nodal coordinates

$$x_k(\boldsymbol{\xi}) = M^c(\boldsymbol{\xi})x_k^c \tag{13}$$

the boundary element S_s is mapped onto a region R_s of standard shape in the $\boldsymbol{\xi}$-space (e.g., a square, or a right triangle).

Accordingly, the symmetric neighbourhood e_ε of $\mathbf{y}$ in the real space is mapped onto a neighbourhood σ_ε of $\boldsymbol{\eta}$ in the $\boldsymbol{\xi}$-space. (Figure 2). It is important to note that, in general, σ_ε is *not a circle*.

Thus, in the space of intrinsic coordinates, expression (12) becomes

$$I = \lim_{\varepsilon\to 0}\left\{\int_{R_s-\sigma_\varepsilon} V_{ikj}(\mathbf{y},\mathbf{x}(\boldsymbol{\xi}))N^a(\boldsymbol{\xi})J(\boldsymbol{\xi})\,d\xi_1 d\xi_2 + N^a(\boldsymbol{\eta})\frac{b_{ikj}(\mathbf{y})}{\varepsilon}\right\} \tag{14}$$

where $dS_x = J(\boldsymbol{\xi})\,d\xi_1\,d\xi_2$. It is worth noting that geometric elements can be of any kind and order.

Following a common practice in the BEM, polar coordinates $(\rho,\,\theta)$ centered at $\boldsymbol{\eta}$ (the image of $\mathbf{y}$) are defined in the $\boldsymbol{\xi}$-space (Figure 2)

$$\begin{cases} \xi_1 &= \eta_1 + \rho\cos\theta \\ \xi_2 &= \eta_2 + \rho\sin\theta \end{cases}$$

so that $d\xi_1\,d\xi_2 = \rho\,d\rho\,d\theta$. Hence, the quantity I in (14) becomes

$$I = \lim_{\varepsilon\to 0}\left\{\int_0^{2\pi}\int_{\alpha(\varepsilon,\theta)}^{\hat\rho(\theta)} F_{ikj}(\rho,\theta)\,d\rho\,d\theta + N^a(\boldsymbol{\eta})\frac{b_{ikj}(\mathbf{y})}{\varepsilon}\right\} \tag{15}$$

where: $F_{ikj}(\rho,\theta) = V_{ikj}\,N^a\,J\,\rho = O(\rho^{-2})$ is the hypersingular integrand, $\rho = \alpha(\varepsilon,\theta)$ is the equation in polar coordinates of the distorted neighbourhood σ_ε (Figure 2), and $\rho = \hat\rho(\theta)$ is the equation in polar coordinates of the external contour of R_s.

Now, let us analyse the singular function $F_{ikj}(\rho,\theta)$. Since it is singular of order ρ^{-2}, we have a (Laurent) series expansion with respect to ρ in the form (subscripts in the expansion are dropped)

$$F_{ikj}(\rho,\theta) = \frac{F_{-2}(\theta)}{\rho^2} + \frac{F_{-1}(\theta)}{\rho} + O(1) \tag{16}$$

Notice that in the BEM both F_{-2} and F_{-1} are just real functions of θ (even when $F_{ikj}(\rho,\theta)$ is complex valued, as in time-harmonic problems). The dependence on θ is crucial for expansion (16)) to actually represent the asymptotic behaviour of $F_{ikj}(\rho,\theta)$, when $\rho \to 0$. Expansion (16) is one of the key ingredients of the present analysis.

Also of basic relevance is the Taylor series expansion for $\alpha(\varepsilon,\theta)$, with respect to ε

$$\rho = \alpha(\varepsilon,\theta) = \varepsilon\,\beta(\theta) + \varepsilon^2\gamma(\theta) + O(\varepsilon^3) \tag{17}$$

Note that, in general, $\rho = \varepsilon\,\beta(\theta)$ is the equation of an ellipse.

A systematic way of obtaining the explicit expressions of $F_{-2}(\theta)$, $F_{-1}(\theta)$, $\beta(\theta)$ and $\gamma(\theta)$ is presented in [6]. It is really an easy task for any kernel function.

Adding and subtracting the first two terms of the series expansion (16) in expression (15), we obtain

$$\begin{aligned}
I = \lim_{\varepsilon\to 0}&\left\{\int_0^{2\pi}\int_{\alpha(\varepsilon,\theta)}^{\hat\rho(\theta)}\left[F_{ikj}(\rho,\theta) - \left(\frac{F_{-2}(\theta)}{\rho^2} + \frac{F_{-1}(\theta)}{\rho}\right)\right] d\rho\, d\theta\right.\\
&\left. + \int_0^{2\pi}\int_{\alpha(\varepsilon,\theta)}^{\hat\rho(\theta)} \frac{F_{-1}(\theta)}{\rho} d\rho\, d\theta + \left[\int_0^{2\pi}\int_{\alpha(\varepsilon,\theta)}^{\hat\rho(\theta)} \frac{F_{-2}(\theta)}{\rho^2} d\rho\, d\theta + N^a(\eta)\frac{b_{ikj}(\mathrm{y})}{\varepsilon}\right]\right\} =\\
&= I_0 + I_{-1} + I_{-2}
\end{aligned} \tag{18}$$

Each term I_0, I_{-1}, and I_{-2} in (18) is now analysed separately.

According to equation (16), in I_0 the integrand is *regular*. Therefore, the limit is straightforward and simply becomes

$$I_0 = \int_0^{2\pi}\int_0^{\hat\rho(\theta)}\left[F_{ikj}(\rho,\theta) - \left(\frac{F_{-2}(\theta)}{\rho^2} + \frac{F_{-1}(\theta)}{\rho}\right)\right] d\rho\, d\theta \tag{19}$$

This double integral can be evaluated by standard quadrature rules.

Now, let us consider I_{-1} and integrate to get

$$\begin{aligned}
I_{-1} &= \lim_{\varepsilon\to 0}\int_0^{2\pi}\int_{\alpha(\varepsilon,\theta)}^{\hat\rho(\theta)}\frac{F_{-1}(\theta)}{\rho} d\rho\, d\theta = \lim_{\varepsilon\to 0}\int_0^{2\pi} F_{-1}(\theta)[\ln|\hat\rho(\theta)| - \ln|\alpha(\varepsilon,\theta)|]d\theta\\
&= \int_0^{2\pi} F_{-1}(\theta)\ln|\hat\rho(\theta)|d\theta - \lim_{\varepsilon\to 0}\int_0^{2\pi} F_{-1}(\theta)\ln|\varepsilon\,\beta(\theta)|d\theta\\
&= \int_0^{2\pi} F_{-1}(\theta)\ln\left|\frac{\hat\rho(\theta)}{\beta(\theta)}\right| d\theta - \lim_{\varepsilon\to 0}\left\{\ln\varepsilon\int_0^{2\pi} F_{-1}(\theta)d\theta\right\}\\
&= \int_0^{2\pi} F_{-1}(\theta)\ln\left|\frac{\hat\rho(\theta)}{\beta(\theta)}\right| d\theta
\end{aligned} \tag{20}$$

Equation (20) shows that I_{-1} is equivalent to a simple *regular* one-dimensional integral. In the derivation of this equation, first we integrated analytically with respect to ρ, then we made use of expansion (17) (limited to the first term), and, finally, we considered the property $\int_0^{2\pi} F_{-1}(\theta)d\theta = 0$.

A similar treatment applies to I_{-2} such that

$$\begin{aligned}
I_{-2} &= \lim_{\varepsilon\to 0}\left\{\int_0^{2\pi}\int_{\alpha(\varepsilon,\theta)}^{\hat\rho(\theta)}\frac{F_{-2}(\theta)}{\rho^2}\, d\rho\, d\theta + N^a(\eta)\frac{b_{ikj}(\mathrm{y})}{\varepsilon}\right\}\\
&= \lim_{\varepsilon\to 0}\left\{\int_0^{2\pi} F_{-2}(\theta)\left[-\frac{1}{\hat\rho(\theta)} + \frac{1}{\alpha(\varepsilon,\theta)}\right] d\theta + N^a(\eta)\frac{b_{ikj}(\mathrm{y})}{\varepsilon}\right\}
\end{aligned}$$

$$= \lim_{\varepsilon \to 0} \left\{ \int_0^{2\pi} \frac{F_{-2}(\theta)}{\varepsilon\,\beta(\theta)} \left(1 - \varepsilon \frac{\gamma(\theta)}{\beta(\theta)}\right) d\theta + N^a(\boldsymbol{\eta}) \frac{b_{ikj}(\mathbf{y})}{\varepsilon} \right\} - \int_0^{2\pi} \frac{F_{-2}(\theta)}{\hat{\rho}(\theta)}\, d\theta$$

$$= \lim_{\varepsilon \to 0} \left(\frac{1}{\varepsilon}\right) \left\{ \int_0^{2\pi} \frac{F_{-2}(\theta)}{\beta(\theta)} d\theta + N^a(\boldsymbol{\eta}) b_{ikj}(\mathbf{y}) \right\} - \int_0^{2\pi} F_{-2}(\theta) \left[\frac{\gamma(\theta)}{\beta^2(\theta)} + \frac{1}{\hat{\rho}(\theta)}\right] d\theta$$

$$= - \int_0^{2\pi} F_{-2}(\theta) \left[\frac{\gamma(\theta)}{\beta^2(\theta)} + \frac{1}{\hat{\rho}(\theta)}\right] d\theta \tag{21}$$

Therefore, I_{-2} is also equivalent to just a one-dimensional *regular* integral. Owing to the higher order of singularity of the integrand (cf. (20)), in this case both terms $\varepsilon\,\beta(\theta)$ and $\varepsilon^2\gamma(\theta)$ must be retained in the expansion (17) for $\alpha(\varepsilon, \theta)$.

The singularity cancellation we have been speaking about is made explicit in (21), where $\{\int_0^{2\pi} [F_{-2}(\theta)/\beta(\theta)]\, d\theta + N^a(\boldsymbol{\eta}) b_{ikj}(\mathbf{y})\} = 0$. The unbounded term $N^a b_{ikj}/\varepsilon$ due to an integral on s_ε (see (9)) is cancelled out by a corresponding unbounded term arising from the integral on $(S_s - e_\varepsilon)$, so that the final result is perfectly bounded and meaningful. This cancellation is strictly related to the nature of the kernels involved.

By collecting the previous results, the following *final formula* for the evaluation of hypersingular integrals in three-dimensional BEM analyses can be given

$$\begin{aligned} I \;=\; & \int_0^{2\pi} \int_0^{\hat{\rho}(\theta)} \left\{ F_{ikj}(\rho, \theta) - \left[\frac{F_{-2}(\theta)}{\rho^2} + \frac{F_{-1}(\theta)}{\rho}\right] \right\} d\rho\, d\theta \\ & + \int_0^{2\pi} \left\{ F_{-1}(\theta) \ln \left|\frac{\hat{\rho}(\theta)}{\beta(\theta)}\right| - F_{-2}(\theta) \left[\frac{\gamma(\theta)}{\beta^2(\theta)} + \frac{1}{\hat{\rho}(\theta)}\right] \right\} d\theta \end{aligned} \tag{22}$$

This formula is the second fundamental result of the present paper. It proves that the quantity I, originally given by a limiting process involving a hypersingular integral plus an unbounded term (see (12), (14), and (16)), is simply equal to a regular double integral plus a regular one-dimensional integral. Hence, it is easily computable.

The terms containing $\hat{\rho}(\theta)$ takes into account the external shape of R_s, while the terms with $\beta(\theta)$ and $\gamma(\theta)$ account for the *distorsion* of σ_ε, that is introduced by the mapping in the originally symmetric neighbourhood e_ε (Figure 2).

Both integrals in (22) are in polar coordinates defined in the local plane, which allows for a standard numerical implementation. Standard Gaussian quadrature rules of low order provide very good accuracy. After [5] and formula (22), the idea that singular integrals are intractable for numerical computations should be definitely abandoned.

Formula (22) is fully general. It holds for any kind of boundary elements employed, provided that the necessary continuity requirements for u_i are satisfied at each collocation point. Of course, a formula formally identical to (22) can be given for any other hypersingular boundary integral equation.

If the singular point is shared by more than one element, formula (22) becomes

$$\begin{aligned} I \;=\; & \sum_m \left\{ \int_{\theta_1^m}^{\theta_2^m} \int_0^{\hat{\rho}^m(\theta)} \left[F^m_{ikj}(\rho, \theta) - \left(\frac{F^m_{-2}(\theta)}{\rho^2} + \frac{F^m_{-1}(\theta)}{\rho}\right) \right] d\rho\, d\theta \right. \\ & \left. + \int_{\theta_1^m}^{\theta_2^m} \left[F^m_{-1}(\theta) \ln \left|\frac{\hat{\rho}^m(\theta)}{\beta^m(\theta)}\right| - F^m_{-2}(\theta) \left(\frac{\gamma^m(\theta)}{(\beta^m(\theta))^2} + \frac{1}{\hat{\rho}^m(\theta)}\right) \right] d\theta \right\} \end{aligned} \tag{23}$$

where the index m refers to one element around the collocation point at a time, and $\theta_1^m \le \theta \le \theta_2^m$ on the m-th element.

4 Numerical Test: Hypersingular Integrals on Curved Elements

A hypersingular kernel function for 3D potential problems $V_3 = -(1/4\pi r^3)[3r_{,3}\,\partial r/\partial n - n_3(\mathbf{x})]$ is integrated over four curved quadratic elements (Figure 3). Formula (23) is used, with $N^a(\boldsymbol{\xi}) \equiv 1$ (therefore $C^{1,\alpha}$-continuity is satisfied). Notice that kernel V_3 is made up of a hypersingular and of a strongly singular term. However, they must be considered together for the cancellation in (21) to occur. The results are reported in Table 1, for Gauss orders from 4 to 10. The results are remarkably stable. Order 6 already provides 7 exact digits. The elastic kernels (cf. (3)) are given by combinations of terms similar to V_3.

Table 1: Numerical evaluation of hypersingular integrals on curved elements with Gaussian formulae (Figure 3).

order n	numerical value
4	-0.3649122
6	-0.3649081
8	-0.3649081
10	-0.3649081

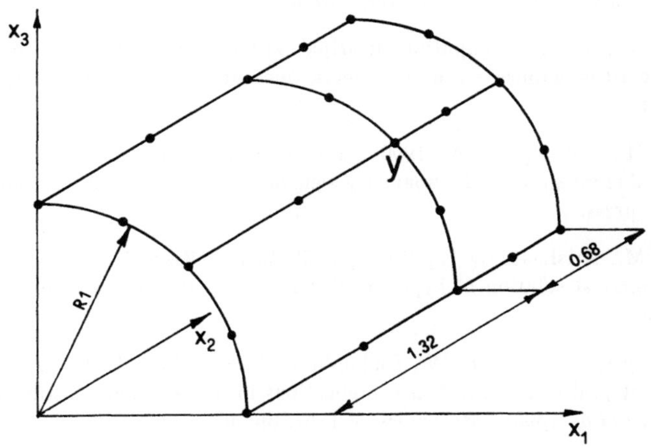

Figure 3: Curved boundary elements and collocation point.

5 Conclusions

In the first part of this paper, it has been shown that theoretical problems in dealing with hypersingular BIE's are only apparent. In fact, no unbounded quantities arise if the limit is properly taken. Equation (9) provides a rigorous, unambiguous form for any hypersingular BIE, even for $\mathbf{y}$ at non-smooth boundary points.

In the second part, a new *direct* approach to the evaluation of integrals with hypersingular kernels has been presented. The method deeply rely on the first theoretical part. Formula (22) (or (23)) shows the regular integrals that need be computed. Since all computations are carried out in the space of intrinsic coordinates, the proposed method can easily deal with boundary elements of any kind. Interestingly, it can be regarded as the extension, with additional relevant work, of the method developed in [5, 4] for strongly singular integrals. Actual computation only requires standard quadrature formulas of low order, as shown by numerical tests.

Acknowledgements

Part of this work was carried out while M. Guiggiani was visiting Iowa State University sponsored by a CNR fellowship. Additional support to M. G. has been provided by MURST. Partial support for F. J. Rizzo was provided by the Office of Naval Research under Contract No. N00014-89-K-0109, Y. Rajapakse program official.

References

[1] Bonnet M., 1989, "Regular boundary integral equations for three-dimensional finite or infinite bodies with or without curved cracks in elastodynamics", *Boundary Element Techniques: Applications in Engineering*, Brebbia C.A. and Zamani N., eds., Computational Mechanics Publications, Southampton, pp. 171–188..

[2] Cruse T. A., and Novati G., 1990, "Traction BIE formulations and applications to non-planar and multiple cracks", forthcoming.

[3] Gray L. J., Martha L. F., and Ingraffea A. R., 1990, "Hypersingular integrals in boundary element fracture analysis", *Int. J. Num. Methods Eng.*, Vol. 29, pp. 1135–1158.

[4] Guiggiani M., 1989, "Computing principal value integrals in three-dimensional time-harmonic elastodynamics—A direct general method", *Proc. ISBEM-89*, East Hartford, Connecticut.

[5] Guiggiani M., and Gigante A., 1990, "A general algorithm for multidimensional Cauchy principal value integrals in the boundary element method", ASME *Journal of Applied Mechanics*, in press.

[6] Guiggiani M., Krishnasamy G., Rudolphi T. J., and Rizzo F. J., "A general algorithm for the numerical solution of hypersingular boundary integral equations", submitted for publication.

[7] Krishnasamy G., Schmerr L. W., Rudolphi T. J., and Rizzo F. J., 1990, "Hypersingular boundary integral equations: Some applications in acoustic and elastic wave scattering", ASME *Journal of Applied Mechanics*, Vol. 57, pp. 404–414.

[8] Martin P. A., and Rizzo F. J., 1989, "On boundary integral equations for crack problems", *Proc. Royal Soc. London*, Vol. A 421, pp. 341–355.

[9] Nishimura N., and Kobayashi S., 1989, "A regularized boundary integral equation method for elastodynamic crack problems", *Computational Mechanics*, Vol. 4, pp. 319–328.

[10] Polch E. Z., Cruse T. A., and Huang C.-J., 1987, "Traction BIE solutions for flat cracks", *Computational Mechanics*, Vol. 2, pp. 253–267.

[11] Rudolphi T. J., 1990, "The use of simple solutions in the regularization of hypersingular boundary integral equations", *Computers and Mathematics with Applications*, Special Issue on BIEM/BEM, in press.

[12] Sládek V., and Sládek J., 1984, "Transient elastodynamic three-dimensional problems in cracked bodies", *Applied Mathematical Modelling*, Vol. 8, pp. 2–10.

Hyperbolic Grid Generation with BEM Source Term

M.H.L. HOUNJET

National Aerospace Laboratory (NLR)
Anthony Fokkerweg 2
1059 CM Amsterdam
The Netherlands

Abstract

A method is presented for the generation of O-type grids about transverse cross-sections of transport type aircraft. The method combines a hyperbolic grid generation scheme with source terms obtained with a boundary element method in such a way that O-type grids around fairly complex shapes with concavities can be generated easily. The components of the method: a boundary element method, a method to generate grids with a boundary element method and the hyperbolic grid generation scheme are described and applications are shown.

1 Introduction

In the development of many modern airplanes aeroelastic analysis is required in the transonic speed range. This implies the development of computer methods to determine the unsteady transonic flow about realistic aircraft configurations. At present some methods have been proposed in the literature or have been put into use recently. All methods apply computational grids on which the flow is described. The methods have in common that the grid generation is a substantial part in the computation of the flow. The computational grids may be generated with recent sophisticated grid generation methods such as the multi-block methods and unstructured grid methods, at the expense however of a few drawbacks: 1) multi- block methods are not easy to use for 'non-grid expert' applicators and 2) multi- block and unstructured grids increase the computation time and the development time considerably. At the Unsteady Aerodynamics and Aeroelasticity Departement of NLR it was decided to develop a mono-block OH -type grid generator for aeroelastic applications to complete aircraft aiming at reducing the aforementioned drawbacks. The grid generator is required in particular to generate grids of acceptable quality about concave areas such as airfoil noses and wing-fuselage junctions and should be easy to use for 'non-grid expert' applicators. Therefore a 2-D investigation was conducted of the generation of O-type grids around transverse cross-sections of transport type aircraft. Research was performed on methods which generate grids in a more natural way by an evolutionary process starting at the boundaries of the configuration and constructing the grid according to an inflation analogy. The research resulted in a BEM method for generating grids by solving the flow about an inflating body. During these developments NLR came to the conclusion that a combination of the BEM grid generator method and the hyperbolic grid generation method [1, 2] would be preferable in terms of computational cost and complexity and this has resulted in the present method. Its essence is a hyperbolic grid generator which uses an integral method for direction and growth control in concave areas where the standard hyperbolic grid generator fails thereby reducing the effort to generate grids of acceptable quality. This paper describes this method and shows results of 2-D examples.

2 BEM grid generation method

2.1 Boundary 'Volume' Method

As pointed out before, the first step in the BEM grid generation method is the calculation of the flow about an inflating body. The procedure of this calculation was taken from a 'panel' method which was developed for solving the inviscid incompressible flow about oscillating blades of an horizontal axis wind turbine [3]. This method was later extented towards the subsonic and supersonic potential flow about oscillating aircraft configurations [4, 5] . The method comprises the calculation of the potential due to constant source and/or doublet panels at the boundaries of the configuration while satisfying the velocity boundary conditions by means of a finite volume discretization on an external one-layer mesh. Therefore the potential has to be calculated not only on collocation points at the panels but also at exterior points on the one-layer mesh. This approach was motivated by the following:

1. Lifting surfaces (zero thickness) can be handled without any modification and without extra cost because the potential (jump) at the lifting surface is known a priori.
2. The efficient evaluation of the helicoidal doublet wake layer influence coefficients.
3. An accurate determination of the suction force on cambered lifting surfaces needed for an accurate prediction of the power of wind turbines.(Zero drag on 2-D lifting surfaces at angle of attack, according to D'Alembert's paradox).
4. No evaluation and storage of velocity influence coefficients, especially in 3-D. The evaluation of potential influence coefficients is sometimes less cumbersone.

The formulation of the Boundary 'Volume' Method (BVM) is described here for the 2-D Laplace equation simulating the external incompressible potential flow about an arbitrary body in a Cartesian coordinate system yz with the y-axis in the direction of the free-stream. The potential ϕ satisfies the Laplace equation:

$$\phi_{yy} + \phi_{zz} = 0$$

and the boundary condition:

$$\phi_y n_y + \phi_z n_z = f$$

where $\vec{n}$ is the normal and f denotes the transpiration at the body surface. In addition, for lifting bodies the so called Kutta condition is applied at the trailing edge:

$$\lim_{\to te}(C_{P_{lower}} - C_{P_{upper}}) = 0$$

where C_P is the pressure coefficient $C_P = 1 - \phi_y^2 - \phi_z^2$.

The problem is solved by an integral equation formulation in which the assumption is made that the solution can be written as the integral of a source or doublet singularity distribution at the body surface.

The discretization is performed on the set of collocation points at the boundary $\vec{r}_{j,0}$ where j denotes the index and on the set of exterior points $\vec{r}_{j,1}$ which are constructed in normal direction at about one half of the mesh spacing measured along the boundary.

The volume integration of the Laplace equation takes place over the number of volumes V_j which are formed by the collocation points (see Figure 1) :

$$\int\int_{V_j} (\phi_{yy} + \phi_{zz}) dv = \int_{S_j} \vec{q}.\vec{n} ds = 0$$

or in discretized form:

$$F_{j+\frac{1}{2},0} + F_{j,\frac{1}{2}} - F_{j-\frac{1}{2},0} - F_{j,0} = 0 \tag{1}$$

F denotes the normal flux through the four sides of the volume. By the boundary condition it follows $F_{j,0} = T_j f_j$ where T denotes the length of the side of the volume at the boundary. The normal fluxes at $j \pm \frac{1}{2}, 0$ are evaluated as follows:

$$F_{j+\frac{1}{2},0} = \frac{\delta_{j,0}\phi\vec{J}.\vec{J} + \delta_{k,0}\phi\vec{K}.\vec{J}}{\vec{J}.\vec{j}}$$

where

$$\vec{j} = \delta_{j,0}\begin{pmatrix} y \\ z \end{pmatrix}, \vec{k} = \delta_{k,0}\begin{pmatrix} y \\ z \end{pmatrix}, \vec{K} = \delta_{j,0}\begin{pmatrix} -z \\ y \end{pmatrix}, \vec{J} = \delta_{k,0}\begin{pmatrix} z \\ -y \end{pmatrix}$$

and

$$\delta_{j,0}g = g_{j+1,0} - g_{j,0}$$

$$\delta_{k,0}g = 0.5(g_{j+1,1} + g_{j,1} - g_{j+1,0} - g_{j,0})$$

where g being some operand. The normal flux at $j, \frac{1}{2}$ is evaluated in a similar way:

$$F_{j,\frac{1}{2}} = \frac{\delta_{j,1}\phi\vec{J}.\vec{K} + \delta_{k,1}\phi\vec{K}.\vec{K}}{\vec{J}.\vec{j}}$$

where

$$\delta_{j,1}g = 0.25(g_{j+1,0} - g_{j-1,0} + g_{j+1,1} - g_{j-1,1})$$

$$\delta_{k,1}g = g_{j,1} - g_{j,0}$$

and the same formula are used for $\vec{j}, \vec{k}, \vec{J}, \vec{K}$.

The expressions for the fluxes are substituted into (1) resulting into a linear system for the potential:

$$\sum_{k=0}^{1}\sum_{i=-1}^{1} C^j_{ik}\,\phi_{i+j\,k} = T_j f_j \tag{2}$$

At the body boundary a constant source σ_j or doublet singularity distribution μ_j is asumed for each panel j .The potential at the collocation point i,k due to each panel j is calculated according to:

$$\phi^j_{ik} = \frac{1}{2\pi}\arctan\frac{Z}{Y}\Big|_{Y_{j-\frac{1}{2}}}^{Y_{j+\frac{1}{2}}}$$

for a constant doublet distribution and:

$$\phi^j_{ik} = \frac{1}{2\pi}\left(|\,Z\,|\arctan\frac{Y}{|\,Z\,|} - Y - Y\log\sqrt{Y^2+Z^2}\right)\Big|_{Y_{j-\frac{1}{2}}}^{Y_{j+\frac{1}{2}}}$$

for a constant source distribution. The transformed coordinates Z and Y are defined by:

$$Z = \vec{n}.\vec{r}_{i,k} - \vec{n}.\vec{r}_{j\pm\frac{1}{2},0}$$

$$Y_{j\pm\frac{1}{2}} = \vec{t}.\vec{r}_{i,k} - \vec{t}.\vec{r}_{j\pm\frac{1}{2},0}$$

where $\vec{t} = (n_z, -n_y)^t$ and $\vec{r}_{j\pm\frac{1}{2},0} = 0.5\,(\vec{r}_{j,0} + \vec{r}_{j\pm1,0})$.

On substitution of the ϕ^j_{ik} in (2) a linear system results for the strength of the source or doublet distributions which is solved by direct solution or iterative methods. After that the potential can be calculated and velocities can be derived by similar finite differences expressions as have been used for (1).

Using the aforementioned procedure several methods have been developed in the past for the solution of the Laplace and Helmholtz equations and also an extention has been made for the hyperbolic equations modeling the steady and unsteady supersonic flow [3, 4, 5].

The BVM method is demonstrated here for the flow about a lifting surface and for the flow about a cylinder for several number of elements and cell aspect ratios. Figure 2 compares the calculated pressure coefficients with the exact solution. A good agreement is obtained. Figure 3a shows the relative error in lift and absolute error in drag as compared to the exact solution for a flat plate at 45 degrees of attack. The forces are obtained by discretizing $\vec{F} = -\int_S C_P\vec{n}+(\vec{q}.\vec{n})\vec{q}ds$ at $j\pm\frac{1}{2}$ and D'Alembert's paradox (no drag) is almost recovered to machine zero for all calculations. The lift converges monotonically for the larger and smaller aspect ratios and has a tendency to oscillate for values in between. A general conclusion about the convergence cannot be draw for this example because the solution has a square root singularity at the leading edge. For engineering purposes an aspect ratio of 0.50 is preferable. Figure 3b shows the maximum relative error in the C_P coefficient as compared with the exact solution for a circular cylinder. At the larger aspect ratio both the source and doublet approaches show about the same convergence which tends to second order. For the lower aspect ratio first order convergence shows up as might be expected. The doublet approach shows a more monotonic convergence as compared with the source approach. For engineering purposes the source approach is preferable with an aspect ratio of 0.50.

2.2 BEM grid generator

The BEM generation method starts from an initial boundary contour discretization $\vec{r}_{j,0}$ (boundary surface grid) and generates the required contours $\vec{r}_{j,k}$ denoted by the contour index k as follows:

1. From the contour grid line discretization $\vec{r}_{j,k}$ an estimate of a new contour grid line discretization $\vec{r}^e_{j,k+1}$ is constructed in normal direction at a distance proportional to the spacing of the contour. The estimate is improved by the following steps to avoid grid folding.
2. The BVM method as outlined in the previous section is applied using the previous contour discretization $\vec{r}_{j,k}$ as collocation points and the estimate $\vec{r}^e_{j,k+1}$ as auxilary points with uniform outflow boundary conditions:

$$f_j = 1 \tag{3}$$

   or non-uniform boundary conditions:

$$f_j = (\vec{r}^e_{j,k+1} - \vec{r}_{j,k}).\vec{n}_{j,k} \tag{4}$$

   which simulate the incompressible flow about a steadily inflating body. The solution provides the flow velocities $\vec{q}_{j,k+\frac{1}{2}}$.
3. Various ways of constructing the final grid contour line discretization $\vec{r}_{j,k+1}$ in the flow direction $\vec{q}$ are possible. One way which has been applied succesfully is:

$$(\vec{r}_{j,k+1} - \vec{r}_{j,k}) = (1-\beta)d\vec{r}^e_{j,k} + \beta\frac{d\vec{r}^e_{j,k}.\vec{q}_{j,k+\frac{1}{2}}}{\mid q_{j,k+\frac{1}{2}} \mid}\,\frac{\vec{q}_{j,k+\frac{1}{2}}.\vec{n}}{f_j}\,\vec{q}_{j,k+\frac{1}{2}} \tag{5}$$

where $d\vec{r}^{e}_{j,k} = (\vec{r}^{e}_{j,k+1} - \vec{r}_{j,k})$ and β is a relaxation parameter. Another way is:

$$(\vec{r}_{j,k+1} - \vec{r}_{j,k}) = (1-\beta)d\vec{r}^{e}_{j,k} + \beta \frac{| d\vec{r}^{e}_{j,k} |}{| q_{j,k+\frac{1}{2}} |} \frac{\vec{q}_{j,k+\frac{1}{2}}.\vec{n}}{f_j} \vec{q}_{j,k+\frac{1}{2}} \tag{6}$$

Choosing $\beta = 1$ will construct the grid along streamlines emanating from the contours. The first fraction reduces the growth in areas where the flow is non-orthogonal and the second fraction reduces the flow in zones where parts of the contour are close together. Equation (5) will reduce the growth more strongly in areas where the flow is non-orthogonal as compared with applying equation (6).

This algorithme with variants of equation (3) , (4) , (5) and (6) has been applied to several problems with concavities and the ability to prevent grid folding has been determined. Examples of grids generated with the method are shown in figures 4a-f . No grid folding shows up in concave areas and in convex areas grid lines seem to cluster a bit. The use of equation (5) in the 'b'-like shape seems to reduce the growth at the upper convex corner to much. Also it can be concluded that the effect of the geometry on flow directions is too global for some geometries and therefore the Laplace equation is replaced by the Helmholtz equation: $\phi_{yy} + \phi_{zz} - \kappa^2\phi = 0$ which controled by κ reduces the global effect of the geometry. An example with $\kappa = 1.0$ is depicted in figure 4g showing a reduction of the flow directions.

The aforementioned method has a few drawbacks:

1. In general, orthogonal grids cannot be generated with the Laplace equation and a BVM method for other linear 'viscous' equations like the biharmonic one is required.
2. The algorithme is explicit and might become unstable for some choices of gridspacing and growth.
3. The computational cost is equivalent to N^2,where N denotes the number of grid points on a contour. To reduce the cost,especially for 3-D applications, the method needs the embedding of multi-grid and clustering techniques [6, 7]. Because accuracy and convergence requirements are much less restrictive in grid generation it should be possible to make the computational cost closely equivalent to N.

Due to these drawbacks it was decided to combine the BEM grid generator method with a hyperbolic grid generator method which is described in the following section. The latter can generate orthogonal grids ,is implicit and efficient, but suffers from grid folding which can be prevented by the embedding of the BEM grid generator.

3 Hyperbolic grid generator with BEM control

The hyperbolic grid generator is based on the mathematical model as published in [2] of which the main lines will be repeated. The mapping from the computational space to the physical domain is based on the equations:

$$\frac{\vec{r}_\eta.\vec{r}_\zeta}{| r_\eta || r_\zeta |} = \cos\theta \tag{7}$$

$$|\vec{r}_\eta \times \vec{r}_\zeta| = V \tag{8}$$

Here (η, ζ) denote the coordinates of the computation space, θ is the angle control term which in [2] has to be specified interactively by the user and now can be automatically specified by the BEM method. V is a volume control term. Next equation (7) and (8) are linearized about the state $\vec{r}^0 = (y^0, z^0)^t$ resulting in:

$$A\vec{r}_\eta + B\vec{r}_\zeta = \vec{f}$$

which can be written as the 2x2 hyperbolic system:

$$\vec{r}_\zeta + B^{-1}A\vec{r}_\eta = B^{-1}\vec{f} \tag{9}$$

where:

$$A = \begin{pmatrix} \frac{-\cos\theta^0 y^0_\eta}{|r^0_\eta|^2} + \frac{y^0_\zeta}{|r^0_\eta||r^0_\zeta|} & \frac{-\cos\theta^0 z^0_\eta}{|r^0_\eta|^2} + \frac{z^0_\zeta}{|r^0_\eta||r^0_\zeta|} \\ z^0_\zeta & -y^0_\zeta \end{pmatrix},$$

$$B = \begin{pmatrix} \frac{-\cos\theta^0 y^0_\zeta}{|r^0_\zeta|^2} + \frac{y^0_\eta}{|r^0_\eta||r^0_\zeta|} & \frac{-\cos\theta^0 z^0_\zeta}{|r^0_\zeta|^2} + \frac{z^0_\eta}{|r^0_\eta||r^0_\zeta|} \\ -z^0_\eta & y^0_\eta \end{pmatrix}$$

and

$$\vec{f} = \left(\cos\theta - \cos\theta^0, V + V^0\right)^t.$$

Equation (9) is integrated in ζ direction by:

$$\left[I + \alpha(B^{-1}A)_k\delta_\eta - \epsilon(\Delta\nabla)^2_\eta\right](\vec{r}_{k+1} - \vec{r}_k) = B^{-1}_k\vec{f}_k - \left[(B^{-1}A)_k\delta_\eta - \epsilon(\Delta\nabla)^2_\eta\right]\vec{r}_k \tag{10}$$

where k denotes the constant ζ contours and ∇, Δ, and δ are backward,forward and central difference operators in η, respectively. B^{-1}, A, and $\vec{f}$ are calculated at the known k level. α is the implicitness parameter. Values of $\alpha > 1$ can be used to inhibit grid folding in concave zones. ϵ is the fourth-order dissipation parameter. Values of $\epsilon > 0$ can be used to smooth initial discontinuities. Extending [1] the right hand side of equation (10) is specified by functions with 'simple' shape and exponential growth or shrinkage:

$$V = \epsilon_c^k V^l + (1 - \epsilon_c^k)V^u$$

and

$$\cos\theta = \epsilon_c^{k+1}(\cos\theta^b)_{k=0}$$

where ϵ_c is the clustering (exponential) parameter, $V^u = (R_{k+1} + R_k)(R_{k+1} - R_k)\frac{\pi}{2N}$ is the volume of a uniform distribution in the far field ,R_k is the radius of equivalent circular bodies having the same circumference or volumes as the grid contours. R_1 is obtained by letting each point j grow by $|(\vec{r}_\eta)_{j,0}|$ AR where AR denotes the average aspect ratio of the grid cells. The other R_k's are determined as follows: For a completely convex contour by:

$$R_{k+1} = R_k + (1 + \epsilon_c)(R_k - R_{k-1})$$

and in other cases by:

$$R_{k+1} = R_k + \frac{(R_k - R_{k-1})R_k}{R_{k-1}}$$

The other volumes are determined by:

$$V^l = \epsilon_b\left(\frac{V^b}{V^s}\right)_{k=0} V^s + (1 - \epsilon_b)V^s$$

$$V^s = \epsilon_u\,|(\vec{r}_\eta)_{j,0}|\,\frac{(R_{k+1} - R_k)R_k}{R_0} + (1 - \epsilon_u)\,|(\vec{r}_\eta)_{j,k}|^2\,AR$$

where ϵ_b is the BEM parameter. Values of $\epsilon_b > 0$ can be applied to activate the BEM grid generation algoritme to the $k = 0$ contour to prevent grid folding in concave zones and to concentrate grid lines somewhat in convex zones. The evaluation of V_b and $\cos\theta^b$ is straightforward. ϵ_u is the uniformity parameter. A value of 1 produces contours at approximately equal distances while a value of 0 produces contours at approximately the contour spacing. The latter is less stable.

The 2x2 equations (10) are solved for each contour by inverting a 9 diagonal matrix for all points j. Besides the abovementioned formulations the computer program has been designed such that the BEM grid generation method can be invoked also from $k\#0$ grid contours which makes it possible to start with an orthogonal grid, limiting of V_s can be applied in concave areas, smoothing of the right hand side can be applied and finally post-elliptic smoothing with control functions [8] is embedded.

Results of the method are presented for several geometries with concaveness in figure 5. The figure compares grids obtained with the hyperbolic grid generator using a large amount of implicitness $\alpha = 3$ to prevent grid folding at the expense of losing orthogonality with grids obtained with the present method $\alpha = 1, \epsilon_b = 0.7$,using equations (3) and (6) with $\kappa = 0.0$. The direction effects of the present grid generator are obvious. In general the present method prevents grid folding and does not destroy the grid spacing in concave corners and clusters grid lines somewhat better at convex corners. It should be noted that no attempts have been made to optimize the control parameters or the surface grid distributions in the examples to obtain a smooth grid. In general smoothness of the grids can be improved by applying elliptic smoothing afterwards.

From the applications it is concluded that generation of O-type grids around tranverse cross-sections of transport type aircraft using the BEM control of the present method is much easier compared to the original hyperbolic grid generator. Therefore the present method will be extented towards 3-D.

4 Conclusion

A demonstration has been given of an integral equation method which is used to generate grids. The method has been embedded in a hyperbolic grid generator method to prevent grid folding in concave areas. All components of the method have been described in detail and results of applications have been shown.

From the applications it is concluded that the BEM control of the present method will strongly reduce the effort to generate grids around tranverse cross-sections of transport type aircraft with concavities.

Its potential for 3-D configurations and use in algebraic grid generation methods for direction control should be investigated.

5 Acknowledgement

This investigation was carried out partly under contract with the Netherlands Agency for Aerospace Programs(NIVR),contractnumber 1904N. Special thanks are due to Dr. H. Schippers at NLR for providing a part of the hyperbolic grid generation methodology.

References

[1] J.L. Steger et al. *Generation of body-fitted coordinates using hyperbolic partial differential equations*,SIAM J.Sci.Stat.Comput.,Vol. 1,No. 4, Dec. 1980, pp. 431-437

[2] J.Q. Cordova et al. *Grid generation for general 2-D regions using hyperbolic equations* ,AIAA-88-0520,Jan. 1988.

[3] M. H. L. Hounjet. *ARSPNS: A method to calculate steady and unsteady potential flow about fixed and rotating lifting and non- lifting bodies*,NLR TR 85114 U ,October 1985

[4] M. H. L. Hounjet. *ARSPNSC: A method to calculate subsonic steady and unsteady potential flow about complex configurations* ,NLR TR 86122 U ,November 1986

[5] M. H. L. Hounjet. *Calculation of unsteady subsonic and supersonic flow about oscillating wings and bodies by new panel methods*, NLR TP89119 U,April 1989

[6] J. W. Slooff. *Requirements and developments shaping a next generation of integral methods*,NLR MP81007 U,March 1981

[7] W. Hackbush et al. *On the fast matrix multiplication in the boundary element method by panel clustering*,Numer.Math.,Vol. 54,pp 463-491, 1989

[8] J. F. Thompson. *A general three-dimensional elliptic grid generation system on a composite block-structure*,Computer Methods in applied Mechanics and Engineering,Vol. 64,pp 377-411,1987

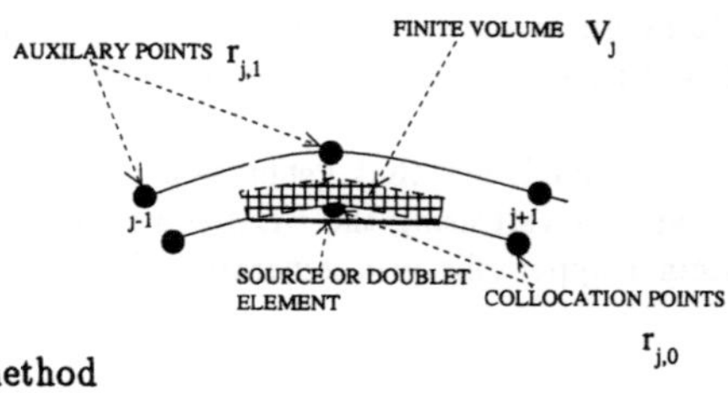

Figure 1. The BVM method

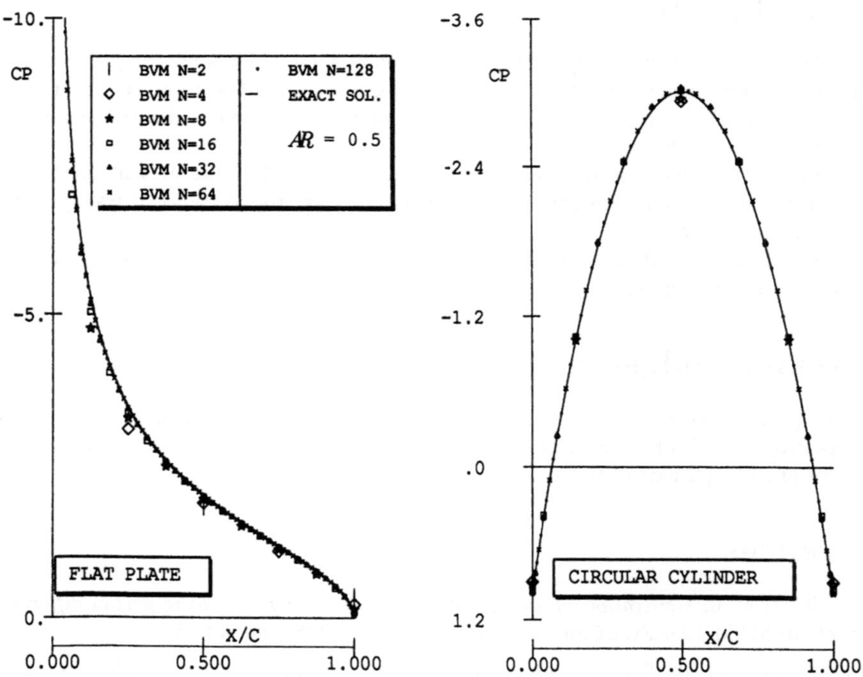

Figure 2. Comparison of BVM solutions with exact data

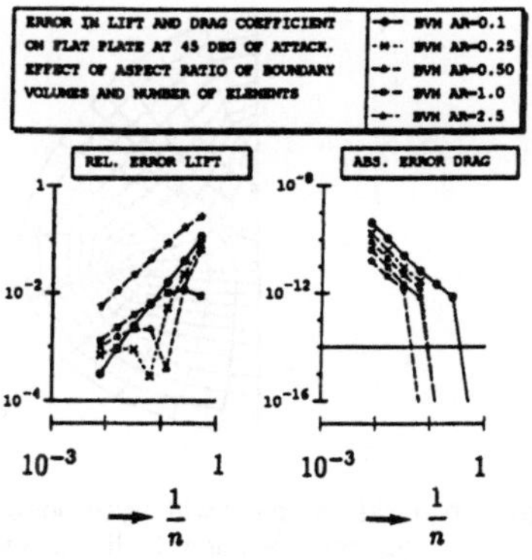

Figure 3a. Convergence characteristics of BVM method of flat plate application.

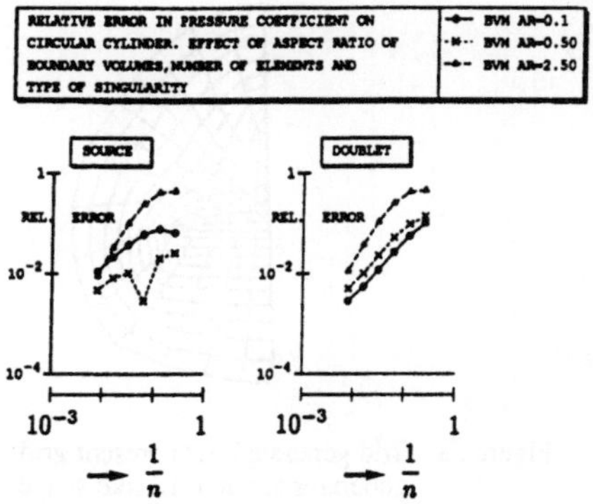

Figure 3b. Convergence characteristics of BVM method of circular cylinder application.

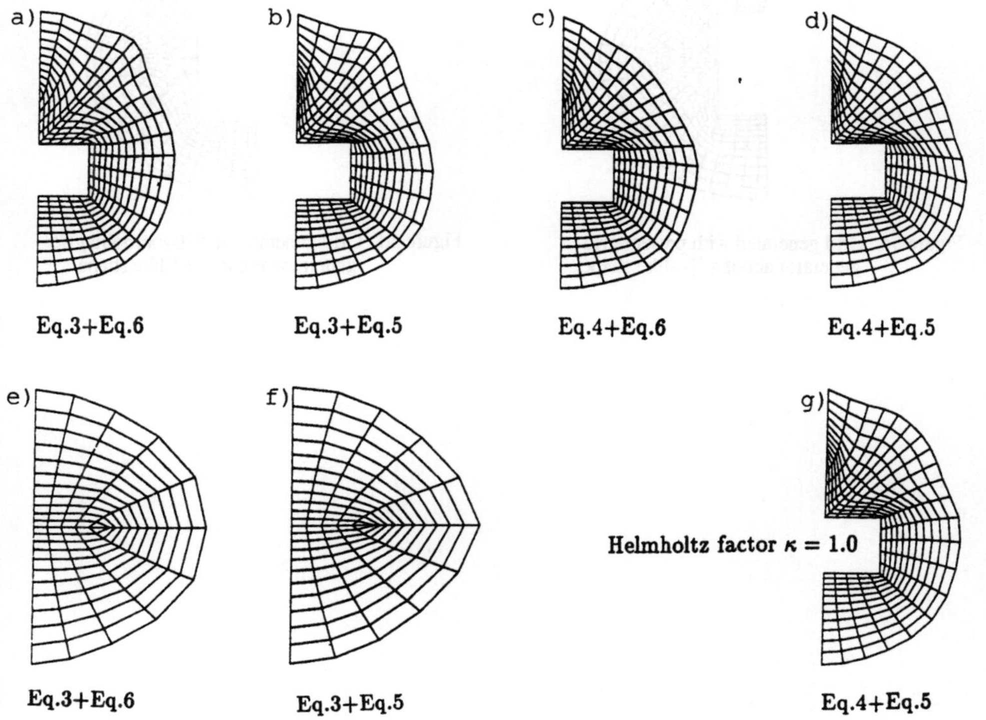

Figure 4a-g. Grids generated by BEM grid generator about 'b' and '-'-like shapes with uniform (Eq.3) or non-uniform (Eq.4) boundary conditions and normal (Eq.6) or stronger (Eq.5) volume reduction.

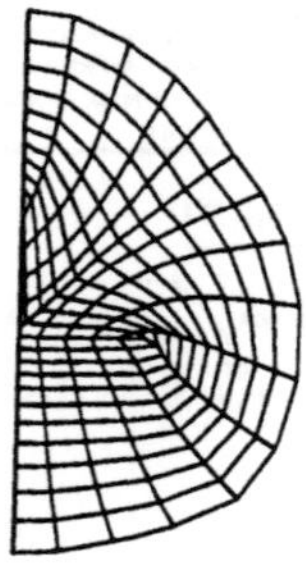

Figure 5a. Grid generated with present grid generator about a 'L'-like shape

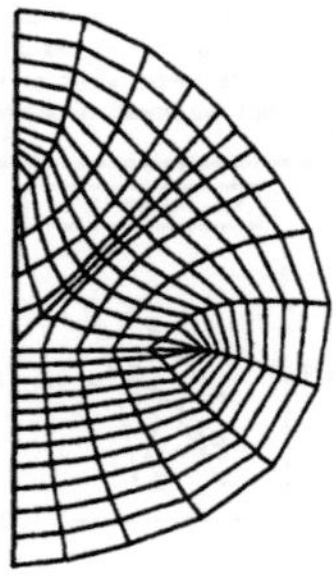

Figure 5b. Grid generated with hyperbolic grid generator about a 'L'-like shape

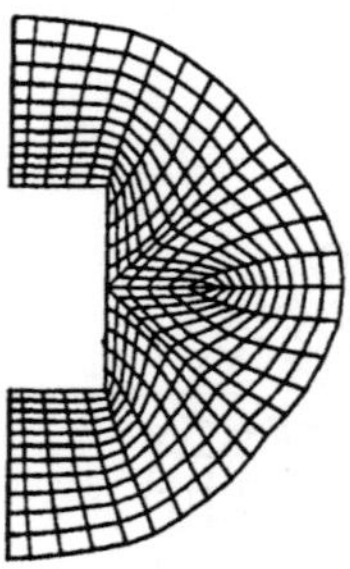

Figure 5c. Grid generated with present grid generator about a ']-'-like shape

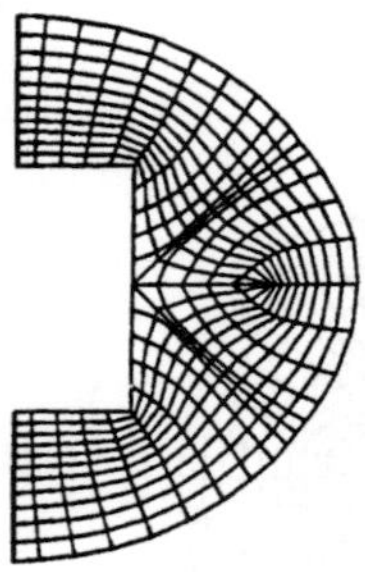

Figure 5d. Grid generated with hyperbolic grid generator about a ']-'-like shape

Solution of Boundary Value Problems by Integral Equations of the First Kind – A Update*

G.C. HSIAO

Department of Mathematical Sciences
University of Delaware, Newark, Delaware 19716, USA

Summary

This paper is concerned with the recent developments in the solution of boundary value problems by integral equations of the first kind. Basic results for weakly singular and hypersingular boundary integral operators will be discussed. Emphases will be given to the mathematical foundation of the method as well as to the physical interpretations of various side conditions derived for the unique solvability of the integral equations of the first kind.

Introduction

It has been almost twenty years since the author's joint paper with MacCamy appeared in [9]. Needless to say, during these years, much progress has been made in the treatment of boundary value problems by the use of integral equations of the first kind. In particular, in recent years, considerable interest has been generated in the mathematical community in applying the method of integral equations of the first kind to obtain numerical solutions of boundary value problems for partial differential equations. On the other hand, in the engineering community, it seems that most of the practitioners still try to avoid the reduction of boundary value problems to integral equations of the first kind, because they generally believe that integral equations of the first kind are ill–posed and hence numerical instability may occur. It is the purpose of this paper to review and to clarify some of the basic results as well as to summarize some of the recent developments in the solution of boundary value problems by integral equations of the first kind.

In addition to Dirichlet problems presented in the earlier paper, this paper will discuss the generalization of the approach to a large class of strongly elliptic boundary value problems in elasticity and fluid mechanics, including the ones with boundary conditions other than those of Dirichlet type. Emphases will be given to the mathematical foundations of the method and to the physical interpretations of various side conditions derived for the unique solvability of integral equations of the first kind. In contrast to the method of integral equations of the second kind, all the essential

* *This paper is dedicated to Professor Richard C. MacCamy on the occasion of his 65th birthday.*

properties of the original partial differential operators necessary for the variational formulation, such as the properties of symmetry and coerciveness, are generally preserved for the corresponding integral operators. Hence from the variational point of view, integral equations of the first kind are more satisfactory than those of the second kind, and their Galerkin approximations can be treated in exactly the same manner as for the partial differential equations.

Model Problems

Throughout the paper, we shall be confined to the exterior boundary value problems in the plane $\mathbb{R}^2$. Let us begin with two basic exterior boundary value problems for the Laplace equation:

$$\Delta u := \frac{\partial^2 u}{\partial x_1^2} + \frac{\partial^2 u}{\partial x_2^2} = 0 \quad \text{in } \Omega^c := \mathbb{R}^2 \backslash \overline{\Omega}, \tag{1}$$

where Ω is a bounded domain with smooth boundary Γ, and $\overline{\Omega} = \Omega \cup \Gamma$ denotes the closure of Ω. Typical boundary conditions are those of *the Dirichlet type (DBC)*: $u|_\Gamma = f$, and of *the Neumann type (NBC)*: $\frac{\partial u}{\partial n}|_\Gamma = g$, where f and g are prescribed functions satisfying appropriate regularity conditions. Here and in the sequel, n always denotes the outward normal to Γ with respect to Ω. Because of the exterior boundary value problem, we also need a proper growth condition at infinity

$$u(x) = \frac{A}{2\pi} \log |x| + \omega + o(1) \quad \text{as } |x| \to \infty, \tag{2}$$

where A and w are both constants; generally A is given. The Dirichlet boundary value problem (*DBVP*) is defined by (1)(2) and the *DBC* for given A and f, while the Neumann boundary value problem (*NBVP*) is defined by (1)(2) and the *NBC* for given A, ω and g. Here g is required to satisfy the *compatibility condition* $\int_\Gamma g ds = A$.

To reduce *DBVP* or *NBVP* to boundary integral equations, we employ direct approach which is based on *Green's representation formula*:

$$u(x) = \int_\Gamma \varphi(y) \frac{\partial}{\partial n_y} \gamma(x,y) ds_y - \int_\Gamma \sigma(y) \gamma(x,y) ds_y + w, \; x \in \Omega^c \tag{3}$$

with $\gamma(x,y)$ being the fundamental solution for the two–dimensional Laplacian,

$$\gamma(x,y) := \frac{-1}{2\pi} \log |x-y|. \tag{4}$$

Here $\varphi := u|_\Gamma$ and $\sigma := \frac{\partial u}{\partial n}|_\Gamma$ are the Cauchy data for the solution u of the Laplace equation and are related according to

$$\begin{pmatrix} \varphi \\ \sigma \end{pmatrix} = \begin{pmatrix} \frac{1}{2}\mathbf{I} + \mathbf{K} & -\mathbf{V} \\ -\mathbf{D} & \frac{1}{2}\mathbf{I} - \mathbf{K}' \end{pmatrix} \begin{pmatrix} \varphi \\ \sigma \end{pmatrix} + \begin{pmatrix} \omega \\ 0 \end{pmatrix} =: \mathbf{C}_{\Omega^c} \begin{pmatrix} \varphi \\ \sigma \end{pmatrix} + \begin{pmatrix} \omega \\ 0 \end{pmatrix}, \tag{5}$$

where $\mathbf{K}, \mathbf{V}, \mathbf{D}$ and $\mathbf{K}'$ are the four basic boundary integral operators:

$$\mathbf{K}\varphi(x) := \int_\Gamma \frac{\partial \gamma}{\partial n_y}(x,y)\varphi(y)ds_y, x \in \Gamma \quad \textit{(Double-layer potential operator)},$$

$$\mathbf{V}\sigma(x) := \int_\Gamma \gamma(x,y)\sigma(y)ds_y, x \in \Gamma \quad \textit{(Simple-layer potential operator)},$$

$$\mathbf{D}\varphi(x) := -\frac{\partial}{\partial n_x}\int_\Gamma \frac{\partial \gamma}{\partial n_y}(x,y)\varphi(y)ds_y, x \in \Gamma \quad \textit{(Hyper-singular potential operator)},$$

$$\mathbf{K}'\sigma(x) := \int_\Gamma \frac{\partial \gamma}{\partial n_x}(x,y)\sigma(y)ds_y, x \in \Gamma \quad \textit{(Adjoint double-layer potential operator)}.$$

The mapping properties of these operators are now well-known (see, e.g., [11],[12]). The operator $\mathbf{C}_{\Omega^c}$ defined by (5) is termed the *Calderòn projector* associated with the Laplacian for the exterior domain Ω^c.

For the *DBVP*, $\varphi = f$ is given. From the first equation of (5) and the growth condition (2), we obtain the boundary integral equations for the unknowns σ and w,

$$\mathbf{V}\sigma - \omega = \left(-\frac{1}{2}\mathbf{I} + \mathbf{K}\right)f, \quad \int_\Gamma \sigma(y)ds_y = A. \tag{6}$$

Existence and uniqueness of solutions of (6) in classical Hölder function spaces as well as in Sobolev spaces have been established in [9],[11]. We remark that in the indirect approach the right hand side of the first equation in (6) is simply replaced by f. This is the *Method of Fichera* introduced in [9], and is a simplified version of a general procedure from [3],[4] for treating higher order elliptic equations in the plane.

On the other hand, for the *NBVP* where $\sigma = g$ is given, the second equation of (5) then leads to the boundary integral equation for the unknown φ,

$$\mathbf{D}\varphi = -\left(\frac{1}{2}\mathbf{I} + \mathbf{K}'\right)g. \tag{7}$$

This equation has constants as the one-dimensional eigenspace. The special right-hand side in (7) satisfies an orthogonality condition in the classical Fredholm alternative, which is also valid for (7), e.g., in the space of Hölder continuous functions on Γ. Therefore (7) always has solutions φ. For the numerical implementations, one may modify (7) in a more convenient form

$$\mathbf{D}\varphi - \omega_b = -\left(\frac{1}{2}\mathbf{I} + \mathbf{K}'\right)g, \quad \int_\Gamma \varphi ds = B \tag{8}$$

by introducing an additional unknown parameter ω_b together with the normalization condition of φ for any given constant B. It is easy to see that because of the special right hand side of (7), the parameter ω_b is in fact equal to zero. Given g and B, it can be shown that (8) is always uniquely solvable for φ and ω_b [7],[8].

Fundamental Problems in Mechanics

As a generalization, we now extend the previous approach to the cases where Γ is either a Lipschitz boundary with corners or Γ is an open arc. For illustrations, we consider here two fundamental problems in mechanics: (a) *Flow Past a Thin Airfoil* in fluid mechanics and (b) *Crack Problems* in linear elasticity.

(a) *Thin Airfoil.* Let Γ be the profile of the thin airfoil Ω with one corner point at the trailing edge TE. Then mathematically the uniform plane flow of an incompressible inviscid fluid past a thin airfoil can be formulated as an exterior boundary value problem for the velocity field $\mathbf{q} = (q_1, q_2)$:

$$\begin{gathered}(\nabla \times \mathbf{q})_3 := \frac{\partial q_2}{\partial x_1} - \frac{\partial q_1}{\partial x_2} = 0 \quad \text{and} \quad \nabla \cdot \mathbf{q} := \frac{\partial q_1}{\partial x_1} + \frac{\partial q_2}{\partial x_2} = 0 \quad \text{in } \Omega^c, \\ \mathbf{q} \cdot \mathbf{n}|_\Gamma = 0, \quad \mathbf{q} - \mathbf{q}_\infty = o(1) \ \text{ as } |x| \to \infty, \\ \lim_{\Omega^c \ni x \to TE} |\mathbf{q}(x)| = |\mathbf{q}|_{|TE} \quad \text{exists at } TE,\end{gathered} \tag{9}$$

where $\mathbf{q}_\infty$ is the given free stream velocity. The last condition in (9) is equivalent to the Kutta–Joukowski condition, which requires bounded and equal pressure at the trailing edge because of Bernoulli's law. In the case of two–dimensional incompressible fluid flow, it is well known that there exists a stream function ψ defined by $\mathbf{q} = (\nabla\psi)^\perp := \left(\frac{\partial\psi}{\partial x_2}, -\frac{\partial\psi}{\partial x_1}\right)$ and that (9) can be reformulated in terms of ψ. Denoting by $u = u(x)$ the *perturbation* or *disturbance stream function* due to the airfoil, we write $\psi(x) = \psi_\infty(x) + u(x)$, where ψ_∞ is the free stream function defined by $\mathbf{q}_\infty$. Then we may reformulate (9) as an exterior *DBVP* for u:

$$\begin{gathered}\Delta u = 0 \quad \text{in } \Omega^c, \qquad u|_\Gamma = -\psi_\infty \quad \text{on } \Gamma, \\ u = \frac{A}{2\pi} \log|x| + \omega + o(1) \quad \text{as } |x| \to \infty.\end{gathered} \tag{10}$$

In this formulation, however, both constants A and ω are unknown. Physically, $-A$ equals to the *circulation* κ around the airfoil which is defined by

$$\kappa := \int_\Gamma \mathbf{q} \cdot dx = -\int_\Gamma \frac{\partial u}{\partial n} ds.$$

The Kutta–Joukowski condition now states that *the circulation around the airfoil should be such that the perturbation velocity* $\mathbf{q} - \mathbf{q}_\infty = (\nabla u)^\perp$ *should be finite and continuous at the trailing edge* TE.

Following the procedure for the model problems, we represent the solution of (10) by (3) with $\varphi = -\psi_\infty|_\Gamma$, and arrive at the boundary integral equations for the unknown function $\sigma = \frac{\partial u}{\partial n}|_\Gamma$ as well as for the unknown constants A and ω such that

$$\mathbf{V}\sigma - \omega = (-\theta_x \mathbf{I} + \mathbf{K})\varphi, \int_\Gamma \sigma ds = A \tag{11}$$

together with the Kutta–Jourkowski condition on the *circulation* related to the unknown constant A. Here θ_x is the solid angle subtended by Γ at $x \in \Gamma$, and $\theta_x = 1/2$ if $x \neq TE$. We seek solution of (11) in the form

$$\sigma = \sigma_0 + A\sigma_1, \quad \omega = \omega_0 + A\omega_1, \tag{12}$$

where $(\sigma_i, \omega_i), i = 0,1$ are the unique solutions of the systems:

$$\mathbf{V}\sigma_0 - \omega_0 = (-\theta_x \mathbf{I} + \mathbf{K})\varphi, \int_\Gamma \sigma_0 ds = 0; \quad \mathbf{V}\sigma_1 - \omega_1 = 0, \int_\Gamma \sigma_1 ds = 1. \tag{13}$$

To determine A, we note that the solutions $\sigma_i, i = 0,1$ generally are *singular* at TE. In fact one can show that σ_i can be decomposed in the form: $\sigma_i = c_i\sigma_s +$ regular term, where c_i's are constants (similar to the stress intensity factors in elasticity); The singular term σ_s is of $O(r^{-\beta})$ as $r = |x - TE| \to 0^+$ with $\beta = -\left(\frac{1}{2-\alpha} - 1\right) = \frac{1-\alpha}{2-\alpha}$, where $0 < \alpha < 1$ and $\alpha\pi$ is the enclosed angle by the tangents to the profile at the trailing edge TE. Thus, the circulation constant A can be determined uniquely by σ_0 and σ_1, namely, $A = -(\lim_{r\to 0} r^\beta \sigma_o)/(\lim_{r\to 0} r^\beta \sigma_1) = -c_o/c_1$ and one can show that $c_1 \neq 0$. We emphasize that in contrast to (6), the constant A can not be prescribed á prioŕi. Otherwise, one may result incorrect flow patterns.

(b) *Crack Problems.* Let Γ be an open smooth arc denoting the crack of the elastic material occupied on the whole plane. Then the displacement vector field $\mathbf{u}$ is governed by the Lamé system

$$\Delta^*\mathbf{u} := \mu\Delta\mathbf{u} + (\lambda + \mu)\,\mathrm{grad}\,\mathrm{div}\,\mathbf{u} = \mathbf{0} \quad \text{in } \Omega^c := I\!R^2 \backslash \overline{\Gamma}, \tag{14}$$

where $\mu > 0$ and $\lambda + \mu > 0$ are the Lamé constants. Here the boundary $\partial\Gamma = \{p_1, p_2\}$ consists of only two points, the tips of the crack. The growth condition at infinity now reads

$$\mathbf{u}(x) = -\gamma(x,0)\mathbf{A} + \boldsymbol{\omega}(x) + O(|x|^{-1}) \quad \text{as} \quad |x| \to \infty. \tag{15}$$

Here $\gamma(x,y)$ is the fundamental displacement tensor of $\Delta^*\mathbf{u} = \mathbf{0}$,

$$\gamma_{ij}(x,y) := \frac{\lambda + 3\mu}{4\pi\mu(\lambda + 2\mu)}\left\{-\log|x-y|\delta_{ij} + \frac{\lambda+\mu}{\lambda+3\mu}\frac{(x_i - y_i)(x_j - y_j)}{|x-y|^2}\right\}, i,j = 1,2, \tag{16}$$

$\mathbf{A}$ is a given constant vector and $\boldsymbol{\omega}(x)$ denotes the rigid motion [13] defined by

$$\boldsymbol{\omega}(x) := \omega_1 \hat{\mathbf{e}}_1 + \omega_2 \hat{\mathbf{e}}_2 + \omega_3(-x_2 \hat{\mathbf{e}}_1 + x_1 \hat{\mathbf{e}}_1), \tag{17}$$

where δ_{ij} denotes the Kronecker delta and $\hat{\mathbf{e}}_i, i = 1, 2$, are the unit vectors along the x_i–axis. The *DBVP* now consists of (14), (15) with given $\mathbf{A}$ and the *DBC*:

$$\mathbf{u}|_{\Gamma_-} = \mathbf{f}_- \quad \text{and} \quad \mathbf{u}|_{\Gamma_+} = \mathbf{f}_+, \tag{18}$$

where the prescribed displacements $\mathbf{f}_\pm$ are required to satisfy the *compatability condition* $[\mathbf{f}]|_{\partial\Gamma} := (\mathbf{f}_+ - \mathbf{f}_-)|_{\partial\Gamma} = \mathbf{0}$. Here $\Gamma_\pm$ denote the $+$ and $-$ sides of Γ.

The *NBVP* then consists of (14), (15) with given $\mathbf{A}$ and $\boldsymbol{\omega}(x)$ together with the *NBC*:

$$T(\mathbf{u})|_{\Gamma_-} = \mathbf{g}_- \quad \text{and} \quad T(\mathbf{u})|_{\Gamma_+} = \mathbf{g}_+, \tag{19}$$

where the jump of the prescribed tractions across $\Gamma, [\mathbf{g}] := \mathbf{g}_+ - \mathbf{g}_-$, is required to satisfy the *compatability condition* $\int_\Gamma [\mathbf{g}] ds = \mathbf{A}$. In the formulation (19), $T(\mathbf{u})$ is the traction operator defined by

$$T(\mathbf{u}) := \lambda(\operatorname{div} \mathbf{u})\mathbf{n} + 2\mu \frac{\partial \mathbf{u}}{\partial n} + \mu \mathbf{n} \times \operatorname{curl} \mathbf{u}. \tag{20}$$

The solution of the *DBVP* or the *NBVP* admits now the *Betti representation formula*[13]:

$$\mathbf{u}(x) = \int_\Gamma (T_y \gamma(x,y))^t [\boldsymbol{\varphi}](y) ds_y - \int_\Gamma \gamma(x,y)[\boldsymbol{\sigma}](y) ds_y + \boldsymbol{\omega}(x), \; x \in \Omega^c, \tag{21}$$

where the Cauchy data are the *jumps* across Γ of the displacement and traction fields, namely, $[\boldsymbol{\varphi}] := \mathbf{u}|_{\Gamma_+} - \mathbf{u}|_{\Gamma_-}$ and $[\boldsymbol{\sigma}] := T(\mathbf{u})|_{\Gamma_+} - T(\mathbf{u})|_{\Gamma_-}$, respectively. Here the notation $(\;)^t$ denotes the transpose of $(\;)$. Again, $[\boldsymbol{\varphi}]$ and $[\boldsymbol{\sigma}]$ are not independent but are related in terms of appropriate boundary integral operators defined on the open arc Γ.

Now if we introduce the boundary integral operators $\mathbf{K}_\Gamma, \mathbf{V}_\Gamma, \mathbf{D}_\Gamma$ and $\mathbf{K}'_\Gamma$ as those in (5) with $\gamma, \frac{\partial}{\partial n_y}, \frac{\partial}{\partial n_x}$ replaced by $\gamma, (T_y)^t, T_x$ accordingly, we may then reduce from (21) to the following *uniquely solvable* boundary integral equations on the arc. For the *DBVP*, we have the system for the unknowns $[\boldsymbol{\sigma}]$ and $\boldsymbol{\omega}$:

$$\begin{aligned} \mathbf{V}_\Gamma[\boldsymbol{\sigma}] - \boldsymbol{\omega} &= -\frac{1}{2}(\mathbf{f}_- + \mathbf{f}_+) + \mathbf{K}_\Gamma[\mathbf{f}], \\ \int_\Gamma [\boldsymbol{\sigma}] ds = \mathbf{A}, \quad & \int_\Gamma [\boldsymbol{\sigma}] \cdot (-y_2, y_1)^t ds_y = A_3. \end{aligned} \tag{22}$$

Here we have included a third normalization condition which means the total moment is given by the constant A_3, while the other two represent the total force is given by the constant vecors $\mathbf{A}$. Thus, we have the same number of equations as unknowns. For the *NBVP*, we have the boundary integral equation for the unknown $[\varphi]$:

$$\mathbf{D}_\Gamma[\varphi] = -\frac{1}{2}(\mathbf{g}_- + \mathbf{g}_+) - \mathbf{K}'_\Gamma[\mathbf{g}] \tag{23}$$

together with the imposed condition $[\varphi]|_{\partial\Gamma} = \mathbf{0}$. In scattering theory, this condition, $[\varphi]|_{\partial\Gamma} = \mathbf{0}$, is termed the *edge condition* which is necessary for the unique solvability for (23) and ensures a local finite energy for the corresponding potential.

It is worth mentioning that in contrast to the closed boundary, neither the *DBVP* nor the *NBVP* can be solved by using a straight forward boundary integral equations of the second kind as one can see from the right hand sides of (23) and (22). In particular, we note that the argument of $\mathbf{K}'_\Gamma$ is $[\boldsymbol{\sigma}]$ instead of $\boldsymbol{\sigma}$ for the *DBVP* and similarly, the argument of $\mathbf{K}_\Gamma$ is $[\varphi]$ but not φ for the *NBVP*.

Basic Properties for **V** and **D**

Neither the differential equation formulation nor the boundary integral equation formulation for the boundary value problems is complete without specifying the appropriate function spaces in which solutions will be sought. This section is concerned mainly with the solution spaces for equations (6) and (8). In what follows, unless stated otherwise, we will assume that the boundary Γ is at least as smooth as a Lipschitz boundary [*l*] (i.e $\Gamma \in C^{0,1}$). This means locally Γ can be represented by a Lipschitz continuous function and such a boundary can contain corner points.

We denote by $C^{m,\lambda}(\Gamma)$ the space of all m-times continuously differentiable Hölder continuous functions in Γ, that is, the supspace of $C^m(\Gamma)$ consisting of functions for which the *mth* order derivatives are Hölder continuous of exponent $\lambda, 0 < \lambda < 1$, in Γ. As usual, we denote by $H^s(\Gamma), 0 < s < 1$, the *Sobolev space* equipped with the norm

$$||\varphi||_s := \left\{ ||\varphi||_0^2 + \int_\Gamma \int_\Gamma \frac{|\varphi(x) - \varphi(y)|^2}{|x-y|^{1+2s}} ds_x ds_y \right\}^{\frac{1}{2}},$$

where $||\varphi||_0$ is the $L^2(\Gamma)$ – norm defined by $||\varphi||_0 = \{\int_\Gamma |\varphi|^2 ds\}^{\frac{1}{2}}$. We recall that $H^s(\Gamma), 0 < s < 1$, may be defined to be the *completion* of the space

$$C_s^0(\Gamma) := \{\varphi \in C^0(\Gamma) : ||\varphi||_s < \infty\}$$

with respect to the norm $||\varphi||_s$. By this we mean that for every $\varphi \in H^s(\Gamma)$, there exists a sequence $\{\varphi_k\} \subset C_s^0(\Gamma)$ such that $\lim_{k\to\infty} ||\varphi - \varphi_k||_s = 0$. We denote by $H^{-s}(\Gamma)$

the dual of $H^s(\Gamma)$ with respect to the $L^2(\Gamma)$ scalar product, i.e. the completion of $L^2(\Gamma)$ with respect to the norm

$$||\sigma||_{-s} := \sup_{||\varphi||_s=1} | < \varphi, \sigma >_0 | = \sup_{||\varphi||_s=1} | \int_\Gamma \varphi(x)\sigma(x)ds|.$$

These are boundary spaces of negative orders whose elements are in fact continuous, linear functionals on the Sobolev space $H^s(\Gamma)$. The essential properties for the operators $\mathbf{V}$ and $\mathbf{D}$ of (5) can be summarized as follows:

(a) *Continuity.* The following operators are continuous:

$$\mathbf{V} : \begin{array}{c} C^{0,\lambda}(\Gamma) \to C^{1,\lambda}(\Gamma) \\ H^{-1/2}(\Gamma) \to H^{1/2}(\Gamma) \end{array}; \quad \mathbf{D} : \begin{array}{c} C^{1,\lambda}(\Gamma) \to C^{0,\lambda}(\Gamma) \\ H^{1/2}(\Gamma) \to H^{-1/2}(\Gamma). \end{array}$$

The *order* of the operator is defined to be the difference of indices of Sobolev spaces in the mapping. Thus we see that the *order* of $\mathbf{V} = -1$ whereas the *order* of $\mathbf{D} = +1$.

(b) *Gårding's Inequality.* Both $\mathbf{V}$ and $\mathbf{D}$ satisfy the Gårding inequality:

$$< \mathbf{V}\sigma, \sigma >_0 \; \geq \alpha||\sigma||^2_{-1/2} - \beta||\sigma||^2_{-1} \quad \forall\sigma \in H^{-1/2}(\Gamma);$$
$$< \mathbf{D}\varphi, \varphi >_0 \; \geq \alpha||\varphi||^2_{1/2} - \beta||\varphi||^2_0 \quad \forall\varphi \in H^{1/2}(\Gamma),$$

where α and β are constants. In the variational formulation, $H^{-1/2}(\Gamma)$ and $H^{1/2}(\Gamma)$ are the *energy spaces* for $\mathbf{V}$ and $\mathbf{D}$ respectively. We remark that Gårding's inequality implies the validity of the classical Fredholm alternative and hence existence follows from uniqueness.

(c) *Stability and Condition Number.* It is now well known that Fredholm integral equations of the first kind are generally ill–posed in the sense that solutions do not depend continuously on the given data, if the corresponding integral operators have *negative orders* such as $\mathbf{V}$ and if the given data are not in appropriate function spaces. The mapping $\mathbf{V} : L^2(\Gamma) \to L^2(\Gamma)$ is *compact* and consequently $\mathbf{V}^{-1}$ is *not bounded* from $L^2(\Gamma)$ into itself. On the other hand, $\mathbf{D}^{-1}$ is *compact* but $\mathbf{D}$, like differential operators, is *unbounded* from $L^2(\Gamma)$ into itself. Thus, if we denote by $\mathbf{V}_h^{-1}$ and $\mathbf{D}_h^{-1}$ the corresponding inverses of the discrete operators defined by the Galekin equations of (6) and (8), it follows that $||\mathbf{V}_h^{-1}|| = O(h^{-1})$ but $||\mathbf{D}_h^{-1}|| = O(1)$. However, in both cases, the L^2–condition number is always of $O(h^{-|\alpha|}) = O(h^{-1})$, where α is the order of the operator under consideration [6],[7].

(d) *Fredholm Operators.* For Γ sufficiently smooth, both $\mathbf{V}$ and $\mathbf{D}$ are Fredholm operators of index zero. We see that $(2\mathbf{V})(2\mathbf{D}) = \mathbf{I} - (2\mathbf{K})^2$ and $(2\mathbf{D})(2\mathbf{V}) = \mathbf{I} - (2\mathbf{K}')^2$. This follows from the fact that the Calerdón projector of (5) is a projection

[2],[14] and both $\mathbf{K}$ and $\mathbf{K}'$ are compact, provided that Γ is sufficiently smooth. The operators $\mathbf{V}$ and $\mathbf{D}$ are also related by the formula: $\mathbf{D}\varphi = -\frac{d}{ds}\mathbf{V}\left(\frac{d}{ds}\varphi\right) \quad \forall\varphi \in H^{1/2}(\Gamma)$ (or $C^{1,\lambda}(\Gamma)$), which will become particularly desirable, if one solves (8) by the Galerkin method.

(e) *Hadamard's Finite Part.* The hyper–singular potential operator $\mathbf{D}$ is sometime expressed in terms of Hadamard's finite part integral, $⨎_\Gamma \frac{\partial^2\gamma}{\partial n_x \partial n_y}(x,y)\varphi(y)ds_y$. More precisely, for $\Gamma \in C^{2,\lambda}$ and $\varphi \in C^{1,\lambda}(\Gamma)$, we can show [14] that

$$
\begin{aligned}
\mathbf{D}\varphi(x) &:= -\lim_{z\to x\in\Gamma, z\in\Omega^c} n_x \cdot \nabla_z \int_\Gamma \frac{\partial}{\partial n_y}\gamma(z,y)\varphi(y)ds_y \\
&= -⨎_\Gamma \frac{\partial^2\gamma}{\partial n_x \partial n_y}(x,y)\varphi(y)ds_y, \quad x \in \Gamma,
\end{aligned}
$$

where the Hadamard's finite part is defined by

$$
⨎_\Gamma \frac{\partial^2\gamma}{\partial n_x \partial n_y}(x,y)\varphi(y)ds_y := \lim_{\epsilon\to 0}\left\{\int_{\Gamma_\epsilon} \frac{\partial^2\gamma}{\partial n_x \partial n_y}(x,y)\varphi(y)ds_y - H(x;\epsilon;\varphi)\right\}
$$

with $H(x;\epsilon;\varphi) = \int_{\Gamma_\epsilon} \frac{d}{ds_x}\frac{d}{ds_y}(\gamma(x,y)\varphi(y))ds_y = \frac{\varphi(x)}{\pi\epsilon} + O(\epsilon^\lambda)$, and $\Gamma_\epsilon := \Gamma \cap \{y : |y-x| \geq \epsilon\}$. We remark that the definition of the finite part integral depends significantly on the geometry of Γ_ϵ and is generally *not invariant* under the change of Γ_ϵ.

Now from the properties of the operators, the following existence and uniqueness results have been established in [9],[11],[12]:

Theorem. *(a) Given $(f,A) \in C^{1,\lambda}(\Gamma)\times I\!R$ (or $H^{1/2}(\Gamma)\times I\!R$), the system of equations (6) has a unique pair of solutions $(\sigma,\omega) \in C^{0,\lambda}(\Gamma)\times I\!R$ (or $H^{-1/2}(\Gamma)\times I\!R$). (b) Given $(g,B) \in C^{0,\lambda}(\Gamma)\times I\!R$ (or $H^{-1/2}(\Gamma)\times I\!R$), the system of equations (8) has a unique pair of solutions $(\varphi,\omega_b) \in C^{1,\lambda}(\Gamma)\times I\!R$ (or $H^{1/2}(\Gamma)\times I\!R$).*

We remark that similar results for equations (22) and (23) are also available[10]. Here the energy spaces for the operators $\mathbf{V}_\Gamma$ and $\mathbf{D}_\Gamma$ are $\tilde{H}^{-1/2}(\Gamma)$ and $\tilde{H}^{1/2}(\Gamma)$ respectively. To conclude this paper, we list here some additional relevant references concerning boundary integral equations of the first kind. These references are [5],[15],[16] and [17] as well as those in [8].

References

1. Costable, M.: Boundary integral operators on Lipschitz domains: elementary results, SIAM J. Math. anal. 19 (1988), 613–626.

2. Costabel, M.; Wendland, W.L.: Strong ellipticity of boundary integral operators, J. Reine Angew. Math., 373(1986), 39–63.

3. Fichera, G.: Linear elliptic equations of higher order in two independent variables and singular integral equations with applications to anisotropic inhomogeneous elasticity, Proceedings of the Symp. Partial Differential Equations and Continuum Mechanics, Ed. R.E. Langer, The University of Wisconsin Press, (1961), 55–80.

4. Fichear,G; Ricci, P.E.: The single layer of potential approach in the theory of boundary value problems for elliptic equations, Lecture Notes in Math., 561, 39–50, Springer–Verlag, Berlin, Heidelberg, New York, 1976.

5. Giroire, J; Nedelec, J.C.: Numerical solution of an exterior Neumann problem using a double layer potential, Math. Comp., 32 (1978), 973–990.

6. Hsiao, G.C.: On the stability of integral equations of the first kind with logarithmic kernels, Arch, Rational Mech. Anal., 94 (1986), 179–192.

7. Hsiao, G.C.: On the stability of boundary element methods for integral equations of the first kind, Boundary Elements IX 1, Ed. C.A. Brebbia, W.L. Wendland, G. Kuhn, Springer–Verlag, 1987, 177–192.

8. Hsiao, G.C.: On boundary integral equations of the first kind, J. Comp. Math. 7 (1989), 121–131.

9. Hsiao, G.C.; MacCamy, R.C.: Solution of boundary value problems by integral equations of the first kind, SIAM Rev., 15 (1973), 687–705.

10. Hsiao, G.C.: Stephan, E.P.; Wendland, W.L.: On the Dirichlet problem in elasticty for a domain exterior to an arc, J. Comp. Appl. Math. 33 (1990), to appear.

11. Hsiao, G.C.; Wendland, W.L.: A finite element method for some integral equations of the first kind, J. Math. Anal. Appl., 58 (1977), 449–481.

12. Hsiao, G.C.; Wendland, W.L.: The Aubin–Nitsche lemma for integral equations, J. of Integral Equations, 3 (1981), 299–315.

13. Hsiao, G.C.; Wendland, W.L.: On a boundary integral method for some exterior problems in elasticity, Proceedings of Tbilisi University, Tbilisi University Press, Tbilisi, 257 (1985), 31–61.

14. Hsiao, G.C.; Wendland, W.L.: *Variational Methods for Boundary Integral Equations and Mathematical Foundations of Boundary Element Methods*, Springer–Verlag, in preparation.

15. LeRoux, M.N.: Méthode d'Elément Finis Pour la Rèsolution Numérique de Problèms Exterieurs en Dimansion 2, R.A.I.R.O. Anal. Numer., 11 (1977), 27–60.

16. MacCamy, R.C.: On a class of two–dimensional Stokes flows, Arch. Rational Mech. and Anal., 21 (1966), 256–268.

17. Stephan, E.; Wendland, W.L.: Remarks to Galerkin and least squares methods with finite elements for general elliptic problems, Lecture Notes Math., 546, 461–471, Springer–Verlag, Berlin, Heidelberg, New York, 1976.

GENESIS-A Mesh-free, Knowledge-based, Nonlinear Boundary Integral Methodology for Compressible, Viscous Flows over Arbitrary Bodies: Theoretical Framework and Basic Physical Principles

Barry Hunt
GE Aircraft Engines, Cincinnati, OH 45215, USA

Summary. Following a discussion of why design engineers need a knowledge–based CFD methodology to complement their existing differential toolkit, the paper proceeds from a universal mathematical identity to develop a generic boundary integral formulation for incompressible, irrotational flow fields. A new formulation known as SAVER is introduced, which determines optimal source and vorticity distributions on the body boundary by a novel relaxation approach. This is then generalized through a modification of the boundary conditions to a methodology known as GENESIS, for analysis or design in compressible, rotational flow. A discussion is presented of how the basic nature of integral methods allows a causal explanation of some important flow phenomena, such as shocks and separations, and facilitates aerodynamic sensitivity analysis. This causality makes integral formulations an ideal basis for the development of a knowledge–based CFD methodology, allowing the designer access to affordable simulations of realistic flows over aeronautical shapes.

1. Introduction: The Need for a Knowledge–Based CFD Methodology

Many of today's aerodynamic flight surfaces were designed on the basis of simple models which adequately represented their primary properties. Examples of these models include: lifting–line and horseshoe vortex theory; actuator disc theory; airscrew theory; one–dimensional shocked nozzle flows; and simple, 2–D integral boundary layer theory, including concepts such as "displacement thickness". Many of these models are intuitive, and *causal*: they are based on the concept of a *force field* generated between an *induced field* and the *field inducers*. These causal models yielded useful predictions, with minimal computational resources, for flow regimes close to "design" conditions. Invariably, these predictions were refined and validated by extensive experiments on scale models and by flight test.

Today's environment is much more complex aerodynamically. This, coupled with increasing demands for performance guarantees, has reduced the usefulness of the earlier models, and of numerous long–held design guidelines and "rules of thumb". The design engineer requires access to more realistic flow models.

The emergence of mainframe computers in the late 1960's led to advances in modeling capability: in particular, Boundary Integral or "panel" methods became almost obligatory design tools, while multi–streamsurface techniques better modeled the aerodynamics of engines. Boundary layer models and compressibility corrections were developed to estimate Reynolds– and Mach–Number effects, and to better model off–design behavior. Throughout the 1970's, Industry conducted significant in–house development work on "interim" technologies, usually involving the iterative coupling of "linear" methods with other computational techniques for nonlinear flow regions. These interim methods represented a *refinement* of the earlier models. They may be referred to as *phenomenological* models since they employ direct representations of the critical physical phenomena expected (e.g. a vortical wake).

A different trend emerged in the 1980's as configuration complexity continued to increase with operation in more demanding flow environments. Priority switched to solving nonlinear partial differential equations (PDE's) offering more realistic representation of compressible and/or viscous flows. Although the airframe industry entered the 1980's with a variety of Boundary Integral Methods (BIM's) in an advanced state of development and application, a rapid shift was underway towards *differential* methods.

A prime reason for this demise of integral methods was the (erroneous) belief that BIM's were intrinsically *linear*, and therefore incapable (at least without coupling with some other type of method) of treating compressible or viscous flows. Such flows are modeled with varying realism by nonlinear PDE's such as the Full Potential Equation, the Euler equations and the Navier–Stokes (N–S) equations. Their numerical solution by a variety of differential methods, primarily of the finite–difference, –element or –volume type, has spawned the growth industry known as *Computational Fluid Dynamics* (CFD). Integral methods have been generally assumed to lie outside the field of CFD, in the sense that they model pre–supposed features of an idealized flow, rather than computing the dynamics of a real fluid.

Although major progress has been made in differential methodologies for the prediction of 3–D flows, they do possess several inherent deficiencies. Among these, we may identify the following: they require *major computational resources*; the user has to generate a *computational grid* upon which the PDE's are discretized; the *complexity of configuration* for which grids can be generated is limited; they can not properly model *discontinuities*; it is very difficult to develop a *design code* to compute a geometric shape which will generate a prescribed pressure distribution. To this list we may add the fact that it is difficult for the applications engineer to derive from a differential method any *understanding* of flow physics, or to perform *sensitivity studies* quantifying interactions between components of a complex configuration.

Observation of the design process indicates a significant change since the 1960's. Until then, design engineers were strong in understanding the flow (usually much simpler then than now), *and* in their ability to apply their experience to steer a design towards the desired properties. The use of differential methods for the complex problems of the 1980's has made the data–generation and result–interpretation processes virtually inaccessible to the design engineer. This has led to the new role of *CFD specialist*, who, while conversant with the emerging technologies of *grid generation, multigrid cycles*, etc., is often totally isolated from the design process. It would be highly desirable to develop computational models capable of handling complexity of both flow and configuration, but based on *cause–and–effect* concepts with which an applications engineer can identify. Such a prospect is offered by the new generation of integral methods now starting to emerge.

The early BIM's placed *surface* distributions of source, and/or vorticity, and/or normal dipoles on the body surface (or some representative internal surface), and on assumed external wake surfaces. The new generation of integral methods also considers the effect of *field* distributions of source and vorticity, simulating the divergence and/or rotation of a compressible and/or rotational flow field. We can divide these new methods into two classes. The first (*Field Integral Methods*, or *FIMs*), perform a direct integration of the extra perturbation velocity field induced by these field distributions. The second constructs "equivalent" surface distributions inducing nominally the same perturbation velocity field as the original field distributions.

While FIMs can have arbitrarily high accuracy, their computational needs can exceed even those of differential methods. They also require a field grid to carry the field distributions (though in this respect they are much less demanding than differential methods). In this paper we shall be concerned primarily with the second class, for which we now introduce the acronym **GENESIS** representing a ***GE****neralized* ***N****onlinear* ***E****xtension of* ***S****urface* ***I****ntegral* ***S****chemes*. At the expense of some reduction in achievable accuracy [which can be restored to that of the FIM by the addition of (small) incremental source/vorticity fields on a (sparse) field grid], GENESIS has none of the stated deficiencies of differential methods.

Integral methods have another inherent advantage over differential methods: their *causal* nature makes them an eminently suitable vehicle upon which to develop a proposed methodology introduced here by the name of Knowledge–Based Computational Fluid Dynamics (KBCFD). In Section 4 we shall discuss some of the implications of such a methodology; first, we give in Section 2 an overview of the principles underlying GENESIS, then in Section 3 we develop the concept of causality for integral methods.

2. Principles of the GENESIS methodology

We shall now outline some general concepts developed by Hunt and Plybon [1], Hunt and Hewitt [2] and Hunt [3]; we confine our attention here to 2–D, steady, compressible, rotational flow. First we define a physical domain Ω with closed inner boundary S and closed outer boundary S_0. We represent the domain *inside* the boundary S by Ω'. We denote by $\hat{n}$ the unit normal vector pointing from S or S_0 into Ω, and by $\hat{s}$ the unit tangent defined by $\hat{s} = \hat{j} \times \hat{n}$, with $\hat{j}$ the unit vector into the paper. We postulate an *arbitrary* vector field $\bar{F}$ in Ω, and an *arbitrary Laplacian* field $\bar{F}'$ in Ω' (i.e. $\nabla \cdot \bar{F}' = 0$ and $\nabla \times \bar{F}' = 0$). A running point Q can lie anywhere in Ω or on S or S_0. We write the value of $\bar{F}$ (and similarly $\bar{F}'$) at a point Q on S in the form $\bar{F} = F_n \hat{n} + F_s \hat{s}$ where $F_n = \hat{n} \cdot \bar{F}$ and $F_s = \hat{s} \cdot \bar{F}$. We represent the normal and tangential components of the jump $\bar{F} - \bar{F}'$ at Q by $\sigma = \hat{n} \cdot (\bar{F} - \bar{F}') = F_n - F_n'$ and $\omega = \hat{s} \cdot (\bar{F} - \bar{F}') = F_s - F_s'$. We define a *vector kernel function*: $\bar{K} = \bar{r}_{PQ}/2\pi r_{PQ}^2$ where $\bar{r}_{PQ}$ is the position vector from Q to an arbitrary fixed point P in Ω or on S or S_0. [The 3–D counterpart is $\bar{r}_{PQ}/4\pi r_{PQ}^3$, the *inverse–square operator* of gravity, electromagnetism and electrostatics.] We can then write the following *universal mathematical identity* expressing the value of the vector $\bar{F}$ at a point P in Ω:

$$\bar{F}_P = \bar{F}_0 + \int_S \sigma\, \bar{K}\, dS + \int_S \omega\, \hat{j} \times \bar{K}\, dS + \int_\Omega (\nabla \cdot \bar{F})\, \bar{K}\, d\Omega + \int_\Omega \hat{j} \cdot (\nabla \times \bar{F})\, \hat{j} \times \bar{K}\, d\Omega \tag{1}$$

where $\bar{F}_0 = \int_{S_0} F_n\, \bar{K}\, dS + \int_{S_0} F_s\, \hat{j} \times \bar{K}\, dS$. In Section 3 we examine the causal nature of (1), in terms of induced fields and their interactions – concepts long familiar to engineers from other fields. First, we consider the linear case where $\nabla \cdot \bar{F}$ and $\nabla \times \bar{F}$ are both zero, then generalize to the nonlinear case.

2.1 Surface Integral Schemes; Source and Vortex Evaluation by Relaxation: SAVER

When $\bar{F}$ is taken as the velocity vector $\bar{V}$ in a fluid flow, and S_0 is "at infinity", $\bar{F}_0$ becomes simply the "onset" freestream velocity $\bar{V}_0$. If the velocity field is *incompressible and irrotational*, (1) then reduces to:

$$\bar{V}_P = \bar{V}_0 + \int_S \sigma\, \bar{K}\, dS + \int_S \omega\, \hat{j} \times \bar{K}\, dS \tag{2}$$

This equation forms the basis of all the various "direct" and "indirect" velocity–based BIM formulations for solving inviscid, incompressible problems for arbitrary bodies. We refer to σ and ω as *surface source* and *vorticity* distributions which *induce* velocity perturbations in accordance with (2). Some of the possible linear formulations, distinguished by their different boundary conditions and different internal fields $\bar{F}'$ in Ω' (i.e. their different σ and ω distributions), are discussed in detail by Hunt [4, 5, 6].

The linear solver in GENESIS employs *both* σ and ω distributions, with either piecewise constant or linear form on flat or curved panels. Analytic expressions are used, in 2–D, for the piecewise integrals in the discretized form of (2). Steps are taken to minimize the numerical errors associated with discontinuities at panel edges. Extending a method proposed by Raj and Gray [7], this solver uses *relaxation* to compute the σ and ω distributions satisfying the boundary conditions prescribed on S for V_n and/or V_s; it has the acronym SAVER, for Source And Vortex Evaluation by Relaxation. SAVER is aimed primarily at *three*–dimensional problems and replaces the usual matrix solution methods. It does *not* require equal numbers of "unknowns" and boundary conditions, and avoids "ill–conditioning" problems. SAVER fits well with the nonlinear part of the methodology outlined below in Section 2.2.

SAVER is based on the observation that if we force the field $\bar{F}'$ in Ω' to be equal to the unperturbed vector $\bar{V}_0$, then by definition of the jump quantities at any boundary point P the vector $\bar{V}_P$ on the external face of S must equal $\bar{V}_0 + \sigma \hat{n} + \omega \hat{s}$. In a standard "*Green's Identity*" approach [4, 5, 6], the source density is fixed as $\sigma = -\hat{n} \cdot \bar{V}_0 + V_{nBC}$, where V_{nBC} is the boundary condition specified for V_n (non–zero for a moving boundary). We thus have: $\vec{V}_P = (\hat{s}\cdot\vec{V}_0 + \omega)\,\hat{s} + V_{nBC}\hat{n}$. The distribution ω, and thus V_s, is obtained by solving a linear set of equations forcing ω to cancel *one component* of the velocity induced by σ at a set of control points on the *interior* surface of S; the assumption is that *mathematical uniqueness* [4] will *automatically* force the other component to be zero. One problem here is that the equation set is ill–conditioned, making iterative techniques inapplicable and a "direct" solution unreliable. Also, "leakage" between control points violates the requirements for "uniqueness", with the result that V_s is in error. Finally, unless special precautions are taken, the solution is not constrained to satisfy the so–called Kutta condition: arbitrary multiples of an "eigensolution" can be present.

SAVER starts from some initial estimate of ω and uses as the initial σ the "Green's Identity" form $\sigma = -\hat{n}\cdot\vec{V}_0 + V_{nBC}$. At each iteration, the overall velocity $\vec{V} = V_n\hat{n} + V_s\hat{s}$ on the external face of S is computed using the current estimates of σ and ω in (2), then ω is *relaxed* towards $V_s - \hat{s}\cdot\vec{V}_0$. Now for lifting airfoil problems, the sum $\omega_u + \omega_l$ of the (signed) values of ω at points on the upper and lower surface must be zero at the trailing edge. At each iteration, a correction $(\omega_u + \omega_l)/2$ is subtracted uniformly from *all* the relaxed ω values, to meet this condition. The computed V_n now differs from the target V_{nBC} by some error, say: $\epsilon_n = V_n - V_{nBC}$. Therefore σ is relaxed towards $-\hat{n}\cdot\vec{V}_0 + V_{nBC} + \epsilon_n$. The cycle is then repeated. At convergence, the *interior* field exactly matches $\vec{V}_0$ at each control point (*both* components are precise, due to the dual relaxation); the computed value V_n on the *external* face of S is very close, but not identical, to the target V_{nBC}. Comparisons with analytical solutions (not presented here) demonstrate that the V_s thus obtained seems consistently more accurate than that from other methods.

SAVER can also be used for "design": a tangential velocity (pressure) boundary condition $V_s = V_{sBC}$ is prescribed for a fixed angle of attack, together with an initial estimate of the required shape S; relaxation then yields V_n, from which the required geometry adjustment is deduced.

2.2 Generalized Nonlinear Extension of Surface Integral Schemes: GENESIS

If we interpret $\vec{F}$ as the velocity $\vec{V}$, the field integrals in (1) represent an induced velocity perturbation, $\vec{V}_F$ say, associated with compressibility and rotationality. In a FIM approach, these field integrals are evaluated directly. In GENESIS, they are converted to "equivalent" boundary integrals. We construct a fictitious vector $\vec{F}$ such that $\nabla\cdot\vec{F} = \nabla\cdot\vec{V} = \Sigma$ and $\nabla\times\vec{F} = \nabla\times\vec{V} = \boldsymbol{\Gamma}$, say, throughout Ω, with $\vec{F} = 0$ at the outer boundary S_0. We let the vector $\vec{F}'$ in Ω' be zero. Introducing the notation $\sigma_F = -F_n$ and $\omega_F = -F_s$ for the *equivalent boundary distributions*, we can write (1) at a point on the external face of S:

$$\vec{V}_F = \int_\Omega \Sigma\,\vec{K}\,d\Omega + \int_\Omega \boldsymbol{\Gamma}\times\vec{K}\,d\Omega \;=\; -\sigma_F\hat{n} - \omega_F\hat{s} + \int_S \sigma_F\,\vec{K}\,dS + \int_S \omega_F\,\hat{j}\times\vec{K}\,dS \qquad (3)$$

For points P in Ω (external to S), we replace $-\sigma_F\hat{n} - \omega_F\hat{s}$ on the r.h.s. of (3) by the field value $\vec{F}_P$. The analysis in [1, 3] shows how (3) converts the general nonlinear problem (1) to a *pseudo–Laplacian problem* comprising (2) with the prescribed normal velocity V_{nBC} replaced by $V_{nBC} + \sigma_{FP}$, where $\sigma_{FP} = -\hat{n}_P\cdot\vec{F}_P$. Denoting by $\vec{V}_L$ the pseudo–Laplace velocity solution, the compressible, rotational field $\vec{V}$ is given by $\vec{V} = \vec{V}_L + \vec{F}$; the tangential component on S is given by $V_s = V_{sL} - \omega_F$. The SAVER algorithm is thus modified to relax ω towards $V_{sL} - \hat{s}\cdot\vec{V}_0$ and σ towards $-\hat{n}\cdot\vec{V}_0 + V_{nBC} + \sigma_F + \epsilon_n$. At convergence, we compute $V_s = V_{sL} - \omega_F$ as the required solution of the *compressible, rotational flow.*

The process for constructing the fictitious vector $\vec{F}$ considers the fields Σ and $\boldsymbol{\Gamma}$ separately, [1, 3]. Estimates of the "physical" Σ and $\boldsymbol{\Gamma}$ at the body boundary S are computed using the current estimate of V_s. For the Euler equations we have $\Sigma = M^2 dV_s/ds$, with the Mach number M an explicit function of V_s;

normal derivatives of Σ can also be obtained explicitly. The value of Γ downstream of a shock is discussed in Section 3.3; the forms of Σ and Γ for a viscous flow will be discussed in a future paper. At each boundary point, the variation of Σ or Γ away from the surface is hypothesized to be the linearly weighted sum of pre-defined shape functions. The weights are computed to match the local value and some number of normal derivatives of Σ or Γ. A vector function F is then constructed, with $\nabla \cdot F = \Sigma$ and $\nabla \times F = 0$ when Σ is being considered, or with $\nabla \times F = \Gamma$ and $\nabla \cdot F = 0$ when Γ is being considered. This is achieved by "marching" along the surface, constructing between one point and the next a piecewise-quadratic function as the solution of a forced simple-harmonic equation. The process starts from an initial plane S_I marking entry to the compressible or rotational zone Ω, and ends at a terminating plane S_T marking its exit. The surfaces S_I, S_T and the body surface S define the boundary over which the boundary integrals are defined; no integral is required over the part of the outer boundary S_0 completing the boundary of Ω, since $F = 0$ on S_0.

A detailed analysis in [3] takes the outer boundary S_0 as the edge $z = \delta$ of a boundary layer. It is shown that the velocity induced by the distributed vorticity in the boundary layer is equal to that induced by "equivalent" distributions σ_F and ω_F on the body boundary S. As $\delta \to 0$, σ_F approaches the value of the streamwise derivative $d(V_{s\delta}\,\delta^*)/ds$, where $V_{s\delta}$ is the streamwise velocity at the edge $z = \delta$, and δ^* the *displacement thickness*. Thus σ_F is equal to the "surface blowing" often applied in an inviscid flow model coupled with a boundary layer calculation [6]. The analysis in [3] suggests that at Reynolds numbers of aeronautical interest, the omission of the further terms in the full definition of σ_F, and the failure to allow for the effect of ω_F, could be responsible for significant errors in the treatment of lifting airfoils.

We shall now discuss BIM causality as an aid in sensitivity analysis, then in interpretation of flow physics.

3. Integral Methods and Causality

3.1 Sensitivity Analysis

The *causal model* provided by a BIM allows a problem to be decomposed into mutually interacting blocks. Within any one block (e.g. a foreplane mounted on a fuselage), a refined flow model may be employed for an accurate computation of the "cause" of its influence on another component, while the "effect" it induces at that other component (e.g. an engine nacelle) may be computed with consistent fidelity by a less refined model. [An electrostatic analogy is the use of point charges to represent the "action at a distance" of charge distributed on an imperfect conductor.] Thus the causal model can offer not only a significant computational saving, but also an insight into the "interference" between the components of a machine: it more or less automates the process of *sensitivity analysis*. For example, it is possible [8] to derive from a BIM the direct contribution from, say, a foreplane and its wake, to, say, the stability derivative n_p (yawing moment due to rate of roll) at a particular flight condition (say, during a coning motion). Similarly, component contributions to flow distortion entering an intake can be directly established.

3.2 Interacting Force Fields

A familiar concept in Electromagnetics is that of a *force field* generated by the interaction between an *induced field* and the *field inducers*. This concept has been only partially exploited in fluid dynamics: a fundamental "law" of aerodynamics is that the sectional lift L on a wing is given by $L = \varrho V_0 \Gamma$, where ϱ is the fluid density, V_0 the freestream speed, and Γ the "circulation" around the wing section. We may speculate that the elemental force dF on a *fluid particle* of elemental volume $d\Omega$ contains a contribution $-\varrho\, V \times \Gamma\, d\Omega$, where V is the *local* velocity and Γ the local *vorticity* vector $\Gamma = \nabla \times V$. [This is the fluid-dynamic analogue of the electro-magnetic force on an elemental electric current in a magnetic field]. We shall now prove a generalized form of this vector relationship.

Consider a fixed control volume Ω with boundary S, the unit normal $\hat{n}$ on S pointing out from Ω. Gauss' Theorems tell us that $\int_\Omega \nabla\cdot\overline{Q}\, d\Omega = \int_S \overline{Q}\cdot\hat{n}\, dS$, and that $\int_\Omega \nabla\theta\, d\Omega = \int_S \theta\,\hat{n}\, dS$, where $\overline{Q}$ and θ are arbitrary vector and scalar fields. The first (vector) theorem leads to a *universal mathematical identity*:

$$\int_S \theta\,\overline{Q}\,(\overline{Q}\cdot\hat{n})\,dS = \int_\Omega \theta\,\nabla\tfrac{1}{2}Q^2\,d\Omega \;-\int_\Omega \theta\,\overline{Q}\times(\nabla\times\overline{Q})\,d\Omega \;+\int_\Omega \theta\,\overline{Q}\,\nabla\cdot\overline{Q}\,d\Omega$$
$$+\int_\Omega (\overline{Q}\cdot\nabla\theta)\,\overline{Q}\,d\Omega \tag{4}$$

First, let us interpret θ as the fluid density ϱ and $\overline{Q}$ as the velocity $\overline{V}$ in a region Ω of a *physical* fluid flow. [In general, ϱ and $\overline{V}$ will not be constant in Ω, and the flow may be compressible and/or rotational, conducting, non–adiabatic, etc.] Then the first integrand in (4) becomes $\varrho\,\overline{V}\,(\overline{V}\cdot\hat{n})\,dS$, so that the l.h.s. of (4) is the *physical momentum flux* outwards across S. This must therefore equal the overall force acting on the particles in Ω, for a *steady* flow. In general, this force will be the sum of *boundary forces* acting on S (i.e., a pressure acting normal to the boundary, plus viscous shear forces, acting tangentially), and *body forces* (i.e. gravity, but possibly also electric and magnetic forces). In the limit as Ω shrinks to an elemental volume $d\Omega$, we denote the overall physical force on this element by $d\overline{F}_r$.

It is easy to show that the sum of the final two integrands in (4) is $\overline{Q}\,\nabla\cdot(\theta\,\overline{Q})$, i.e. in this physical flow field $\overline{V}\,\nabla\cdot(\varrho\,\overline{V})$, or *zero* if this flow satisfies mass conservation [i.e. if $\nabla\cdot\overline{V} = -(\overline{V}\cdot\nabla\varrho)/\varrho$]. Suppose we now introduce a function $\wp$ defined in $d\Omega$ by $\wp = const. - \varrho_c\,Q^2/2$, where ϱ_c is the density at the geometric centroid c of the elemental region $d\Omega$. Then the first integrand on the r.h.s. of (4) may be replaced by $-\nabla\wp$, so that, for an infinitesimal element of volume $d\Omega$, (4) may be written:

$$d\overline{F}_r = -\nabla\wp\,d\Omega \;-\; \varrho\,\overline{V}\times\overline{\Gamma}\,d\Omega \tag{5}$$

Thus, for this general (steady) flow, the force $d\overline{F}_r$ acting on $d\Omega$ can be expressed as the sum of a *pseudo–pressure force* on its boundary and a *pseudo–body force* acting on its particles. The pseudo–pressure $\wp$ is the pressure which would pertain if the flow had uniform density ϱ_c and satisfied the incompressible Bernoulli equation $p = const. - \varrho_c\,Q^2/2$. The pseudo–body force is $-\varrho\,\overline{V}\times\overline{\Gamma}$ per unit volume.

[An example of (5) is a *Rankine vortex* Ω of radius b, with its axis at the origin of an (r,ϕ) coordinate system, in an otherwise stationary medium with constant density ϱ. The velocity inside Ω is the same as that of a solid body rotating with constant angular velocity ω; that is, $V_\phi = \omega\,r$ and $V_r = 0$. Outside Ω, the "induced" velocity field is irrotational: the product $V_\phi\,r$ remains equal to the value ωb^2 at $r = b$. The field inside Ω is rotational, the vorticity being of uniform magnitude 2ω; the vector product $-\varrho\,\overline{V}\times\overline{\Gamma}$ in (5) is thus of magnitude $2\varrho\omega^2 r$ and is oriented radially inwards. The vector $-\nabla\wp$ in (5) is equal by definition to $\varrho\,\nabla V^2/2$ and thus has magnitude $\varrho\omega^2 r$, oriented radially outwards. If we consider an element $d\Omega$ at radius r inside the vortex, and assume that $-\varrho\,\nabla V^2/2$ and $-\nabla\wp$ remain uniform within $d\Omega$, then (5) shows that the overall force $d\overline{F}_r$ acting on $d\Omega$ is $\varrho\omega^2 r\,d\Omega$, radially inwards. Now the mass of the fluid particles within $d\Omega$ is $\varrho\,d\Omega$; the force *required* to maintain their centripetal acceleration $\omega^2 r$ is therefore equal to $\varrho\omega^2 r\,d\Omega$, exactly matching the effective force $d\overline{F}_r$.]

Let us now hypothesize a *fictitious incompressible* flowfield in which $\overline{Q}$ in (4) is defined, point by point, to be equal in value to the velocity $\overline{V}$ of a compressible, rotational flow. This velocity field $\overline{Q}$ is considered to be the sum of some externally imposed velocity field and a perturbation velocity field *induced by* a distributed vorticity distribution $\overline{\Gamma}$ defined by $\overline{\Gamma} = \nabla\times\overline{Q} = \nabla\times\overline{V}$ and by a distributed source distribution Σ defined by $\Sigma = \nabla\cdot\underline{Q} = \nabla\cdot V$. We again consider an infinitesimal control volume $d\Omega$. By definition, the distributed source Σ is injecting mass into $d\Omega$; we hypothesize that in $d\Omega$ both the injected fluid and the resulting "mixture" have the constant density ϱ_c from the physical flow defined above. We now let θ in (4) be equal throughout $d\Omega$ to this *constant* value ϱ_c . Thus for this case the final integral in the universal identity (4) vanishes. The first integrand in (4) is now $\varrho_c\, V\,(V\cdot\hat{n})\, dS$; we represent by dF_f the overall momentum flux of the fictitious fluid, outwards from $d\Omega$. We now introduce the same function $\wp$ defined earlier, so that the first integrand on the r.h.s. of (4) may again be replaced by $-\nabla\wp$. Thus, for this fictitious field, (4) can be written for the element $d\Omega$:

$$dF_f = -\nabla\wp\, d\Omega - \varrho_c\, V\times\Gamma\, d\Omega + \varrho_c\, V\,\Sigma\, d\Omega \tag{6}$$

The elemental forces dF_r and dF_f in (4) and (6) are *not* equal when the density ϱ is not constant in $d\Omega$ in the real flow: the control volume $d\Omega$ in the fictitious flow requires an additional, non–physical body force equal to $\varrho_c\, V\,\Sigma$ per unit volume; this is required to take the velocity of the injected mass $\varrho_c\Sigma$ from zero to the local value V. This elemental force acts parallel to V; it can be interpreted as a *drag* force. If the fictitious incompressible flow is constructed using field sources such that the velocity vector V is everywhere equal to that of a physical compressible flow satisfying the appropriate physical laws, then the fluid particles in that fictitious flow will experience this extra body force. This extra force is the analogue of the electrostatic force on a charged particle in an electric field.

3.3 Interpretation of shock waves and shock–induced separation.

Relation (6) offers causal explanations of certain phenomena associated with shock waves. We consider here the case of a shock emanating normally from a curved boundary S with local surface curvature κ . We know that when a flow with non–uniform profile passes through a shock, the flow downstream is non–uniformly rotational. The field vorticity evaluated on the body boundary on the downstream face of the shock has the (signed) magnitude $\Gamma = -\sigma_s\,\kappa\,(M_1^2-1)$, where the local Mach number M_1 entering the shock satisfies $M_1^2 = \frac{2}{\gamma-1}\left[\frac{V_1^2}{V_{max}^2-V_1^2}\right]$, with $V_{max} = \frac{V_o}{M_o}\left[M_o^2+\frac{2}{\gamma-1}\right]^{\frac{1}{2}}$ the thermodynamically maximum possible speed and V_1 the local velocity entering the shock, for a perfect gas with specific heats ratio γ . Since the flowspeed must decrease across a compression shock, the *Rankine–Hugoniot* normal–velocity jump $\sigma_s = V_1\left[\frac{\gamma-1}{\gamma+1}\,\frac{V_{max}^2}{V_1^2} - 1\right]$ must be negative. V_1^2 must exceed $\frac{2}{\gamma+1}\frac{V_o^2}{M_o^2}\left[1+\frac{\gamma-1}{2}M_o^2\right] = \frac{\gamma-1}{\gamma+1}\,V_{max}^2$, corresponding to $M_1 = 1$, for a shock to occur. Were V_1 to approach V_{max} , then σ_s would approach a limit $-2V_{max}/(\gamma+1)$, but M_1^2 and Γ would become indefinitely large.

This vorticity Γ *induces* a perturbation, V_Γ say, which at the boundary S is of opposite sign to the velocity of an irrotational flow. It is thus evident from (6) that the particles carrying Γ will experience a "lift" force contribution $-\varrho\, V_\Gamma\times\Gamma$ acting away from the surface, *which would not be present in a potential flow*; if V_1 is sufficiently large, this force contribution will lift the flow particles from the surface immediately upon exiting the shock. Downstream of the shock [1], Γ increases along a streamline, in proportion to the local *static pressure* p . On a lifting airfoil, p rises quite sharply near the trailing edge; therefore both Γ and the V_Γ induced by it will also increase sharply in magnitude. Since ϱ also increases, the magnitude of the force $\varrho\, V_\Gamma\times\Gamma$ will progressively increase downstream of the shock. There is thus the motivation for the flow to separate at some distance downstream of the shock, even in an inviscid flow. Since the direction of rotation of the vorticity Γ is the same as that in the physical boundary layer, it follows that the influence of viscosity will merely be to enhance this *essentially inviscid* shock–induced separation effect.

We note also a further interaction here between the "cause" and the "effect": the vorticity Γ behind the shock will induce a (potentially unbounded) perturbation $\vec{V}_\Gamma$, acting so as to *reduce* the value V_1 entering the foot of the shock, while the "sink" σ_s representing the shock will induce a (bounded) perturbation increasing V_1. If the velocity entering the shock attempts to become too large during a computation, the induced velocity field will be strongly reduced by $\vec{V}_\Gamma$, until equilibrium is attained: this "damping", *which is absent from a potential flow model*, can become arbitrarily large, as indicated earlier.

Similar arguments offer a causal explanation, without direct recourse to the Second Law of Thermodynamics, of why *expansion shocks* never appear in a physical flow. In addition to the element $d\Omega$ discussed earlier, we now also consider a second element $d\Omega'$ located at a point where $\nabla \cdot \vec{V}$ has the value Σ'. We represent the net divergence $\Sigma' d\Omega'$ of the particle by $d\Sigma'$, and consider this to act as an elemental *source* which *induces* an elemental perturbation velocity $d\vec{V}' = \vec{r}\, d\Sigma'/4\pi r^3$ at the first element $d\Omega$, where $\vec{r}$ is the position vector drawn from $d\Omega'$ to $d\Omega$. We can see that the elemental force $d\vec{F}$ exerted by $d\Omega'$ on $d\Omega$ is: $d\vec{F} = \varrho\, \vec{r}\, d\Sigma\, d\Sigma'/4\pi r^3$ which may be recognized as the analogue of the *inverse square law* of gravity, or of *Coulomb's law* in electrostatics. It is clear that particles $d\Omega'$ and $d\Omega$ carrying "source" intensities $\nabla \cdot \vec{V}$ of opposite sign will be attracted towards each other, while the force will be one of repulsion for like signs. In a real fluid flow, $\nabla \cdot \vec{V}$ is directly related to the gradient of the density ϱ; this is related to $\vec{V}$ by the laws of thermodynamics, and is affected by mechanical dissipation and thermal conduction. These phenomena generally become significant only in the interior of a shock wave. We can speculate that the extra forces of attraction/repulsion between fluid particles, implied by this dissipation and conduction, *sharpen* compressive shock waves, and *suppress* the formation of expansion shocks.

Suppose we include in the definition of Σ only the "ideal" Euler term $M^2 \partial V/\partial s$, with M the local Mach number and $\partial V/\partial s$ the rate of change of speed along the streamline [1]. We can expect Σ to induce a steepening negative slope in the computed velocity, in the vicinity of points where the flowspeed decreases through $M = 1.0$, representing the formation of a compression shock. It will however also give rise to a steepening positive slope in the velocity, in the vicinity of the sonic point where the flowspeed *increases* through $M = 1.0$, representing a (non–physical) expansion shock: the "ideal" Euler equations contain no information to prevent the Second Law of Thermodynamics from being violated by the computation.

Suppose now we also include in the definition of Σ appropriate terms representing the irreversible effects of mechanical dissipation and thermal conduction, in addition to the "ideal" Euler term. For the purposes of shock capture, it is permissible [1] to ignore, in the dissipative terms in the equations of momentum and energy, the first coefficient of viscosity μ and retain only the second coefficient λ factoring Σ. The dissipation terms can then be expressed as a nonlinear local multiple of the reciprocal of a Reynolds number *Re* based on λ. The resulting effect will be that the compression shock is "sharpened" by the attraction of additional positive and negative source elements in its front and rear parts. The expansion shock will be *suppressed* by the repulsion between elements carrying extra source intensities of *equal* sign.

These additional forces are negligible in most of the flow field, and are effective only when $\partial V/\partial s$ is very large. Unfortunately, such gradients can only be detected if the grid carrying the field source is very fine – on the scale of $1/Re$. Since such grids are infeasible, the interplay between dissipation and conduction, which controls the internal structure and jump properties of a physical shock, will be unrealistic on any practicable grid. It would not be possible to simulate real shocks, i.e. ones with a Rankine–Hugoniot jump.

This is the same problem as that faced by differential methods. In those methods, the standard fix is to resort to an unrepresentatively low *Re*, or to throw physics out of the window entirely and rely on some sort of "artificial" viscosity related arbitrarily to discretization errors in the approximations of the physical derivatives, or to the use of some type of "upwinding" in the estimation of those derivatives. It would appear desirable to remove this arbitrariness from the computation, by using our empirical knowledge of flow behavior. We now go on to discuss the concept of a knowledge–based approach to CFD.

4. Integral Methods as a Basis for Knowledge–Based CFD

Integral methods can be applied *implicitly*, to *capture* phenomena such as shock waves or flow separations. They can also be used *explicitly*. A phenomenon can be modeled as an *explicit discontinuity* (shock) or *near–discontinuity* (shear layer); axiomatic or empirical knowledge of its properties is then bestowed upon it, without the various differential–method artifices. Such a process is common in other "knowledge–based" (KB) technologies, and is known as *hypothesize and test*: *hypothesize* the presence of a feature, *test* the hypothesis, then iteratively *adjust* it until some criteria are satisfied, or reject the hypothesis. Such a process has been common for many years in vortex flow computations by "inviscid" panel methods [6], where vortex sheets are hypothesized to emerge from the edges of lifting wings, or from smooth surfaces, into an otherwise potential flow. The mathematical justification lies in extending the surface S_0 defined for the "Laplace" problem in (2) to include the boundaries of these vortex sheets *with prescribed strength and location*. The "onset" velocity $\bar{V}_0$ then includes the contribution from these sheets. The strength and trajectory are *iterated* to simulate physical constraints. If this results in vortex sheets of zero strength, then their induced velocity fields will be evaluated as zero. This happens automatically, for example, if a panel method designed for lifting wings is applied to a symmetric wing at zero angle of attack.

The infeasibility of solving the N–S equations for realistic configurations by differential methods motivates the development of such KB simulations. However, care must be exercised in their application and interpretation: the success of such an approach depends strongly on the quality of the "knowledge". For example, at high Re and low angle of attack α, we can be confident that the trailing edge of a lifting wing will be a very good estimate of the separation line. Similarly, empirical observation shows that under fairly repeatable conditions a spiraling vortex sheet will emerge from a swept leading edge. If a reputable boundary layer scheme were coupled with the inviscid model, the computed separation line could be expected to agree closely, if not exactly, with the hypothesized separation line.

Expectations are lower for smooth–surface separation. In fact, for forebody separation, it is possible to obtain stable vortex–sheet solutions for a range of hypothesized separation–line locations; none of these is necessarily close to a physical solution: physical constraints absent from the model are being violated. For example, the physical solution may have a boundary–layer reversal at another location. We can strengthen confidence in the KB model by identifying such possibilities and testing for them in the computation.

Similar arguments can be extended to the representation of shocks by *source sheets* in a non–dissipative flow model. For example, we can have some confidence in the conditions under which a shock wave will emanate from a line of angular discontinuity. Although there are generally no *a priori* criteria for predicting a shock wave on a *smooth* surface, a rapid steepening of the computed velocity gradient would indicate the approximate location of an incipient shock wave. This offers a starting point for iterating the location and strength σ_s of an explicit source surface simulating an hypothesized shock, until the Rankine–Hugoniot (R–H) condition relating σ_s and the computed shock–entry speed V_1 is satisfied.

The surface S_0 defined earlier can now be further extended to include the boundaries of *source sheets of prescribed strength and location*, embedded in an *incompressible* field. The "onset" flow velocity $\bar{V}_0$ now includes a contribution from these sheets. Suppose we have a source sheet impinging normally upon a body boundary S, with prescribed strength σ_s on S and with some pre–defined rate of decay with normal distance. It would be simple to use this modified BIM to compute the velocity on S, including the value V_1 at entry to the fixed source sheet. If we interpret V_{max} as a non–physical function, then for *any* chosen σ_s and location of the sheet there would be some value of V_{max} for which that σ_s and the computed V_1 would satisfy *formally* the R–H condition. This would be a "solution" in a kinematic sense only: this fictitious incompressible flow would require some extraneous force to hold the source sheet in position.

Suppose we now modify the problem and "switch on" compressibility by introducing a field source $\Sigma = M^2 \partial V/\partial s$, with V_{max} again taking on its physical meaning. For any "reasonable" *fixed* shock location, there would still be a σ_s satisfying the R–H relation between that σ_s and the computed V_1. The flow would show a tendency to generate auxiliary field source or sink distributions Σ in the vicinity of the fixed source sheet, in an attempt to shift its strength and location towards "physical" values. Nonetheless, thermodynamic properties computed from the velocity $\bar{V}$ and from the entropy jump implied by V_1 would nominally satisfy all the physical conservation laws in the vicinity of the "shock".

This flow model would, however, also attempt to generate a non–physical *expansion shock* near the sonic point. Paraphrasing an argument by Nixon and Liu [9] we may speculate that only when a physically correct shock location has been attained will the non–dissipative Euler equations lose their tendency to form an expansion shock. We can conjecture a causal relationship between the shock position and the conditions at the sonic point: *the shock location needs to be adjusted until the sonic line acts like a nozzle throat for each streamtube traversing it.* This "design condition" will be met when the sonic point, and the point at which the streamwise mass–flux derivative $\partial(\varrho V_s)/\partial s$ vanishes, coincide. The degree of violation of this constraint can be expressed in terms of a required correction to the velocity field in the vicinity of these two points. This can then be equated to a required change in the perturbation contribution induced by the shock source surface, which can then be interpreted as a required shift in its location with σ_s temporarily frozen. The strength σ_s is then re–evaluated as a function of the new entry speed V_1. This process is analogous to that outlined in Section 3.1 for aerodynamic sensitivity analysis. This speculative shock–relaxation process is currently under investigation; the findings will be reported at a later date.

Acknowledgments. The author has engaged numerous individuals in discussions relating to some of the speculative concepts introduced in this paper, and would like to express particular thanks to his colleagues *Art Adamson, George Converse, Jim Keith* and *Ron Plybon* at GE Aircraft Engines, and to *Carson Yates* (NASA LaRC), *David Nixon* (Nielsen Engineering & Research Inc.), *George Dulikravich* (Pennsylvania State University) and *James Wu* (Georgia Institute of Technology).

References

1. Hunt, B.; Plybon, R.C.: Generalization of the B.I.M. to Nonlinear Problems of Compressible Fluid Flow: The No–mesh Alternative. Pt I: Maths; Pt II: Physics. Boundary Element Methods in Engineering, (to appear, Proc. ISBEM–89, Eds. Annigeri, B.S., Tseng, K.); Springer Verlag.
2. Hunt, B.; Hewitt, B.L.: The Indirect B.I. Formulation for Elliptic, Hyperbolic, and Nonlinear Fluid Flows. Developments in Boundary Element Methods – IV, (Eds. Banerjee, P.K., Watson, J.O.); Elsevier Applied Science Publishers.
3. Hunt, B.: Generalized Nonlinear Boundary Integral Methods for Compressible, Viscous Flows over Arbitrary Bodies: Knowledge–Based CFD. Japan/USA Boundary Elements Symposium, Palo Alto, 5–7 June, 1990. To appear in Engineering Analysis with Boundary Elements.
4. Hunt, B.: The Panel Method for Subsonic Aerodynamic Flows: A Survey of Mathematical Formulations and Numerical Models and an Outline of the new BAe scheme. VKI L. S. 1978–4.
5. Hunt, B.: The Mathematical Basis and Numerical Principles of the B.I.M. for Incompressible Potential Flow over 3D Aerodynamic Configurations. Numerical Methods in Applied Fluid Dynamics (Ed. Hunt, B.); Academic Press, London, 1980.
6. Hunt, B.: Recent and Anticipated Advances in the Panel Method: The Key to Generalised Field Calculations? VKI Lecture Series 1980–5.
7 Raj, P; Gray, R.B.: Computation of Three–Dimensional Potential Flow Using Surface Vorticity Distribution. J. Aircraft Vol. 16, No. 3, 79–4039 (1979).
8. Hunt, B.: The Role of Computational Fluid Dynamics in High–Angle–of– Attack Aerodynamics. AGARD Lecture Series 121.
9. Nixon, D.; Liu, Y: The Mechanisms of Determining Shock Locations in One and Two Dimensional Transonic Flows. J. Appl. Mech. 53 (1986) 203–205.

An Iterative Boundary Element Analysis of Helically Symmetric MHD Equilibria

H. Igarashi and T. Honma
Department of Electrical Engineering, Faculty of Engineering,
Hokkaido University, Kita 13, Nishi 8, Kita-ku, Sapporo, 060 Japan

Summary

This paper presents an iterative scheme based on boundary element methods for calculating helically symmetric magnetohydrodynamic (MHD) equilibrium configuration. A boundary integral equation is derived from the MHD equilibrium equation in the helically symmetric system. It is shown that the numerical solutions by the present scheme are in good agreement with the exact solutions. Moreover, the numerical stability is reported when the Picard iteration is used in this scheme.

Introduction

The axisymmetric magnetohydrodynamic (MHD) equilibria of the nuclear fusion plasmas have been well studied by means of finite difference [1] and finite element [2] methods. On the other hand, in the stellarator-type fusion machines, the equilibrium configuration has the three dimensional structure. Hence, we need to analyze those configurations with large storage memory and extremely long CPU time in comparison with the axisymmetric calculations. For this reason, the stellarator configuration is often approximated as being helically symmetric in the limit of large aspect ratio [3, 4]. In this approximation, it is assumed that the equilibrium configuration depends on only two variables r and $\zeta = \theta - hz$, where (r, θ, z) are the cylindrical coordinates and h is the helical wave number.

We obtained a vacuum solution to the helically symmetric MHD problem by the use of boundary element methods [5] and compared it with the finite and boundary element solutions for the plasmas enclosed by a circular shell [6]. In this paper, we report the convergence and accuary of the linear and nonlinear solutions to the above problem when using an iterative boundary element methods. Moreover, the numerical stability of the Picard iteration, which is the most standard technique of the iteration, is reported in this problem.

In this paper, after the helically symmetric MHD equilibrium equation is introduced, we derive a boundary integral equation for the helically symmetric system from the general form defined in the system which has a symmetry. Moreover, we compare the boundary element solution with the exact solution for a helically symmetric MHD plasma model with a circular cross-section. Finally, we report the numerical stability of the Picard iteration in both the linear and nonlinear cases.

MHD Equilibrium Equation

The MHD equilibrium is described by

$$\mathbf{J} \times \mathbf{B} = \nabla p, \quad \nabla \times \mathbf{B} = \mu_0 \mathbf{J}, \quad \nabla \cdot \mathbf{B} = 0, \tag{1}$$

where $\mathbf{J}$ is the current density, $\mathbf{B}$ is the magnetic field, p is the plasma pressure and μ_0 is the permeability of the free space. Here, we assume that the system has a symmetry in the u^3-direction in the general curvilinear coordinates (u^1, u^2, u^3),

$$\mathbf{B} = \mathbf{B}(u^1, u^2), \; \mathbf{J} = \mathbf{J}(u^1, u^2), \; \mathbf{A} = \mathbf{A}(u^1, u^2), \; p = p(u^1, u^2), \; g_{ij} = g_{ij}(u^1, u^2), \tag{2}$$

where $\mathbf{A}$ is the magnetic vector potential and g_{ij} is the metric tensor. From the relation $\mathbf{B} \cdot \nabla A_3 = 0$ derived from eq. (2), we can see that the magnetic lines lie on the surfaces of A_3 =const. and these surfaces are referred to as magnetic surfaces. We can derive the MHD equilibrium equation [7] for A_3 from eqs. (1) and (2) as follows :

$$-g_{33} \nabla \cdot \left(\frac{\nabla A_3}{g_{33}} \right) + \frac{g_{33}}{\sqrt{g}} \left\{ \frac{\partial}{\partial u^1} \left(\frac{g_{23}}{g_{33}} \right) - \frac{\partial}{\partial u^2} \left(\frac{g_{13}}{g_{33}} \right) \right\} B_3 = \mu_0 g_{33} p' + B_3 B_3'. \tag{3}$$

In this equation, $g \equiv \det(g_{ij})$ and the prime denotes the differentiation with respect to A_3. In addition, the functions p and B_3 can be proved to be constant on the magnetic surfaces.

On the other hand, the relation among $\mathbf{J}$, $\mathbf{B}$ and p can be derived from eqs. (1) and (2) using the basis vector $\mathbf{e}_3 \equiv \sqrt{g}\,(\nabla u^1 \times \nabla u^2)$ as follows :

$$\mu_0 \mathbf{J} = B_3' \mathbf{B} + \mu_0 p' \mathbf{e}_3. \tag{4}$$

From this relation, it can be seen that when $B_3' = p' = 0$, eq. (3) corresponds to the vacuum field and when $p' = 0$ and B_3'=const., it corresponds to the force-free field.

In the following, we derive the MHD equilibrium equation in the helically symmetric system from eq. (3). In this case, we use the helical polar coordinates (r, ζ, z) [8]. The second term in the left hand side of eq. (3) does not vanish in the coordinates while it vanishes in the well-known axisymmetric equilibria. In the (r, ζ, z) coordinates, the position vector $\mathbf{r}$ can be represented using the basis vectors $(\mathbf{i}, \mathbf{j}, \mathbf{k})$ in the Cartesian system as follows:

$$\mathbf{r} = r\cos(\zeta + hz)\,\mathbf{i} + r\sin(\zeta + hz)\,\mathbf{j} + z\,\mathbf{k}. \tag{5}$$

From eq. (5) and relations $\mathbf{e}_i = \partial\mathbf{r} / \partial u^i$, $g_{ij} = \mathbf{e}_i \cdot \mathbf{e}_j$, the metric tensor g_{ij} in (r, ζ, z) coordinates is given by

$$\left[g_{ij}\right] = \begin{bmatrix} 1 & 0 & 0 \\ 0 & r^2 & hr^2 \\ 0 & hr^2 & 1+h^2r^2 \end{bmatrix}, \tag{6}$$

and $g = r^2$. In addition, the relation of vector components between in (r, ζ, z) and in (r, θ, z) coordinates are $A_3 = hrA_\theta + A_z$, $B_3 = hrB_\theta + B_z$. We can derive the helically symmetric MHD equilibrium equation from the above relations as follows:

$$\frac{1}{r}\left\{\frac{\partial}{\partial r}\left(rK\frac{\partial\Psi}{\partial r}\right) + \frac{h^2}{r}\frac{\partial^2\Psi}{\partial\zeta^2}\right\} = \frac{2K^2}{h}f(\Psi) - Kf(\Psi)f'(\Psi) - \mu_0 h^2 p'(\Psi) \equiv F(\Psi), \tag{7}$$

where $K \equiv h^2/(1+h^2r^2)$, $\Psi \equiv A_3$, $f(\Psi) \equiv B_3$.

In eq. (7), the profiles of functions $f(\Psi)$ and $p(\Psi)$ are usually specified in accordance with some model of the plasma. In this paper, we express these functions in the form

$$f(\Psi) = \sum_{m=0}^{\infty} f_m \Psi^m, \quad p(\Psi) = \sum_{m=0}^{\infty} p_m \Psi^m. \tag{8}$$

On the other hand, we choose the boundary condition

$$\Psi = 0, \qquad \text{(on wall)}. \tag{9}$$

Under the boundary condition (9) at the circular wall of radius a, the exact solution to eq. (7) for $f_m = p_m = 0$ ($m \geqq 2$) can be expressed as [9]

$$\Psi = -\frac{f_0}{f_1} - \frac{\mu_0 p_1}{f_1^2}\left(1+h^2r^2+\frac{2h}{f_1}\right) + h\frac{f_0 f_1^2+\mu_0 p_1(f_1+h^2a^2f_1+2h)}{f_1^3\{ahJ_1(f_1a)+J_0(f_1a)\}}\left\{rJ_1(f_1r)+\frac{1}{h}J_0(f_1r)\right\}, \quad (10)$$

where J_0 and J_1 are the zeroth- and first-order Bessel functions of the first kind, respectively.

Boundary Element Formulation

In this section, we derive a boundary integral representation of the helically symmetric MHD equilibrium equation (7). For this purpose, firstly, we introduce a boundary integral equation described in the (u^1, u^2, u^3) coordinates for the system which has a symmetry in u^3-direction.

Let us consider the identity

$$G\nabla\cdot(K\nabla\Psi) - \Psi\nabla\cdot(K\nabla G) = \nabla\cdot(KG\nabla\Psi) - \nabla\cdot(K\Psi\nabla G), \quad (11)$$

and, in eq. (11), we choose the function G which satisfies the relation

$$\nabla\cdot(K\nabla G) + \frac{4\pi}{\sqrt{g}}\delta(u^1-u^1_i)\,\delta(u^2-u^2_i) = 0. \quad (12)$$

Integrating eq. (11) over one period λ of region V enclosed by boundary ∂V (for $-\lambda/2 \leqq u^3 \leqq \lambda/2$) yields

$$\lambda\int_\Omega GF(\Psi)\sqrt{g}\,du^1du^2 + \lambda C_i\Psi_i = \int_{\partial V} KGq\,dS - \int_{\partial V} K\Psi\frac{\partial G}{\partial n}\,dS, \quad (13)$$

where Ω is the surface of $u^3=0$ inside the region V, C_i is a constant which equals 2π on ∂V and 4π inside V, $\partial/\partial n$ is the normal differential operator of ∂V and $q \equiv \partial\Psi/\partial n$. Note that the surface integrals over the end of the region V in eq. (13) cansel by periodicity. Moreover, we can rewrite eq.(13) on the surface of $u^3=0$ as follows :

$$\int_\Omega GF(\Psi)\sqrt{g}\,du^1du^2 + C_i\Psi_i = \oint_C KGqP\,ds - \oint_C K\Psi Q(G)\,ds, \quad (14)$$

$$P \equiv \left\{\left(\frac{du^1}{ds}\right)^2(g_{11}g_{33}-g_{13}^2)+2\frac{du^1}{ds}\frac{du^2}{ds}(g_{12}g_{33}-g_{13}g_{23})+\left(\frac{du^2}{ds}\right)^2(g_{22}g_{33}-g_{23}^2)\right\}^{1/2}, \quad (15)$$

$$Q \equiv -\frac{1}{\sqrt{g}}\frac{du^1}{ds}\left\{(g_{13}g_{23}-g_{12}g_{33})\frac{\partial}{\partial u^1}+(g_{11}g_{33}-g_{13}^2)\frac{\partial}{\partial u^2}\right\}$$

$$+\frac{1}{\sqrt{g}}\frac{du^2}{ds}\left\{(g_{22}g_{33}-g_{23}^2)\frac{\partial}{\partial u^1}+(g_{13}g_{23}-g_{12}g_{33})\frac{\partial}{\partial u^2}\right\}. \quad (16)$$

where C is the intersection of ∂V and the surface of $u^3=0$ and ds is the differential line element on C. Equation (14) is the boundary integral equation for the u^3-symmetric system.

If we choose the (r, ζ, z) coordinates in eq. (14), we can obtain the integral form of the helically symmetric equilibrium equation (7). Namely, using eqs. (6) and (9), we easily get

$$\int_{\Omega} G F(\Psi)\, r\, dr\, d\zeta + C_i \Psi_i = \oint_C KGq \left\{1 + \left(hr\frac{dr}{ds}\right)^2\right\}^{\frac{1}{2}} ds\,. \tag{17}$$

In this case, the fundamental solution G in eq. (17) is given by [10]

$$G(r,\zeta | r_i,\zeta_i) = -\frac{2}{h^2}\left(\log r_> + \frac{1}{2}h^2 r_>^2\right)$$
$$- 4rr_i \sum_{m=1}^{\infty} I_m'(mhr_<)K_m'(mhr_>)\cos[m(\zeta-\zeta_i)], \tag{18}$$

where I_m and K_m are the modified Bessel functions of the first and second kind of order m, $r_> \equiv max(r, r_i)$ and $r_< \equiv min(r, r_i)$, (r, ζ) and (r_i, ζ_i) are the field and source point. Because the convergence of the infinite series in G is very slow when (r, ζ) is near (r_i, ζ_i), we exactly evaluate the component of the slowest convergence in the series and sum up finite terms of the other components [5].

Moreover, to discretize eq. (17), the stream function Ψ and its normal derivative q are assumed to be constant on boundary elements Γ_j and in triangle domain elements $\triangle_k$. With the above assumptions, eq. (17) becomes

$$\sum_{k=1}^{M} F(\Psi_k)\int_{\triangle_k} G_{ik}\, r\, dr\, d\zeta + C_i\Psi_i = \sum_{j=1}^{N} q_j \int_{\Gamma_j} KG_{ij}\left\{1 + \left(hr\frac{dr}{ds}\right)^2\right\}^{\frac{1}{2}} ds\,. \tag{19}$$

In this paper, M and N in eq. (19) are taken to be 684 and 36, respectively. Moreover, the unknown variables q_j on Γ can be obtained by iteratively solving eq. (19). In this paper, we choose the Picard iteration [11] to solve it, *i.e.*, we successively solve the simultaneous equation derived from eq. (19),

$$\mathbf{A}\,\mathbf{q}^{n+1} = \mathbf{B}(\Psi^n), \tag{20}$$

where n denotes the number of iterative steps.

Numerical Results

In this section, using the present scheme, we calculate MHD equilibria for the model in which the plasma is enclosed by a perfectly conducting wall with circular cross-section of radius 1 with its center at the origin. Moreover, the helical wave number h is taken to be 1 and the parameters in eq. (8) are set as follows:

(Case A) $f_0 = f_1 = 1, \mu_0 p_1 = 0, f_m = p_m = 0\ (m \geqq 2)$,

(Case B) $f_0 = f_1 = \mu_0 p_1 = 1, f_m = p_m = 0\ (m \geqq 2)$,

(Case C) $f_0 = f_1 = \mu_0 p_1 = 1, f_2 = \mu_0 p_2 = 0.2, f_m = p_m = 0\ (m \geqq 3)$.

Note that the problem comes to be nonlinear in case C while it is linear in cases A and B. In the cases A and B, we can compare the boundary element solutions with the exact solution (10).

Figures 1 - 6 show the numerical results for the above three cases A - C. Figures 1, 3 and 5 are equilibrium profiles as a function of radius r and Figs. 2, 4 and 6 are the changes of the value of Ψ at $r=0$ with the number of iteration. The exact solution is also shown in Figs. 1 and 3. From Figs. 1 to 4, it can be seen that the final solutions obtained after several iterative steps are in good agreement with the exact solution for the linear cases A and B. Besides, from Figs. 5 and 6, we can see that the numerical solution also converges after several steps for the nonlinear case C.

Moreover, we evaluate the dependence of convergence on the value of parameters in eq. (8) for linear and nonlinear cases. In the linear case, f_1 increases under the condition that the other parameters are set to be the same as the case A. On the other hand, in the nonlinear case, f_2 increases under the condition that the other parameters are set to be the same as the case C.

Here, it is known that the convergence of the Picard iteration depends on the inhomogeneous terms and, in particular, it diverges for $|\beta| > \beta_0$ when the equation $L(\Psi) = \alpha + \beta\Psi$ is solved [11 - 13] (where β_0 is the smallest eigenvalue of L).

Figure 7 shows the change of relative error E with the number of iteration, where $E \equiv (\Psi_{\text{exact}} - \Psi_{\text{numerical}}) / \Psi_{\text{exact}}$. In this calculation, f_1 is set to be near the value for which the iteration becomes divergent from convergent. From Fig. 7, we can see that the marginal value of f_1 is about 3.1 and, as mentioned above, this value approximately corresponds to the smallest eigenvalue of the equation

$$\frac{1}{r}\left\{\frac{\partial}{\partial r}\left(rK\frac{\partial \Psi}{\partial r}\right) + \frac{h^2}{r}\frac{\partial^2 \Psi}{\partial \zeta^2}\right\} = \frac{2K^2}{h} f_1 \Psi - K f_1^2 \Psi. \tag{21}$$

Here, the smallest eigenvalue $f_1^{\,1}$ of eq. (21) satisfies $J_1(f_1^{\,1}) + J_0(f_1^{\,1}) = 0$.

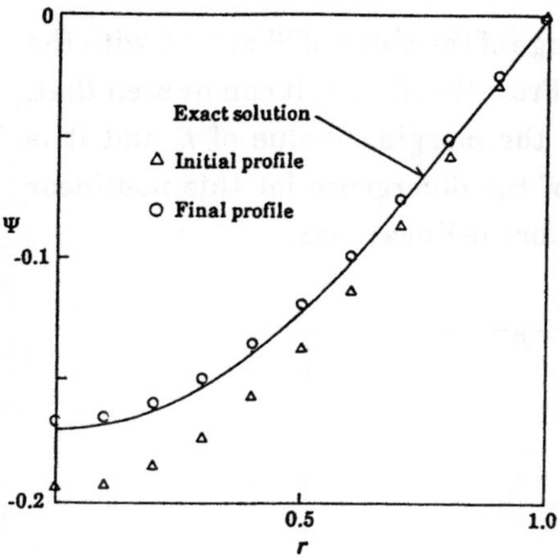

Fig. 1 Equilibrium profile for case A

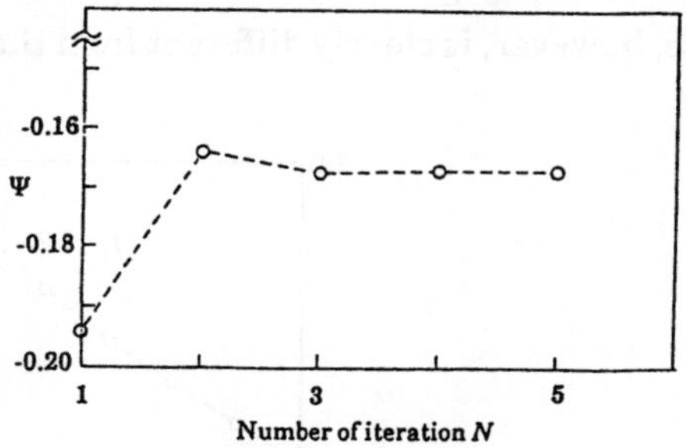

Fig.2 Convergence for case A

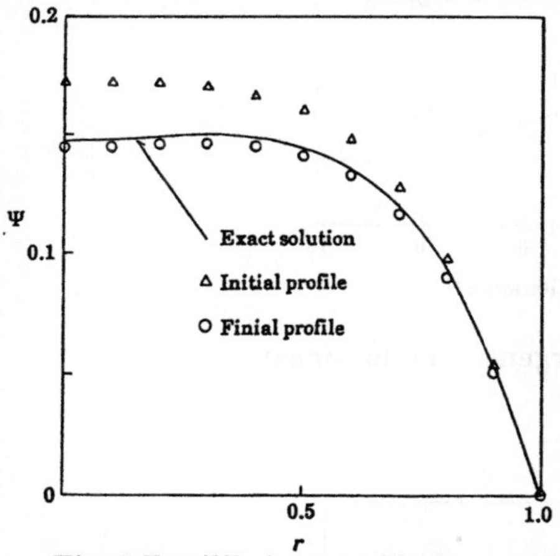

Fig. 3 Equilibrium profile for case B

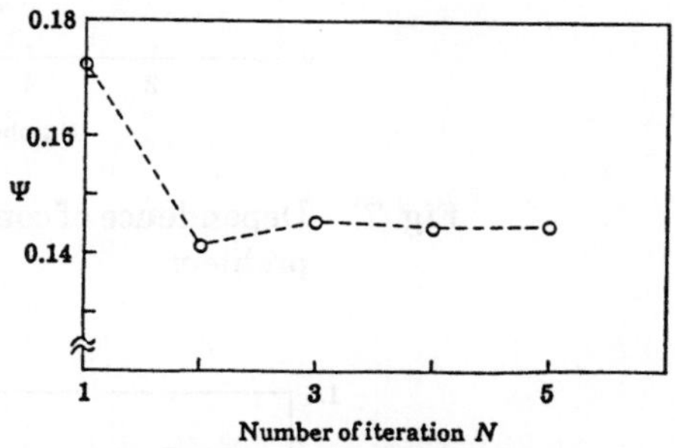

Fig.4 Convergence for case B

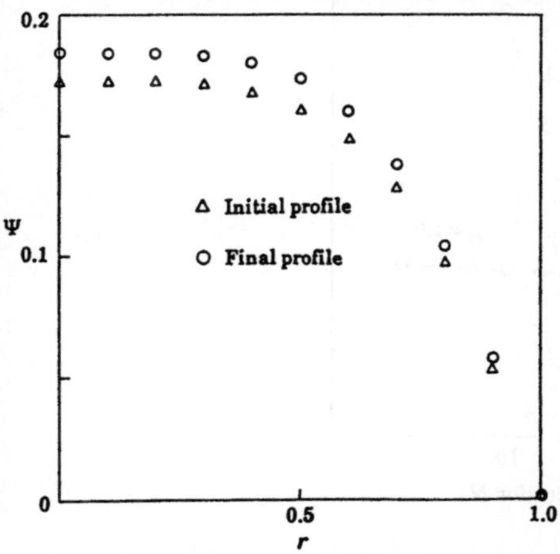

Fig. 5 Equilibrium profile for case C

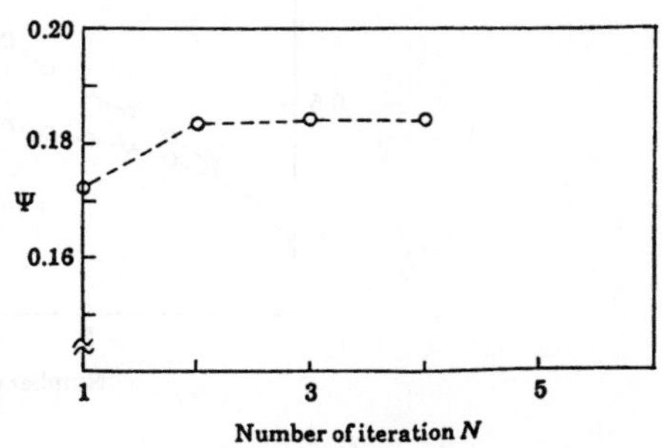

Fig.6 Convergence for case C

On the other hand, Fig. 8 shows the the change of the value of Ψ at $r=0$ with the number of iteration for the nonlinear case. From this figure, it can be seen that, as is seen in the linear case, there exists the marginal value of f_2 and it is approximately given by 1.0. The process of the divergence for this nonlinear case, however, is clearly different from that for the linear case.

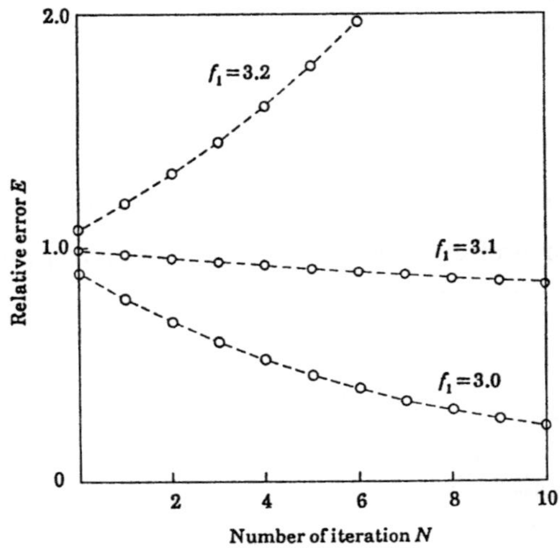

Fig. 7 Dependence of convergence on f_1 for linear problem

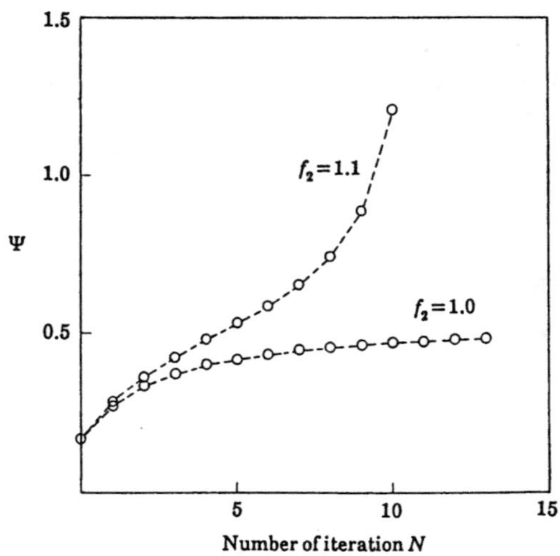

Fig. 8 Dependence of convergence on f_2 for nonlinear problem

The above situation does not depend on the scheme solving eq. (7) and may be the same as in the finite difference and finite element analyses. Besides, the above marginal values increase, *e.g.*, using the Marder-Weitzner iteration [11-12].

Conclusions

In this paper, we have introduced a boundary integral formulation for the helically symmetric MHD equilibria by reducing the u^3-symmetric boundary integral equation. Moreover, we have compared the iterative boundary element solution with the exact solution for a helically symmetric MHD plasma model. Furthermore, we have reported the numerical stability of the present scheme, in which the Picard iteration is used, for the linear and nonlinear cases. In conclusion, the numerical solutions can be obtained in a good accuracy when the parameters in the inhomogeneous term are within the stable region of the Picard iteration.

References

1. Dnestrovskii, Y. N. ; Kostomarov, D. P. Numerical simulation of plasmas. Berlin, Heidelberg : Springer-Verlag, 1986.
2. Gruber, R. ; Rappaz, J. Finite element methods in linear ideal magnetohydrodynamics. Berlin, Heidelberg : Springer-Verlag, 1985.
3. Markel, P. ; Nührenberg, J. : HASE - A quasi-analytical 2D MHD equilibrium code. Comp. Phys. Comm. **31** (1984) 115-122.
4. Monticello, D. A. ; Dewar, R. L. ; Furth, H. P. ; Reiman, A. : Heliac parameter study. Phys. Fluids **27** (1984) 1248-1252.
5. Igarashi, H. ; Honma, T. : An equilibrium analysis of helically symmetric plasmas using boundary element method. to be published in IEEE Trans. Mag. **MAG-26** (1990).
6. Igarashi, H. ; Honma, T. : BE and FE analysis of helically symmetric MHD equilibrium configuration. to be published in Int. J. App. Electromagnetics in Materials.
7. Edenstrasser, J. W. : Unified treatment of symmetric MHD equilibria. Plasma Phys. **24** (1980) 299-313.
8. Freidberg, J. P. : Stability of a finite β, $l=2$ stellarator. Phys. Fluids **16** (1973) 1349-1358.
9. Correa, D. ; Lortz, D. : A class of helically symmetric MHD-equilibria. Nucl. Fusion **13** (1973) 127-129.

10. Gardner, H. J. ; Dewar, R. L. ; Sy, W. N-C. : The free boundary equilibrium problem for helically symmetric plasmas. J. Comp. Phys. 74 (1988) 477-487.

11. Blum, J. Numerical simulation and optimal control in plasma physics with application to tokamaks. Paris : John Wiley & Sons, 1989.

12. Marder, B. ; Weitzner, H. : A bifurcation problem in E-layer equilibria. Plasma Phys. 12 (1970) 435-445.

13. Miller, G. ; Faber, V. ; White, Jr. A. B. : Finding plasma equilibria with magnetic islands. J. Comp. Phys. 79 (1988) 417-435.

The Generalized Boundary Element Approach to Viscous Flow Problems by Using the Time Splitting Technique

K. Kakuda and N. Tosaka

Department of Mathematical Engineering,
College of Industrial Technology, Nihon University,
Narashino, Chiba 275, Japan.

Summary

We present a new attempt by means of the generalized boundary element approach to solve an unsteady-state problem of viscous fluid flow. This approach is based on the well-known Fractional Step (FS) scheme which is one of the time splitting techniques. The fundamental equations are split into the advection-diffusion-type equation and the linear Euler-type ones. The advection-diffusion-type equation is transformed into the integral representation with the fundamental solution for the Iaplace operator. The Poisson equation which is derived by applying some manipulations to the Euler-type equations is also solved by using the generalized boundary element method. Numerical results of the driven cavity flow demonstrate the accuracy and applicability of our method.

Introduction

The numerical solutions of viscous fluid flows which are governed by the Navier-Stokes equations have been performed by many researchers by using the finite difference method [1,2] or the finite element method[3] based on the time splitting technique. In addition to the two numerical methods, the integral equation method has been also applied to the flow problems. Wu and his-workers [4,5] presented the numerical solution procedure based on the integral equation representation in terms of the velocity and vorticity as the field variables. Onishi, Kuroki and Tanaka [6] proposed a boundary element formulation in terms of the stream function and vorticity for the two-dimensional viscous flows.

We have been also developing the boundary-domain-type element method [7,8] and the generalized boundary element method [9] to solve the flow problems in terms of the velocity and pressure. The integral equations obtained from the first approach are discretized by not only boundary elements but also internal elements. The final system of equations with a full coefficient matrix was solved effectively by using the Newton-Raphson iterative procedure. On the other hand, the second approach is based on the boundary integral equation formulation by using the fundamental solution on each subdomain in the whole domain. The final system of equations with a sparse coefficient matrix was solved implicitly by using a simple iterative procedure.

In this paper, we present a new approach by means of the generalized boundary elements based on the time splitting technique to solve an incompressible viscous fluid flow governed by the Navier-Stokes equations. This technique is based on the well-known FS method. In order to stabilize our computational scheme for the flow problems at large Reynolds number, we adopt the Navier-Stokes equations [10] with the rotation-type form as the advection term. The fundamental equations are split into the advection-diffusion-type equation and linear Euler-type equation. As the first step of the computational scheme, the integral representation derived by applying the generalized boundary element method to the advection-diffusion-type equation can be solved explicitly. As the second step, by the use of the generalized boundary element method we solve the Poisson equation in terms of a scalar potential derived by taking some manipulations for the linear Euler-type equation. Numerical results of the driven cavity flow problem are demonstrated through a comparison with the other existing ones [2,3].

Statement of the Problems

Let Ω be a bounded domain in Euclidean space with a piecewise smooth boundary Γ . The unit outward normal vector to Γ is denoted by $\boldsymbol{n}$. Also, $\mathcal{T}$ denotes a closed time interval.

The flow of an incompressible viscous fluid is governed by the following Navier-Stokes equations and continuity equation :

$$\frac{\partial \boldsymbol{V}}{\partial t} + (\boldsymbol{V} \cdot \nabla)\boldsymbol{V} = -\nabla p + \frac{1}{Re}\nabla^2 \boldsymbol{V} \quad in\ \boldsymbol{T} \times \Omega \tag{1}$$

$$\nabla \cdot \boldsymbol{V} = 0 \quad in\ \boldsymbol{T} \times \Omega \tag{2}$$

in which Re is the Reynolds number, $\boldsymbol{V}$ is the velocity vector, and p is the pressure.

Here, we adopt the Navier-Stokes equations [10] with the rotation-type form as the advection term of equation (1)'. And, by applying the semi-implicit scheme to time derivative of the obtained equations, the governing equations are given as follows :

$$\frac{\boldsymbol{V}^{n+1} - \boldsymbol{V}^n}{\Delta t} - \boldsymbol{V}^n \times \boldsymbol{\omega}^n = -\nabla H^{n+1} + \frac{1}{Re}\nabla^2 \boldsymbol{V}^n \quad in\ \Omega \tag{3}$$

$$\nabla \cdot \boldsymbol{V}^{n+1} = 0 \quad in\ \Omega \tag{4}$$

where Δt is the time step, $\boldsymbol{\omega}^n$ is the vorticity vector$(= \nabla \times \boldsymbol{V}^n)$ at n-th time step, and H^{n+1} is the Bernoulli's function defined by

$$H^{n+1} = p^{n+1} + \frac{\| \boldsymbol{V}^n \|^2}{2} \tag{5}$$

Defining an auxiliary variable $\bar{\boldsymbol{V}}$, and applying the Fractional Step scheme to equations (3) and (4), we can obtain the following two types of equation :

1. Advection-diffusion-type equation

$$\frac{\overline{\boldsymbol{V}} - \boldsymbol{V}^n}{\Delta t} - \boldsymbol{V}^n \times \boldsymbol{\omega}^n = \frac{1}{Re}\nabla^2 \boldsymbol{V}^n \tag{6}$$

2. Linear Euler-type equations

$$\left.\begin{aligned} \boldsymbol{V}^{n+1} &= \overline{\boldsymbol{V}} - \Delta t \nabla H^{n+1} \\ \nabla \cdot \boldsymbol{V}^{n+1} &= 0 \end{aligned}\right\} \tag{7}$$

Here, by taking the rotation of the first equation in (7) and by making use of the Helmholtz resolution, we obtain the following equation :

$$\boldsymbol{V}^{n+1} = \overline{\boldsymbol{V}} + \nabla\phi \tag{8}$$

in which ϕ is the scalar potential. Moreover, by taking the divergence of equation (8) and by taking into consideration the continuity equation in (7), we obtain the following Poisson equation in terms of ϕ :

$$\nabla^2\phi = -\nabla \cdot \overline{\boldsymbol{V}} \tag{9}$$

And also, by substituting equation (8) into (7), we derive the following relation between H^{n+1} and ϕ :

$$H^{n+1} = -\frac{1}{\Delta t}\phi \tag{10}$$

Generalized Boundary Element Approach

In this stage, we describe the integral representations for our fundamental differential equations (6) and (9) of the viscous fluid flow.

From the integral identity of equation (6) for the arbitrary scalar function ϕ^* and with the aid of the divergence theorem, we can derive the following integral representation :

$$\begin{aligned} C(\boldsymbol{y})\frac{1}{Re}\boldsymbol{V}^n(\boldsymbol{y}) \;\; &= \;\; -\int_{\Omega_1}\left[\frac{\overline{\boldsymbol{V}} - \boldsymbol{V}^n}{\Delta t} - \boldsymbol{V}^n \times \boldsymbol{\omega}^n\right]\phi^* d\Omega \\ &+ \int_{\Gamma_1}\frac{1}{Re}(\boldsymbol{n}\cdot\nabla)\boldsymbol{V}^n\phi^* d\Gamma \;\; - \int_{\Gamma_1}\frac{1}{Re}\boldsymbol{V}^n(\boldsymbol{n}\cdot\nabla)\phi^* d\Gamma \end{aligned} \tag{11}$$

Here, the scalar function ϕ^* can be chosen as a fundamental solution which satisfies the following differential equation :

$$\nabla^2\phi^* = -\delta(\boldsymbol{x} - \boldsymbol{y}) \tag{12}$$

The fundamental solutions are given as follows:

$$\phi^*(\boldsymbol{x},\boldsymbol{y}) = \frac{1}{2\pi}\ln\frac{1}{r} \quad \text{(for two-dimensional operator)} \tag{13}$$

and

$$\phi^*(\boldsymbol{x},\boldsymbol{y}) = \frac{1}{4\pi r} \qquad \text{(for three-dimensional operator)} \tag{14}$$

By applying the boundary element discretization to equation (11), we obtain the following local matrix form :

$$_i\boldsymbol{M}\frac{_i\overline{\boldsymbol{V}} - {_i\boldsymbol{V}^n}}{\Delta t} = {_i\boldsymbol{G}}_i\boldsymbol{V}^n_{,n} - {_i\boldsymbol{H}}_i\boldsymbol{V}^n + {_i\boldsymbol{A}^n} \tag{15}$$

Moreover, equation (15) is also reduced as follows :

$$_i\overline{\boldsymbol{M}}\frac{_i\overline{\boldsymbol{V}} - {_i\boldsymbol{V}^n}}{\Delta t} = {_i\boldsymbol{V}^n_{,n}} - {_i\overline{\boldsymbol{H}}}_i\boldsymbol{V}^n + {_i\overline{\boldsymbol{A}}^n} \tag{16}$$

where $_i\bar{\boldsymbol{M}}$ is the lumped mass matrix (= $_i\boldsymbol{G}^{-1}{_i\boldsymbol{M}}$).

Taking into consideration of the equilibrium conditions of $_i\boldsymbol{V}^n_{,n}$ on each subdomain and setting up equation (16) for all subdomains, we can obtain the final system of equations as follows :

$$\overline{\boldsymbol{V}} = \boldsymbol{V}^n + \Delta t \boldsymbol{C}^{-1}\boldsymbol{F}^n \tag{17}$$

where $\boldsymbol{C}$ is the diagonal coefficient matrix and $\boldsymbol{F}^n$ denotes the known vector which is consisted of the velocity and vorticity at n -th time step.

As the vorticity vector is piecewise constant in a subdomain, the weighted residual integral is expressed as

$$\boldsymbol{\omega}^n = \frac{1}{\theta}\int_{\Gamma_i} \boldsymbol{n} \times \boldsymbol{V}^n d\Gamma \tag{18}$$

where θ takes the area of Ω_i and the volume of Ω_i for 2D case and 3D case, respectively.

On the other hand, we apply the boundary element method in a subdomain Ω_i to solve the Poisson equation (9). The solution ϕ has the integral representation with the fundamental solution (13) or (14) :

$$c(\boldsymbol{y})\phi(\boldsymbol{y}) = -\int_{\Omega_i} \overline{\boldsymbol{V}} \cdot \nabla\phi^* d\Omega + \int_{\Gamma_i} \overline{\boldsymbol{V}} \cdot \boldsymbol{n}\phi^* d\Gamma$$

$$+ \int_{\Gamma_i} (\boldsymbol{n} \cdot \nabla\phi)\phi^* d\Gamma - \int_{\Gamma_i} \phi(\boldsymbol{n} \cdot \nabla\phi^*) d\Gamma \tag{19}$$

The boundary element discretization of equation (19) is given as follows:

$${}_i\boldsymbol{H}_i\boldsymbol{\phi} = {}_i\boldsymbol{G}_i\boldsymbol{\phi}_{,n} + {}_i\boldsymbol{Q}_i\overline{\boldsymbol{V}} \tag{20}$$

Setting up equation (20) for all subdomains, the final system of equations is obtained as follows:

$$\boldsymbol{B}\boldsymbol{\phi} = \overline{\boldsymbol{F}} \tag{21}$$

where $\boldsymbol{B}$ is the sparse coefficient matrix and $\bar{\boldsymbol{F}}$ denotes the known vector with respect to the anxiliary vector $\bar{\boldsymbol{V}}$.

The computational scheme based on our method is given as follows :

1. Give an initial velocity vector of $\boldsymbol{V}^n$ and calculate the auxiliary vector $\bar{\boldsymbol{V}}$ by equations (17) and (18).
2. Calculate the potential function $\boldsymbol{\phi}$ from equation (21).
3. Calculate $\boldsymbol{V}^{n+1}$ and H^{n+1} at (n+1)-th time step from the weighted residual statement of equations (8) and (10), respectively, and go to 1.

Numerical Examples

In the following, we demonstrate an unsteady-state flow in a square cavity driven by a lid sliding at a uniform velocity. We perform the numerical calculations for $Re = 10^3$. The fixed time step Δt is equal to 0.002 .

The velocity vector fields at each time step are shown in

Fig.1. Fig.2 shows the vorticity contours corresponding to the velocity fields at each time step. Figs. 3 and 4 show the horizontal velocity profiles along vertical centreline and the vertical velocity profiles along horizontal centreline, respectively, at each time step obtained from the proposed method. The agreement between our numerical results and the other ones appears satisfactory as shown in Fig.5.

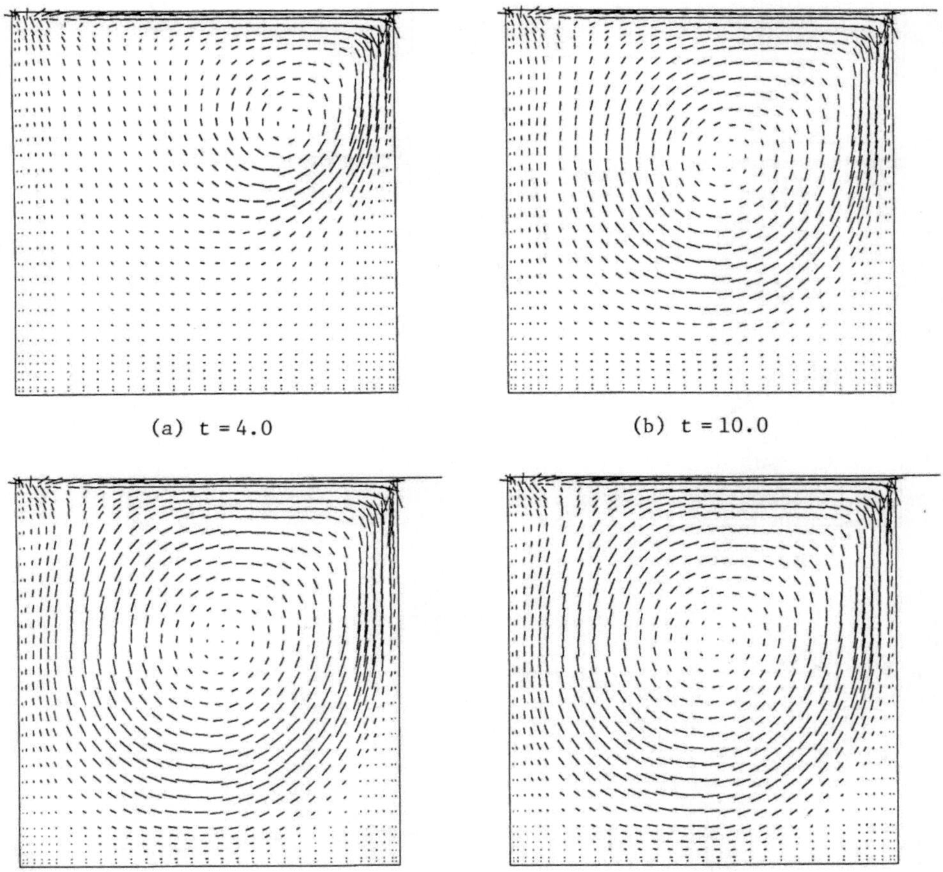

(a) t = 4.0 (b) t = 10.0

(c) t = 20.0 (d) t = 40.0

Fig.1 Velocity vector fields for Re=10^3
(Δt=0.002)

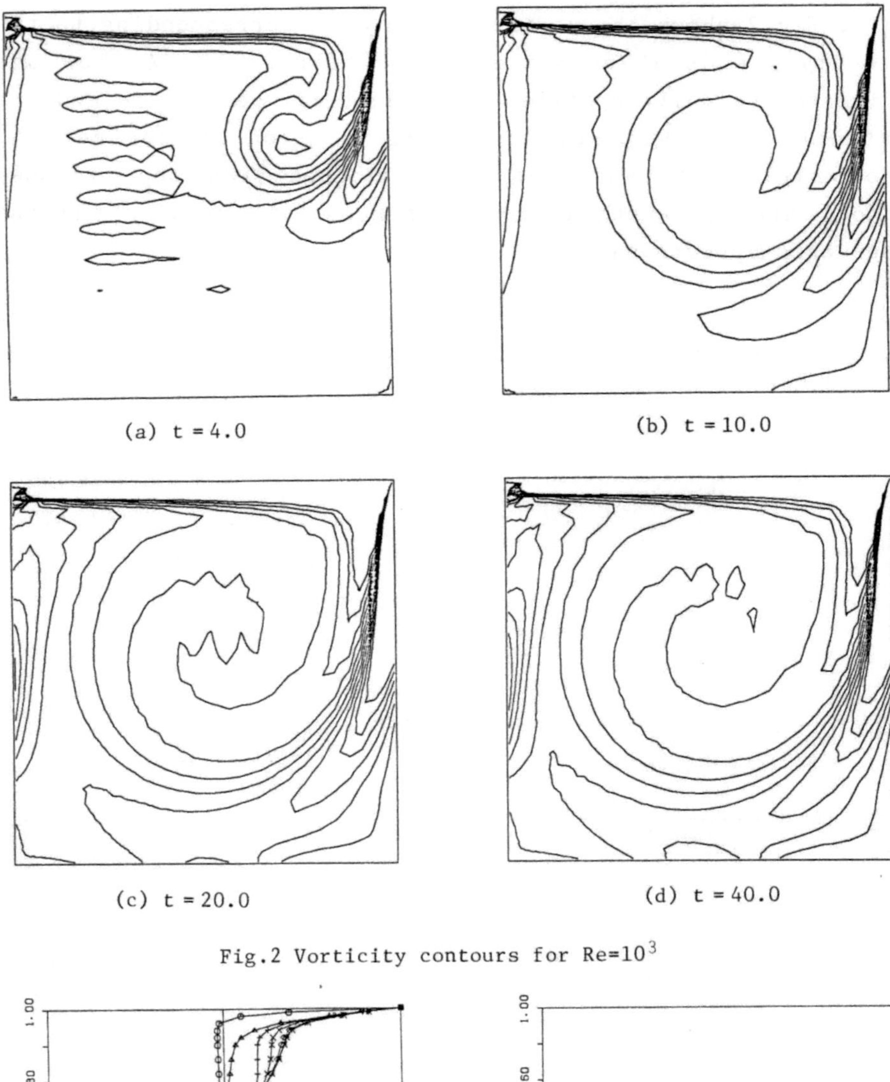

(a) t = 4.0

(b) t = 10.0

(c) t = 20.0

(d) t = 40.0

Fig.2 Vorticity contours for $Re=10^3$

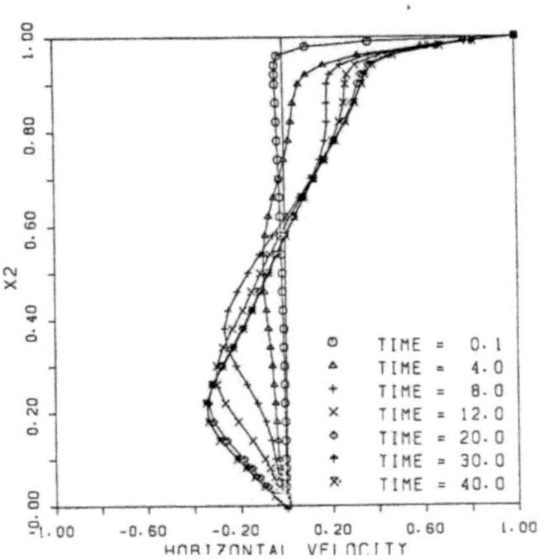

Fig.3 Horizontal velocity profiles along vertical centreline at each time step

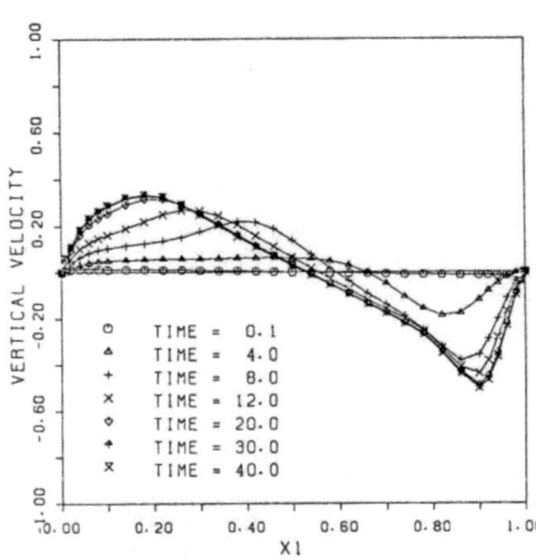

Fig.4 Vertical velocity profiles along horizontal centreline at each time step

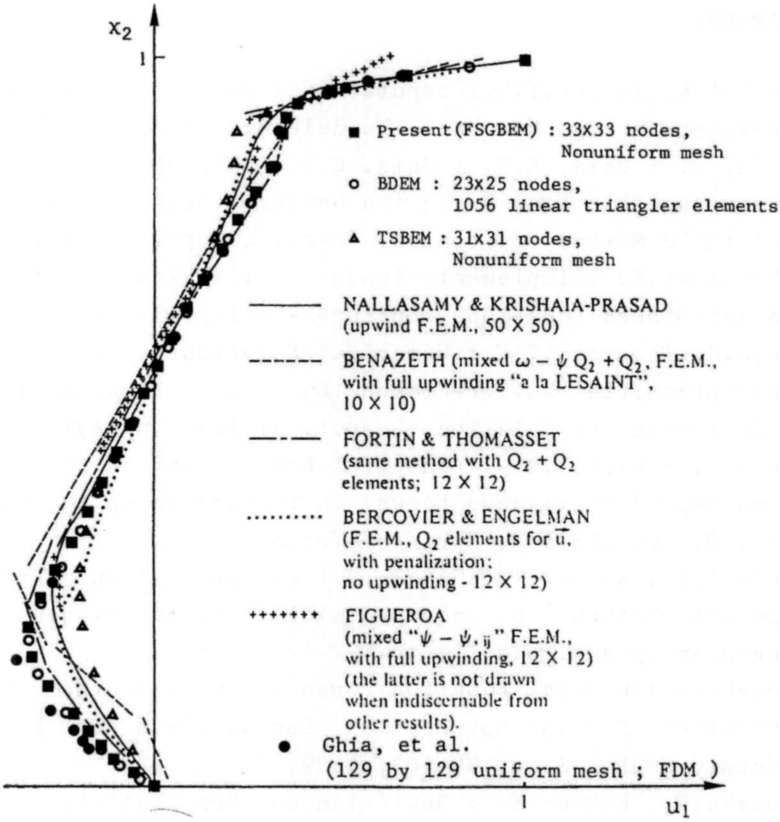

Fig.5 Comparison of horizontal velocity profiles along vertical centreline (Re = 10^3)

Conclusions

We have presented an approach by means of the generalized boundary elements based on applying the fractional step scheme to an incompressible viscous fluid flow problem. The integral representations were constructed by using only a fundamental solution for the Laplace operator. The numerical results of the driven cavity flow problem demonstrated the accuracy and the applicability of the proposed method through a comparison with the other existing results.

References

1. Peyret,R.;Taylor,T.D.:Computational Methods for Fluid Flow, Springer-Verlag, New York, Heidelberg, Berlin, 1983.
2. Ghia, U. ; Ghia, K.N. ; Shin, C.T.: High-Re Solutions for Incompressible Flow Using the Navier-Stokes Equations and a Multigrid Method, J. Comput. Phys., 48, pp.387-411, 1982.
3. Thomasset,F. : Implementation of Finite Element Methods for Navier-Stokes Equations, Springer-Verlag, 1981.
4. Wu,J.C.;Thompson,J.F.: Numerical Solution of Time-Dependent Incompressible Navier-Stokes Equations Using an Integro-Differential Formulation, Comput. Fluids,1,pp.197-215,1973.
5. Wu,J.C. ; Rizk,Y.M. : Integral-Representation Approach for Time-Dependent Viscous Flows, in Lecture Notes in Physics, Vol.90, pp.558-564, Springer-Verlag, 1978.
6. Onishi,K ; Kuroki,T.; Tanaka,M.: An Application of Boundary Element Method to Incompressible laminar Viscous Flows, Engineering Analysis, 1, pp.122-127, 1984.
7. Tosaka,N.:Integral Equation Formulations with the Primitive Variables for Incompressible Viscous Fluid Flow Problems, Comput. Mech., 4, pp.89-103, 1989.
8. Tosaka,N.; Kakuda,K. : Newtonian and Non-Newtonian Unsteady Flow Problems, in Chapter 5 of Boundary Element Methods in Nonlinear Fluid Dynamics, Developments in Boundary Element Methods 6 (Eds., P.K. Banerjee and L.Morino), pp.151-181, Elsevier Applied Science, 1990.
9. Kakuda, K. ; Tosaka, N. : The Generalized Boundary Element Approach to Burgers' Equation, Int. J. Num. Meth. Engng., Vol.29, pp.245-261, 1990.
10. Kanai,E. ; Tanahashi,T. : GSMAC-A New Finite Element Method for Unsteady Incompressible Viscous Flow Problems (1st Report, A Stable Method at High Reynolds Numbers), (in Japanese), JSME Journal, 53, pp.683-691, 1987.

Sample Point Boundary Element Error Analysis

N. Kamiya and K. Kawaguchi
Department of Mechanical Engineering, Nagoya University,
Nagoya, 464-01, Japan

Summary

This paper considers a method of construction of adaptive boundary element mesh for the problem governed by the two-dimensional Laplace equation. The method relies on an estimation of the discretization error on each boundary element appearing in the boundary integral equation through the magnitude of inconsistency of the intermediate solution except at the boundary nodes for the prescribed discretization. In this regard, a new concept "sample point error analysis" is presented and introduced into ordinary h-version of the adaptive boundary mesh refinement scheme.

Keywords
Boundary element method, Adaptive mesh, Error analysis, Collocation method

Introduction

Automated adaptive schemes for FEM have progressed recently [1] but investigation to the such adaptive schemes for BEM started more recently. The most popular BEM employs collocation on the specified boundary nodes to derive a system of algebraic equation. This paper is devoted to a new development of error estimation and related mesh refinement, which should be appropriate for the proper concept of the collocation boundary element method.

Adaptive Boundary Elements

Alarcon and co-workers[2, 3] proposed a p-version adaptive boundary elements using hierarchy polynomial interpolation functions based on error estimation through the residual of the discretized equation of the boundary integral equation. Assuming that *a priori* error characteristics is proportional to square of the element length, Rencis *et al.*[4] utilized h-version mesh refinement. Another method proposed by Rencis and associates[5] is *a posteriori* error estimation from the analysis using initial assumed mesh by the higher order interpolation. Rank[6] defined an error norm similar to that in FEM from the residual of the boundary integral equation on the other points than nodes and employed it to h-version and h/p-combined version. Some other methods are shown in the references cited in [2, 3, 4, 5, 6]. We can mention that some of these adaptive schemes rely on the error criteria for FEM, which is not necessarily applicable to BEM, and that the other requires very complicated numerical computation.

The integral equation does not hold rigorously on the boundary points other than the nodes and yields error, of which dimension is of potential, representing magnitude of inconsistency of the solution. As one of the marked property of the direct BEM, one can deal with the mixed problem by using the potential and outward normal derivative referred to flux as the unknowns. Therefore, the fact that two variables and errors of different dimensions appear is of great importance for appropriate error estimation. We will relate the errors on the whole elements to the solution inconsistency on the source point, i.e., the latter yields from combination of the errors of each element multiplied by the fundamental solutions, whereby the two errors of different dimensions are unified to the magnitude of identical dimension, of potential.

The error distribution on the element is not known in advance. If, however, we assume the error distribution appropriately, the error level on each element can be specified approximately, which will help to find a strategy to the adaptive mesh refinement. In this report, we will develop a new active error estimation device called "Sample point error analysis" based on the direct BEM and apply it to the h-version with the linear elements. The final error indicator is defined by the value of each error multiplied by the corresponding fundamental solutions on the element, which is referred to as "Extended error indicator".

Error of Boundary Integral Equation

Consider the following boundary-value problem in the two-dimensional domain Ω and its boundary Γ :

$$\text{Governing equation:} \quad \nabla^2 u = 0 \quad \text{in } \Omega \tag{1}$$

$$\text{Boundary condition:} \quad \left.\begin{array}{ll} u = \bar{u} & \text{on } \Gamma_{\bar{u}} \\ q = \bar{q} & \text{on } \Gamma_{\bar{q}} \end{array}\right\} \tag{2}$$

where u and $q(= \partial u/\partial n)$ denote the potential and flux respectively, and n is the unit outward normal on the boundary. Using the fundamental solution u^* of the Laplace equation, we can formulate the boundary integral equation for an arbitrary source point p_i on the boundary.

$$cu(p_i) = \int_\Gamma [q(p_j)u^*(p_i, p_j) - u(p_j)q^*(p_i, p_j)]d\Gamma \tag{3}$$

which is mathematically identical to the original differential equation (1). c is a constant determined by the geometrical condition of the boundary on the point p_i .

Ordinary BEM employs the collocation scheme on the selected boundary nodes to construct required numbers of algebraic equations. Equation (3) is identical to equation (1) in mathematical sense and no error owing to transformation is included.

Adopting the approximate solution $\hat{u}$ and $\hat{q}$ in equation (3), we obtain

$$c\hat{u}(p_i) = \int_\Gamma [\hat{q}u^* - \hat{u}q^*]d\Gamma \tag{4}$$

From equations (3) and (4), we can derive the relation between the error on the source point and that of each element,

$$\text{(Left-hand side)} \;=\; r(p_i) \equiv cu(p_i) - c\hat{u}(p_i) \tag{5}$$

$$\text{(Right-hand side)} \;=\; \int_\Gamma [(q-\hat{q})u^* - (u-\hat{u})q^*]d\Gamma \tag{6}$$

$$r(p_i) = \int_\Gamma [e_q u^* - e_u q^*]d\Gamma \tag{7}$$

$$e_u \equiv u - \hat{u}\ ,\ e_q \equiv q - \hat{q} \tag{8}$$

Equation (7) indicates that the source point error is related to the observation point errors (errors on the whole element) multiplied by the corresponding fundamental solution.

Sample Point Error Analysis

By using equation (7), we may analyze the error on each element provided that the magnitude of r is specified approximately. If we take a source point p_i' on the arbitrary point other than initial nodes on the element (referred to "Sample point") and substitute $\hat{u}, \hat{q}$ obtained by equation (4) into the boundary integral equation, it does not hold on that point.

$$c\hat{u}(p_i') \neq \int_\Gamma [\hat{q}u^* - \hat{u}q^*]d\Gamma \equiv c\acute{u}(p_i') \tag{9}$$

$\acute{u}(p_i')$ defined here generally differs from the value specified by interpolation of the approximate solution : $\hat{u}(p_i')$, and will be shown later by some numerical examples to predict well the actual behavior of the solution to be obtained on $\Gamma_{\bar{q}}$.

The magnitude of r on the source point is thought to be representable by the following known information ;

$$r(p_i') \equiv \begin{cases} c\acute{u}(p_i') - c\hat{u}(p_i') & (p_i' \in \Gamma_{\bar{q}}) \\ c\bar{u}(p_i') - c\acute{u}(p_i') & (p_i' \in \Gamma_{\bar{u}}) \end{cases} \tag{10}$$

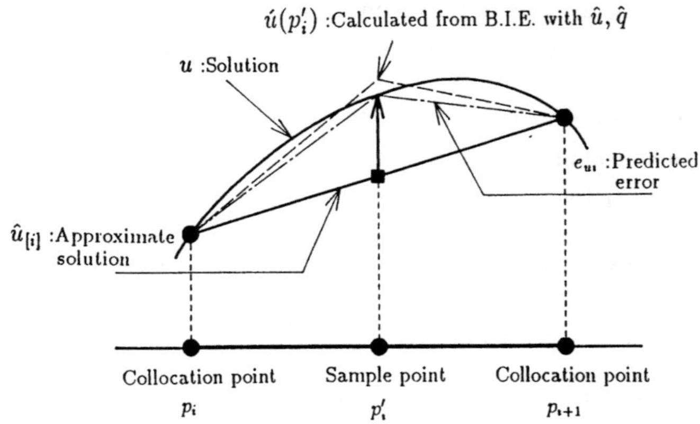

Figure 1: Error on boundary element

The error distributions e_{uj}, e_{qj} on the element j are modeled provisionally by the linear relation in terms of the intrinsic coordinate ξ $(-1 \le \xi \le 1)$ as shown in Figure :1;

$$\left.\begin{aligned} e_{uj}(\xi) &= e_{u[j]}\Psi(\xi) \\ e_{qj}(\xi) &= e_{q[j]}\Psi(\xi) \end{aligned}\right\} \tag{11}$$

where

$$\Psi(\xi) = \begin{cases} (1+\xi) & (-1 \le \xi \le 0) \\ (1-\xi) & (0 < \xi \le 1) \end{cases} \tag{12}$$

Equation (7) is discretized by boundary elements to give the following equation for the sample point p_i'

$$r(p_i') = \sum_{j=1}^{n}\left\{\int_{\Gamma_j} e_{q[j]}\Psi u^* d\Gamma - \int_{\Gamma_j} e_{u[j]}\Psi q^* d\Gamma\right\} \tag{13}$$

where

$$\left.\begin{aligned} e_{q[j]} &= 0 \quad (j \in \Gamma_{\bar{q}}) \\ e_{u[j]} &= 0 \quad (j \in \Gamma_{\bar{u}}) \end{aligned}\right\} \tag{14}$$

For the flux-specified boundary, since the potential is unknown, the error on the element becomes

$$e_{uj}(0) = e_{u[j]}\Psi(0) = e_{u[j]} \tag{15}$$

that is, the following equation holds for the middle point ($\xi = 0$) on the element,

$$ce_{u[i]} \simeq r(p_i') \tag{16}$$

Therefore, $e_{u[j]}$ on the boundary where the potential is unknown can be easily determined. $e_{q[j]}$ is, on the other hand, determined by solving the linear simultaneous equations system reduced by substitution of the given $e_{u[j]}$.

The relation between the source point error r and the error on each element is equation (7) and is interpreted as the following equation :

$$r(p_i') = \tilde{e}_q + \tilde{e}_u \tag{17}$$

by defining the "extended error indicator" $\tilde{e}_q$ and $\tilde{e}_u$. These new quantities have same dimension, potential, and consequently make it possible to estimate errors gathered from whole elements, some are on $\Gamma_{\bar{q}}$ and the others on $\Gamma_{\bar{u}}$. Similar equations to equation (17) are constructed to find the error influence $\tilde{e}_{q[i,j]}$ and $\tilde{e}_{u[i,j]}$ on the i-th sample point from the j-th element error.

Boundary Element Refinement

The absolute value of the above-defined error influence $\tilde{e}_{q[i,j]}$ and $\tilde{e}_{u[i,j]}$

$$\eta_{[i,j]} = |\tilde{e}_{q[i,j]}| + |\tilde{e}_{u[i,j]}| \tag{18}$$

indicates the influence intensity of the j-th element error on the error of the source point i. $\eta_{[i,j]}$ is a matrix of $n \times n$ elements for n sample points. An easy and simple way of the refinement employed here is based on the relative magnitude of $\eta_{[i,j]}$ compared with its average.

$$\eta_{ave} = \frac{1}{n^2} \sum_{i,j=1}^{n} \eta_{[i,j]} \tag{19}$$

i.e., the element i is divided into two smaller elements when η is larger than η_{ave}.

The accurate solution is thought to be obtained when r, representing inconsistency of the solution on the selected sample points, becomes sufficiently small.

$$|r(p_i')| \leq \epsilon \quad (\text{ for } i = 1, 2, \cdots n \ ) \tag{20}$$

Numerical Examples

(1)The first example is heat transfer problem in a transformer coil shown in Figure :2, which was already considered by Rencis *et al.*[5]. Computation starts using only 12 corner points A to L as shown in Figure :2(a). After 5 iterations we obtained the result shown in Figure :2(b). Figure :2(c) shows the result of adaptive mesh refinement by the method presented in [5], which requires least minimum numbers of boundary nodes for estimation of interpolation error in principle. On the other hand, the present scheme has merit that it requires only 12 nodal points to define the original geometry. Although both methods differ from each other, the final results are qualitatively similar.

(2)Figure :3 is the second example of the two-dimensional steady potential flow around a circular obstacle between parallel rigid walls. Initial geometrical data are 5 corner points and additional three taken on a circular boundary, which are also sufficient for correct specification of the given boundary conditions. Figure :3(b) shows the result of flux obtained by 7 iterations.

References

[1] I. Babuška, O. C. Zienkiewicz, J. Gago, and E. R. de A. Oliveira, editors. *Accuracy Estimates and Adaptive Refinements in Finite Element Computations.* John Wiley & Sons, 1986.

[2] E. Alarcon and A. Reverter. P-adaptive boundary elements. *Int. J. Num. Meth. Eng.*, 23:801–829, 1986.

[3] M. Cerrolaza, M. S. Gomez-Lera, and E. Alarcon. Elastostatics p-adaptive boundary elements for micros. *Software for E. W. S.*, 4:18–24, 1988.

[4] R. L. Mullen and J. J. Rencis. Adaptive mesh refinement techniques for boundary element methods. *Advanced Topics in Boundary Element Analysis*, 235–255, 1985.

[5] J. J. Rencis and K. Y. Jong. A self-adaptive h-refinement technique for the boundary element method. *Comp. Meth. Appl. Mech. Eng.*, 73:295–316, 1989.

[6] E. Rank. Adaptive h,p and hp versions for boundary integral element methods. *Int. J. Num. Meth. Eng.*, 28:1335–1349, 1989.

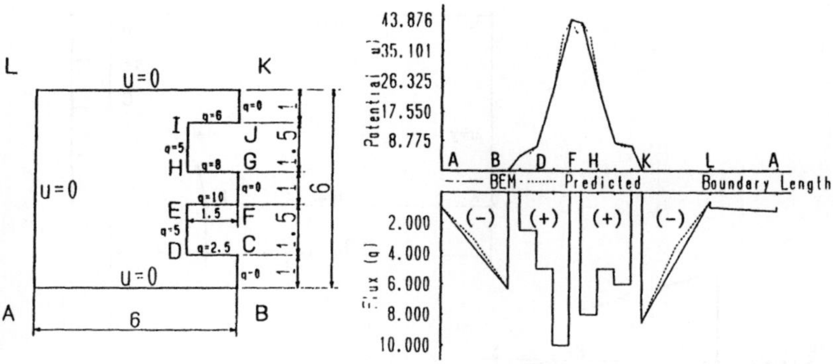

(a) Boundary condition and initial solution

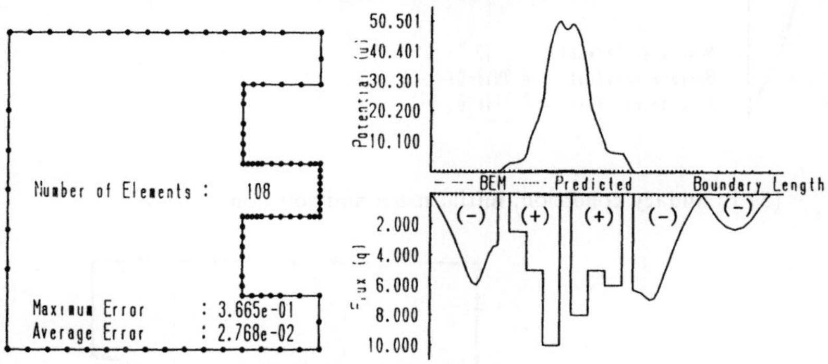

(b) After 5 iterations

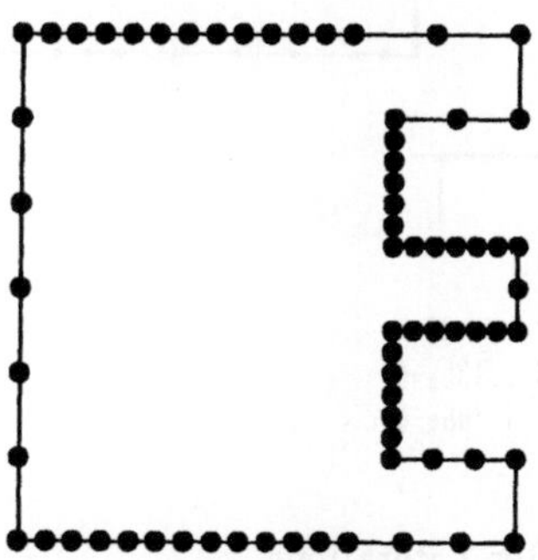

(c) Adaptive mesh for example 1 (Rencis *et al.*) Error 5%

Figure 2: Example 1

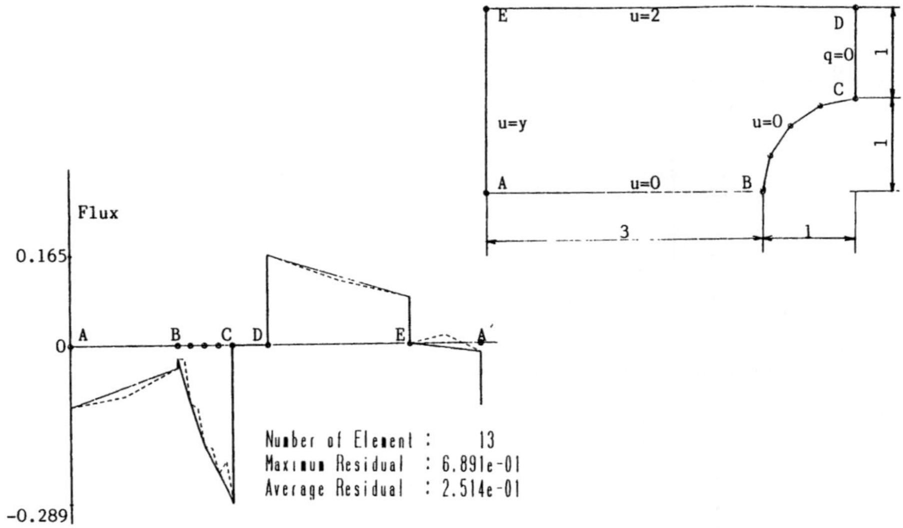

(a) Boundary condition, initial mesh and solution

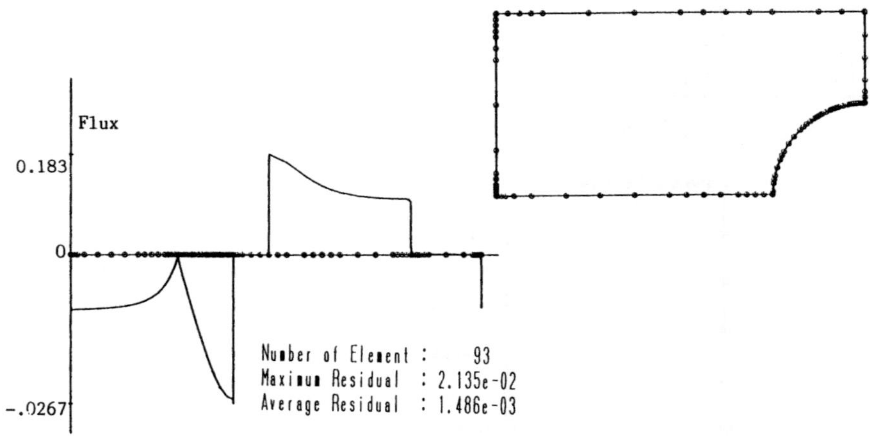

(b) After 7 iterations

Figure 3: Example 2

Boundary Formulations for Nonlinear Thermal Response Sensitivity Analysis

James H. Kane and Hua Wang
Mechanical & Aeronautical Engineering Department
Clarkson University, Potsdam, New York 13699

Abstract

Implicit differentiation of the discretized boundary integral equations governing the conduction of heat in solid objects, subjected to both nonlinear boundary conditions, and with temperature dependent material properties, is shown to generate an accurate and economical approach for the computation of shape sensitivities for this class of problems. Several iterative strategies are presented for the solution of the resulting sets of nonlinear equations and the computational performances examined. Multi-zone analysis and zone condensation strategies are demonstrated to provide substantive computational economies in this process for models with either localized nonlinearities or regions of geometric insensitivity to design variables. Nonlinear example problems are presented that have closed form solutions. Exact analytical expressions are compared with the sensitivities computed using this boundary element formulation.

Introduction

Several papers have been written on the implicit differentiation approach to continuum shape design sensitivity analysis utilizing boundary element analysis (BEA) formulations. Applications in acoustics [1], solid mechanics [2,3], heat transfer [4], and the coupling of these phenomena [5,6] have appeared. Separate treatment of nonlinear heat transfer with zone condensation [7], DSA of problems with nonlinear boundary conditions [8,9] and temperature dependent conductivity [10] have appeared. In this paper, a unified treatment of the implicit differentiation approach to the computation of shape sensitivities for objects with temperature dependent conductivity, convection, and radiation boundary conditions is presented. Two examples are presented to demonstrate the accuracy and efficiency of this formulation.

Thermal BEA Formulation with Nonlinear Boundary Conditions

The boundary integral equation governing the thermal response of a medium with constant conductivity [4,7,8,9] can be written for any location of the source point of the fundamental solution. A singular boundary element formulation is obtained by locating this source point at each of the nodes present in the BEA model, producing a square system of algebraic equations.

$$[F]\,\{t\} = [G]\,\{q\} \tag{1}$$

$\{t\}$ is a column vector of nodal point temperatures and $\{q\}$ is a column vector of nodal point normal heat flux components. The $\{t\}$ vector has an entry for each node in the overall problem, while $\{q\}$ may have additional entries if jumps in the normal component of the heat flux occurs at any node. The matrix $[F]$ is square, and $[G]$ is either square of rectangular.

In a well posed boundary value problem, half of the temperature and normal heat flux components will be specified (and therefore known) and the other half will be unknown. Transferring all known values to $\{q\}$, placing all unknown temperature and normal heat flux components in $\{t\}$, exchanging corresponding columns of the respective rectangular matrices, and performing the indicated matrix-vector multiplication on the right hand side, a solvable system of equations can be produced.

$$[A]\{x\} = \{b\} \tag{2}$$

This matrix equation is usually solved by the triangular factorization of the matrix $[A]$ using Gauss elimination with partial pivoting, followed by forward reduction of $\{b\}$ and backward substitution to obtain the unknown response vector $\{x\}$. The notation has been generalized to mean that $\{t\}$ is the vector of unknown boundary response quantities, while $\{q\}$ denotes the vector of specified boundary conditions in the problem.

Convection boundary conditions relate the normal heat flux q and temperature T, on the surface of an object, to the free stream (or bulk) temperature T_{oo} and a convection coefficient h.

$$q = h\,(T - T_\infty) \tag{3}$$

Considering the ith equation in the BEA system equations, substitution of the convection law for a convection boundary condition at node k yields the following relationships.

$$f_{i1}\,T_1 + \cdots + f_{ik}\,T_k + \cdots + f_{in}\,T_n = g_{i1}\,q_1 + \cdots + g_{ik}\,q_k + \cdots + g_{in}\,q_n \tag{4}$$

or

$$f_{i1}\,T_1 + \cdots + f_{ik}\,T_k + \cdots + f_{in}\,T_n = g_{i1}\,q_1 + \cdots + g_{ik}\,h_k\,(T_k - T_\infty) + \cdots + g_{in}\,q_n$$

In this expression, the temperature at node k appears on both the left and right hand side. Different approaches can be taken to solve these equations. One consists of leaving T_k on both sides and iterating to find its correct value.

$$f_{i1}\,T_1 + \cdots + f_{ik}\,T_k^{(j+1)} + \cdots + f_{in}\,T_n = g_{i1}\,q_1 + \cdots + g_{ik}\,h_k\,(T_k^{(j)} - T_\infty) + \cdots + g_{in}\,q_n \tag{5}$$

In a second approach, the terms that multiply T_k on the right side of Equation (4) are brought to the left hand side.

$$f_{i1}\,T_1 + \cdots + (f_{ik} - g_{ik}\,h_k)\,T_k + \cdots + f_{in}\,T_n = g_{i1}\,q_1 + \cdots + g_{ik}\,(-\,h_k T_\infty) + \cdots + g_{in}\,q_n \tag{6}$$

For cases where h is not a function of T, the second method is preferred because it results in a linear problem that can be solved without the need for iteration. When h is a function of T, both approaches can be attempted, however, there are many cases in which the approach characterized by Equation (5) does not converge.

The approach characterized by Equation (5) leaves $[A]$ unchanged from one iteration to the next, thus allowing for the $[L]\,[U]$ factorization, formed in the first iteration, to be reused in all subsequent iterations. The second method involves changing $[A]$ in each iteration. Although this latter approach may seems noncompetitive, algorithms that employ left side modification strategies have been demonstrated [7] to converge in just a few iterations, while algorithms that only modify the right side of the BEA system equations converge at a much slower rate and may actually diverge. It has also been shown [7] that zone condensation techniques reduce the computational effort associated with the re-factorization of partially modified left hand side matrices to the point where the overall algorithm becomes superior to other approaches for this class of problems.

The law for q on the surface of an object due to radiation can be manipulated into a form exactly like a temperature dependent convection coefficient.

$$q = \sigma\,E\,(T^4 - T_r^4) \tag{7}$$

where

$$E = V\left\{\frac{1}{\varepsilon} + \frac{1}{\varepsilon_r} - 1\right\}^{-1}$$

and

$$q = \left\{\sigma E\,(T^2 + T_r^2)\,(T + T_r)\right\}(T - T_r) = h_r\,(T - T_r) \tag{8}$$

s is the Stefan-Boltzmann constant, T_r is the temperature of the known external radiation source, V is the radiation view factor, ε is the surface emissivity of the object being analyzed, and ε_r is the emissivity of the radiation source. It is thus possible to solve problems involving radiation using a nonlinear BEA code that allows for a temperature dependent convection coefficient that behaves like the cubic polynomial.

Temperature Dependent Conductivity and the Kirchhoff Transformation

The governing differential equation and boundary conditions for steady 2D temperature dependent conductivity problems that can be treated in a linear fashion are

$$\frac{\partial}{\partial x_1}\left(k(T)\,\frac{\partial T}{\partial x_1}\right) + \frac{\partial}{\partial x_2}\left(k(T)\,\frac{\partial T}{\partial x_2}\right) = \nabla \cdot [k(T)\,\nabla T] = 0 \tag{9}$$

$$T = \bar{T} \text{ on } \Gamma_1 \text{ and } q = -k(T)\,\frac{\partial T}{\partial n} = \bar{q} \text{ on } \Gamma_2 \tag{10}$$

where k(T), n, Γ_1, and Γ_2 are the temperature dependent thermal conductivity, unit outward surface normal vector, portion of the surface with specified temperatures, and portion of the surface with specified normal heat flux, respectively. Note that the over-bar is used to denote specified quantities. This problem can be transformed as shown below.

Let $$k(T) = \frac{\partial f}{\partial T}\,;\; f = \int_{T_0}^{T} k(T)\,dT \tag{11}$$

$$\nabla \cdot [k(T)\,\nabla T] = \nabla \cdot \left[\frac{\partial f}{\partial T}\frac{\partial T}{\partial x_1}\,\mathbf{e_1} + \frac{\partial f}{\partial T}\frac{\partial T}{\partial x_2}\,\mathbf{e_2}\right] = \nabla^2 f = 0 \tag{12}$$

and

$$f = \int_{T_0}^{\bar{T}} k(T)\,dT = \bar{f} \text{ on } \Gamma_1\,; \text{ and } \frac{\partial \bar{f}}{\partial n} = \frac{\partial \bar{f}}{\partial T}\frac{\partial \bar{T}}{\partial n} = +k(\bar{T})\,\frac{\partial \bar{T}}{\partial n} = -\bar{q} \text{ on } \Gamma_2 \tag{13}$$

The kirchhoff transformation thus changes the problem into a standard Laplace problem in the transformed variable f. Once this problem is solved, the primal variables T and q can be obtained using Equation (13) and the inverse transformation symbolized below.

$$T = k^{-1}(f) \tag{14}$$

An example transformation and its inverse are given below.

$$\left.\begin{aligned} &k = k_o \exp\left\{\beta\left(\frac{T - T_o}{T_o}\right)\right\};\; f = \frac{k_o\,T_o}{\beta}\left[\exp\left\{\beta\left(\frac{T - T_o}{T_o}\right)\right\} - 1\right] \\ &\text{and} \\ &T = T_o + \frac{T_o}{\beta}\ln\left\{\frac{\beta f}{k_o\,T_o} + 1\right\} \end{aligned}\right\} \tag{15}$$

A standard isoparametric BEA formulation can be employed to solve the transformed problem that is analogous to the constant conduction problem, with the exception that the fundamental solution do not contain the conductivity and g is the normal gradient of f.

$$[F]\{f\} = [G]\{g\} \tag{16}$$

where

$$[F]^{(E,P)} = \int_{-1}^{+1} g^*[H]\, J\, da\ ; \text{ and } [G]^{(E,P)} = \int_{-1}^{+1} f^*[H]\, J\, da \tag{17}$$

and

$$f^* = -\frac{\ln(R)}{2\pi}\ ; \text{ and } g^* = \frac{\partial f^*}{\partial n} = \frac{(x_1 - d_1)n_1 + (x_2 - d_2)n_2}{2\pi R^2} \tag{18}$$

$[H]$ is a row vector of element interpolation functions on element E, a is an element intrinsic coordinate, and J is the Jacobian of the transformation from the element's intrinsic coordinate system to the actual coordinate system. $\{f\}$ and $\{g\}$ are column vectors of node point transformed temperatures and normal heat flux components respectively in the overall algebraic system equations shown in Equation (16).

A convection boundary condition with constant convection coefficient can be incorporated by employing Equation (13) and utilizing the transformations given, for example, by Equation (15).

$$g = \frac{\partial f}{\partial n} = -q = -h(T - T_\infty) \tag{19}$$

$$g = -h\left\{\frac{T_0}{\beta}\ln\left(\frac{\beta f}{k_0 T_0} + 1\right) - (T_\infty - T_0)\right\} \tag{20}$$

Thus, the transformed convection boundary condition, even with a constant h, manifests itself as a nonlinear boundary condition. Problems with temperature dependent h and radiation can also be treated via the Kirchhoff transformation. In this case, the nonlinear primal problem is transformed into a transformed nonlinear problem.

Considering the ith transformed BEA system equation, substitution of the boundary condition shown in Equation (20) at node k yields the following general relationship.

$$f_{i1}f_1 + \cdots + f_{ik}f_k + \cdots + f_{in}f_n = g_{i1}g_1 + \cdots + g_{ik}g(f_k) + \cdots + g_{in}g_n \tag{21}$$

In this expression, $g(f_k)$ represents the general relation between g_k and f_k. Separating $g(f_k)$ into a parts that are respectively independent and dependent on f_k yields

$$\begin{aligned} f_{i1}f_1 + \cdots + f_{ik}f_k + \cdots + f_{in}f_n &= \\ g_{i1}g_1 + \cdots + g_{ik}h_k(T_\infty - T_0) &- g_{ik}h_k\left\{\frac{T_0}{\beta}\ln\left(\frac{\beta f_k}{k_0 T_0} + 1\right)\right\} + \cdots + g_{in}g_n \end{aligned} \tag{22}$$

or

$$f_{i1}f_1 + \cdots + f_{ik}f_k + \cdots + f_{in}f_n = g_{i1}g_1 + \cdots + g_{ik}c_k + g_{ik}z(f_k) + \cdots + g_{in}g_n \tag{23}$$

In this expression, the transformed temperature at node k appears on both the left and right hand side. As discussed in the context of the constant conductivity problem with nonlinear boundary conditions, different approaches can be taken to solve these equations. One consists of leaving f_k on both sides and iterating to find its correct value. In a second approach, the terms that are dependent on f_k on the right side of Equation (23) are brought to the left hand side. These two approaches are characterized as shown below.

$$f_{i1}f_1 + \cdots + f_{ik}f_k^{(j+1)} + \cdots + f_{in}f_n = g_{i1}g_1 + \cdots + g_{ik}c_k + g_{ik}z(f_k^{(j)}) + \cdots + g_{in}g_n \tag{24}$$

and

$$f_{i1}f_1 + \cdots + [f_{ik} - g_{ik}\{f_k^{(j)}\}^{-1}z(f_k^{(j)})]\, f_k^{(j+1)} + \cdots + f_{in}f_n = g_{i1}g_1 + \cdots + g_{ik}c_k + \cdots + g_{in}g_n \tag{25}$$

In these expressions, the superscript (j) and (j+1) are employed to indicate that the value of f_k is associated with the jth and j+1th iteration respectively. The approach characterized by Equation (24) leaves the left hand side of the overall boundary element system equations unchanged from one iteration to the next, thus allowing for the triangular factorization of [A], formed in the first iteration, to be reused in all subsequent iterations. The second method involves changing [A] in each iteration.

Solution Procedures and Design Sensitivity Analysis

Letting f be t and g be q, in the constant conductivity problem, both the constant and temperature dependent conductivity nonlinear problems can be discussed in a unified format. Solution of the nonlinear set of equations is accomplished by an iterative technique consisting of an initial guess at the surface (transformed) temperature at nodes associated with nonlinear boundary conditions, computation of z_k, and assembly of the matrix equations corresponding to Equation (25). The left hand side matrix is then factored and forward reduction and back substitution is done to obtain an updated response vector. This response is then used to repeat the entire process until it converges. The overall process can be characterized by an expression analogous to Equation (2).

$$[A]^{(M-1)}\{x\}^{(M)} = \{b\}^{(M-1)} \tag{26}$$

Superscript notation is used to denote that the (transformed) temperatures obtained in the M-1th iteration are used to construct [A] and {b} to predict the response in iteration M.

An effective nonlinear DSA formulation is developed by performing implicit differentiation of *the converged* Equation (25) with respect to the Lth design variable X_L.

$$\left\{ f_{i1}f_1+\cdots+[f_{ik}-g_{ik}f_k^{-1}z(f_k)]\,f_k+\cdots+f_{in}f_n = g_{i1}g_1+\cdots+g_{ik}c_k+\cdots+g_{in}g_n \right\}_{,L} \tag{27}$$

or

$$f_{i1}f_{1,L}+\cdots+[f_{ik}-g_{ik}f_k^{-1}z(f_k)]\,f_{k,L}+\cdots+f_{in}f_{n,L} = g_{i1,L}g_1+\cdots+g_{ik,L}c_k+\cdots+g_{in,L}g_n$$

$$-\ \{f_{i1,L}f_1+\cdots+[f_{ik}-g_{ik}f_k^{-1}z(f_k)]_{,L}\,f_k+\cdots+f_{in,L}f_n\} \tag{28}$$

where

$$[f_{ik}-g_{ik}f_k^{-1}z(f_k)]_{,L} = f_{ik,L}-\left\{ g_{ik,L}f_k^{-1}z(f_k) + [g_{ik}f_k^{-1}z_{,L}(f_k) - g_{ik}f_k^{-2}z(f_k)]\,f_{k,L}\right\} \tag{29}$$

Observe that Equation (28) has exactly the same coefficients on the left hand side as Equation (25). This fact is significant, in that it allows the converged left hand side matrix (and its [L] [U] factorization), evolved in a previous nonlinear analysis, to be reused in the sensitivity analysis process. Note further that the DSA equations remain nonlinear, with $f_{k,L}$ appearing on both the left and right sides of the equation. Thus an iterative scheme must be employed in their solution.

$$f_{i1}f_{1,L}^{(j+1)}+\cdots+[f_{ik}-g_{ik}f_k^{-1}z(f_k)]\,f_{k,L}^{(j+1)}+\cdots+f_{in}f_{n,L}^{(j+1)} = g_{i1,L}g_1+\cdots+g_{ik,L}c_k+\cdots+g_{in,L}g_n$$

$$-\left\{ f_{i1,L}f_1+\cdots+\left(f_{ik,L}-\{g_{ik,L}f_k^{-1}z(f_k) + [g_{ik}f_k^{-1}z_{,L}(f_k) - g_{ik}f_k^{-2}z(f_k)]\,f_{k,L}^{(j)}\}\right)f_k+\cdots+f_{in,L}f_n\right\} \tag{30}$$

The dramatic difference in the type on nonlinearity present in this DSA formulation can be quantified by contrasting it with the previous analysis step. During the analysis phase of the nonlinear thermal problem, $[A]^{(M-1)}$ is evolved to its *correct* state, with all (transformed) temperatures used to make the z_k functions at their converged values. The resulting sensitivity equations can be interpreted in a matrix sense.

$$\frac{\partial}{\partial X_L}\Big([A]\{x\} = \{b\}\Big) \Rightarrow [A]\{x\}_{,L} = \Big(\{b\}_{,L} - [A]_{,L}\{x\}\Big) \tag{31}$$

Notice that the left hand side matrix shown in this equation is indeed the converged $[A]^{(M-1)}$, formed and factored during the previous nonlinear thermal analysis step. The other terms shown on the right hand side of Equation (30) make up $\{b\}_{,L}$ and $[A]_{,L}\{x\}$. Symbolically the solution of Equation (31) is characterized as

$$[A]\{x\}_{,L}^{(M)} = \{c\}_{,L}^{(M-1)} \tag{32}$$

Thus, Equation (32) is far different from the nonlinear equation set solved in the previous (transformed) thermal analysis. The left hand side matrix present in Equation (45) is *correct*, having been constructed using the converged (transformed) thermal response. Also, this left hand side matrix has already been factored and can be saved from the last iteration of the previous thermal analysis and reused in the iterative process to evolve $f_{,L}$ to its converged state. These iterations therefore involve only forward reduction and back substitution operations.

Extensive formulae for c_k and $z(f_k)$ and their sensitivities are given in References [7-10], for temperature dependent h and radiation boundary conditions and temperature dependent conductivities characterized by exponential, linear and power laws. Listed below are formulae for an exponential conductivity law and constant h.

$$c_k = h(T_\infty - T_0)\,;\ z(f_k) = \frac{-h\,T_0}{\beta}\ln\left(\frac{\beta\,f_k}{k_0T_0}+1\right);\ \text{and}\ z_{,L}(f_k) = \frac{-h}{k_0}\left(\frac{\beta\,f_k}{k_0T_0}+1\right)^{-1} \tag{33}$$

For the case of constant conductivity and radiation boundary conditions, the quantities become

$$c_k = 0\,;\ z(T_k) = \sigma E(T_k^2+T_r^2)(T_k+T_r)\,;\ \text{and}\ z_{,L}(T_k) = \sigma E(3T_k^2+2T_kT_r+T_r^2) \tag{35}$$

Example Problems

The hollow circular cylinder with constant conductivity shown in Figure 1 is presented first. The inner radius is maintained at 200 degrees, while the outer radius is subjected to a radiation boundary condition. A quarter symmetry model with 40 nodes and 20 three-node quadratic elements is employed with adiabatic conditions along the straight sides. The design variable in this example is the outer radius 'b' of the cylinder. The node point geometric sensitivity is a linear function of their radial position as shown. The solution of this nonlinear problem was differentiated to produce an exact sensitivity solution. Table 1

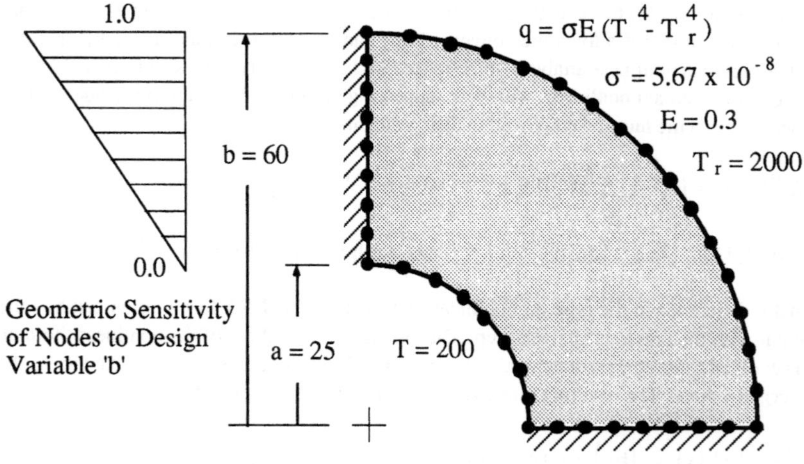

Figure 1. Hollow Circular Cylinder with a Radiation Boundary Condition

contains a comparison of the exact and computed sensitivities at a number of sample points on the surface of the BEA model. The agreement between the exact and computed response sensitivities is excellent.

Table 1. Response Sensitivities for Hollow Cylinder with Radiation Boundary Conditions.

R	R,L	T,L exact	T,L computed	q,L exact	q,L computed
25.0	0.0	0.0	0.0	-1.5656	-1.5611
29.375	0.125	2.4372	2.4398	-1.6302	na
33.75	0.25	3.4842	3.4864	-1.6106	na
38.125	0.375	3.7069	3.7087	-1.5571	na
42.5	0.5	3.4204	3.4218	-1.4901	na
46.875	0.625	2.8107	2.8112	-1.4198	na
51.25	0.75	1.9928	1.9929	-1.3508	na
55.625	0.875	1.0404	1.0403	-1.2851	na
60.0	1.0	0.0024	0.0023	-1.2234	-1.2240

na - not available because only normal components of the heat flow vector are computed

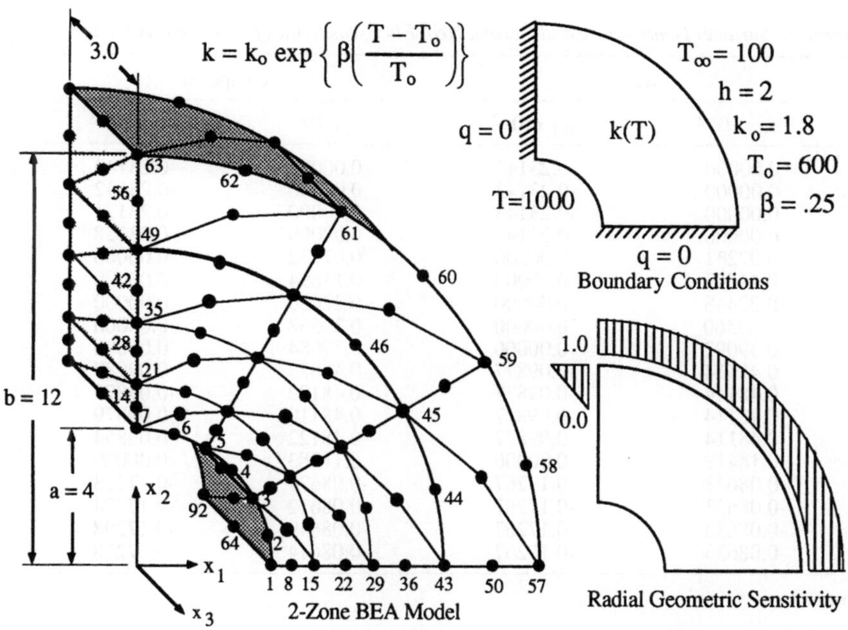

Figure 2. Cylinder with Temperature Dependent Conductivity and Convection

The three dimensional, two zone BEA model shown in Figure 2 was used to demonstrate the nonlinear DSA formulation on a problem with a material with an exponential conductivity conductivity law and a convection boundary condition on the outer radius with a constant h. This problem was solved iteratively as described in this paper. The analysis required 7 iterations while the DSA required 9 iterations. CPU statistics and acomparison of exact and computed surface sensitivities are given in Tables 2 and 3. Again, the accuracy of the approach is demonstrated to be excellent. Note that all computed response sensitivities have less than one percent error. The timings shown in this example are a bit misleading. This example was run on a computer system that required that the scratch files containing all matrix coefficients reside on SCSI (small computer system interface) disk. Most 3D BEA is done on systems with higher performance disk storage systems. On such systems, the computer resources consumed in the I/O of these matrix coefficients will be a much smaller percentage of the overall computational effort.

Table 2. CPU Statistics for Second Example

Operation	Time in Analysis	Time in DSA
Preliminaries	1.4	0.0
Integration*	21.7	9.1
Assembly*	18.6 (7)	22.3 (9)
Factorization	12.1 (7)	0.0 (0)
Reduction & Substitution	0.5 (7)	0.6 (9)
Recovery	0.5	0.6
Total	54.8	32.6

* - CPU times slow due to SCSI disk

Table 3. Surface Temperature and Normal Heat Flux Sensitivities for Second Example

	exact		computed	
Point	$t,_L x 10^{+2}$	$q,_L x 10^{+2}$	$t,_L x 10^{+2}$	$q,_L x 10^{+2}$
1*	0.00000	-0.23143	0.00000	-0.23149
2	0.00000	-0.23143	0.00000	-0.23142
3	0.00000	-0.23143	0.00000	-0.23171
4	0.00000	-0.23143	0.00000	-0.23158
8	0.07231	0.00000	0.07232	0.00000
15	0.14693	0.00000	0.14694	0.00000
22	0.22448	0.00000	0.22450	0.00000
29	0.30560	0.00000	0.30558	0.00000
36	0.39092	0.00000	0.39084	0.00000
43*	0.48114	-0.09877	0.48083	-0.09987
44	0.48114	-0.09877	0.48102	-0.09879
45	0.48114	-0.09877	0.48110	-0.09879
46	0.48114	-0.09877	0.48122	-0.09854
50	0.18419	0.00000	0.18383	0.00000
57*	-0.08633	-0.17267	-0.08677	-0.17353
58	-0.08633	-0.17267	-0.08612	-0.17224
59	-0.08633	-0.17267	-0.08649	-0.17298
60	-0.08633	-0.17267	-0.08614	-0.17228

* - corner

Zone Condensation

Multi-zone BEA [7] is accomplished by breaking up an entire model into zones and writing the governing boundary integral relationship for each zone. By evaluating this expression at the load points corresponding to node point locations for the zone in question, one can generate a matrix system of equations for each zone. The individual zone matrix relations can be put together for use in an overall analysis by considering the conditions of (transformed) temperature compatibility and thermal energy conservation of the (transformed) normal heat flux components at zone interfaces. As detailed in [7], the resulting matrix equation is actually hypermatrix equation with matrices for its entries. The concept of condensation of degrees of freedom in the thermal BEA context [7] has also been demonstrated by considering the matrix equations for a single zone. Reordering degrees of freedom and partitioning Equation (1) into blocks that correspond to master degrees of freedom and blocks that correspond to degrees of freedom that could be condensed, one can arrive at the matrix equation shown below.

$$\begin{bmatrix} [F_{MM}] & [F_{MC}] \\ [F_{CM}] & [F_{CC}] \end{bmatrix} \begin{Bmatrix} \{f_M\} \\ \{f_C\} \end{Bmatrix} = \begin{bmatrix} [G_{MM}] & [G_{MC}] \\ [G_{CM}] & [G_{CC}] \end{bmatrix} \begin{Bmatrix} \{g_M\} \\ \{g_C\} \end{Bmatrix} \qquad \begin{matrix} (36a) \\ (36b) \end{matrix}$$

Solving the matrix Equation (36b) for $\{f_C\}$, and substituting the result into Equation (36a), collecting terms, and performing considerable further manipulations yields a zone condensation approach for (transformed) temperature dependent conductivity problems. The details of this condensation process has been documented elsewhere [7]. Only the final relations are provided below.

$$[M_1]\{f_M\} = [M_2]\{g_M\} + \{v_C\}\ ;\ \{v_C\} = [G_{MC}]\{g_C\} - [F_{MC}][F_{CC}]^{-1}[G_{CC}]\{g_C\} \tag{37}$$

$$\{f_C\} = [F_{CC}]^{-1}\Big([G_{CM}]\{g_M\} + \{v_1\} - [F_{CM}]\{f_M\}\Big);\ \{v_1\} = [G_{CC}]\{g_C\} \tag{38}$$

and

$$[M_1] = [F_{MM}] - [F_{MC}][F_{CC}]^{-1}[F_{CM}]\ ;\ \text{and}\ \ [M_2] = [G_{MM}] - [F_{MC}][F_{CC}]^{-1}[G_{CM}] \tag{39}$$

Equation (37) is called a condensed BEA zone matrix equation, while Equation (38) is called a BEA zone matrix expansion equation. The condensation procedure presented above is an exact formulation, in that, no terms have been neglected, nor has any approximation been made in order to write these equations. Note that whenever $[F_{CC}]^{-1}$ appears in these equations, it always pre-multiplies either a column vector or rectangular matrix. Thus, the use of the matrix inversion notation is purely symbolic. In the computer implementation of this approach, no matrix inversion is ever actually performed. Instead, the triangular factorization of $[F_{CC}]$ is performed once, and subsequently these factors are used to solve matrix equations by forward reduction and backward substitution of a right hand side vector or group of vectors.

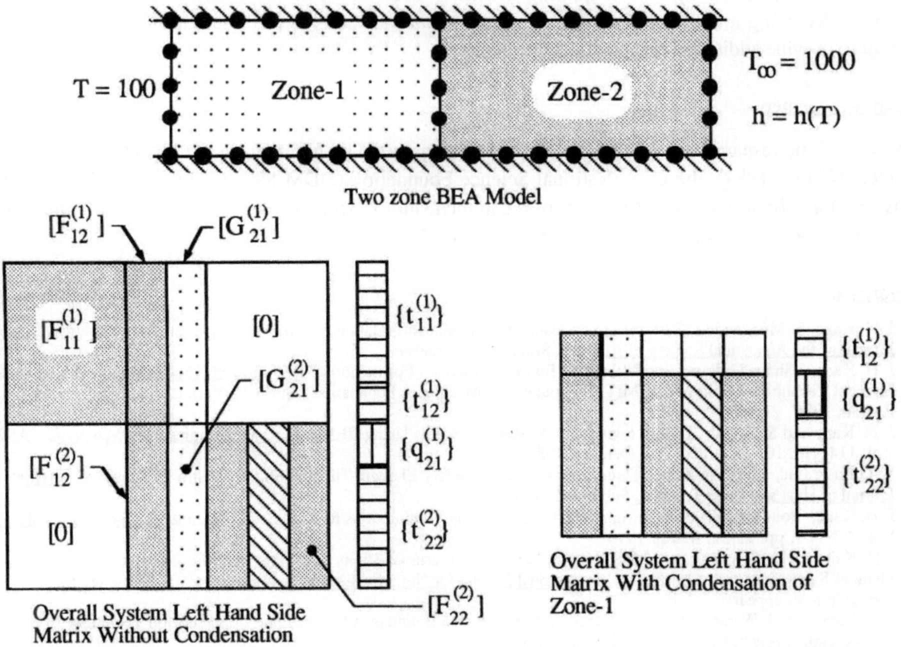

Figure 3. System Matrices Associated with Multi-zone Models and Condensation

A natural way to combine substructuring with multi-zone BEA capability is to allow for the possible condensation of degrees of freedom that appear exclusively in any particular zone. In this case, the partitions to be eliminated by the condensation process coincide exactly with certain partitions already present in the multi-zone BEA procedure. The impact of the zone condensation technique, when employed in nonlinear heat transfer, can be explained by considering the two zone example problem shown in Figure 3. Note that the nonlinear boundary condition is confined exclusively to zone two. This two zone boundary element model produces the sparse blocked left hand side matrix shown. In this figure, the changing entries in the left hand side matrix due to the nonlinearities are highlighted using diagonal crosshatching. A second matrix is shown that corresponds to the overall left hand side matrix for the case where boundary element zone-1 has been condensed. Comparison of these two matrices shows very clearly why the iterative process evolving this class of nonlinear problem to its converged solution can be performed in a more economical fashion when condensation is employed.

The advantage of the multi-zone BEA and zone condensation techniques also extend to the DSA process described in this paper. For example, in the two zone problem described above, the process of one condensation step followed by 7 factorizations of a reduced size left hand side matrix required only 3.1 seconds, as opposed to the 12.1 seconds required in the problem without condensation. Similar reductions in CPU requirements were obtained in the DSA process. The uncondensed left hand side matrix required 22028 double precision words of computer storage while the left hand side matrix with zone-1 condensed could be held in only 5464 double precision words of memory.

Conclusions

Nonlinear example problems have been presented to demonstrate that implicit differentiation can generate an accurate and economical approach for the computation of shape sensitivities for nonlinear thermal problems. A unified formulation was given treating nonlinear boundary conditions and temperature dependent conductivity using a common notation. Multi-zone analysis and zone condensation strategies was also shown to provide additional computational economy.

Acknowledgement

Portions of the research discussed herein have been supported by grants from the NASA Lewis Research Center (NAG 3-1089), the U. S. National Science Foundation (DDM-8996171) to Clarkson University. Any opinions, findings, and conclusions or recommendations expressed in this publication are those of the authors and do not reflect the views of these other organizations.

References

1. J. H. Kane, S. Mao, and G. C. Everstine, 'A boundary element formulation for acoustic shape sensitivity analysis,' The Journal of the Acoustical Society of America, Submitted for review
2. J. H. Kane, " Shape Optimization Utilizing a Boundary Element Formulation," BETECH 86, Proceedings, 1986 Boundary Element Technology Conference, MIT, Computational Mechanics Publications, Springer-Verlag, Southampton and Boston, 1986
3. J. H. Kane and S. Saigal, "Design Sensitivity Analysis of Solids Using BEM," Journal of Engineering Mechanics, ASCE, Vol. 114, No. 10, pp. 1703-1722, October, 1988
4. K. Guru Prasad, and J. H. Kane, 'Three Dimensional Boundary Element Thermal Shape Sensitivity Analysis,' International Journal of Heat and Mass Transfer, Submitted for review
5. J. H. Kane, 'Boundary Element Design Sensitivity Analysis Formulations for Coupled Problems,' Engineering Analysis, Vol. 7, No. 1, pp. 21-32, March 1990
6. J. H. Kane, B. L. K. Kumar, and M. Stabinsky, "Transient Thermoelasticity and Other Body Force Effects in Boundary Element Shape Sensitivity Analysis," International Journal for Numerical Methods in Engineering, Accepted for publication, to appear.
7. J. H. Kane and H. Wang, 'Nonlinear Thermal Analysis with a Boundary Element Zone Condensation Technique,' Computational Mechanics, Accepted for publication.
8. J. H. Kane and H. Wang, 'Boundary Element Shape Sensitivity Analysis Formulations for Thermal Problems with Nonlinear Boundary Conditions,' AIAA Journal, Accepted for publication, to appear
9. J. H. Kane and H. Wang, 'A Boundary Element Shape Design Sensitivity Analysis Formulation for Thermal Radiation Problems,' ISBEM89 Proceedings, Springer-Verlag, 1990
10. J. H. Kane and H. Wang, 'Boundary Formulations for Shape Sensitivity of Temperature Dependent Conductivity Problems,' Numerical Heat Transfer, Submitted for review.

Non-Linear Analysis of the Flow Around Partially or Super-Cavitating Hydrofoils by a Potential Based Panel Method

S. A. KINNAS and N. E. FINE

Department of Ocean Engineering
Massachusetts Institute of Technology
Cambridge, MA 02139, USA

Abstract

The problem of analyzing the flow around a partially or super-cavitating hydrofoil in an ideal fluid is addressed by employing a low order perturbation potential based panel method. The cavity surface is determined as a part of the solution in an iterative manner. As a first iteration in determining the final cavity surface, the foil beneath the cavity is used in the case of partially cavitating hydrofoils, and the cavity shape from linear theory is used in the case of super-cavitating hydrofoils. The numerical scheme is shown to be very robust and to converge to the final cavity shape quicker than a previous numerical scheme based on a surface vorticity velocity based panel method.

1 Introduction

Cavitation is very often unavoidable and its accurate prediction is therefore a very important aspect in estimating the hydrodynamic performance of marine propellers, water pumps or high speed hydrofoils. The main intricacy in predicting the flow around a cavitating lifting surface is the fact that the extent and shape of the cavity are not known *a priori* and have to be determined as a part of the solution. An additional difficulty arises at the trailing edge of finite extent cavities where a cavity termination model must be specified.

Analytical methods for predicting the cavitating flow around hydrofoils are limited to two-dimensions and to simple geometries, while linear theory has enabled the study of cavitating flows around general shape hydrofoils and around lifting surfaces in steady as well as unsteady flows. It is outside the scope of this work, however, to review the vast literature on analytical or numerical techniques based on linear cavity theory. [1]

One of the main defects of the linear cavity theory is its inability to predict the correct behavior of the cavity shape with changes in the hydrofoil thickness. This is especially true in the case of partially cavitating round nosed hydrofoils, as discovered first by Tulin and Hsu [1] and later by Uhlman [3]. The first of those authors applied a linear short cavity theory on the fully wetted non-linear solution and the second applied a non-linear boundary element method with an iterative scheme for determining the exact shape of the cavity. A method for accounting, within the linear cavity theory, for the non-linear leading edge effects (and thus predict the correct effect of the hydrofoil thickness on the cavity) has been developed by Kinnas [2].

[1] Extensive related literature may be found in [1] and [2].

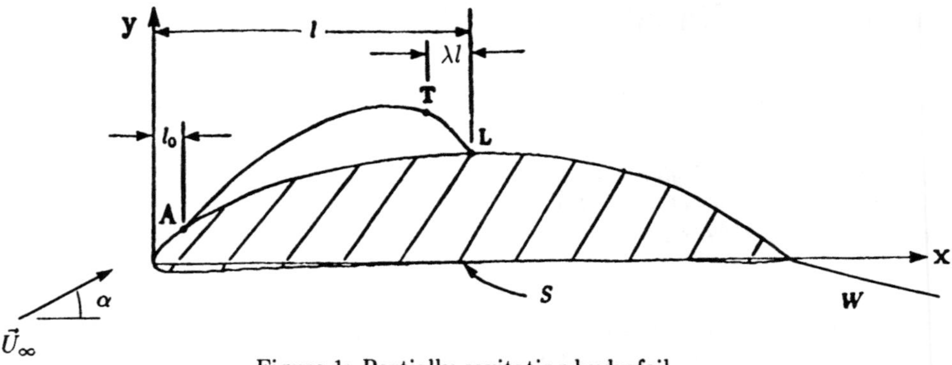

Figure 1: Partially cavitating hydrofoil.

Velocity based boundary element methods have been applied for the analysis of the flow around super-cavitating hydrofoils by Pellone and Rowe [4], who applied a source based formulation, for partially [3] and super-cavitating [5] hydrofoils by Uhlman, who applied a surface vorticity formulation, and for partially cavitating hydrofoils by Lemonnier and Rowe [6], who applied a numerically optimum mixed source and vorticity formulation. All of the previous authors placed the singularities on the exact cavity surface, the shape of which was determined in an iterative manner.

In the present work we employ a potential based formulation for analyzing, in the context of non-linear theory, the flow around partially and super-cavitating hydrofoils. The method is shown to converge to the non-linear solution at a faster rate than the surface vorticity formulation [3]. This can partly be attributed to the less singular nature of the influence coefficients in the potential based versus the velocity based formulation in the case of thin hydrofoil sections. The fast rate of convergence to the non-linear cavity solution is a very desirable characteristic of an iterative method, especially when applied to the computation of three dimensional flows for which computer effort is an important consideration.

2 The Partially Cavitating Hydrofoil

2.1 Formulation

Consider a partially cavitating hydrofoil subject to a uniform inflow, $\vec{U}_\infty$, with ambient pressure, p_∞, as shown in Figure 1. The cavity detaches at point A (arbitrary in this work) on the suction side of the foil and ends at point L. The pressure on the cavity surface is constant and equal to p_c between A and T. In the transition zone between T and L a cavity termination model must be imposed, as will be described later in this section. The length of the cavity l is defined as the distance between the leading edge of the foil and the trailing edge L of the cavity measured parallel to the chord, as shown in Figure 1. The position of the upper cavity detachment point, l_0, is likewise defined by the horizontal distance from the leading edge of the foil to the point A. Unless otherwise noted, all length scales are normalized on the chord length c.

Assuming that the fluid is inviscid and incompressible, and that the resulting flow is irrotational, we can express the total velocity flow field, $\vec{q}$, in terms of either the total potential, Φ, or the perturbation potential, ϕ, as:

$$\vec{q} = \nabla\Phi = \vec{U}_\infty + \nabla\phi. \tag{1}$$

The total and perturbation potentials are related as follows:

$$\phi(x, y) = \Phi(x, y) - \Phi_{in}(x, y) \tag{2}$$

where the inflow velocity potential Φ_{in} corresponds to the uniform inflow of magnitude U_∞ at an angle of attack α:

$$\Phi_{in}(x, y) = U_\infty(x \cos\alpha + y \sin\alpha). \tag{3}$$

The perturbation potential ϕ will satisfy Laplace's equation in the domain outside the cavity and hydrofoil

$$\nabla^2\phi = 0. \tag{4}$$

In addition, the following boundary conditions on ϕ are applied:

1. Kinematic boundary condition

The flow is required to be tangent to the wetted hydrofoil surface as well as to the cavity surface. If $\hat{n}$ is the unit normal to the hydrofoil or cavity surface directed into the fluid domain, then the kinematic boundary condition is given by:

$$\frac{\partial\phi}{\partial n} = -\frac{\partial\Phi_{in}}{\partial n} = -\nabla\Phi_{in} \cdot \hat{n} \quad \text{on the wetted foil and cavity surface.} \tag{5}$$

2. Dynamic boundary condition on the cavity (from A to T)

The pressure is required to be constant on the cavity surface from A to T and equal to p_c. Defining the cavitation number, σ, as follows:

$$\sigma = \frac{p_\infty - p_c}{\frac{1}{2}\rho U_\infty^2}, \tag{6}$$

and applying Bernoulli's equation, we find that the magnitude of the total velocity on the cavity, q_c, must be constant and given by $q_c = U_\infty\sqrt{1+\sigma}$. The dynamic boundary condition can be written in terms of the perturbation potential by using equation (2):

$$\frac{\partial\phi}{\partial s_c} = q_c - \frac{\partial\Phi_{in}}{\partial s_c}, \tag{7}$$

where s_c is the cavity arclength measured from the cavity detachment point A. Integrating (7) yields a more convenient form of the dynamic boundary condition:

$$\phi(s_c) - \phi(0) = q_c s_c - \Phi_{in}(s_c) + \Phi_{in}(0)\ ; \quad \text{on the cavity surface.} \tag{8}$$

3. Kutta condition

$$\nabla\phi = \mathit{finite}\ ; \quad \text{at the trailing edge} \tag{9}$$

4. Condition at infinity

$$\nabla\phi \longrightarrow 0 \ \text{ at infinity} \tag{10}$$

5. Cavity termination model and closure condition

A cavity termination model is applied for which there exists a transition zone (between T and L in Figure 1) in which the cavity velocity varies continuously from its constant value on the cavity to a value matching the wetted foil velocity downstream of the cavity. In this model, as introduced in [6], the dynamic boundary condition (7) is extended to include the transition zone, as follows:

$$\frac{\partial \phi}{\partial s_c} = q_c + (q_L - q_c)\, f(s_f) - \frac{\partial \Phi_{in}}{\partial s_c} \tag{11}$$

where

$$f(s_f) = \begin{cases} 0 & s_f < s_T \\ \left[\frac{s_f - s_T}{s_L - s_T}\right]^{\nu} & s_T \leq s_f \leq s_L \end{cases} \tag{12}$$

where q_L is the total velocity at the trailing edge of the cavity, s_f is the arclength of the foil beneath the cavity measured from the cavity leading edge, s_L refers to the value of s_f corresponding to the cavity trailing edge, and s_T to the beginning of the transition region, where

$$s_T = (1 - \lambda) s_L. \tag{13}$$

For cavities which begin at the leading edge of the foil, λ is the fraction of the cavity length l which comprises the transition zone. The parameters ν and λ are arbitrary, but in the future will be taken from experimental information at the trailing edge of the cavity. Their effect on the cavity shape will be shown in Section 2.3.

Equation (11) can be integrated to yield the perturbation potential on the cavity surface:

$$\phi(s_c) - \phi(0) = q_c s_c + (q_L - q_c) \int_0^{s_c} f(s_f) ds_c - \Phi_{in}(s_c) + \Phi_{in}(0). \tag{14}$$

Since the cavity surface is not known, the new arclength, s_c, must be approximated by the arclength from the previous iteration, s. The dynamic boundary condition then becomes

$$\phi(s) - \phi(0) = q_c s + (q_L - q_c) \int_0^{s} f(s_f) ds - \Phi_{in}(s) + \Phi_{in}(0). \tag{15}$$

In addition, we will assume that the cavity height vanishes at its trailing edge:

$$h(s_L) = 0. \tag{16}$$

The objective of this work is, for given cavity detachment point A and cavity length l, to solve equation (4), subject to the conditions (5), (15), (9), (10) and (16) and to determine the cavity shape and the corresponding cavitation number. The solution will be obtained numerically by employing a low-order potential based panel method, as described in the following section.

2.2 Numerical Implementation

According to Green's theorem the perturbation potential, ϕ_p, at any point p in the domain outside and on the cavity and foil can be expressed as:

$$2\pi\phi_p = \int_S \left[-\phi\frac{\partial logR}{\partial n} + \frac{\partial\phi}{\partial n}logR\right] dS - \int_W \Delta\phi_w \frac{\partial logR}{\partial n} dS; \quad \text{for } p \text{ outside } S/W \quad (17)$$

$$\pi\phi_p = \int_S \left[-\phi\frac{\partial logR}{\partial n} + \frac{\partial\phi}{\partial n}logR\right] dS - \int_W \Delta\phi_w \frac{\partial logR}{\partial n} dS; \quad \text{for } p \text{ on } S/W \quad (18)$$

where S is the surface of the wetted foil or the cavity and W is the surface of wake. as shown in Figure 1. R is the distance from the surface element dS to the point p.

Equation (17) may be regarded as a representation of the perturbation potential in the fluid domain in terms of a dipole distribution of strength ϕ on the foil/cavity surface S, a source distribution of strength $\frac{\partial\phi}{\partial n}$ on S, and a dipole distribution of constant strength $\Delta\phi_w$ on the wake surface W. Notice that ϕ, as given by equation (17), already satisfies the condition at infinity, equation (10).

The perturbation potential and its normal derivatives on the wetted foil or cavity surface are related via the integral equation (18). According to the boundary conditions described in the previous section, $\partial\phi/\partial n$ is known on the wetted foil (*Neumann* boundary condition), while on the cavity $\partial\phi/\partial n$ is unknown and ϕ is known (*Dirichlet* boundary condition). The cavity surface, however, is not known and must be determined as a part of the solution. Note that the potential on the cavity is known only up to the unknown constant, q_c, which also must be determined as part of the solution.

To invert equation (18) numerically we discretize the cavity and foil surface into straight panels whose vertices lie on S. A full cosine spacing is implemented between the foil leading edge and the cavity trailing edge and between the cavity trailing edge and the foil trailing edge. The continuous source and dipole distributions on each panel are approximated by constant strength distributions. Equation (18) is finally applied. in its discretized form, at the midpoints of the panels.

The Kutta condition (9) is numerically implemented by employing Morino's condition [7]:

$$\Delta\phi_w = \phi_T^+ - \phi_T^- \quad (19)$$

where ϕ_T^+ and o_T^- are the potentials at the upper and lower trailing edge panels, respectively.

As already mentioned, the cavity surface is not known and has to be determined iteratively. As a first iteration the cavity panels are placed on the foil underneath it. At each iteration the edges of the cavity panels are relocated on the *current* cavity surface which was computed at the end of the previous iteration.

Denoting N_c the number of panels on the cavity, and N_w the number of panels on the wetted foil. the total number of panels, N, is the sum $N = N_c + N_w$. Applying equation (18) at the panel midpoints provides N equations. On the fully wetted part of the foil, the source strengths are known via equation (5) but the N_w values of the dipole strengths (potentials) are unknown. On the cavity, the potentials are expressed in terms of the unknown cavity velocity q_c via equation (15), but the N_c values of the source strengths are unknown. Therefore, the unknowns are: N_w potentials on the wetted foil. N_c sources on the cavity, and the cavity velocity q_c. The additional equation is provided by numerically implementing the cavity closure condition (16).

At this point we define, as shown in Figure 2, as $\hat{n}$ and $\hat{s}$ the normal and tangential unit vectors to the current cavity surface, respectively, as $h_c(s)$ the distance between the

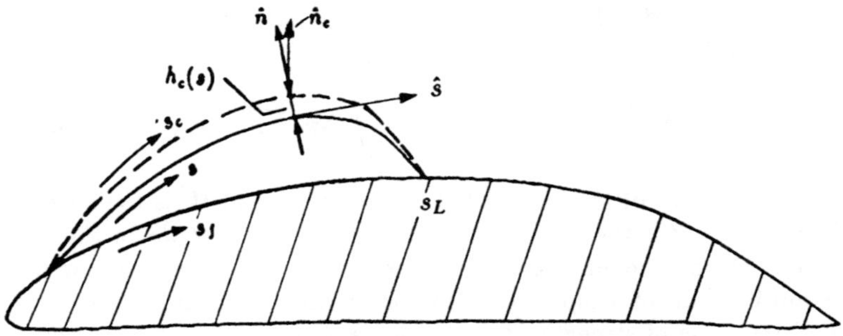

Figure 2: Cavity shapes at the current and next iteration.

cavity surfaces from the next and current iterations (taken along the n direction), and as $\hat{n}_c$ the unit normal vector to the cavity surface at the next iteration. We denote the arclength along the current cavity surface as s, and the arclength along the next cavity surface as s_c. In addition, s_L is the value of s at $x = L$.

The cavity closure condition will thus become:

$$h_c(s_L) = \int_0^{s_L} \frac{dh_c}{dt} dt = 0 \tag{20}$$

where t is a dummy variable for s. We may write this in a more convenient form by imposing flow tangency on the cavity surface, which may be expressed in the following form:

$$\nabla\Phi \cdot \hat{n}_c = 0 \qquad \text{on the cavity.} \tag{21}$$

It can be shown that the cavity normal, $\hat{n}_c$, may be decomposed into components which are normal and tangent to the current cavity surface:

$$\hat{n}_c = -\frac{dh_c}{ds}\hat{s} + \hat{n} + O(h_c^2). \tag{22}$$

Substituting equation (22) in equation (21), expressing $\nabla\Phi$ in terms of its (n, s) components and by using equation (11), we finally get:

$$-\left[q_c + (q_L - q_c)f(s_f)\right]\frac{dh_c}{ds} + \frac{\partial\phi}{\partial n} + \frac{\partial\Phi_{in}}{\partial n} = 0 \tag{23}$$

from which we can solve for dh_c/ds. The resulting expression may be inserted in equation (20) to give the final form of the cavity closure condition:

$$\int_0^{s_L} \frac{\partial\phi}{\partial n}\frac{dt}{[q_c - (q_L - q_c)f(s_f)]} = -\int_0^{s_L} \frac{\partial\Phi_{in}}{\partial n}\frac{dt}{[q_c - (q_L - q_c)f(s_f)]}. \tag{24}$$

Equation (24) provides an additional equation between the unknown source strengths on the cavity surface. Once the solution is found and $\partial\phi/\partial n$ is known, the additional cavity thickness $h_c(s)$ can be determined by integrating equation (24). The new position of the cavity surface and the new arclength, s_c, can then be determined and used in the next iteration.

N	σ	V/c^2
80	1.0235	.01412
100	1.0120	.01367
120	1.0073	.01344
150	1.0029	.01340
200	0.9979	.01335
250	0.9970	.01329

Table 1: Convergence of the cavitation number and cavity volume. with total number of panels, of the present method at the first iteration.

The value of q_L is not known, and is approximated by $\partial\Phi_{in}/\partial s$ for the first iteration and, for successive iterations, is approximated by $\partial\Phi/\partial s$ from the previous iteration, extrapolated to the cavity trailing edge from its values on the wetted foil.

The value of $\phi(0)$ in equation (15) corresponds to the value of the perturbation potential at the detachment point of the cavity A. It is also unknown and in the present numerical scheme is expressed via a cubic extrapolation in terms of the unknown potentials on the wetted panels in front of the cavity.

2.3 Numerical Validation and Results

In this section, the present method is applied to a NACA16006 thickness form at $\alpha = 4°$ and for $l = .50$, $l_0 = 0$, $\nu = .50$, and $\lambda = .05$.

Table 1 displays the rapid convergence of the cavitation number and cavity volume with number of panels for the first iteration.

The convergence of the cavity shapes with number of iterations is shown in Figure 3 for the same foil with $N = 100$. It appears that the cavity shape has practically converged in the second iteration and that even the cavity shape from the first iteration (where the cavity panels are located on the foil beneath the cavity) is close to the converged result. In Figure 3, the pressure distribution on the cavity and foil is also shown as computed by employing a wetted flow analysis [8] on the foil with the cavity from the second iteration. This pressure (plotted with open circles in Figure 3) is shown superimposed on the pressure distribution computed by the present method (plotted with a solid line). The pressure on the cavity appears to satisfy the imposed dynamic boundary condition and this provides a good check on the present scheme.

The convergence of cavitation number and cavity volume with number of iterations for the above example is shown in Figure 4 in comparison with a surface vorticity velocity based panel method developed by Uhlman [3]. While the cavitation numbers computed by the two methods appear to converge at comparable rates, the convergence of cavity volume is much quicker in the potential based panel method.

The cavity shapes predicted by the present method at the first and last iterations are compared to linear theory results in Figure 5. Here, it is seen that the linear theory overpredicts the cavity shape substantially. However, the linear theory with the leading edge corrections [2] predicts a cavity shape which is closer to the converged non-linear result, especially near the leading edge. Note that a comparison of the different cavity solutions, although useful in a qualitative sense, is not quantitatively useful, since each

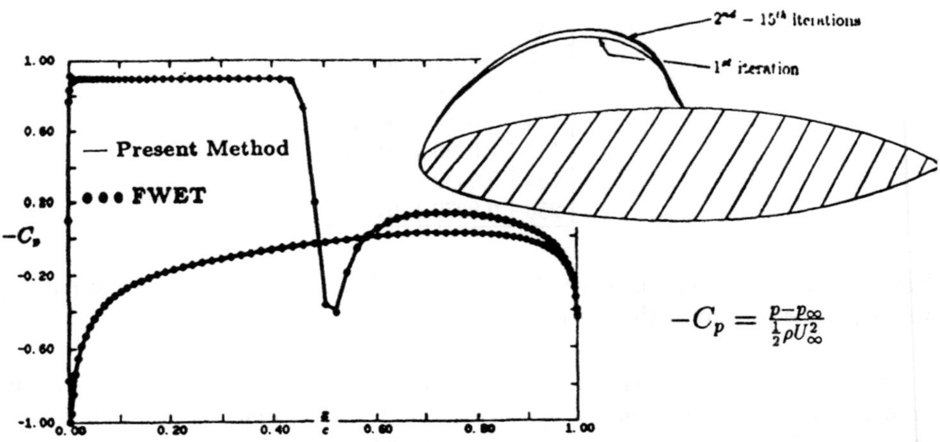

Figure 3: Convergence of cavity shape with number of iterations and pressure distribution on the cavity and foil at the second iteration. The pressure is compared to the pressure computed by a fully wetted analysis (FWET) of the foil and cavity from the second iteration.

incorporates its own cavity termination model.

Figure 6 shows the effect of varying the length of the transition zone. λ, on the cavity shape at the last iteration. While varying λ from .06 to .24, with all other parameters frozen, decreased the volume by over 12%, for the same perturbation of λ the cavitation number increased by only 3%. The parameter ν was also varied in a numerical experiment which showed the effect to be relatively minor.

The effect of thickness on cavity volume is demonstrated in Figure 7, where the cavity volume is plotted against the parameter $\frac{\alpha}{\sigma}$ for a NACA16 foil with thicknesses of 6%, 9%, and 12%. The results appear to be in agreement with those presented by Uhlman [3]. For this example, the detachment point and closure model are the same as in the above examples.

3 The Super-Cavitating Hydrofoil

3.1 Formulation and Numerical Implementation

The non-linear method for predicting the cavity shape and cavitation number for partially cavitating hydrofoils, described in the previous sections, has been extended for the analysis of super-cavitating hydrofoils. The formulation of the problem of super-cavitation is very similar to that of partial cavitation, with several key differences. First, the cavity shape predicted by linear theory, which has been shown to be very close to non-linear results [5], is used as the first iteration[2]. Note also that for super-cavitation there are two points at which the cavity detaches from the foil: one on the upper side, near the foil leading edge, and one on the lower side, near the foil trailing edge.

An additional difference is the necessity to integrate the dynamic boundary condition (7) along both the upper and lower cavity surfaces in order to form a Dirichlet boundary condition on ϕ. As a result, this dynamic boundary condition, unlike the one for partial

[2]The linear cavity shape is obtained by employing a source and vorticity formulation [9]

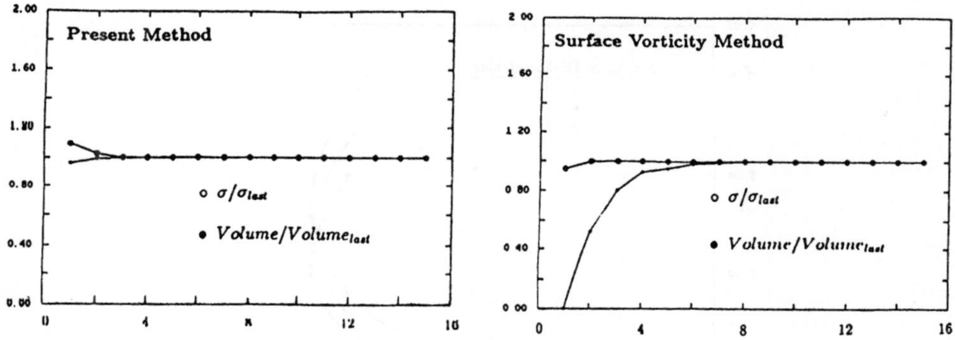

Figure 4: Convergence with number of iterations for the cavitation number and cavity volume by using the present method and Uhlman's surface vorticity method.

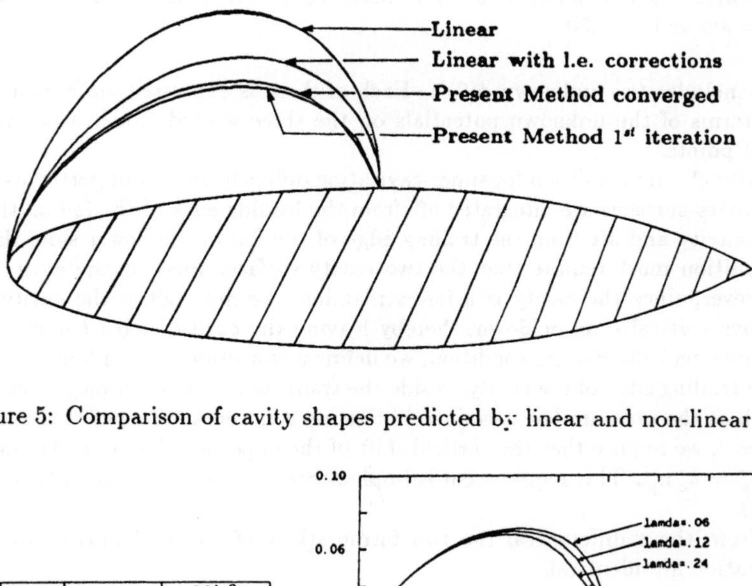

Figure 5: Comparison of cavity shapes predicted by linear and non-linear theory.

λ	ν	σ	V/c^2
.06	.50	0.90113	0.015894
.12	.50	0.91333	0.014948
.24	.50	0.92780	0.013980
.12	.75	0.91027	0.015184
.12	1.00	0.90788	0.015370

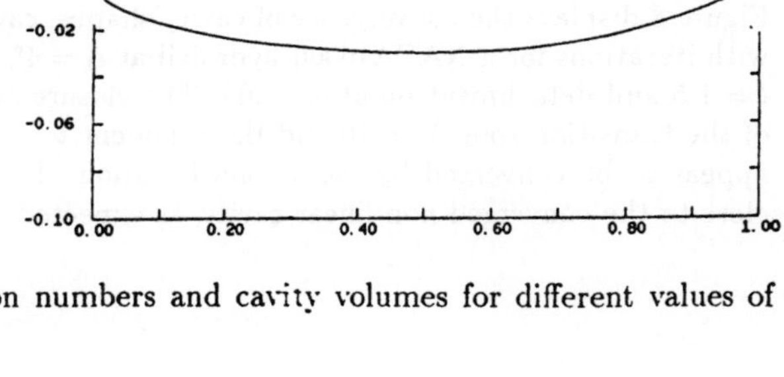

Figure 6: Cavity shapes, cavitation numbers and cavity volumes for different values of the parameters λ and ν.

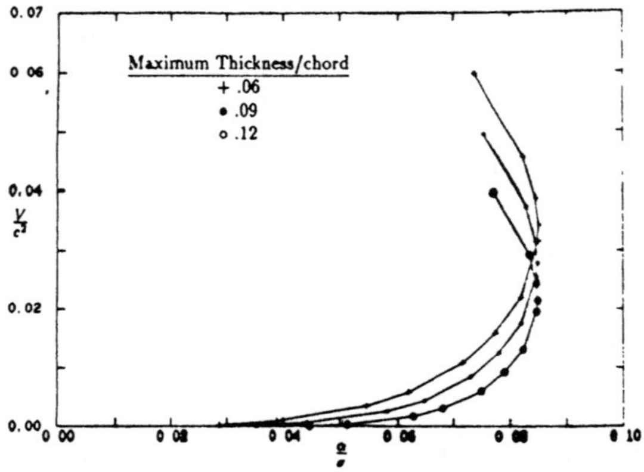

Figure 7: Cavity volume versus α/σ for NACA16006, NACA16009 and NACA16012 at $\alpha = 4°$, $\lambda = .05$ and $\nu = .50$.

cavitation, includes *two* unknown $\phi(0)$'s. Each of these is expressed via a cubic extrapolation in terms of the unknown potentials on the three wetted panels adjacent to the detachment points.

The cavity closure condition for super-cavitation differs from that of partial cavitation. Since the cavity surfaces are integrated aft from the leading edge of the foil on the upper side of the cavity and aft from the trailing edge of the foil on the lower side, the cavity closure condition must require that the two cavity surfaces meet at the cavity trailing edge. However, since the cavity is a free streamline, we must allow the cavity trailing edge to move vertically up or down, thereby leaving the cavity length l unchanged. In order to implement the closure condition, we define a transition zone of length $\lambda \cdot l$ which cuts off the trailing edge of the cavity. Inside the transition zone we employ a termination model, such as the pressure law described in Section 2.1. and at the end of the transition zone, at $x = l$, we require that the vertical shift of the upper and lower cavity surfaces be equal, $h_c^+ n_y^+ = h_c^- n_y^-$. This requirement is implemented in the closure condition, utilizing equation 23.

Aside from these differences, the two formulations of the problems of partial and super-cavitation are identical.

3.2 Results

Figure 8 displays the convergence of cavity shapes, cavitation number, and cavity volume with iterations for a NACA16006 hydrofoil at $\alpha = 4°$ super-cavitating with cavity length $l = 1.5$ and detachment point $l_0 = .05$. The closure model is characterized by the length of the transition zone $\lambda = .10$ and the exponent $\nu = .50$. All of the computed quantities appear to be converged by the second iteration. Furthermore, the linear cavity is very close to the converged non-linear cavity, as expected.

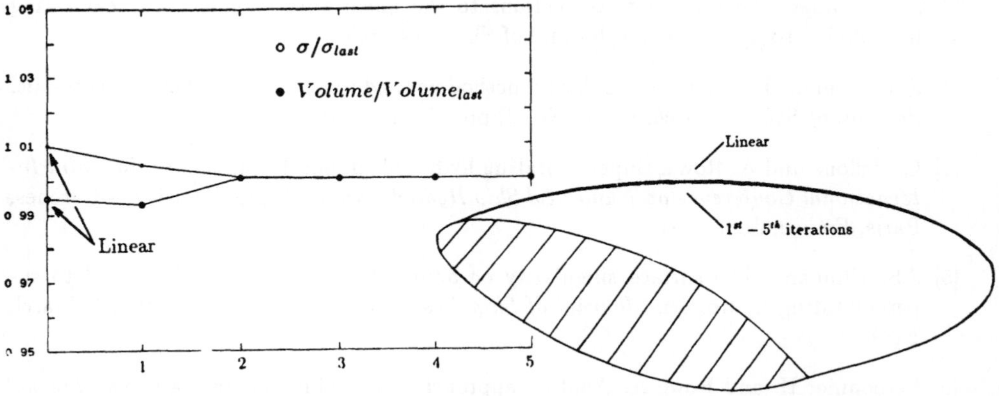

Figure 8: Convergence of cavity shapes, cavitation number, and cavity volume with number of iterations for a super-cavitating hydrofoil.

4 Conclusions

The results obtained by the perturbation potential based panel method for the analysis of cavitating hydrofoils are very promising. In particular, the cavity shapes from the first and final iterations are very close to each other, and even the results from intermediate iterations appear to satisfy the nonlinear boundary conditions with ample accuracy (see Figure 3). These results indicate that the method is well suited for application to three dimensional lifting surfaces in unsteady cavitating flow, since regridding of the cavity surface, if necessary at all, will be minimal. This is an extremely important characteristic of a three dimensional method, since we are seeking a numerical solution which may be obtained with a reasonable amount of computer effort.

In the future, the closure model described in this paper will be supplemented with an open cavity model in order to model the viscous wake behind the cavity. The thickness of the wake will be determined by experiments to be performed at the MIT Marine Hydrodynamics Lab Variable Pressure Water Tunnel, together with the parameters λ and ν.

Acknowledgments

This research has been supported by the Applied Hydromechanics Research Program administered by the Office of Naval Research (Contract: N00014-90-J-1086).

References

[1] M.P. Tulin and C.C. Hsu. New applications of cavity flow theory. In *13th Symposium on Naval Hydrodynamics*, Tokyo, Japan, 1980.

[2] S.A. Kinnas. Leading edge corrections to the linear theory of partially cavitating hydrofoils. To appear in the Journal of Ship Research.

[3] J.S. Uhlman. The surface singularity method applied to partially cavitating hydrofoils. *Journal of Ship Research*, vol 31(No. 2):pp. 107–124, June 1987.

[4] C. Pellone and A. Rowe. Supercavitating hydrofoils in non-linear theory. In *Third International Conference on Numerical Ship Hydrodynamics*, Basin d'essais des Carènes, Paris, France, June 1981.

[5] J.S. Uhlman. The surface singularity or boundary integral method applied to supercavitating hydrofoils. *Journal of Ship Research*, vol 33(No. 1):pp. 16–20, March 1989.

[6] Lemonnier H. and Rowe A. Another approach in modelling cavtating flows. *Journal of Fluid Mechanics*, vol 195, 1988.

[7] Luigi Morino and Ching-Chiang Kuo. Subsonic potential aerodynamic for complex configurations : a general theory. *AIAA Journal*, vol 12(no 2):pp 191–197, February 1974.

[8] J.E. Kerwin, S.A. Kinnas, J-T Lee, and W-Z Shih. A surface panel method for the hydrodynamic analysis of ducted propellers. *Trans. SNAME*, 95, 1987.

[9] S.A. Kinnas and N.E. Fine. Analysis of the flow around supercavitating hydrofoils with midchord and face cavity detachment. To appear in the Journal of Ship Research.

A Discussion of BEM with Reference to Trusses

E. Korach, S. Miccoli, and G. Novati

Structural Engineering Department,
Politecnico di Milano (Technical University),
Piazza Leonardo da Vinci, 32
I 20133 Milano, Italy

Summary

BEM-like approaches are here illustrated with reference to trusses. The matrix equations arrived at represent the counterpart to the weak (weighted residual) version of the displacement and traction boundary integral equations for the continuum. The double-integral bilinear forms and the self-adjointness of the integral operators involved in the continuum formulation reduce, in the present context, to matrix bilinear forms and to matrix symmetry, respectively. The choice of the sources to be adopted at the truss "boundary" in order to obtain a symmetric coefficient matrix is discussed. A condensation of degrees of freedom is introduced to simulate the boundary field discretization typical of a continuum BE model. It is shown that with such forced reduction of degrees of freedom the symmetry of the key matrix operators can be preserved only operating in the spirit of a Galerkin approach.

A Boundary Formulation for Truss-Analysis

Let us consider the actual truss to be analyzed, highlighted in Fig. 1, embedded in a larger structure. We define the truss "boundary" Γ as the set of nodes, through which the actual truss interacts with its complementary portion in the larger structure. Besides, let the remaining nodes and all the bars of the actual truss be collected in a set Ω_I, while Ω_C denotes the set containing all bars of the complementary structure and all of its nodes, except those belonging to Γ (see Fig. 1). Finally Ω will denote the union of these three disjoint sets: $\Omega = \Gamma \cup \Omega_I \cup \Omega_C$.

In the spirit of the indirect BEM (see e. g. Banerjee and Butterfield [1]), we introduce "sources" in Ω acting at the nodes of Γ. Namely we will deal with static sources represented by (external) forces acting on the nodes of Γ and kinematic sources represented by nodal discontinuities which, when active, make the two structures separate in the deformed state. At difference from the approach usually followed to introduce the BEM for the continuum, the notion of fundamental solution is not operatively exploited here; in fact we obtain instead the influence matrices involved in the representation formulae by manipulating the stiffness matrices of the actual truss ($\Omega_I \cup \Gamma$) and the complementary one ($\Omega_C \cup \Gamma$), conceived as separate structures. As for the notation adopted in what follows, upper-case symbols will denote matrices, bold lower-case symbols will indicate column vectors, and superscript ^{T} transposition.

Let us first define the stiffness matrices of the actual and complementary truss. Consider the actual truss $\Omega_I \cup \Gamma$ and define $\mathbf{u}_{\Gamma_I}$ and $\mathbf{u}_{\Omega_I}$ as the displacements of the nodes in Γ (conceived as boundary nodes of Ω_I) and of the nodes of Ω_I, respectively. Assuming

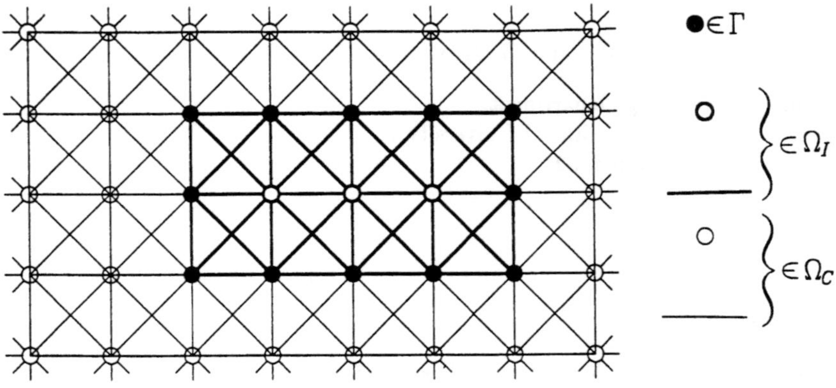

Figure 1: Structure Ω surrounding actual truss: sets of boundary nodes Γ, internal truss Ω_I, and complementary truss Ω_C.

no loads on the Ω_I-nodes and denoting by $\mathbf{p}_I$ the external forces acting on the Γ-nodes, the usual displacement matrix-equation reads

$$\hat{K}_I \begin{Bmatrix} \mathbf{u}_{\Gamma_I} \\ \mathbf{u}_{\Omega_I} \end{Bmatrix} = \begin{Bmatrix} \mathbf{p}_I \\ \mathbf{0} \end{Bmatrix}, \tag{1}$$

where the stiffness matrix $\hat{K}_I$ is symmetric and positive semidefinite (since the external constraints to the actual structure have been disregarded so far). With reference to the complementary truss $(\Omega_C \cup \Gamma)$, the equation analogous to (1) reads

$$\hat{K}_C \begin{Bmatrix} \mathbf{u}_{\Omega_C} \\ \mathbf{u}_{\Gamma_C} \end{Bmatrix} = \begin{Bmatrix} \mathbf{0} \\ -\mathbf{p}_C \end{Bmatrix}, \tag{2}$$

where $\mathbf{u}_{\Omega_C}$ and $\mathbf{u}_{\Gamma_C}$ have obvious meaning, vector $-\mathbf{p}_C$ collects the external forces acting on the Γ-nodes of this structure, and $\hat{K}_C$ is symmetric and positive definite (since the constraints of Ω have been taken into account).

The influence matrices transforming the "sources" at the Γ-nodes into the relevant "effects" at the same nodes are now generated in terms of $\hat{K}_I$ and $\hat{K}_C$ imposing appropriate matching conditions between the actual and the complementary structure. We will denote these matrices by G^{ij} $(i, j = u, p)$ according to the following rule: the first superscript i indicates the nature of the effect, being u or p for kinematic or static effects respectively; the second superscript j is the work-conjugate of the source, being u or p in the case of static or kinematic sources, respectively.

At first let us focus on the case in which static sources $\mathbf{f}$ are present at the Γ-nodes of Ω. This situation arises when the following matching conditions are enforced:

$$\mathbf{p}_I = \mathbf{p}_C + \mathbf{f}, \quad \mathbf{u}_{\Gamma_I} = \mathbf{u}_{\Gamma_C}. \tag{3}$$

The sign conventions used in writing the first of (3) are clarified in Fig. 2. We define $\mathbf{u}_\Gamma \overset{\text{def}}{=} \mathbf{u}_{\Gamma_I} \equiv \mathbf{u}_{\Gamma_C}$, collect all the nodal displacements in the vector $[\mathbf{u}_{\Omega_C} \;\; \mathbf{u}_\Gamma \;\; \mathbf{u}_{\Omega_I}]^T$, and

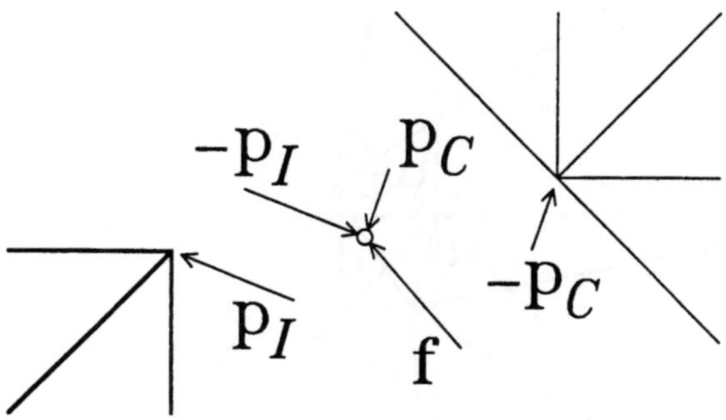

Figure 2: Static discontinuity $\mathbf{f}$ acting on a Γ-node; the nodal equilibrium is enforced evidentiating resultant forces acting on the node from the Ω_I- and Ω_C-bars.

denote by K_C and K_I the stiffness matrices expanded to the dimension of such global displacement vector. By assembling (1) and (2), account taken of (3), we obtain

$$(K_C + K_I)\begin{Bmatrix} \mathbf{u}_{\Omega_C} \\ \mathbf{u}_\Gamma \\ \mathbf{u}_{\Omega_I} \end{Bmatrix} = \begin{Bmatrix} \mathbf{0} \\ \mathbf{f} \\ \mathbf{0} \end{Bmatrix}, \tag{4}$$

where matrix $K_C + K_I$ is the global stiffness matrix which we will call K in the sequel. In order to obtain the first influence matrix G^{uu}, a Boolean matrix B is now introduced such that

$$\mathbf{u}_\Gamma = B\begin{Bmatrix} \mathbf{u}_{\Omega_C} \\ \mathbf{u}_\Gamma \\ \mathbf{u}_{\Omega_I} \end{Bmatrix}, \quad \begin{Bmatrix} \mathbf{0} \\ \mathbf{f} \\ \mathbf{0} \end{Bmatrix} = B^T\mathbf{f}. \tag{5}$$

In fact, solving (4) and making use of (5), gives

$$\mathbf{u}_\Gamma = BK^{-1}B^T\mathbf{f} \stackrel{\text{def}}{=} G^{uu}\mathbf{f}. \tag{6}$$

Being a diagonal submatrix of K^{-1}, G^{uu} is symmetric and positive definite.

Analogously, account taken of (1) and (2), we are able to compute static effects due to the sources $\mathbf{f}$

$$\mathbf{p}_I = BK_IK^{-1}B^T\mathbf{f} \stackrel{\text{def}}{=} G_I^{pu}\mathbf{f}, \quad \mathbf{p}_C = -BK_CK^{-1}B^T\mathbf{f} \stackrel{\text{def}}{=} G_C^{pu}\mathbf{f}, \tag{7}$$

thus defining matrices G_I^{pu} and G_C^{pu}. These matrices transform the loads $\mathbf{f}$ into the resultants $\mathbf{p}_I$ and $\mathbf{p}_C$ acting on the internal and the complementary structure. From the definition of G_C^{pu} and G_I^{pu}, we obtain the "jump relation":

$$G_I^{pu} - G_C^{pu} = BKK^{-1}B^T = I. \tag{8}$$

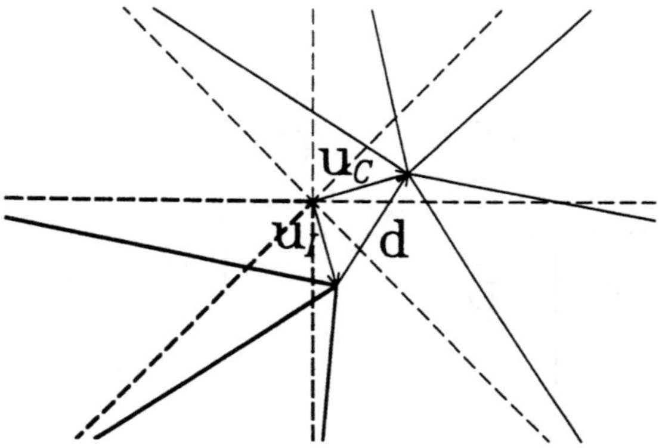

Figure 3: Displacement discontinuity at a Γ-node; the undeformed configuration of the Ω_I and Ω_C bars is denoted by a dashed line.

Dually to what done before, instead of imposing external loads $\mathbf{f}$ on the Γ-nodes, we impose relative displacements $\mathbf{d}$ at the same nodes as in Fig. 3; therefore the matching conditions to enforce are

$$\mathbf{p}_I = \mathbf{p}_C, \quad \mathbf{u}_{\Gamma_I} = \mathbf{u}_{\Gamma_C} - \mathbf{d}. \tag{9}$$

Note that in the second equation of (9), the components of $\mathbf{u}_{\Gamma_I}$ ($\mathbf{u}_{\Gamma_C}$) relevant to a single node of Γ denote the absolute displacements of the endpoints of the bars belonging to Ω_I (Ω_C) and converging into that node in the undeformed state. We define $\mathbf{p} \stackrel{\text{def}}{=} \mathbf{p}_I \equiv \mathbf{p}_C$. Since both $\mathbf{u}_{\Gamma_I}$ and $\mathbf{u}_{\Gamma_C}$ are unknown, but the difference $\mathbf{d}$ between the two quantities is assigned, assembling (1) and (2), we have free choice whether to include $\mathbf{u}_{\Gamma_I}$ or $\mathbf{u}_{\Gamma_C}$ in the vector of primary unknown displacements. If we choose to include $\mathbf{u}_{\Gamma_I}$, assembling (1) and (2) account taken of the matching conditions (9), we have

$$(K_C + K_I) \begin{Bmatrix} \mathbf{u}_{\Omega_C} \\ \mathbf{u}_{\Gamma_I} \\ \mathbf{u}_{\Omega_I} \end{Bmatrix} = -K_C \begin{Bmatrix} \mathbf{0} \\ \mathbf{d} \\ \mathbf{0} \end{Bmatrix}. \tag{10}$$

Solving (10) for $\mathbf{u}_{\Gamma_I}$, we obtain the influence matrix G_I^{up}:

$$\mathbf{u}_{\Gamma_I} = -BK^{-1}K_C B^T \mathbf{d} \stackrel{\text{def}}{=} G_I^{up}\mathbf{d}. \tag{11}$$

If vector $\mathbf{u}_{\Gamma_C}$, instead of $\mathbf{u}_{\Gamma_I}$, were considered among the primary displacement unknowns, the same steps would lead to the generation of matrix G_C^{up}:

$$\mathbf{u}_{\Gamma_C} = BK^{-1}K_I B^T \mathbf{d} \stackrel{\text{def}}{=} G_C^{up}\mathbf{d}. \tag{12}$$

From the definition of these influence matrices, one obtains the jump relation

$$G_C^{up} - G_I^{up} = BK^{-1}KB^T = I. \tag{13}$$

The tractions due to the displacement discontinuities are obtained solving (10) and substituting the displacement vector into the expanded version of eq. (1) (eq. (2) could be used instead); this leads to:

$$\mathbf{p} = -BK_I K^{-1} K_C B^T \mathbf{d} \stackrel{\text{def}}{=} G^{pp}\mathbf{d}. \tag{14}$$

It can be easily shown that this matrix is symmetric; in fact, substituting $K_I = K - K_C$, in the definition of G^{pp} we we have $G^{pp} = B(K_C K^{-1} K_C - K_C)B^T$; it could be also proved that this matrix is negative semidefinite.

Beside the jump relations already evidenced, the following other links among the influence matrices are worth noting:

$$G_C^{up} = (G_I^{pu})^T, \quad G_I^{up} = (G_C^{pu})^T; \tag{15}$$

$$G^{pp} = G_I^{pu}(G^{uu})^{-1} G_I^{up}, \quad G^{pp} = G_C^{pu}(G^{uu})^{-1} G_C^{up}. \tag{16}$$

Equations (15) are trivially verified on the basis of the definitions of the matrices involved; the non trivial proof of (16) is here omitted for brevity. Notice that once established the first of (16), the second is simply proved by taking the transpose of the previous and using equations (15).

Mechanical Interpretation of the Links among Influence Matrices

The matrix relations obtained so far only by means of algebraic properties, are analyzed in this section from a mechanical point of view. Equations (8) and (13), which we here rewrite

$$G_I^{pu} - G_C^{pu} = I, \quad G_C^{up} - G_I^{up} = I, \tag{17}$$

are "jump relations". For the first of (17) we have in fact $\mathbf{p}_I = G_I^{pu}\mathbf{f}$, and $\mathbf{p}_C = G_C^{pu}\mathbf{f}$. Subtracting both sides of the previous equations, we obtain $\mathbf{f} = \mathbf{p}_I - \mathbf{p}_C = (G_I^{pu} - G_C^{pu})\,\mathbf{f}$, which implies the desired jump relation. The mechanical justification of the second jump relation is identical.

The symmetry of matrices G^{pp} and G^{uu} and (15) stem from Betti's reciprocity theorem. To show the symmetry of G^{uu} we consider two distinct elastic states, $\mathcal{A}$ and $\mathcal{B}$ respectively, due to sets of nodal loads $\mathbf{f}^{\mathcal{A}}$ and $\mathbf{f}^{\mathcal{B}}$. In this case Betti's theorem, which reads $\left(\mathbf{u}_\Gamma^{\mathcal{A}}\right)^T \mathbf{f}^{\mathcal{B}} = \left(\mathbf{f}^{\mathcal{A}}\right)^T \mathbf{u}_\Gamma^{\mathcal{B}}$, expressing displacements by means of influence matrices, becomes:

$$\left(\mathbf{f}^{\mathcal{A}}\right)^T (G^{uu})^T \mathbf{f}^{\mathcal{B}} = \left(\mathbf{f}^{\mathcal{A}}\right)^T G^{uu}\mathbf{f}^{\mathcal{B}};$$

the symmetry of matrix G^{uu} is thus proved. For $\mathcal{A} \equiv \mathcal{B}$ the expression of the elastic energy (always positive) is $\mathcal{L} = \frac{1}{2}\mathbf{u}_\Gamma{}^T\mathbf{f} = \frac{1}{2}\mathbf{f}^T G^{uu}\mathbf{f}$; thus matrix G^{uu} is positive definite.

To prove the first of (15), let us consider now two elastic states, $\mathcal{A}$, caused by a set of kinematic actions $\mathbf{d}^{\mathcal{A}}$, and $\mathcal{B}$, caused by a set of static actions $\mathbf{f}^{\mathcal{B}}$. For the state $\mathcal{A}$ we have

$$\mathbf{u}_{\Gamma_I}^{\mathcal{A}} = G_I^{up}\mathbf{d}^{\mathcal{A}}, \; \mathbf{u}_{\Gamma_C}^{\mathcal{A}} = G_C^{up}\mathbf{d}^{\mathcal{A}}, \; \mathbf{p}^{\mathcal{A}} = G^{pp}\mathbf{d}^{\mathcal{A}}.$$

For the state $\mathcal{B}$ we have

$$\mathbf{u}_\Gamma^{\mathcal{B}} = G^{uu}\mathbf{f}^{\mathcal{B}}, \; \mathbf{p}_I^{\mathcal{B}} = G_I^{pu}\mathbf{f}^{\mathcal{B}}, \; \mathbf{p}_C^{\mathcal{B}} = G_C^{pu}\mathbf{f}^{\mathcal{B}}.$$

Due to the kinematic discontinuities, Betti's theorem is no more applicable to the whole truss Ω. Instead, we obtain a generalized version by writing Betti's theorem two times: once for the complementary and once for the internal truss:

$$\left(\mathbf{u}_{\Gamma_C}^{\mathcal{A}}\right)^T\left(-\mathbf{p}_C^{\mathcal{B}}\right)=\left(-\mathbf{p}^{\mathcal{A}}\right)^T\mathbf{u}_{\Gamma}^{\mathcal{B}},\quad \left(\mathbf{u}_{\Gamma_I}^{\mathcal{A}}\right)^T\mathbf{p}_I^{\mathcal{B}}=\left(\mathbf{p}^{\mathcal{A}}\right)^T\mathbf{u}_{\Gamma}^{\mathcal{B}}.$$

Adding both sides of the previous equations we have $\left(\mathbf{u}_{\Gamma_C}^{\mathcal{A}}\right)^T\left(-\mathbf{p}_C^{\mathcal{B}}\right)+\left(\mathbf{u}_{\Gamma_I}^{\mathcal{A}}\right)^T\mathbf{p}_I^{\mathcal{B}}=0$. Summing and subtracting to the first side of the previous equation $\left(\mathbf{u}_{\Gamma_C}^{\mathcal{A}}\right)^T\mathbf{p}_I^{\mathcal{B}}$ we obtain the generalized Betti's theorem:

$$\left(\mathbf{u}_{\Gamma_C}^{\mathcal{A}}\right)^T\mathbf{f}^{\mathcal{B}}+\left(\mathbf{d}^{\mathcal{A}}\right)^T\left(-\mathbf{p}_I^{\mathcal{B}}\right)=0. \tag{18}$$

In term of influence matrices it reads:

$$\left(\mathbf{d}^{\mathcal{A}}\right)^T(G_C^{up})^T\mathbf{f}^{\mathcal{B}}=\left(\mathbf{d}^{\mathcal{A}}\right)^T G_I^{pu}\mathbf{f}^{\mathcal{B}},$$

thus proving the sought relation. Substituting in the above demonstration $\left(\mathbf{u}_{\Gamma_I}^{\mathcal{A}}\right)^T\mathbf{p}_C^{\mathcal{B}}$ to $\left(\mathbf{u}_{\Gamma_C}^{\mathcal{A}}\right)^T\mathbf{p}_I^{\mathcal{B}}$, we can write the generalized Betti's theorem as $\left(\mathbf{u}_{\Gamma_I}^{\mathcal{A}}\right)^T\mathbf{f}^{\mathcal{B}}+\left(\mathbf{d}^{\mathcal{A}}\right)^T\left(-\mathbf{p}_C^{\mathcal{B}}\right)=0$, and prove in this way the second of (15).

By a similar way we can state that if two elastic states, $\mathcal{A}$ and $\mathcal{B}$, are both caused by a set of kinematic actions $\mathbf{d}^{\mathcal{A}}$ and $\mathbf{d}^{\mathcal{B}}$, the generalized Betti's theorem may be written as $\left(\mathbf{d}^{\mathcal{A}}\right)^T\left(-\mathbf{p}^{\mathcal{B}}\right)=\left(-\mathbf{p}^{\mathcal{A}}\right)^T\left(\mathbf{d}^{\mathcal{B}}\right)$, which leads to the symmetry of matrix G^{pp}:

$$\left(\mathbf{d}^{\mathcal{A}}\right)^T(-G^{pp})\,\mathbf{d}^{\mathcal{B}}=\left(\mathbf{d}^{\mathcal{A}}\right)^T(-G^{pp})^T\left(\mathbf{d}^{\mathcal{B}}\right).$$

From the above generalized Betti's theorem, identifying $\mathcal{A}\equiv\mathcal{B}$, it follows that the expression of the internal work for the whole truss Ω, under a kinematic action $\mathbf{d}$, is $\mathcal{L}=\frac{1}{2}\mathbf{d}^T(-\mathbf{p})$. The internal work cannot be negative, but if $\mathbf{d}$ represents a rigid body motion of the Γ nodes, then the elastic solution which fulfils $\mathbf{u}_{\Gamma_C}-\mathbf{u}_{\Gamma_I}=\mathbf{d}$ is the one expressed by $\mathbf{u}_{\Gamma_C}=\mathbf{0}$ and $\mathbf{u}_{\Gamma_I}=-\mathbf{d}$; this means that the complementary truss remains immobile and the internal one rigidly moves. This solution is strain- and stress-free, and thus the internal work vanishes. Expressing this work as $\mathcal{L}=\frac{1}{2}\mathbf{d}^T(-G^{pp})\,\mathbf{d}$, we deduct that matrix G^{pp} must be negative semidefinite; further the zero eigenvalue is associated with the rigid body motions of the Γ-nodes.

We can prove the first of equations (16) by mechanical considerations. Let us consider two different elastic states in Ω: the first caused by a kinematic discontinuity $\mathbf{d}^{\mathcal{A}}$, the second by a static load $\mathbf{f}^{\mathcal{B}}$, related to the first by $\mathbf{f}^{\mathcal{B}}=(G^{uu})^{-1}G_I^{up}\mathbf{d}^{\mathcal{A}}$. Note that the two states coincide in terms of the $\mathbf{u}_{\Gamma_I}$ displacements; in fact for the first state we have $\mathbf{u}_{\Gamma_I}^{\mathcal{A}}=G_I^{up}\mathbf{d}^{\mathcal{A}}$, and for the second $\mathbf{u}_{\Gamma}^{\mathcal{B}}=G^{uu}\mathbf{f}^{\mathcal{B}}=G^{uu}\left(G^{uu}\right)^{-1}G_I^{up}\mathbf{d}^{\mathcal{A}}=G_I^{up}\mathbf{d}^{\mathcal{A}}$. If the displacements of the two solutions coincide in all the boundary nodes of Ω_I, and the internal nodes are unloaded, then the two solutions must coincide in Ω_I; thus $\mathbf{p}^{\mathcal{A}}=\mathbf{p}_I^{\mathcal{B}}$. This equation implies

$$G^{pp}\mathbf{d}^{\mathcal{A}}=G_I^{pu}\left(G^{uu}\right)^{-1}G_I^{up}\mathbf{d}^{\mathcal{A}}.$$

or (16) since the arbitrariness of $\mathbf{d}^{\mathcal{A}}$.

Symmetric Boundary Solution Methods

Let us assume that the jumps **f** and **d** are present simultaneously at the Γ-nodes; by combining equations (6), (7), (11), (12), and (14), the displacements and resultant forces at the Γ nodes due to static and kinematic discontinuities (or jumps) applied in Ω at the same nodes, can be expressed for the actual $(\Omega_I \cup \Gamma)$ and the complementary $(\Omega_C \cup \Gamma)$ trusses, according to the following "representation formulae":

$$\left\{\begin{matrix} \mathbf{u}_{\Gamma_I} \\ \mathbf{p}_I \end{matrix}\right\} = \begin{bmatrix} G^{uu} & G_I^{up} \\ G_I^{pu} & G^{pp} \end{bmatrix} \left\{\begin{matrix} \mathbf{f} \\ \mathbf{d} \end{matrix}\right\}, \quad \left\{\begin{matrix} \mathbf{u}_{\Gamma_C} \\ \mathbf{p}_C \end{matrix}\right\} = \begin{bmatrix} G^{uu} & G_C^{up} \\ G_C^{pu} & G^{pp} \end{bmatrix} \left\{\begin{matrix} \mathbf{f} \\ \mathbf{d} \end{matrix}\right\}. \tag{19}$$

These formulae are equivalent to those presented in Hartmann et al. [2] with reference to BIE approaches for plates, and in Maier et al. [3] in conjunction with a BE discretization of the continuum.

A reinterpretation of formulae (19) can be suggested. Consider two distinct elastic states $\mathcal{A}$ and $\mathcal{B}$ pertaining to the actual $(\Omega_I \cup \Gamma)$ and the complementary $(\Omega_C \cup \Gamma)$ trusses respectively, conceived as separate structures. The jumps $\mathbf{f} \equiv -(\mathbf{p}_C^{\mathcal{B}} - \mathbf{p}_I^{\mathcal{A}})$ and $\mathbf{d} \equiv (\mathbf{u}_{\Gamma_C}^{\mathcal{B}} - \mathbf{u}_{\Gamma_I}^{\mathcal{A}})$, subordinated by the above two states, identically satisfy equation (19) when the l. h. sides are interpreted as the displacements and external force resultants at the Γ-nodes relevant to the actual (state $\mathcal{A}$) and complementary (state $\mathcal{B}$) trusses. Note that equations (16) express a linear dependence between the rows of the block-matrices in (19): their rank is half their order. Note also that these matrices are almost symmetric, due to the symmetry of G^{uu} and G^{pp}, and to the relations

$$(G_I^{up})^T = G_I^{pu} - I, \quad (G_C^{up})^T = G_C^{pu} + I, \tag{20}$$

which stem from (15) and (17). The lack of symmetry is localized in the diagonal terms of submatrices G_I^{up}, G_I^{pu}, for the first influence matrix, and of submatrices G_C^{up}, G_C^{pu} for the second one.

With reference to the boundary conditions to enforce on the actual truss, let us define Γ_u as the subset of the Γ-nodes subject to imposed displacements $\bar{\mathbf{u}}$ and $\Gamma_p = \Gamma \setminus \Gamma_u$ as the complementary subset acted on by external forces $\bar{\mathbf{p}}$. Besides, let **s** be the vector of unknown displacements at the Γ_p-nodes and **r** the unknown reaction forces on the nodes of Γ_u. The nodal quantities related to the Γ-nodes of the actual truss are partitioned as follows

$$\mathbf{u}_{\Gamma_I} = \left\{\begin{matrix} \mathbf{s} \\ \bar{\mathbf{u}} \end{matrix}\right\}, \quad \mathbf{p}_I = \left\{\begin{matrix} \bar{\mathbf{p}} \\ \mathbf{r} \end{matrix}\right\}, \tag{21}$$

where the subvectors on the r. h. s. of (21) pertain to the nodes of Γ_p and Γ_u respectively.

In the spirit of the indirect method, our aim is to compute sources **f** and **d** in Ω so that the first formula in (19) satisfies (21); this means:

$$\left\{\begin{matrix} \mathbf{s} \\ \bar{\mathbf{u}} \\ \bar{\mathbf{p}} \\ \mathbf{r} \end{matrix}\right\} = \begin{bmatrix} {}_pG^{uu} & {}_pG_I^{up} \\ {}_uG^{uu} & {}_uG_I^{up} \\ {}_pG_I^{pu} & {}_pG^{pp} \\ {}_uG_I^{pu} & {}_uG^{pp} \end{bmatrix} \left\{\begin{matrix} \mathbf{f} \\ \mathbf{d} \end{matrix}\right\}, \tag{22}$$

were the influence matrices G^{ij} have been subdivided in block rows, according to the partition of $\mathbf{u}_{\Gamma_I}$ and $\mathbf{p}_I$ in (21). The subscript u or p located on the left of each submatrix in (22) denotes the subset (Γ_u or Γ_p) to which the effect given by that influence matrix is to be referred; e. g. ${}_pG^{uu}$ is the influence matrix which expresses displacements at the Γ_p-nodes due to static sources located in all the nodes of Γ. We rewrite (22) separating in the vector on the l. h. side data from unknowns:

$$\left\{\begin{matrix}\bar{\mathbf{u}}\\ \bar{\mathbf{p}}\end{matrix}\right\} = \begin{bmatrix} {}_uG^{uu} & {}_uG_I^{up} \\ {}_pG_I^{pu} & {}_pG^{pp}\end{bmatrix}\left\{\begin{matrix}\mathbf{f}\\ \mathbf{d}\end{matrix}\right\}, \quad \left\{\begin{matrix}\mathbf{s}\\ \mathbf{r}\end{matrix}\right\} = \begin{bmatrix} {}_pG^{uu} & {}_pG_I^{up} \\ {}_uG_I^{pu} & {}_uG^{pp}\end{bmatrix}\left\{\begin{matrix}\mathbf{f}\\ \mathbf{d}\end{matrix}\right\}. \tag{23}$$

We make use of the first of (23) to obtain the auxiliary sources $\mathbf{f}$ and $\mathbf{d}$ which, according to the indirect approach, when substituted in the second of (23) determine the value of the actual unknowns $\mathbf{r}$ and $\mathbf{s}$. The system to be solved is underdetermined: if n is the number of Γ-nodes, then (in the case of bidimensional problems) the vector of given data $[\bar{\mathbf{u}}\ \bar{\mathbf{p}}]^T$ has $2n$ components, while the unknown vector $[\mathbf{f}\ \mathbf{d}]^T$ has $4n$ components. This indetermination of the primary unknowns $[\mathbf{f}\ \mathbf{d}]^T$ does not jeopardize the uniqueness of the actual unknowns $\mathbf{r}$ and $\mathbf{s}$: in fact altough we obtain ∞^{2n} solutions from the first of (23), once substituted in the second equation, they all lead to the same result, since the matrices in (19) have rank $2n$.

For instance, analogously to what usually done in the standard formulation of the BEM for the continuum, we can set $\mathbf{d} = 0$ and obtain $\mathbf{f}$ by inverting the non-symmetric matrix $\begin{bmatrix} {}_uG^{uu} \\ {}_pG_I^{pu}\end{bmatrix}$. To generate a symmetric indirect method, we set to zero the static source $\mathbf{f}$ at the Γ_p-nodes and the kinematic source $\mathbf{d}$ at the Γ_u-nodes. Calling $\mathbf{f}_u$ and $\mathbf{d}_p$ the subvectors of $\mathbf{f}$ and $\mathbf{d}$ corresponding to the Γ_u- and Γ_p-nodes, respectively, we have from (23):

$$\left\{\begin{matrix}\bar{\mathbf{u}}\\ \bar{\mathbf{p}}\end{matrix}\right\} = \begin{bmatrix} {}_uG^{uu}{}_u & {}_uG_I^{up}{}_p \\ {}_pG_I^{pu}{}_u & {}_pG^{pp}{}_p\end{bmatrix}\left\{\begin{matrix}\mathbf{f}_u\\ \mathbf{d}_p\end{matrix}\right\}, \quad \left\{\begin{matrix}\mathbf{s}\\ \mathbf{r}\end{matrix}\right\} = \begin{bmatrix} {}_pG^{uu}{}_u & {}_pG_I^{up}{}_p \\ {}_uG_I^{pu}{}_u & {}_uG^{pp}{}_p\end{bmatrix}\left\{\begin{matrix}\mathbf{f}_u\\ \mathbf{d}_p\end{matrix}\right\}. \tag{24}$$

In the previous equation the subscript u or p located on the right of each submatrix indicates their partition in column-blocks: precisely, it denotes the subset of the Γ-nodes where the sources relevant to the influence matrices are located; e. g. ${}_pG^{uu}{}_u$ is the influence matrix which expresses the displacements at the Γ_p-nodes due to static sources located on Γ_u-nodes. The coefficent matrix of the first equation in (24) is symmetric. To prove this one exploits the partitioned form of (20) and the symmetry of matrices G^{uu} and G^{pp}.

To obtain a direct formulation, for which primary unknowns are represented by $\mathbf{s}$ and $\mathbf{r}$, we impose that the complementary truss Ω_C be undisturbed under the sources $\mathbf{f}$ and $\mathbf{d}$ acting in Ω: i. e. $\mathbf{u}_{\Gamma_C} = \mathbf{0}$ and $\mathbf{p}_C = \mathbf{0}$; this implies

$$\mathbf{d} = -\mathbf{u}_{\Gamma_I}, \quad \mathbf{f} = \mathbf{p}_I. \tag{25}$$

If we enforce the above conditions by means of the second representation formula of (19), account taken of (25) and changing the sign of the second block row, we have:

$$\mathbf{0} = \begin{bmatrix} G^{uu} & -G_C^{up} \\ -G_C^{pu} & G^{pp}\end{bmatrix}\left\{\begin{matrix}\mathbf{p}_I\\ \mathbf{u}_{\Gamma_I}\end{matrix}\right\}. \tag{26}$$

By partitioning vector $[\mathbf{p}_I\ \mathbf{u}_{\Gamma_I}]^T$ as in (21), subdividing in two block-columns the G^{ij} influence matrices, and separating data from unknowns, we obtain:

$$\begin{bmatrix} G^{uu}{}_u & -G_C^{up}{}_p \\ -G_C^{pu}{}_u & G^{pp}{}_p\end{bmatrix}\left\{\begin{matrix}\mathbf{r}\\ \mathbf{s}\end{matrix}\right\} = -\begin{bmatrix} G^{uu}{}_p & -G_C^{up}{}_u \\ -G_C^{pu}{}_p & G^{pp}{}_u\end{bmatrix}\left\{\begin{matrix}\bar{\mathbf{p}}\\ \bar{\mathbf{u}}\end{matrix}\right\}. \tag{27}$$

This is a $4n \times 2n$ non-homogeneous system of equations. However, as pointed out above, the matrix of the second formula of (19) has rank $2n$, thus equation (27) has a unique solution. To solve (27) one can therefore indifferently choose a set of $2n$ linearly independent equations. Under general boundary conditions the choice leading to symmetry, as already discussed in [2,3] is the following: within the first block of equations in (27), we select those relevant to nodes of Γ_u; within the second block the equations expressing the resultant forces at the nodes of Γ_p are selected. The symmetric equation system arrived at is:

$$\begin{bmatrix} {}_uG^{uu}{}_u & -{}_uG^{up}_{C\,p} \\ -{}_pG^{pu}_{C\,u} & {}_pG^{pp}{}_p \end{bmatrix} \begin{Bmatrix} \mathbf{r} \\ \mathbf{s} \end{Bmatrix} = - \begin{bmatrix} *.tex_uG^{uu}{}_p & -{}_uG^{up}_{C\,u} \\ -{}_pG^{pu}_{C\,p} & {}_pG^{pp}{}_u \end{bmatrix} \begin{Bmatrix} \bar{\mathbf{p}} \\ \bar{\mathbf{u}} \end{Bmatrix}. \tag{28}$$

Some remarks are worthwhile on the BEM-like solution methods described so far. By the use of the representation formulae (19) we were able to illustrate, in an unitary context, both the indirect and direct method. It is worth noting that according to relations (17) the coefficient matrices of both methods—see the first of (24) and the matrix on the l. h. s. of (28)—coincide, except for the sign of some submatrices; further, in view of (15), the second matrix of (24) is the transpose of the matrix on the r. h. side of (28), if we disregard the signs of some submatrices. Both approaches are effective in computing the exact solution for the actual truss. At difference from the usual matrix displacement method (MDM), BEM-like methods outlined in the present work utilize as primary unknowns quantities relevant solely to the "boundary" nodes; moreover such quantities are of mixed type (i. e. both static and kinematic) while in the case of the MDM unknowns are represented by the nodal displacements of the whole actual structure (i. e. $\Omega_I \cup \Gamma_p$).

Fictitious Reduction of Boundary Degrees-Of-Freedom

As pointed out in the previous section all the BEM-like methods described so far are "exact", since no approximations are introduced. We will now introduce in this discrete context the concept of boundary modelling typical of the BEM methods for the continuum, and thus the concept of approximate solutions. For simplicity we restrict this discussion to the case of the direct method. Besides, as in the previous section, reference is to trusses in 2D.

Let the actual boundary quantities $\mathbf{u}_{\Gamma_I}$ and $\mathbf{p}_I$ be approximated as

$$\mathbf{u}^r_{\Gamma_I} = H_u \mathbf{u}^*_{\Gamma_I}, \quad \mathbf{p}^r_I = H_p \mathbf{p}^*_I. \tag{29}$$

The above equation are to be understood in the following sense: the vector $\mathbf{u}^*_{\Gamma_I}$ and $\mathbf{p}^*_I$, collect $2r$ modelling parameters each—the displacements and resultant forces at r nodes, say—with $r < n$; H_u and H_p are $2n \times 2r$ "interpolation" matrices of rank $2r$, which give displacements and tractions in all the Γ-nodes, in terms of modelling parameters. Thus we are trying to approximate vectors $\mathbf{u}_{\Gamma_I}$ and $\mathbf{u}_{\Gamma_C}$, which belong to the $2n$-dimensional euclidean vector space $\mathbf{R}^{2n}$, by vectors $\mathbf{u}^r_{\Gamma_I}$ and $\mathbf{p}^r_I$, which belong to a $2r$-dimensional subspace of $\mathbf{R}^{2n}$. When this approximation is sustituted in (26), a system of $4n$ equations in $4r$ unknowns is obtained:

$$\mathbf{0} = \begin{bmatrix} G^{uu} H_p & -G^{up}_C H_u \\ -G^{pu}_C H_p & G^{pp} H_u \end{bmatrix} \begin{Bmatrix} \mathbf{p}^*_I \\ \mathbf{u}^*_{\Gamma_I} \end{Bmatrix}. \tag{30}$$

Now we have to "project" the above $4n$ equations on $\mathbf{R}^{4r}$ to recover a square system of equations. One could keep the $4r$ rows of (30) which correspond to equations written at the r nodes chosen as modelling parameters; however this choice would destroy the symmetry properties of (26). This approach is the counterpart to the nodal pointwise enforcement (collocation) of the integral equations for a BE model of a continuum.

If we instead introduce weighting vectors $\mathbf{w}_u = H_u \mathbf{w}_u^*$ and $\mathbf{w}_p = H_p \mathbf{w}_p^*$, and premultiply (30) by $\left[\mathbf{w}_p^T \quad \mathbf{w}_u^T\right]$, we obtain, for arbitrary $\mathbf{w}_p^*$ and $\mathbf{w}_u^*$

$$\mathbf{0} = \begin{bmatrix} H_p^T G^{uu} H_p & -H_p^T G_C^{up} H_u \\ -H_u^T G_C^{pu} H_p & H_u^T G^{pp} H_u \end{bmatrix} \begin{Bmatrix} \mathbf{p}_I^* \\ \mathbf{u}_{\Gamma_I}^* \end{Bmatrix}. \tag{31}$$

This procedure is the counterpart of the Galerkin weighted residual method applied to the BE model of a continuum. Note that in (31) the two diagonal submatrices are obviously symmetric, while, for the off-diagonal submatrix, the following relation, derived using (15) and (17), holds:

$$\left(H_p^T G_C^{up} H_u\right)^T = H_u^T G_I^{pu} H_p = H_u^T G_C^{pu} H_p + H_u^T H_p. \tag{32}$$

If H_p is chosen so that $H_u^T H_p = I$, i. e. if the $\mathbf{p}_I^*$ are generalized forces, exactly the same symmetry properties of (26) are recovered. Following the same steps outlined in the previous section, we would be able to generate a symmetric direct method, like the one described by (28), also for this Galerkin approximation of (26). It could be easy to show that in general this is the only way by which we are able to generate a symmetric method from (30).

Although the Galerkin approach preserves symmetry of the underlying formulation, other properties get lost. For example the rank of (31) is $4r$ and not $2r$ since for the matrices involved in it, an equation equivalent to the second of (16) does not hold:

$$H_u^T G^{pp} H_u \neq H_u^T G_C^{pu} H_p \left(H_p^T G^{uu} H_p\right)^{-1} H_p^T G_C^{up} H_u. \tag{33}$$

Therefore the choice of the equations to select from (31) in generating a square system does affect the solution.

References

[1] Banerjee, P. K.; Butterfield, P. K.: *Boundary Element Methods in Engineering Science.* McGraw-Hill Book Co. 1981.

[2] Hartmann, F.; Katz, C.; Protopsaltis, B.: Boundary elements and symmetry. *Ingeniuer-Archiv*, 55 (1985) 440–449.

[3] Maier, G.; Novati, G.; Sirtori, S.: On symmetrization in boundary element elastic and elastic-plastic analysis. In Kuhn, G.; Mang, H. (eds.): *Discretization Methods in Structural Mechanics*, Proc. IUTAM Symposium Vienna 1989, 191–200.

Time-Harmonic Elastic-Wave Scattering: The Role of Hypersingular Boundary Integral Equations

G. Krishnasamy, F. J. Rizzo

University of Illinois
Department of Theoretical and Applied Mechanics
Urbana, Illinois 61801, USA

Summary

Some numerical data for scattering of elastic waves from cracks are presented using a hypersingular boundary integral formula. Then it is shown how the appropriate hypersingular formulas, needed for a formulation for elastodynamic scattering from any void shape, valid at all frequencies, may be derived from the hypersingular formula for cracks.

Introduction

The purpose of this paper is to present some of the authors' experience in formulating and solving the problem of scattering of time-harmonic elastic waves from voids and cracks, using boundary integral equations (BIE's) with solution via the boundary element method (BEM). Specifically, to fix ideas, think first of a flattened "ellipsoidal void" model of a crack. This crack-like model, since it contains finite volume inside of the ellipsoidal surface, has none of the mathematical difficulties associated with a zero volume model or true crack, wherein the crack surfaces occupy essentially the same place (cf. [1]). Thus for the crack-like model we may use a conventional BIE for time-harmonic scattering from voids (e.g. [2],[3]). For the zero-volume model, however, the conventional BIE formulation degenerates and the preferred approach seems to be, and the one taken here is, to employ a hypersingular integral formula (e.g. [4],[5],[6]) wherein the unknown is the crack-opening displacement.

A BIE-BEM attack on crack problems with each model presents certain difficulties. In the void model there are difficulties with the model itself not containing the crack-edge singularity, although the absence of this is probably not important in the far field. It is essential though in computing the stress intensity factor at the crack edge. Nevertheless, for many problems the void model is a better representation of the physical situation. If the void model is very flat (little volume) there are numerical difficulties associated with two surfaces being very close together. This requires fine discretization and high-order Gaussian quadrature over and above that demanded by the wave phenomena at a given frequency. Finally, although the kernel in the BIE's are only Cauchy-singular (rather than hypersingular) for the void model, the BIE's suffer from the well-known fictitious eigenfrequency difficulty (FED) (e.g. [7], [8] [9]) wherein the BIE's (but not the physical problem) have a multiplicity of solutions at a spectrum of discrete frequencies.

In the zero-volume model the above difficulties are not present, but instead one is faced with a hypersingular kernel in the BIE's such that the integrals are usually regularized in some way before computation is attempted. Interpretation of such integrals before regularization as Hadamard finite-part integrals is now becoming common, but computations for crack problems seem to follow regularization or an analytical evaluation of representative finite-part integrals over elements, rather than use a direct numerical attack on the finite part integrals (e.g. [10]). This paper uses a recently developed approach to hypersingular BIE's, and some new data from our formulas are presented for scattering from circular and elliptical cracks.

One of the main points in this paper, however, is to observe that one way, and we believe the best way (cf. [11] for similar work in acoustics) to remove the FED from the crack-like formulation, is to append a hypersingular BIE to the conventional one, as suggested by Burton and Miller [7] for acoustics, and thus remove the FED altogether. Following procedures of acoustics, such a hypersingular BIE could be derived separately and regularized separately, perhaps with the aid of so-called special solutions (cf. [12]). However, the necessary hypersingular BIE is derivable from the hypersingular BIE used for the zero-volume crack model discussed above, and the derivation is shown here for the first time. In any case, the *vector* hypersingular equation for *non-crack scattering*, however obtained for purposes for the FED, seems to be fairly rare in the literature—the authors are sure of only one reference, i.e. [13].

Therefore hypersingular BIE's play two roles in this paper: 1) as an ingredient in the Burton and Miller procedure for removal of the FED for scattering from voids and void-models of cracks, and 2) to formulate the zero- volume true-crack model. Although the purpose and the model giving rise to hypersingular BIE's in each case is quite different (one a closed surface containing finite volume, the other open coincident surfaces), we show that having the BIE's for cracks 2), those for voids 1) are easily obtained without a separate derivation.

Thus in the next section we present the hypersingular BIE formula for a zero- volume crack model and present some crack-opening-displacement results for a circular-and elliptical-crack model impinged upon by a plane longitudinal wave of a particular frequency. The reader should compare the complexity of the hypersingular BIE used here with the Cauchy-singular BIE as found in [2] or, more recently in the context of the non-zero volume crack-like model, in [3].

Then, in the following section, we obtain the necessary hypersingular BIE needed to remove the FED from non-zero volume cracklike scatterers (or any other non-zero volume scatterers) via the Burton and Miller formulas; and this BIE is obtained directly from the one for cracks.

Ongoing work with these ideas include verifying the Burton and Miller formulation, free of the FED, for scattering of elastic waves, from any shape, scattering of elastic waves from non-planar and multiple cracks, and comparison of data in the near and far field from both crack-like and zero-volume crack models.

Hypersingular formula for cracks

When solving for the scattered field from a stress-free crack in an infinite elastic medium, a physical crack with two (coincident) surfaces S^+ and S^- is envisioned with Δu the crack opening displacement defined over S where S is either S^+ or S^- as desired (cf. [4]), (see Fig. 1). If C_{ijkl} is the tensor which describes the elastic medium, Δu is sufficiently smooth at ξ_o a point on S (see [4]), G_{km} and G^D_{km} are the static and dynamic Greens functions, and the scattered field satisfies the radiation condition, then the boundary integral equation which gives the traction on the crack surface S^+ is

$$-t^s_p(\xi^+_0) = t^i_p(\xi^+_0) = -C_{pqmr}n_q(\xi^+_0)\left\{-\int_S\left[\frac{\partial T^D_{im}}{\partial\xi_r} - \frac{\partial T_{im}}{\partial\xi_r}\right]\Delta u_i(x)dS\right.$$

$$-\int_S \frac{\partial T_{im}}{\partial \xi_r}[\Delta u_i(x) - \Delta u_i(\xi_0) - \Delta u_{i,p}(\xi_0)(x_p - \xi^+_{0p})]dS$$

$$\left.+\Delta\sigma_{ij}(\xi_0)\int_S \frac{\partial G_{im}}{\partial x_r}n_j(x)dS + I_c(\xi^+_0)\right\}. \tag{1}$$

Here the superscript 'i' and 's' corresponds to the incident and scattered field, $n(x)$, the normal to S is same as that of S^+, $\Delta u_i(\xi_0) = u_i(\xi^+_0) - u_i(\xi^-_0)$, $T_{im} = C_{ijkl}G_{km,l}n_j$, $\Delta\sigma_{ij} = C_{ijkl}\Delta u_{k,l}$ and $I_c(\xi_0)$ includes all terms which involve line integrals over the crack edge ([14] Eqs. (21) and (22)). Here the frequency dependence of the G^D_{km} and $\underline{u}$ are suppressed and $u_i = u^i_i + u^S_i$. Given the incident field $\underline{u}^i$, $\underline{t}^i(\xi_0)$, one can solve the above equation for Δu_i through the BEM. The BEM involves discretizing the geometry by elements, approximating Δu_i over the geometry and forming a system of algebric equation which is then solved for Δu_i [1]. To be consistent with theory which resulted in Eq. (1), it is proper to use nonconforming or spline elements to approximate Δu_i so that Δu_i is sufficiently smooth at the collocation point [4]. Also the appropriate behavior of the solution near the crack tip is built into the elements bordering the crack edge.

The normalized crack opening displacement for a penny shaped crack due to a normal incident plane wave, $K_T a = 3.2$ and 4.4 where K_T is the transverse wave number and a is the radius of the crack, is shown in Figs. 2-3. This was obtained by modeling a penny-shaped crack by 25 nonconforming eight-node elements with three concentric rings of eight elements each and one square element at the center. The results thus obtained compare very well with theory. The use of nonconforming elements result in discontinuities at element edges which are usually small, and these can be avoided by using spline elements. The normalized crack-opening displacement due to a plane wave at normal incidence on an ellipticа crack with an aspect ratio of $\sqrt{2}$ is shown in Fig. 4 for a $K_T a = 4.5$. These results compare well with [14].

Hypersingular formula for voids

So far we have shown that the regularized hypersingular boundary integral equation, Eq. 1, is useful to solve scattering problems involving cracks. Here we will extend this equation for a void. One can think of a void as the limiting form of a volume trapped by a crack whose edges merge. A 2-D equivalent is shown in Fig. 5. This results in an edge of zero length and so $I_c(\xi_o^+) = 0$. Since we are interested in the equation for a void enclosed by the S^+ surface involving u_i corresponding to S^+, we can write Eq. 1 as

$$t_p^i(\xi_o^+) = -C_{pqmr} n_q(\xi_o^+) \Bigg\{ -\int_{S^+} \left[\frac{\partial T_{im}^D}{\partial \xi_r} - \frac{\partial T_{im}}{\partial \xi_r} \right] u_i(x) dS$$

$$-\int_{S^+} \frac{\partial T_{im}}{\partial \xi_r} [u_i(x) - u_i(\xi_o^+) - u_{i,p}(\xi_o^+)(x_p - \xi_{op}^+)] dS$$

$$+\sigma_{ij}(\xi_o^+) \int_{S^+} \frac{\partial G_{im}}{\partial x_r} n_j(x) dS$$

$$+\int_{S^-} \left[\frac{\partial T_{im}^D}{\partial \xi_r} - \frac{\partial T_{im}}{\partial \xi_r} \right] u_i(x) dS$$

$$+\int_{S^-} \frac{\partial T_{im}}{\partial \xi_r} [u_i(x) - u_i(\xi_o^-) - u_{i,p}(\xi_o^-)(x_p - \xi_{op}^+)] dS$$

$$-\sigma_{ij}(\xi_o^-) \int_{S^-} \frac{\partial G_{im}}{\partial x_r} n_j(x) dS \Bigg\}. \tag{2}$$

For the closed surface formed by S^-, ξ_o^+ is a point outside the volume enclosed by S^-, and so for a stress free surface the reciprocal theorem gives

$$0 = \int_{S^-} T_{im}^D(x, \xi_o^+) u_i(x) dS. \tag{3}$$

Taking the gradient of the above equation and regularizing gives

$$0=\int_{S}\left[\frac{\partial T_{im}^{D}}{\partial \xi_{r}}-\frac{\partial T_{im}}{\partial \xi_{r}}\right]u_{i}(x)dS$$

$$+\int_{S^-}\frac{\partial T_{im}}{\partial \xi_{r}}[u_{i}(x)-u_{i}(\xi_{o}^{-})-u_{i,p}(\xi_{o}^{-})(x_{p}-\xi_{op}^{+})]dS$$

$$+u_{i}(\xi_{o}^{-})\int_{S^-}\frac{\partial T_{im}}{\partial \xi_{r}}dS$$

$$+u_{i,p}(\xi_{o}^{-})\int_{S^-}\frac{\partial T_{im}}{\partial \xi_{r}}(x_{p}-\xi_{op}^{+})dS. \tag{4}$$

It can be shown by the use of Stokes' theorem that the last two terms above can be written

$$u_{i}(\xi_{o}^{-})\int_{S^-}\frac{\partial T_{im}(x,\xi_{o}^{+})}{\partial \xi_{r}}dS=-u_{i}(\xi_{o}^{-})\oint_{c}\varepsilon_{jrq}C_{ijkl}\frac{\partial G_{km}}{\partial x_{l}}dx_{q} \tag{5}$$

and

$$u_{i,p}(\xi_{o}^{-})\int_{S^-}\frac{\partial T_{im}}{\partial \xi_{r}}(x_{p}-\xi_{op}^{+})dS=-\int_{S^-}\frac{\partial G_{im}}{\partial x_{r}}\sigma_{ij}(\xi_{o}^{-})n_{j}(x)dS$$

$$-\sigma_{kl}(\xi_{o}^{-})\oint\varepsilon_{rlq}G_{km}dx_{q}$$

$$-u_{i,p}(\xi_{o}^{-})\oint\varepsilon_{jrq}C_{ijkl}\frac{\partial G_{km}}{\partial x_{l}}(x_{p}-\xi_{op}^{+})dx_{q}$$

$$+u_{i,r}(\xi_{o}^{-})\int_{S^-}T_{im}(x,\xi_{o}^{+})dS. \tag{6}$$

where C is the boundary to the surface S^-. Since S^- is a closed surface there is no boundary and all the integrals over C are zero. The point ξ_o is outside the closed surface S^- and so the last integral of Eq. 6 is zero. On substituting Eqs. 5 and 6 in Eq. 4 we have

$$0=\int_{S^-}\left[\frac{\partial T_{im}^{D}}{\partial \xi_{r}}-\frac{\partial T_{im}}{\partial \xi_{r}}\right]u_{i}(x)dS$$

$$+\int_{S^-}\frac{\partial T_{im}}{\partial \xi_{r}}[u_{i}(x)-u_{i}(\xi_{o}^{-})-u_{i,p}(\xi_{o}^{-})(x_{p}-\xi_{op}^{+})]dS$$

$$-\int_{S^-}\frac{\partial G_{im}}{\partial x_{r}}\sigma_{ij}(\xi_{o}^{-})n_{j}(x)dS. \tag{7}$$

Hence it is clear from the above Eq. that the last three terms of Eq. 2 add up to zero and the resulting equation for the void is

$$t_p^i(\xi_o^+) = C_{pqmr} n_q(\xi_o^+) \left\{ - \int_{S^+} \left[\frac{\partial T_{im}^D}{\partial \xi_r} - \frac{\partial T_{im}}{\partial \xi_r} \right] u_i(x) dS \right.$$

$$- \int_{S^+} \frac{\partial T_{im}}{\partial \xi_r} [u_i(x) - u_i(\xi_o^+) - u_{i,p}(\xi_o^+)(x_p - \xi_{op}^+)] dS$$

$$\left. + \sigma_{ij}(\xi_o^+) \int_{S^+} \frac{\partial G_{im}}{\partial x_r} n_j(x) dS \right\}. \qquad (8)$$

It is interesting to note that if the crack were to be bent the other way, with S^+ to the inside and S^- outside, then the resulting equation is valid for the equivalent interior problem.

The regularized hypersingular BIE, Eq. 8, can now be combined with the conventional BIE to result in an equation, for scattering from any void shape, which is free from FED (cf. [13]).

Acknowledgement

Partial support for this work was provided by the Solid Mechanics Program of the US Office of Naval Research, Y. Rajapakse program official, and by the National Science Foundation, O. W. Dillon program official. Thanks are due Y. Liu of the University of Illinois for several valuable conversations.

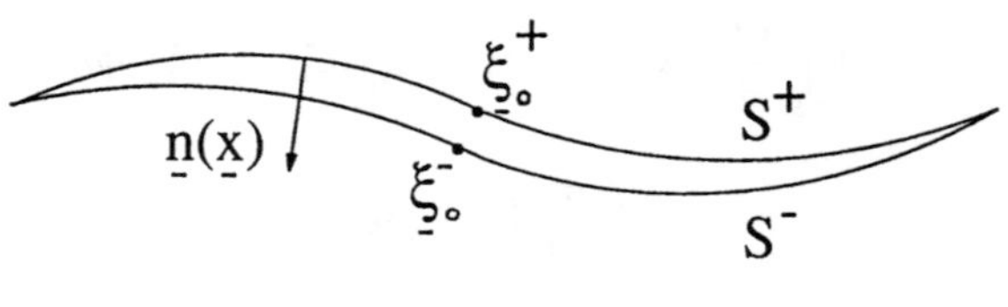

Fig. 1 A 2-D crack.

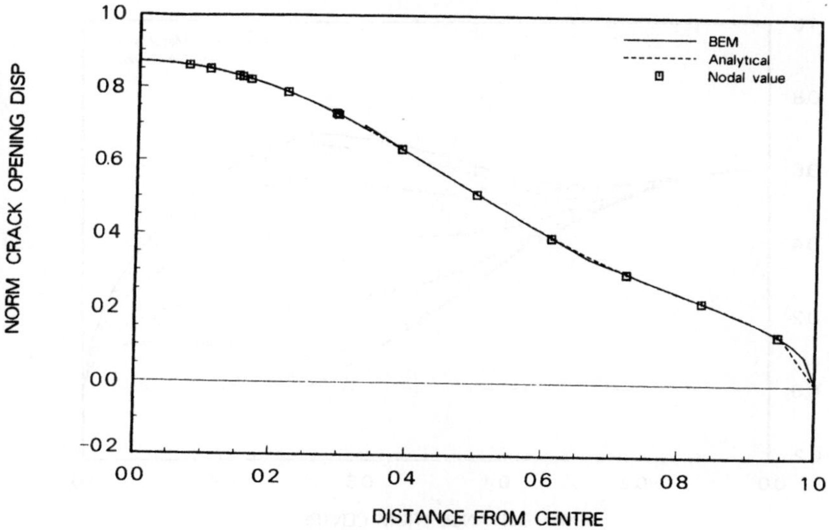

Fig. 2 Normalized crack opening displacement for penny shaped crack and $K_T a = 3.2$

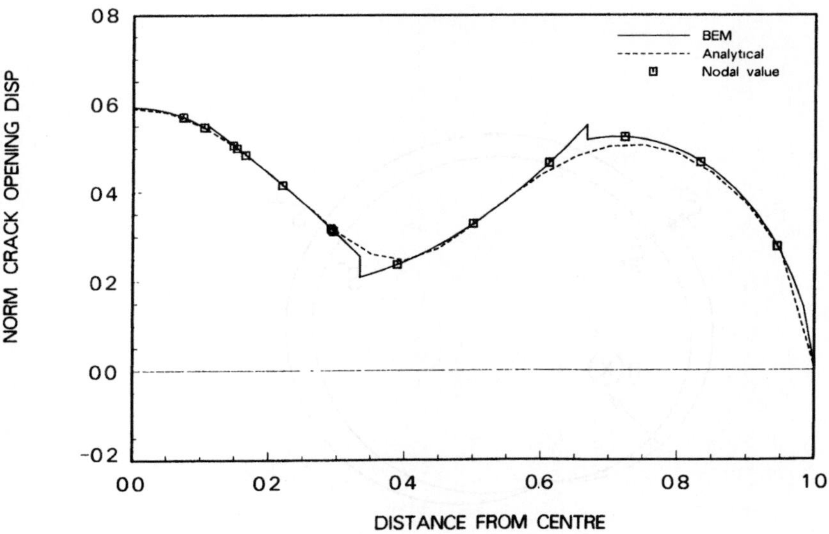

Fig. 3 Normalized crack opening displacement for penny shaped crack and $K_T a = 4.4$

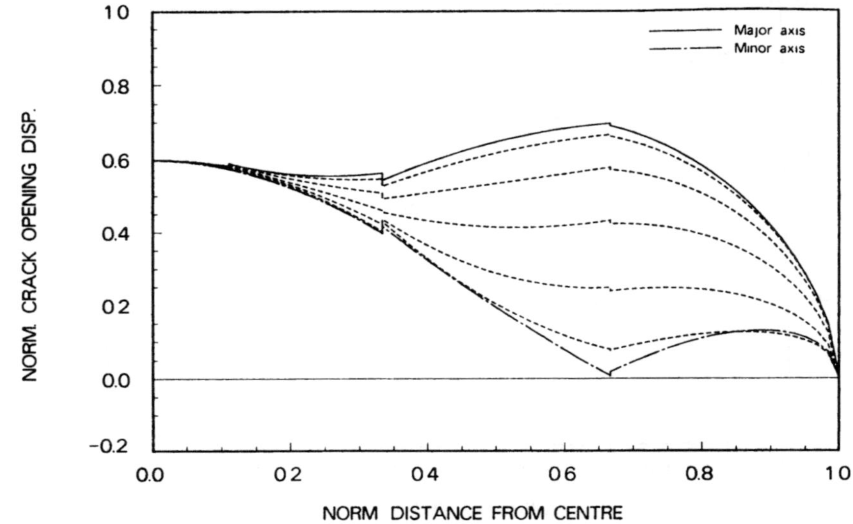

Fig. 4 Normalized crack opening displacement for an elliptical crack and $K_T a = 4.5$

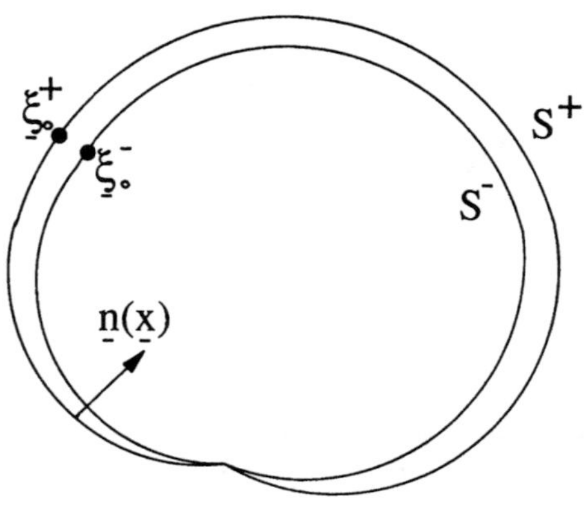

Fig. 5 A 2-D crack with its edge merged

References

1. Cruse, T. A., Van Buren, W.: Three-dimensional elastic stress analysis of fracture specimen with an edge crack. Int. J. Fracture Mech. 7 (1971) 1-15.
2. Rizzo, F. J., Shippy, D. J., Rezayat, M.: A boundary integral equation method for radiation and scattering of elastic waves in three dimensions. International Journal for Numerical Methods in Engineering 21 (1985) 115-129.
3. Schafbuch, P. J., Thompson, R. Bruce, Rizzo, F. J.: Application of the boundary element method to elastic wave scattering by irregular defects. To appear in J. of Nondestructive Evaluation.
4. Krishnasamy, G., Schmerr, L. W., Rudolphi, T. J., Rizzo, F.J.: Hypersingular boundary integral equations: Some applications in acoustic and elastic wave scattering. ASME J. of Appl. Mech. 57 (1990) 404-414.
5. Gray, L. J., Martha, Luiz F., Ingraffea, A. R.: Hypersingular integrals in boundary element fracture analysis. Int. J. of Num. Meth. in Engg. 29 (1990) 1135-1158.
6. Polch, E. Z., Cruse, T. A., Huang, C. J.: Traction BIE solutions for flat cracks. Comput. Mech. 2 (1987) 253-267.
7. Burton, A. J., Miller, G. F.: The application of integral equation methods to the numerical solution of some exterior boundary-value problems. Proc. Roy. Soc. Lond. A 323 (1971) 201-210.
8. Martin, P. A.: Identification of irregular frequencies in simple direct integral-equation methods for scattering by homogeneous inclusions. To appear in Wave Motion (1989).
9. Gonsalves, I. R., Shippy, D. J., Rizzo, F. J.: The direct boundary integral equation method for the three-dimensional elastodynamic transmission problem. To appear in Computers and Mathematics with Applications.
10. Kutt, H. R.: The numerical evaluation of principal value integrals by finite-part integration. Numer. Math. 24 (1975) 20-210.
11. Chien, C. C., Rajiyah, H., Atluri, S. N.: An effective method for solving the hypersingular integral equations in 3-D acoustics. J. Acoust. Soc. Am. 88(2) (1990) 918-936.
12. Liu, Y., Rudolphi, T. J.: Some identities for fundamental solutions and their applications to non-singular boundary element formulations. To appear in J. Engg. Analysis.
13. Jones, D. S.: Boundary integrals in elastodynamics, IMA J. Appl. Math. 34(1) (1985) 83-97.
14. Budreck, D. E., Achenbach, J. D.: Scattering from three-dimensional planar cracks by the boundary integral equation method. ASME J. of Appl. Mech. 55 (1988) 405-411.

A Numerical Solution of II Kind Fredholm Equations: A Naval Hydrodynamics Application

F.Lalli†, E.Campana‡, U.Bulgarelli †.

†INSEAN, *Italian Ship Model Basin*,
Via di Vallerano 139, 00128 Rome (Italy).

‡IBM ECSEC, *European Center for Scientific and Engineering Computing*,
Via del Giorgione 159, 00142 Rome (Italy).

Summary

The aim of the present work is to describe a numerical method, based on the simple layer integral formulation, whose main applications are in naval hydrodynamics, for ship hulls design purposes. The solution of the 2D fully nonlinear problem is obtained by means of an iterative procedure while in the 3D case a linear problem is solved. Under some hypotheses, viscosity effects at the free surface are considered. All the obtained numerical results have been tested with analytical solutions and experimental results.

I. Introduction

In 1977 an important contribution to the numerical approach for the ship wave problem was given by Dawson[3], who proposed a very simple and effective numerical procedure, based on the simple layer potential formulation; may be this method could be considered as the natural extension to naval hydrodynamics of the method of Hess and Smith. The traditional approach in naval hydrodynamics was based on the use of a very complicated Green function, satisfying the linearized free surface boundary conditions. Two important advantages are therefore connected with Dawson idea: simple layer potential is very easy to treat computationally, and the distribution of panels on the free surface allows one to consider different boundary conditions; at last, the extension to the nonlinear case is also possible, without introducing any substantial variation in the procedure, which can maintain the original simplicity also for the nonlinear problem. This method has become quite popular for its simplicity and effectiveness; in the wake of Dawson, several numerical papers appeared then in literature. Some Authors deal with the linear problem, discussing about consistency of the different kinds of linearization, others on the enforcement of radiation condition, while in[6] general criteria are suggested to study the numerical properties (stability, numerical damping and dispersion) of the discrete schemes.

Many Authors have pointed out the limitations of the linear formulation: starting from Dawson solution, the subsequent aim is suggested to be the solution of the nonlinear problem. Someone deal with the 2nd order problem, obtained by means of Taylor expansion. At the same time some researchers started to work on the fully nonlinear problem . In [4,5] moving panels methods are proposed, in which the exact nonlinear conditions are applied on the free surface, computed step by step, but the features of the numerical method were not stressed and no convergency problem was mentioned. Nevertheless severe difficulties will appear as soon as any attempt to update iteratively the linear solution is made; these difficulties grow, of course, with the increase of Froude number.

The range of applicability of moving panel methods depends on the particular numerical scheme adopted. We propose an algorithm which enlarges this range[7], developing a new dynamic boundary condition in which, under some hypotheses, the effects of viscosity at the free surface are taken into account.

II. Mathematical formulation

We consider the mathematical formulation governing the 2D steady state potential flow, due to the motion of a submerged non-lifting body $\mathcal{B}$ in a fluid of infinite depth. The extension for three dimensional cases can be easily obtained. The fluid domain $\mathcal{D}$ is bounded on the upper part by a free boundary $\mathcal{S}$, whose cartesian equation is $y = \eta(x)$, and it is unbounded in the other directions. The frame of reference is assumed to be fixed with the body: the x axis is oriented as the uniform stream $\mathbf{U} = (U, 0)$, the y axis is positive upwards and the undisturbed free surface, is given by $y = 0$ (see fig.1). Thus:

$$\nabla^2\phi(x,y) = 0, \quad inside\ \mathcal{D} - \overline{\mathcal{B}} \tag{1}$$

$$\phi_n(x,y) = 0 \qquad on \quad \partial\mathcal{B} \tag{2}$$

$$\phi_x(x,y)\eta_x(x) = \phi_y(x,y) \qquad on \quad \mathcal{S} \tag{3}$$

$$\eta(x) = \frac{1}{2g}[U^2 - \nabla\phi(x,y)\cdot\nabla\phi(x,y)] \qquad on \quad \mathcal{S} \tag{4}$$

$$\lim_{x\to-\infty} |\nabla\phi| = U \tag{5}$$

where $\phi(x,y) = Ux + \varphi(x,y)$.

For computational purposes, we need to obtain a unique boundary condition for $\mathcal{S}$, by eliminating $\eta(x)$ between (3) and (4):

$$\frac{1}{2}\nabla\phi(x,\eta)\cdot\nabla(\nabla\phi(x,\eta)\cdot\nabla\phi(x,\eta)) + g\phi_y(x,\eta) = 0 \tag{6}$$

The problem described so far can be revisited considering a curvilinear abscissa l defined along the free boundary which, in the steady state case, is a streamline of the flow. Therefore the set of boundary conditions (3),(4),(6) can be rewritten as follows[7]:

$$\phi_l(x,\eta)\eta_l(x) = \phi_y(x,\eta) \tag{7}$$

$$\eta(x) = \frac{1}{2g}[U^2 - \phi_l^2(x,\eta)] \tag{8}$$

$$\phi_l^2(x,\eta)\phi_{ll}(x,\eta) + g\phi_y(x,\eta) = 0 \tag{9}$$

Furthermore, by introducing viscosity effects [7], the following dynamic condition can be obtained:

$$\eta(x) = \frac{1}{2g}[U^2 - (1+\lambda(x)\ )\ \phi_l^2(x,\eta)\ ] + \Lambda(x,\eta) \tag{10}$$

where:

$$\lambda = -4\delta K \quad ; \quad \Lambda = 2\frac{\nu}{g\delta}\int_{-\infty}^{l} K\phi_{l'}\, dl'$$

in which K is the free surface curvature and δ is the thickness of the superficial boundary layer. The unified free surface boundary condition is obtained from (7) and (10):

$$[1+\lambda(x)]\ \phi_l^2(x,\eta)\ \phi_{ll}(x,\eta) + g\phi_y(x,\eta) = \mu(x,\eta)\phi_l^2(x,\eta) \tag{11}$$

in which we have posed $\mu = 2\ [\delta K_l\phi_l + (\nu/\delta)K]$. We observe that the unified condition (6) and the inviscid dynamical condition (4) can be obtained respectively from (11) and (10), in the limiting case of $\delta \to 0$.

The fully nonlinear free surface flow will be computed by means of a numerical scheme implying an iterative procedure: the 1st step will be the solution of the linearized problem.

The simplest linear formulation is obtained assuming the flow to be a small perturbation of the uniform stream; neglecting the nonlinear terms both in (3) and in (4) the Neumann-Kelvin formulation is obtained. This kind of linearization can give reasonable results not only when the body velocity is small enough, but also if the body is either very slender or deeply submerged. Dawson[3] proposed another type of linearized formulation which seems to be more realistic for bodies moving near the free surface or, at least, for floating ones. As basis fluid motion he assumed the free surface flow past the body with the Froude number equal to zero.

III. Numerical method and results

The numerical solution of the mathematical model given in the 1st paragraph is computed by means of a simple layer formulation:

$$\varphi(x,y) = \int_{\partial B} \sigma(\xi_1,\xi_2)\log r d\Sigma + \int_S \sigma(\xi_1,\xi_2)\log r d\Sigma \tag{12}$$

where $r = \sqrt{(x-\xi_1)^2 + (y-\xi_2)^2}$.

To describe the discretization technique, we consider now the 2-D linear problem of the free surface flow past a submerged circular cylinder (a and h are respectively the radius and the depth). For the linear case the method of images is used; the analytical solutions of this problem are given in[1]. Discretizing the Neumann Kelvin condition we obtain an integral equation of the 2nd type. For the numerical solution we discretize a finite part of the undisturbed free surface (where the simple layer potential is defined) by means of the collocation method. A piecewise constant variation is assumed for the unknown function $\sigma(x)$.

$$\sum_{k=1}^{N} \sigma_k [\frac{d}{dx} I_k(x)]_{x=x_i} - \frac{\pi}{Fr^2}\sigma(x_i) = -\frac{4x_i(x_i^2 - 3h^2)}{(x_i^2+h^2)^3} \qquad i = 1,...,N \tag{13}$$

where the Froude number is defined as $U/\sqrt{2ag}$, the variables have been nondimensionalized with respect to U and the diameter. Indicating with L_k the generic linear boundary element of the free surface:

$$I_k(x) = \frac{d}{dx}\int_{L_k} \log \mid x - \xi_1 \mid d\xi_1$$

The derivative d/dx which appears in (13) has been implemented by a 2nd order finite differences scheme[3] of upwind type, in order to enforce the radiation condition (5); the numerical behavior of this scheme has been discussed in [6], where it has been shown that the numerical damping and dispersion are respectively of 5th and 1st order. The scheme converges linearly, that is, is of 1st order.

The term $I_k(x)$ can also be derived analytically[7]: in this case, the radiation condition (5) must be imposed explicitly; numerical experiments have shown that the following conditions must be satisfied at the 1st node upstream, to avoid numerical oscillations of the free surface before the body: $\varphi_{xx}(x_1,0) = 0$, $\varphi_y(x_1,0) = 0$. The numerical free surface profile is plotted in fig.2a and 2b, in comparison with the analytical solution given in [1].

In fig.2a the numerical solution obtained by means of finite differences implementation is plotted, while fig. 2b refers to the method implying analytical derivative of the term $I_k(x)$. In fig. 2a, the numerical properties of the scheme can be observed: the wave amplitude is retained with good accuracy, but for the wavelength the behavior is not so good. Such behavior can be improved with the second method described before, as shown in fig.2b. In fig. 3 the same comparison is shown in the 3D case of a submerged dipole using the analitycal derivative scheme.

In the non linear problem the body surface $\partial\mathcal{B}$ and part of the free surface $\mathcal{S}$, which is an unknown of the problem, are discretized by means of linear elements; as in the linear case, the simple layer density σ is assumed to be constant on every boundary element. During the iterative procedure the free boundary $\mathcal{S}$ is 'followed', step by step, by updating its discretization. At the 1st step, to initialize the procedure, the potential flow and the shape of $\mathcal{S}$ are computed with the linear Dawson formulation.

Details on the iterative scheme can be found in [7]. Fig.1 shows the test case hydrofoil [2,7] used for the computations performed in this work. In our computations, we have arranged 256 panels on the body, (using a cosine type stretching) and 700 panels on the free surface, with 60 elements per wave length; 1/3 of the discretized free surface lies upstream with respect to the leading edge of the body.

The introduction of viscosity effects improves the numerical behavior of the algorithm without causing any significant change in the values of the wave resistance; in fact, as shown in fig.4, the wave amplitude is very lightly damped. In the same figure the typical nonlinear effect of wave steepness can be observed; moreover, with respect to the linear case, a wave length reduction does appear.

The numerical wave profiles obtained by means of the method including viscous effects, for different depths and Froude numbers, are compared in fig.5 with the experimental data taken from [2] and with the Dawson linear solution. The agreement with the measurements is quite good and the improvement with respect to the linear solution is significant. The importance of nonlinear effects in the wave resistance problem is also observed in fig.6. Finally fig.7 is shown the numerical linear solution of the free surface flow past a mathematical ship hull (Wigley model).

Acknowledgements

This work was supported by the Italian Ministry of Merchant Marin in the frame of the INSEAN research plan 1986.

We would like to acknowledge Prof. P. Bassanini, University of Roma, and Prof. G. Monegato, University of Torino, for their kind and helpful suggestions and criticisms.

References

1) Havelock T. H., The Wave Pattern of a Doublet in a Stream, *Proc. Royal Society*, A, Vol.121, 1928.

2) Salvesen N., On Second Order Wave Theory for Submerged Two-Dimensional Bodies,*6th Sym. on Naval Hydro.*, Washington, 1966.

3) Dawson C. W., A Practical Computer Method for Solving Ship-Wave Problems,*2nd Int. Conf. on Numerical Ship Hydro.*, Berkeley, 1977.

4) Daube O., Dulieu A., A Numerical Approach of the Nonlinear Wave Resistance Problem,*3rd Int. Conf. on Numerical Ship Hydro.*, Paris, 1981.

5) Rong H., Liang X., Wang H., A Numerical Method for Solving Nonlinear Ship-Wave Problem,*ITTC*, Kobe, 1987.

6) Sclavounos P. D., Nakos D. E., Stability Analysis of Panel Methods for Free Surface Flows with Forward Speed,*17th Sym. on Naval Hydro.*, The Hague, 1988.

7) Lalli F., Campana E., Bulgarelli U., Numerical Simulation of Fully Nonlinear Steady Free Surface Flow, to be published on *Int. Jou. Num. Methods in Fluid.*.

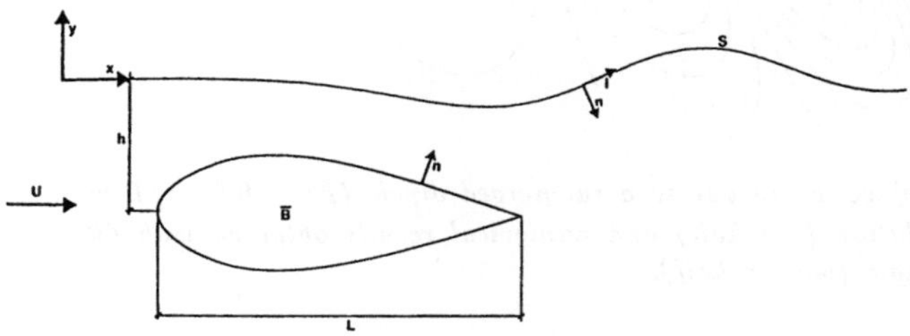

Fig.1:*Definition sketch of the problem.*

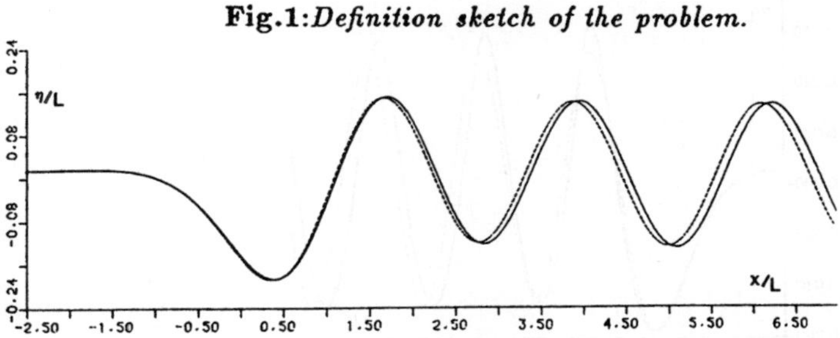

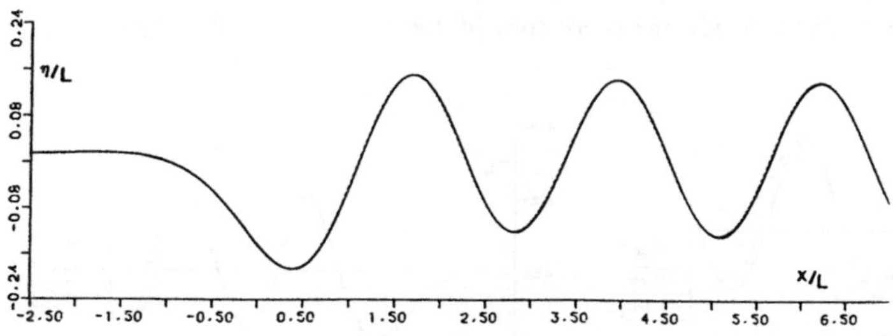

Fig.2 :*2-D surface waves due to a submerged dipole* ($Fr = 0.6$, $h/L = 1.5$)*: analytical solution (solid line) and numerical results obtained with 60 panels per wave length (dashed lines). (a): finite differences scheme, (b): analytical derivative scheme.*

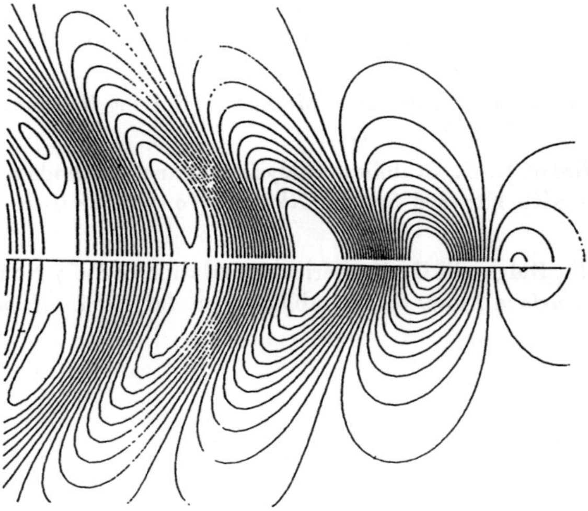

Fig.3 *:3-D surface waves due to a submerged dipole ($Fr = 0.9$, $h/L = 1.3$): analytical solution (top half) and numerical results obtained with 60 panels per wave length (bottom half).*

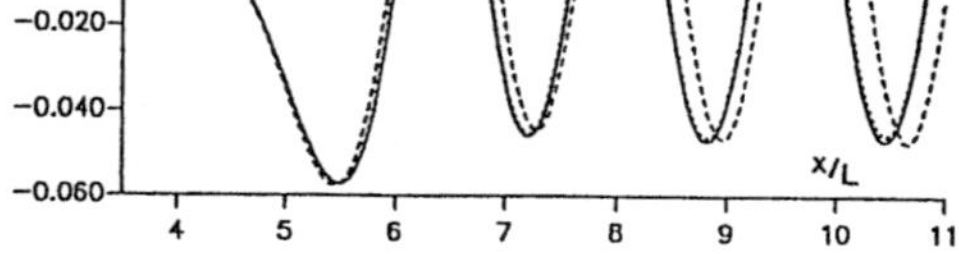

Fig.4: *Computed wave profiles: linear (dashed), inviscid non linear (solid) and non linear with viscous correction (dotted). $Fr = 0.59092$, $h/L = 1.37615$*

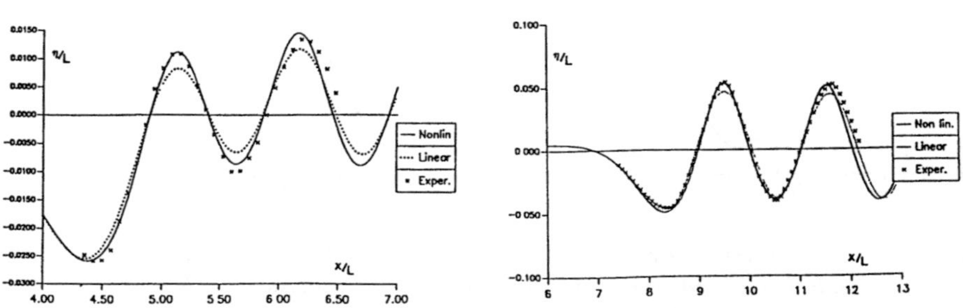

Fig.5: *A comparison between the experimental free surface profiles*[2] *(×) and the numerical solutions : linear (dotted) and nonlinear with viscous correction (solid). (a) $Fr = 0.42208$, $h/L = 1.14678$. (b) $Fr = 0.59092$, $h/L = 1.37615$.*

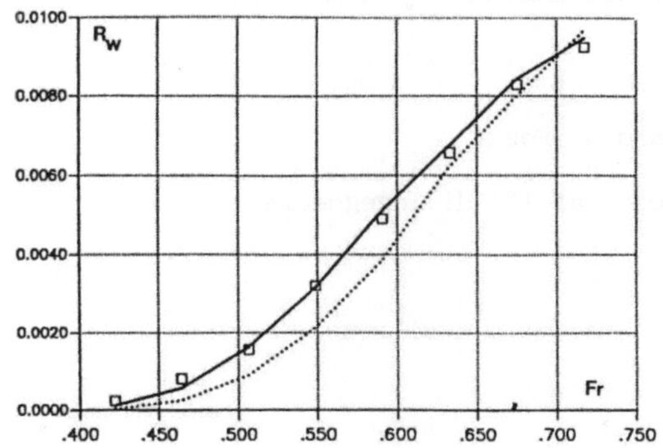

Fig.6:*Experimental (□) and numerical wave resistance: linear (dotted) and nonlinear (solid).*

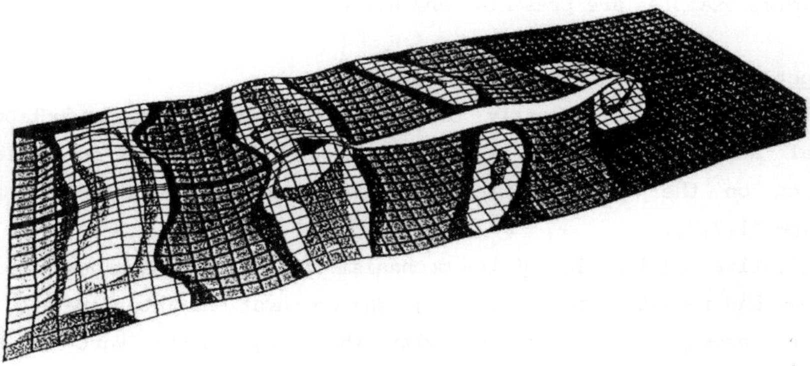

Fig.7:*3D wave pattern due to the steady motion of the Wigley hull, Fr = 0.3 .*

Resistance of a Grooved Surface to Parallel and Cross Flow Calculated by B.E.M.

P. Luchini, F. Manzo and A. Pozzi

Istituto di Gasdinamica, Facoltá di Ingegneria,
University of Naples,
ITALY

Summary

A study is described of both parallel and cross flow in the viscous sublayer generated by a fluid streaming along a grooved surface, with the aim of clarifying the phenomena that underlie the reduction of turbulent drag by such surfaces. A quantitative characterization of the effectiveness of different groove profiles in retarding secondary cross flow is given in terms of the difference of two "protrusion heights." The development of a B.E.M. computer code for the analysis of general profiles is illustrated, and several examples are presented and discussed.

Introduction

One of the methods currently being investigated for drag reduction in internal and external flows is shaping the wall with grooves (or riblets, depending on the way one wishes to see them) cut along the main flow direction [1,2,3].

A qualitative explanation of the mechanism of drag reduction near grooved surfaces has recently emerged [4,5,6]: the corrugations interfere with the secondary cross flow associated with the longitudinal vortices which randomly appear in the turbulent flow, and somehow manage to dampen these vortices and therefore the level of turbulence itself; the consequent reduction in the rate of turbulent diffusion makes for a lower eddy viscosity and the reduction of drag.

Bechert and Bartenwerfer [5] remark that the typical size of corrugations which appear to be experimentally effective is of the same order of magnitude as the height of the viscous sublayer of the turbulent stream. Within the viscous sublayer convective and pressure-gradient terms in the Stokes-Navier equations are negligible compared to the viscous terms, and therefore the flow can be studied in the much simpler framework of the Stokes equations. They argued that the mean longitudinal velocity profile, which is asymptotically linear in the adjacent shear layer, appears as if it originated from an equivalent plane wall located at a distance below the riblet tips which they call "protrusion height", and were able to calculate the protrusion height of a number of riblet configurations for which the Laplace equation can be solved by conformal mapping.

We pursue a further step: since solutions of the Stokes equations superpose linearly, we can calculate the behaviour in the viscous sublayer of flows which have not only a longitudinal but also a transverse, time-varying, component. We shall thus be able to prove the intuitive conjecture that grooved surfaces offer a greater resistance to transverse than they do to longitudinal flow, and to characterize this different resistance quantitatively in terms of a longitudinal and a transverse protrusion height.

To this end we shall present the development of a boundary-element numerical algorithm that calculates the two protrusion heights of an arbitrary corrugated wall.

Formulation of the problem

Our aim in this paper is to study how flow in the viscous sublayer is modified when the plane wall is replaced by a corrugated wall, with corrugations not exceeding the thickness of the viscous sublayer itself. Mathematically, the problem may be stated as follows: we wish to study the Stokes flow of a viscous fluid alongside a cylindrical infinite corrugated wall (represented by the equation $y = y_0(x)$ with y_0 periodic) in the presence of a given shear, or velocity gradient, in the far field.
It is easy to observe that, all quantities being independent of the longitudinal coordinate z, the equation for the longitudinal velocity w decouples from the system and is just the Laplace equation

$$\Delta_2 w = 0 \qquad (1)$$

with boundary conditions $w[x,y_0(x)] = 0$ and $w_y(x,\infty) = \text{constant}$. Eq.(1) with these conditions is the problem that was studied by Bechert et al.[4,5].

The transverse problem for the remaining unknowns u, v and p, which we introduce in this paper, may be reformulated in terms of the streamfunction ψ, defined so that $\psi_y = u$ and $-\psi_x = v$, and of the vorticity $\omega = u_y - v_x$, as the biharmonic equation

$$\Delta_2 \Delta_2 \psi = 0 \qquad (2)$$

with b.c.s $\psi_x[x,y_0(x)] = \psi_y[x,y_0(x)] = 0$, $\psi_{yy}(x,\infty) = \text{constant}$.
By suitably choosing a reference length and velocity, we can nondimensionalize eqs.(1) and (2) in such a way that the period of the corrugations is 2π and the imposed velocity gradient at infinity is unity.

The concept of protrusion height

Through the Fourier-series representation of the general solution of eq.(1) it is possible to show that for $y \to \infty$ the longitudinal velocity approaches the linear behaviour $w \simeq a_0 + y$ with exponential accuracy, and thus imitates the velocity profile produced by a plane wall located at $y = -a_0$. Bechert defines the protrusion height $h_{\parallel}$ as the distance of the riblet tips, which he locates at $y = 0$, from this virtual origin of the velocity profile, i.e., $h_{\parallel} = a_0$.
From the standpoint of dimensional analysis the protrusion height is a length, and therefore depends only on the chosen reference length and not on the reference velocity. The ratio of the protrusion height to the period of the corrugations, which we shall call normalized protrusion height $\bar{h}_{\parallel}$, is an absolute parameter depending only on the geometry of the wall corrugations and neither on their size nor on the actual speed of the driving fluid stream. As far as the main flow is concerned, the corrugated wall is equivalent to a plane wall located at a distance below the riblet tips which is given by the normalized protrusion height times the period.
We showed that a similar, but numerically different, protrusion height $h_{\perp}$ may be defined for the cross flow as well [7]. In fact, the Fourier-series expansion of the general periodic solution of eqs.(2) behaves for $y \to \infty$ as $\psi \simeq A_0 + B_0 y + y^2/2$ (with exponential accuracy). The parallel velocity component $u = \psi_y \simeq B_0 + y$ thus imitates the linear profile generated by a plane wall located at $y = -B_0$. A transverse protrusion height, different from the longitudinal one, may thus be defined as $h_{\perp} = B_0$.
For all effects concerning cross flow in the driving turbulent shear layer, the corrugated wall is equivalent to a plane wall located at a distance below the riblet tips equal to the transverse protrusion height. If this virtual plane wall turns out to lie above the one seen by the longitudinal flow, that is if $h_{\perp}$ is smaller than $h_{\parallel}$, secondary cross flow will experience a higher viscous dissipation, just as if it flowed in a narrower duct, than the main longitudinal flow, and the level of near-wall turbulence will presumably be reduced. The difference of the two protrusion heights $\Delta h = h_{\parallel} - h_{\perp}$ gives the distance between the two virtual plane walls

that the cross flow and the longitudinal flow respectively see, and therefore yields a quantitative characterization of whether and how much the corrugated wall impedes the cross flow more than it does the longitudinal flow.

The boundary-element numerical algorithm

In order to calculate the two protrusion heights numerically for a general wall profile, we have developed a boundary-element computer code which solves the Laplace and the biharmonic equation in a half-plane-like domain bounded by a periodic wall.

All boundary-element algorithms for the Laplace equation are based on Green's formula,

$$f(\underline{r}') = \oint \left[f(\underline{r})\frac{\partial G}{\partial n}(\underline{r},\underline{r}') - G(\underline{r},\underline{r}')\frac{\partial f}{\partial n}(\underline{r}) \right] ds \qquad (3)$$

which gives the value at any point $\underline{r}'$ of a general solution f in terms of the values taken by f and its normal derivative on the boundary of the solution domain (spanned by the curvilinear abscissa s). In eq.(3) the Green function G is, by definition, any one solution of the Poisson equation $\Delta_2 G = \delta(\underline{r} - \underline{r}')$; different formulations will result from different choices of the Green function.

In the present case, in which we are dealing with a periodic corrugated wall, it is useful to enforce periodicity directly by choosing a periodic Green function. We may then take a single period as solution domain and apply eq.(3) to a boundary formed by one period of the wall, two straight lines parallel to the y-axis, say $x = 0$ and $x = 2\pi$, and a line joining these two at $y = +\infty$; if both f and G are periodic the contributions to eq.(3) from the two lines $x = 0$ and $x = 2\pi$ cancel each other and an obvious simplification ensues. A periodic solution of the Poisson equation suitable for this purpose may easily be determined by conformal mapping techniques. More than one choice is possible; of these the simplest is probably $G(\underline{r},\underline{r}') = (4\pi)^{-1}\log[\cosh(y-y') - \cos(x-x')]$. We have thus eliminated the contributions to eq.(3) from the lateral boundaries of the solution domain. We can, in addition, eliminate the contribution from the line at infinity if we choose a Green function which vanishes for $y \to \infty$. This effect can be obtained by subtracting from the previous Green function its asymptotic behaviour, i.e. $(4\pi)^{-1}(y - y' - \log 2)$. The result is

$$G(\underline{r},\underline{r}') = (4\pi)^{-1}\{\log[2\cosh(y-y') - 2\cos(x-x')] - y + y'\} \qquad (4)$$

Adopting this Green function we can use eq.(3) with the line integral extended over one wall period alone. We shall also find useful to have x as the integration variable along the wall, and therefore we rewrite eq.(3) as

$$f(x',y') = \int_0^{2\pi} \left\{ f[x,y_0(x)]\frac{\partial G}{\partial n}\frac{ds}{dx} - G(x-x',y-y')\varphi(x) \right\} dx \qquad (5)$$

where $\varphi(x) = \dfrac{\partial f}{\partial n}[x,y_0(x)]\dfrac{ds}{dx}$.

According to the general philosophy of B.E.M.s, we now particularize eq.(5) to $y' \to y_0(x')$ (with some care needed in taking the limit from the interior) and interpret the result as an integral equation relating the two functions $f[x,y_0(x)]$ and $\varphi(x)$, either one of which may be the unknown.

For the discretization of the integral equation we have devised a piecewise-polynomial technique which can be enacted at low or high orders of approximation with essentially the same ease.

The two main ingredients of this technique are the representation of f and φ through piecewise polynomials and the adoption of a Gaussian integration formula for all the required integrals. Let the interval $(0,2\pi)$ be divided into N, generally disuniform, subintervals (x_i,x_{i+1}). In each subinterval we assume every unknown ,say φ, to be represented by a polynomial of order M-1, and approximate the integral of this polynomial times the Green function by an M-point Gaussian formula, i.e. by the sum of the values taken by the integrand at M purposely chosen points, multiplied by suitable weights. The key property of Gaussian integration, which is also exploited in other numerical techniques such as the spectral-element method for partial differential equations, is that it is exact for polynomials up to order 2M-1, so that we are effectively approximating the Green function through a polynomial of order M in each subinterval. At the same time, we do not need to deal with the polynomial representation of φ explicitly, because we can simply adopt as variables the values of φ at the M Gaussian integration points in each subinterval, and never let the M coefficients of the polynomial appear at all.

The discretization of eq.(5) along the axis x' is achieved by first recasting the equation in weak form, that is multiplying it by a test function T(x') and integrating over $(0,2\pi)$, and then discretizing this new integral in the same way as the previous one. Then, requiring the equation to be satisfied for T being any piecewise polynomial of order M-1 over the chosen partition into N intervals gives a finite linear system of order MN as the discrete representation of the integral equation.

Actually, whereas the above technique would work for a generic integral equation, in the case of eq.(5) there are two additional complications.

One is that the kernel, either G or $\partial G/\partial n$, is singular at $x = x'$, so that the possibility of representing it locally by a polynomial fails, and at the same time the value of the kernel for $x = x'$, required in the integration formula, is infinite. This difficulty has been eliminated by isolating the singular contribution to the kernel of each integral and integrating numerically the regular part only, and analytically the product of the singular part with the polynomials that represent φ and T over the relevant subinterval. This calculation need be done only once, and the result, obtained at first as a bilinear function of the coefficients of the polynomial representations of φ and T, may be recast once and for all as a bilinear function of the values taken by these polynomials at the Gaussian points.

The second complication is that the integral equation itself is singular, in the sense that it admits a nonzero solution with a zero known term and conversely is not guaranteed to have finite solutions unless the known term satisfies a condition. When it does have a solution, this is not unique unless an additional condition is imposed. This behaviour, analogous to that of a linear system with a coefficient matrix of rank deficient by one, is a consequence of the well known property of the solutions of the Laplace equation that the integral over a closed boundary of $\partial f/\partial n$ must be zero. The reason why this is a complication is that the matrix obtained as the discrete representation of the integral equation will be nearly singular but, because of discretization errors, not quite so. The problem is then that of obtaining a solution which satisfies an additional condition, as the exact solution of the continuous problem does, and only approximately satisfies the nearly-singular linear system obtained from the discretization.

In particular, in the physical problem we are concerned with, the additional condition that must be satisfied by φ corresponds to the imposition of the velocity gradient at infinity; the contributions of the sides to the boundary integral of $\partial w/\partial n$ cancel each other and the contribution of infinity, where $\partial w/\partial y$ is constant and equal to 1, is 2π. We thus obtain the condition that

$$\int_0^{2\pi} \varphi \, dx = \int_{wall} \frac{\partial w}{\partial n} \, ds = -2\pi. \tag{6}$$

The simplest approach to the above problem is dictated by the ordinary theory of rank-deficient linear systems: simply drop one of the equations (which are linearly dependent) and replace it by the additional condition. This operation yields a new non-singular system and certainly works also when, owing to discretization errors, the system is not exactly singular, the effect being in this case that the dropped equation will not be satisfied exactly. However, eliminating the singularity in this manner introduces an unwarranted asymmetry, since any one equation chosen to be dropped will correspond to a particular point on the wall which is not otherwise special.

A more symmetric approach, suggested by analogy with a related variational problem, is to add a constant λ to the r.h.s. of eq.(5) and regard λ as a new unknown to be determined simultaneously with $\varphi(x)$ under the additional constraint (6). Doing so effectively de-singularizes the system; for the solution is unique, owing to the explicitly imposed additional constraint, and exists for no matter which known term, the difference being that when the known term is compatible with the original equation λ turns out to be zero in the solution whereas when the known term is not compatible it does not. The discretization of this modified integral equation yields a finite system of MN + 1 equations (one of which is the additional constraint 6) in MN + 1 unknowns (one of which is λ) which is definitely non-singular and may be solved by any standard method. After the solution λ will turn out to have a value the smaller the better is the approximation of the original integral equation by its discretized counterpart.

Application to flow over grooved surfaces

In order to apply the above technique to problem (1), longitudinal flow over a grooved surface, we need only insert the boundary condition $w[x,y_0(x)] = 0$ into eq.(5), that is solve the homogeneous problem for φ under the constraint (6). Having done so, we can determine the longitudinal protrusion height by applying eq.(5) again, but this time in the limit for $y' \to +\infty$. Since G tends to $(y' - y)/2\pi$ and w to $y' + h_{\|}$ in this limit, we easily obtain

$$\hbar_{\|} = h_{\|}/2\pi = -(2\pi)^{-2}\int_0^{2\pi} y_0(x)\varphi(x)\, dx \tag{7}$$

which can be discretized, consistently with the other integrals, by piecewise Gaussian integration.

For problem (2), transverse flow, one possibility would be to set up a boundary-element formulation based on the Green function of the biharmonic equation (a 2×2 matrix itself) and thus obtain a system of 2MN equations in 2MN unknowns; however, we can instead reduce the problem to two coupled Laplace equations, which we can solve by the successive inversion of two MN×MN matrices one of which is in common with the previous problem. It is well known that if f and g are harmonic functions, $\psi = yf + g$ is a solution of the biharmonic equation. In terms of f and g the boundary conditions relevant to eq.(2) may be written as

$$y_0(x)f[x,y_0(x)] + g[x,y_0(x)] = 0 \quad ; \quad y_0\frac{\partial f}{\partial n} + \frac{\partial g}{\partial n} + \hat{n}\cdot\hat{y}\ f = 0 \qquad (8)$$

(where $\hat{n}\cdot\hat{y}$ is the product of the two unit vectors corresponding to the outward normal and the y axis). Let us now assume $f[x,y_0(x)]$ as main unknown. The first of eqs.(8) directly gives us $g[x,y_0(x)]$ in terms of f. Once we have the matrix relating the discrete representations of f and $\partial f/\partial n$ (as well as g and $\partial g/\partial n$) from the solution of eq.(5), we can formally insert their expressions into the second of eqs.(8) and thus get a combined linear system to determine f from. In the process we impose the desired behaviour at infinity of ψ through conditions of type (6) requiring that $\partial f/\partial n = 0.5$ and $\partial g/\partial n = 0$ there.

Finally, the transverse protrusion height is given by a formula similar to eq.(7).

Performance of the numerical algorithm

In order to test the performance of our algorithm, with particular regard to the use of higher approximations, we have considered two geometries: a cosinusoidal wall with a height equal to the period, and an array of parabolic grooves, again with a height equal to the period.

For the cosinusoidal profile a uniform spacing has been used. Fig. 1 reports, on a bilogarithmic scale, the error in the calculation of $h_{\parallel}$ and $h_{\perp}$ versus number of discretization points. (The error being calculated with respect to a value obtained with a number of discretization points higher than all those appearing in the plot.) Although the curves do not display the change in slope that one would expect in going from lower to higher orders of approximation, the error does decrease with increasing M, smoothly for $h_{\parallel}$ and somewhat more irregularly for $h_{\perp}$, losing roughly a factor of 20 in going from M=1 to M=5.

The parabolic groove profile, with its pointed corners, constitutes a much tougher test for the numerical algorithm, because near the corners the solution is non-analytic and thus not representable by polynomials, but is interesting in the applications. Nevertheless, although larger than that obtained for the cosinusoidal profile, the absolute error generated in the range of, say, $50 \leqslant MN \leqslant 100$ is as low as needed for practical applications (provided a non-uniform discretization is used with a much closer spacing near the corners). Fig.2 shows the errors of the computations performed with different values of M. Not surprisingly, the higher approximations do not perform very well in this case, and in fact turn out to worsen slightly with increasing M. It is interesting to observe, however, that M=2 performs significantly better than M=1, and in fact better than all the others, so that this is definitely the order of approximation to be preferred.

Conclusions

The aim of our research has been twofold: to understand physically and substantiate quantitatively the intuitive notion that a grooved surface offers a greater resistance to cross than to parallel flow, which underlies the generally accepted explanation of why a grooved surface can reduce turbulent drag, and to develop a B.E.M. computer code for the analysis of such surfaces.

On the numerical side, we have obtained an algorithm which can easily work at high orders of approximation and, when the surface profile is smooth, offers a significantly better performance than an ordinary piecewise-constant-panel method (to which it more or less reduces for M=1). For pointed profiles performance is not as good, as may easily be expected, but still M=2 is about an order of magnitude better than M=1.

As far as the physical problem is concerned, three groove geometries have been considered: sinusoidal, triangular and parabolic, each for values of the ratio of height to period s varying between zero and one. The results comparing these three profiles have been reported in [7] and are not reproduced here for space limitations. In all three cases it could be noticed that for $s \simeq 0$ the curves of $\bar{h}_{\parallel}$ and $\bar{h}_{\perp}$ are tangent to each other while $\overline{\Delta h}$ goes to zero quadratically, in accordance with the theoretical analysis performed in the same paper, whereas for $s \rightarrow \infty$ the three curves tend to the analytically calculated limit values. The rates at which the limit is approached are, however, different. In fact, at $s = 1$ the parabolic profile already attains 85% of the limit value of the protrusion height difference $\overline{\Delta h}$, whereas the triangular profile attains 72% and the sinusoidal profile only 53%.

In addition, it is interesting to observe that in all three cases the increase of all three parameters $h_{\|}$, $h_{\perp}$ and Δh is monotonic, so that intermediate values higher than the limit never occur. It was also noticed that the curve of $h_{\perp}$ approaches the limit and becomes flat appreciably earlier than the curve of $h_{\|}$, appearing to indicate that parallel flow penetrates deeper into the grooves and thus "sees" the bottom longer than cross flow.

In general our results confirm the trend of pointed profiles to provide better results, for equal depth, than smooth ones, in accordance with the conclusions reached in [4] on the basis of their analysis of the parallel protrusion height alone.

References

1. McLean, J.D., George-Falvy, D.N. and Sullivan, P.P.: Flight-test of turbulent skin-friction reduction by riblets. Proc. Turbulent Drag Reduction by Passive Means, Royal Aeronautical Society, 15-17 Sept. 1987, London.
2. Sawyer, W.G. and Winter, K.G.: An Investigation of the effect on turbulent skin friction of surfaces with streamwise grooves. Proc. Turbulent Drag Reduction by Passive Means, Royal Aeronautical Society, 15-17 Sept. 1987, London.
3. Walsh, M.J.: Riblets as a viscous drag reduction technique. AIAA J. 21 (1983) 485-486.
4. Bechert, D.W., Hoppe, G. and Reif,W.E.: On the drag reduction of the shark skin. AIAA Paper 85-0546 (1985).
5. Bechert, D.W. and Bartenwerfer, M.: The viscous flow on surfaces with longitudinal ribs. J. Fluid Mech. 206 (1989) 105-129.
6. Baron, A., Quadrio, M. and Vigevano L.: Riduzione della resistenza di attrito in correnti turbolente e altezza di protrusione di pareti scanalate. Proc. X AIDAA Conference, Pisa, Oct. 16-20 1989.
7. Luchini, P., Manzo, F. and Pozzi, A.: Resistance of a grooved surface to parallel and cross flow. Proc. X AIMETA Conference, Pisa 2-5 Oct. 1990, pp. 769-774 (1990).

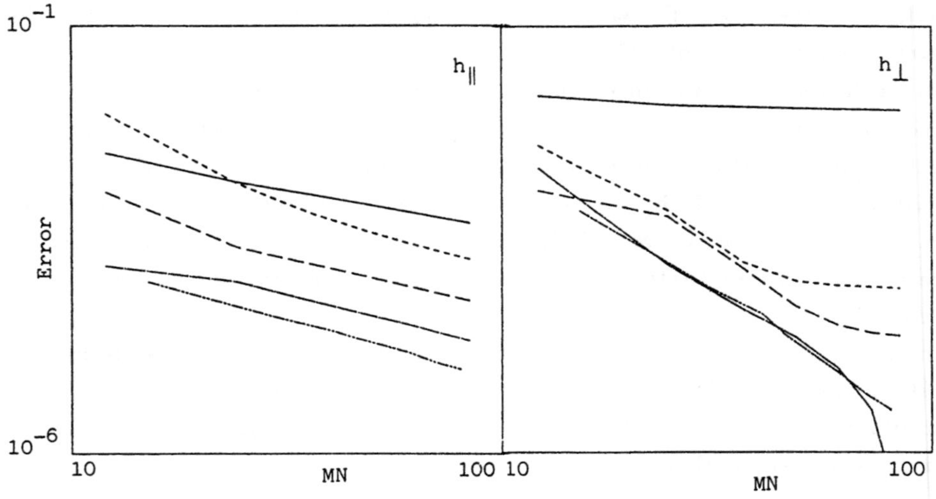

Fig.1. Bilogarithmic plot of the error in the calculation of $h_{\parallel}$ (left) and $h_{\perp}$ (right) versus number of discretization points for a cosinusoidal groove profile. Different line styles denote the order of approximation as follows. Solid: M=1; short-dashed: M=2; long-dashed: M=3; dash-and-dot: M=4; dash-and-double-dot: M=5.

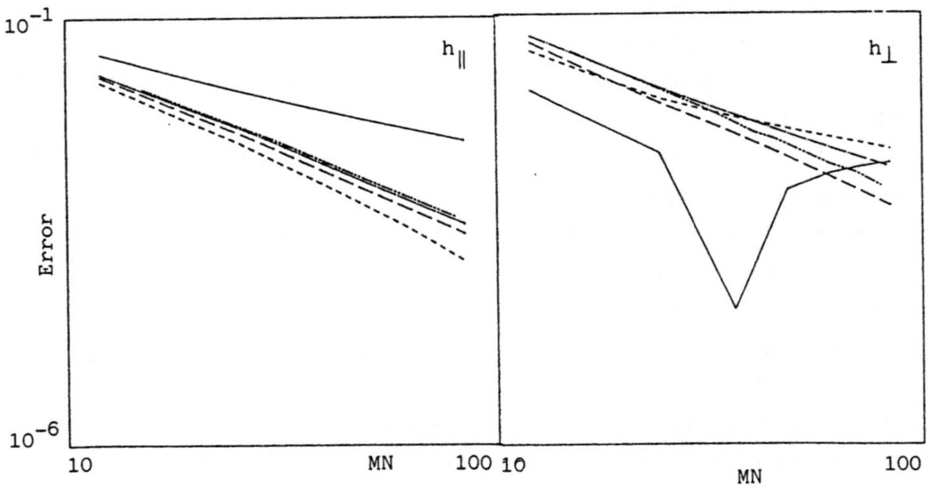

Fig.2. Bilogarithmic plot of the error in the calculation of $h_{\parallel}$ (left) and $h_{\perp}$ (right) versus number of discretization points for a parabolic groove profile with pointed corners. Line styles are as in Fig. 1.

Indirect Evaluation of Surface Stress in the Boundary Element Method

E. Lutz
Cornell University Department of Computer Science and Program of Computer Graphics
486 Engineering Theory Center
Ithaca NY 14850 USA

L.J. Gray
Mathematical Sciences Section
Oak Ridge National Laboratories,
Oak Ridge, Tennessee 37831 USA

A.R.Ingraffea
Cornell University Department of Civil and Environmental Engineering and Program of Computer Graphics,
Hollister Hall
Ithaca NY 14850 USA

ABSTRACT
Evaluation of stress 'on the surface' is a difficult step in the boundary element method for elasticity because it requires evaluation of hypersingular integrals. We describe an indirect method that avoids the hypersingular integral by integrating over a far surface not including the singular point or its local geometry. The derivation makes no demands for smoothness of the local surface, hence can be applied on curved surfaces and at edges. As with other derivations of hypersingular integrals, it requires continuous displacement gradients in the neighborhood of the singular point.

Introduction

For any point $\mathbf{x}$ in the interior or exterior of an elastic solid bounded by surface S, Somigliana's identity states that if a surface displacement and traction functions $\mathbf{u}(\mathbf{y})$ and $\mathbf{p}(\mathbf{y})$ for surface points $\mathbf{y}$ are a solution to the elasticity equations on the body, then

$$\gamma(\mathbf{x})\sigma_{ij}(\mathbf{x}) = \int_S \left[D_{ijk}\mathbf{p}_k(\mathbf{y}) - S_{ijk}\mathbf{u}_k(\mathbf{y})\right] dS(\mathbf{y})$$

where

$$\gamma(\mathbf{x}) = \begin{cases} 1 & \text{if } \mathbf{x} \text{ is on the interior of the region} \\ 0 & \text{if } \mathbf{x} \text{ is on the exterior of the region} \end{cases} \qquad (1)$$

where S_{ijk} and D_{ijk} are third order tensors. This is an exact relationship, allowing a direct evaluation of interior stresses if the surface functions are given and the integrals can be evaluated. As will be discussed below, we consider this boundary integral (and consequently the interior/exterior coefficient $\gamma(\mathbf{x})$) to be *strictly undefined* when $\mathbf{x}$ is precisely on the boundary.

Unfortunately, evauluation of these integrals is complicated by the fact that the D_{ijk} and S_{ijk} tensors are dominated by r^{-2} and r^{-3}, where $r = |\mathbf{x} - \mathbf{y}|$. As $\mathbf{x}$ is taken to the boundary, this "produces a higher order singularity of a type that has not yet been effectively resolved." (Brebbia and Dominguez [1], pg 163).

If surface tractions and (differentiable) displacements are known, the traction can be converted to a normal derivative of displacement, and consequently a full stress tensor, by manipulation of the elasticity relations (Brebbia, Telles, and Wrobel [2]). Since the stress tensor for infinitesimal displacements depends 'only' on the displacement gradient, and not on either the constant or higher order terms of the displacement field, this result is the correct one if the surface functions are accurate. That is, evaluating the full boundary integrals of (1) *with correct surface functions* contributes nothing at all to the stress value other than what is already present in the gradient.

Rank deficiency in the BEM matrix for a cracked body has motivated several efforts to evaluate some global stress integral, such as (1) or a related expression involving the displacement discontinuity across the crack, in order to obtain an additional, independent equation at each point on the crack surface. Examples of this may be found in Cruse and van Buren [4], Polch and Cruse [12], Weaver [15], and Gray, Martha, and Ingraffea [7].

In this paper, we derive an 'indirect' formula for the hypersingular and singular parts of the integral. This gives the hypersingular stress integrals over a singular patch in terms of non-singular integrals over the remaining surface. The process may be interpreted as simultaneously subtracting (a) a rigid body translation and (b) a linear displacement field based on the local traction and tangential derivatives. This is entirely complementary to other 'direct' derivations of the values of the hypersingular integrals (see section 1) , but may be easier to apply because it (a) avoids term-by-term analysis of the internal structure of the kernels and (b) makes no assumptions about the local surface geometry.

Section 1 reviews prior efforts to understand the hypersingular integrals by direct analysis of the kernel expressions. Section 2 presents the new method of indirect analysis. Section 3 gives a computational example.

1 Direct Analyses of Hypersingular Kernels

Gray, Martha, and Ingraffea [7], and Cruse and Novati [3], and Krishnasawamy, Schmerr, Rudolphi, and Rizzo [8] present three different analyses whose common goal is to show that the hypersingular integrals exist and are computable as the point $\mathbf{x}$ in (1) approaches a surface point $\bar{\mathbf{x}}$. The derivations appear to be quite different, but there are

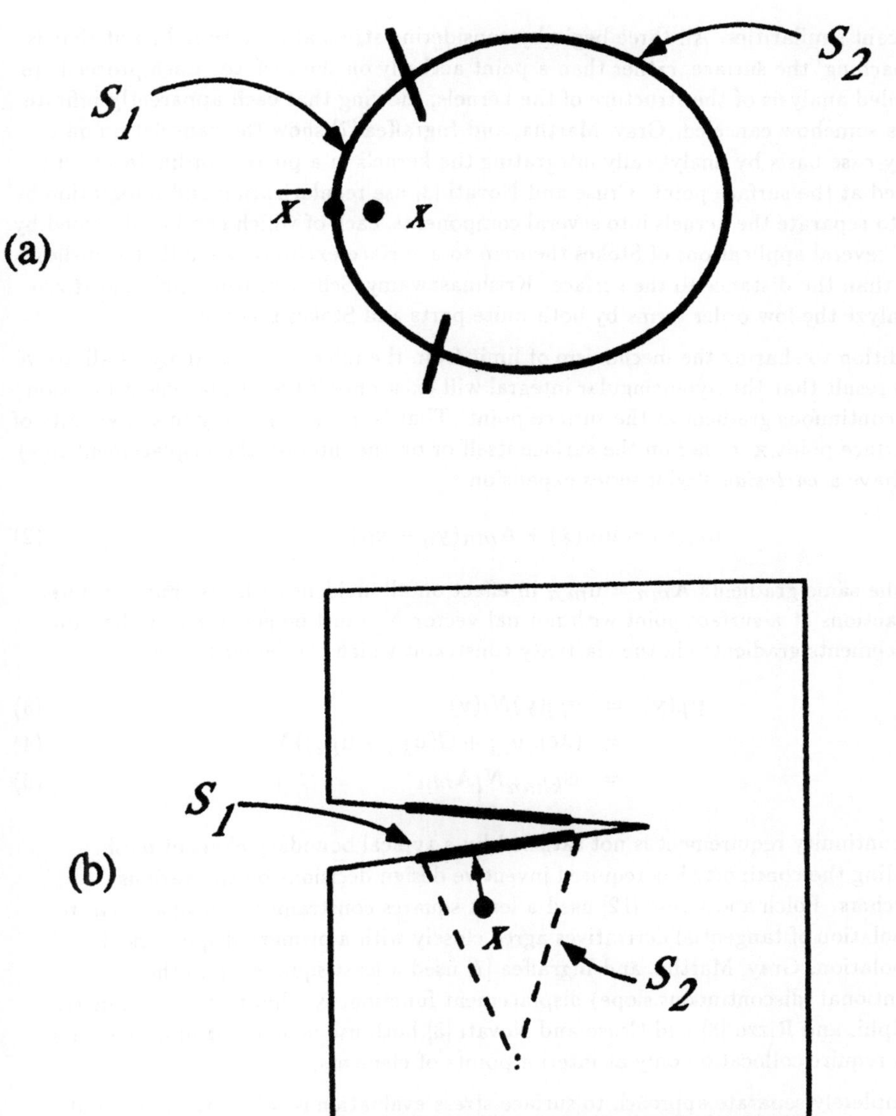

Fig. 1: Surface and point where hypersingular integral is to be computed. (a) General body, with surface divided into two parts. (b) Cracked body. Internal point x approaches a crack face from below.

significant similarities. All three begin by considering stress at an internal point that is 'approaching' the surface, rather than a point actually on the surface. Each proceeds to a detailed analysis of the structure of the kernels, showing that each apparently infinite term is somehow canceled. Gray, Martha, and Ingraffea [7] show the cancelation on a case-by-case basis by analytically integrating the kernels in a polar coordinate system centered at the surface point. Cruse and Novati [3] use regularization and integration by parts to separate the kernels into several components, each of which can be addressed by one of several applications of Stokes theorem to a surface exclusion zone that vanishes faster than the distance to the surface. Krishnasawamy, Schmerr, Rudolphi, and Rizzo [8] analyze the low order terms by both finite parts and Stokes theorem.

In addition to sharing the mechanism of limit-from-the-interior, these analyses all arrive at the result that the hypersingular integral will exist only if the displacement function has a continuous gradient at the surface point. That is, for any point $\mathbf{y}$ in the vicinity of the surface point $\bar{\mathbf{x}}$, either on the surface itself or on the interior, the displacement $\mathbf{u}(\mathbf{y})$ must have a *cartesian* Taylor series expansion

$$\mathbf{u}_m(\mathbf{y}) = \mathbf{u}_m(\bar{\mathbf{x}}) + \mathbf{A}_{mn}(\mathbf{y}_n - \bar{\mathbf{x}}_n) + \cdots \tag{2}$$

with the same gradients $\mathbf{A}_{mn} = \mathbf{u}_{m,n}$ in effect on all incident surfaces. Furthermore, the tractions at a surface point with normal vector N_l must be governed by the same displacements gradients via the elasticity constraint which can be written as

$$\begin{aligned} \mathbf{p}_k(\mathbf{y}) &= \sigma_{kl}(\mathbf{y})N_l(y) && (3)\\ &= (\lambda\delta_{kl}\mathbf{u}_{i,i} + G(\mathbf{u}_{k,l} + \mathbf{u}_{l,k}))N_l && (4)\\ &= \mathbf{E}_{klmn}N_l\mathbf{A}_{mn} && (5) \end{aligned}$$

This continuity requirement is not satisfied by a typical boundary element mesh. Providing the continuity has required inventive design decisions by the various researchers. Polch and Cruse [12] used a least squares constraint to make a separate interpolation of tangential derivatives agree closely with a primary displacement interpolation. Gray, Martha, and Ingraffea [7] used a least squares fit to the conventional (discontinuous slope) displacement function. Krishnasawamy, Schmerr, Rudolphi, and Rizzo [8] and Cruse and Novati [3] both use non-conforming elements, which require collocation only at interior points of elements.

A completely separate approach to surface stress evaluation is taken by Ghosh and Muhkerjee[5] and Okada, Rajiyah, and Atluri [11]. They derive kernel functions for boundary integrals in which tangential displacement gradients, rather than displacements themselves, appear as primary unknown quantities. These methods have the benefit of milder kernel singularities than those appearing in the direct differentiation of conventional kernels. The kernels in Ghosh and Muhkerjee[5] are only as singular as the original displacement kernel, i.e. $\ln(r^{-1})$ in 2D and r^{-1} in 3D. The kernels in Okada, Rajiyah, and Atluri [11] have the same singularity as the traction kernels, i.e. r^{-1} and r^{-2} in 2D and 3D, respectively. There is limited practical experience with these kernels in 2D, and none at all for 3D or for crack surfaces.

2 Indirect Computation of Hypersingular Integrals

As shown in Fig. 1a, we consider a surface divided into two parts S_1 and S_2, with surface point $\bar{\mathbf{x}}$ placed strictly in the interior of S_1. There must be a finite (nonzero) distance from $\bar{\mathbf{x}}$ to the nearest point of S_2. That is, S_1 is a finite-sized surface patch, and will *not* shrink to zero size during the analysis. Dividing the integrals in (1) into two parts, we define the hypersingular integral over S_1 as $I(S_1,\mathbf{x})$, with

$$I(S_1,\mathbf{x}) = \int_{S_1} \left[D_{ijk}\mathbf{p}_k(\mathbf{y}) - S_{ijk}\mathbf{u}_k(\mathbf{y})\right] dS(\mathbf{y}) \quad (6)$$

$$= \gamma(\mathbf{x})\sigma_{ij}(\mathbf{x}) - \int_{S_2} \left[D_{ijk}\mathbf{p}_k(\mathbf{y}) - S_{ijk}\mathbf{u}_k(\mathbf{y})\right] dS(\mathbf{y}) \quad (7)$$

Now assume the displacement has the Taylor series expansion (2) about surface point $\bar{\mathbf{x}}$ and consider the constant and linear parts as separate solutions:

Case 1: Constant part (pure translation): When $\mathbf{u}(\mathbf{y}) \equiv \mathbf{u}(\bar{\mathbf{x}})$ for all $\mathbf{y}$, the stress state and surface tractions are zero. Hence the hypersingular integral of the constant term must vanish over the whole surface, and we obtain

$$\int_{S_1} S_{ijk}\, dS(\mathbf{y}) = -\int_{S_2} S_{ijk}\, dS(\mathbf{y}) \quad (8)$$

Case 2: Linear part (uniform stress): If the entire body has displacement $\mathbf{u}_m(\mathbf{y}) = \mathbf{A}_{mn}(\mathbf{y}_n - \bar{\mathbf{x}}_n)$, the stress state at any interior point $\mathbf{y}$ is entirely determined by $\sigma_{ij}(\mathbf{y}) = \mathbf{E}_{ijmn}\mathbf{A}_{mn}$, and surface traction at the point with normal $N_l(\mathbf{y})$ is $\mathbf{p}_k = \mathbf{E}_{klmn}\mathbf{A}_{mn}N_l$. Hence, renaming clashing subscripts and defining a new tensor quantity $\mathbf{W}_{ijmn}$ for the ij integral of the mn gradient over S_2,

$$\int_{S_1} \left[D_{ijk}\mathbf{E}_{klmn}\mathbf{A}_{mn}N_l - S_{ijm}\mathbf{A}_{mn}(\mathbf{y}_n - \bar{\mathbf{x}}_n)\right] dS(\mathbf{y}) \quad (9)$$

$$= \left[\gamma(\mathbf{x})\mathbf{E}_{ijmn} - \int_{S_2} \left[D_{ijk}\mathbf{E}_{klmn}N_l - S_{ijm}(\mathbf{y}_n - \bar{\mathbf{x}}_n)\right] dS(\mathbf{y})\right] \mathbf{A}_{mn} \quad (10)$$

$$= (\gamma(\mathbf{x})\mathbf{E}_{ijmn} - \mathbf{W}_{ijmn})\mathbf{A}_{mn} \quad (11)$$

The quantities $\mathbf{W}_{ijmn}$ are entirely determined by elasticity constants, the position of $\mathbf{x}$, and the geometry of the far surface.

Case 3: Higher order part: If $\mathbf{u}$ has higher order components, the remainder (after cases 1 and 2 are used for the constant and linear parts) has a leading second order term, which reduces the r^{-3} singularity to r^{-1}, which can be handled by conventional means. Combining cases, the contribution of the local surface patch S_1 when the local displacement field has a continuous Taylor expansion is

$$I(S_1,\mathbf{x}) = -\mathbf{u}_m(\bar{\mathbf{x}})\int_{S_2} S_{ijk}\, dS(\mathbf{q}) \quad (12)$$

$$+ \ (\gamma(\mathbf{x})\mathbf{E}_{ijmn} - \mathbf{W}_{ijmn})\mathbf{A}_{mn} \quad (13)$$

$$+ \ \text{weakly singular higher order contribution} \quad (14)$$

So long as the geometry and surface function modeling in a BEM program provides clearly defined $\mathbf{u}_m(\bar{\mathbf{x}})$ and $\mathbf{A}_{mn}$, the right hand sides of (12), (13), and (14) provide nonsingular expressions for the constant, linear, and higher order parts of the hypersingular integrals.

This is formally valid only at a strictly interior or exterior point, and not for the surface point $\bar{\mathbf{x}}$ itself. However, by the assumption that all points on S_2 are a finite distance from $\bar{\mathbf{x}}$, we can consider points $\mathbf{x}$ that are approaching $\bar{\mathbf{x}}$ in the limit. The integrals over S_2 will be the same whether approaching from the inside or outside. The $\gamma(\mathbf{x})$ term provides the expected discontinuity between the stresses at points in the interior and exterior neighborhoods of $\bar{\mathbf{x}}$.

Remark 1: Other Green's Functions The development made no reference to the internal structure of the kernels. The general outline also applies to Laplace's equation, with some simplification due to the fact that the Laplace integrals have the normal derivative quantity at $\bar{\mathbf{x}}$ itself independent of the tangential derivatives of the primary quantity, rather than as a linear combination.

Remark 2: Coupling of Linear Terms For the linear terms, the expression with $\mathbf{W}_{ijmn}$ has contributions of both the D_{ijk} and S_{ijk} integrals. Except for the special case of zero traction, one does not obtain separate expressions for the constant term D_{ijk} and linear term S_{ijk} integrals on the near surface. This is consistent with results obtained by Gray and Lutz [6], where the apparently infinite terms that appear when one looks at the internal structure of the kernels are shown to cancel only by simultaneous integration of the two kernels.

Remark 3: Infinite Terms The indirect analysis avoids any reasoning about how infinite terms pair with each other to be canceled. This is a higher-order analogy to the common practice of applying a rigid body motion argument to obtain the 'diagonal' term of the conventional BEM matrix as a sum of off-diagonals, obtaining the integral of the singular normal derivative P_{ik} kernel as

$$\int_{S_1} P_{ik}\, dS(\mathbf{q}) = -\gamma(\mathbf{x})\mathbf{I} - \int_{S_2} P_{ik}\, dS(\mathbf{q}) \tag{15}$$

Indirect arguments say that there is a close enough relationship between local and global quantities that the locals can be inferred from the globals, rather than by detailed consideration of the local geometry.

Remark 4: Corners and Curved Surfaces No particular assumptions were made about the surfaces other than that $\bar{\mathbf{x}}$ is on S_1 but not S_2. By considering the effect of changing one surface but not the other, it can be seen that the singular integrals over S_1 and S_2 are actually dependent only on $\bar{\mathbf{x}}$ and the boundary between the two surfaces. Any pair of surfaces with the same boundary will produce the same geometric integrals. As a practial matter, the definition of gradient values at a corner is a more difficult problem than the question of what the integrals are at the corner.

3 Implementation

The indirect computation of hypersingular integrals is being applied to 3D crack modeling in the the FRANSYS modeling system described by Martha [10]. The exterior of the solid and (both) crack surfaces are meshed by triangular and quadrilateral elements. The two crack surfaces have topologically distinct elements that happen to coincide geometrically. The conventional displacement BIE equation

$$\mathbf{u}_i(\bar{\mathbf{x}}) = \lim_{\mathbf{x}\to\bar{\mathbf{x}}} \int_S U_{ik}(\mathbf{x},\mathbf{y})\mathbf{p}_k(\mathbf{y}) - P_{ik}(\mathbf{x},\mathbf{y})\mathbf{u}_k(\mathbf{y})\, dS(\mathbf{y}) \tag{16}$$

is enforced (a) at all non-crack points, (b) on one crack face, and (c) on the crack front, with the limit taken as $\mathbf{x}$ approaching the surface from the interior.

This equation cannot be applied on the second crack face because it is not independent. Instead, for a point where the surface normal is N_j, the traction condition

$$\mathbf{p}_i(\bar{\mathbf{x}}) = \sigma_{ij} N_j \tag{17}$$

$$= \lim_{\mathbf{x}\to\bar{\mathbf{x}}} \int_S D_{ijk}(\mathbf{x},\mathbf{y}) N_j \mathbf{p}_k(\mathbf{y}) - S_{ijk}(\mathbf{x},\mathbf{y}) N_j \mathbf{u}_k(\mathbf{y})\, dS(\mathbf{y}) \tag{18}$$

is enforced, with the left side value provided by the boundary condition traction on the crack face. (At the analytic level, this combination of conditions is exactly as used previously in Gray, Martha, and Ingraffea [7] and Cruse and van Buren [4].)

To avoid difficulties enforcing gradient continuity across element boundaries and (particularly) at corners, nonconforming displacements are used on the crack face and edges where the crack intersects the surface. That is, adjacent elements are not required to have identical displacements along shared edges. Each element must have as many interior collocations as it has nodes. On a 4-noded element, collocations are at the positions of a 2-by-2 Gauss rule; on a 3 noded element, they are at barycentric coordinates $(\frac{1}{5}, \frac{1}{5}, \frac{3}{5})$ and its 2 rotations.

For an interior point approaching a collocation point on the crack face, as in Fig. 1b, both the singular displacement integrals and hypersingular traction integrations are required over both the lower and upper surface. If we take the lower element as surface S_1 in the indirect integration formulas, we cannot take the entire remainder of the solid (which includes the mating crack element) as S_2 because the mating element violates the condition that S_2 be a nonzero distance from $\bar{\mathbf{x}}$. However, there is nothing that requires S_2 to be related to the actual solid; the integrands over S_2 are just constant and cartesian linear terms multiplied by kernels. Any surface that mates with S_1 to enclose volume can be used. In particular, we can choose any point off of the element and form a ruled surface this apex point and the edges of the element, as suggested by the triangular S_2 in the figure. For a 3D element, this enclosed volume is a 3- or 4-sided pyramid.

This strategy provides the necessary smoothness to the interpolating functions at the integration points. Because collocation points can be quite near element edges, extremely difficult near-singular integrations can occur both over neighboring elements and over the closure surface for the hypersinuglar integrals. Some assistance in these integrations is provided by Gaussian quadrature formulas that incorporate the distance between the singular point and an integration element as a parameter of the orthogonal polynomial constructon, as described in Lutz, Wawrzynek, and Ingraffea [9].

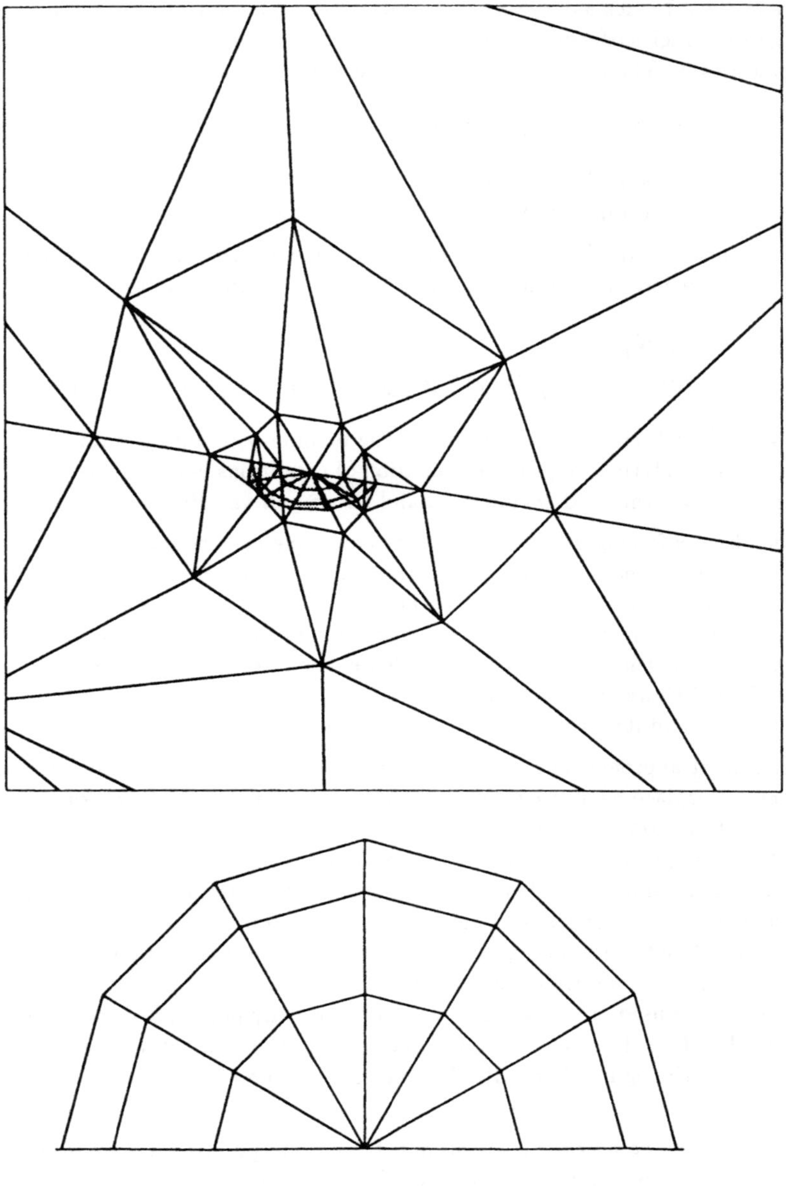

Fig. 2: Semicircular surface crack. (a) Crack intruding through planar face. (b) discretization of crack surface.

3.1 Computational Results

Fig. 2a shows a semi-circular crack intruding through a planar face of a large (effectively infinite, relative to crack size) block. Fig. 2b shows the mesh on the face in more detail. The analytic solution for internal pressure p=1 and radius a=4 in Tada, Paris, and Irwin [14] gives a uniform mode I crack-tip stress-intensity factor $K_I = 2p\sqrt{a/\pi} = 2.26$. Values obtained from the boundary element method, with displacements converted to stress intensity factors using displacement correlation as described by Sousa [13], vary between 2.22 and 2.28 along the crack front. Considering the coarse mesh and linear elements, these results are surprisingly accurate.

4 Conclusions

We have shown that only non-singular integrations are required to compute hypersingular integrals arising in surface stress integrals. The derivation applies to flat, curved, and corner points of a surface, so long as (a) the local displacement gradient is known at the singular point and (b) a closure surface can be computed around the singular patch. Both the existence proof and the implementation are greatly simplified by the absence of manipulation of internal structure of the kernels.

Computational results using a non-conforming mesh on a 3D crack face are promising, but significant integration difficulties remain for near-singular integrals.

Acknowledgements: Computations were carried out in the facilities of the Cornell University Program of Computer Graphics. Financial support was provided by the Unisys Corporation and by the Applied Mathematical Sciences Program, Office of Energy Research, US Department of Energy under contract DE-AC05-84OR21400. Thanks to Dave Potyondy for providing the geometry and mesh for the semicircular crack. Thanks to D. Potyondi and S. Muhkerjee for comments on drafts of the paper.

References

[1] C.A. Brebbia and X.Y. Dominguez. *Boundary Elements: An Introductory Course.* Computational Mechanics Publications (McGraw-Hill), 1988.

[2] C.A. Brebbia, J.C.F. Telles, and L.C. Wrobel. *Boundary Element Techniques - Theory and Applications in Engineering.* Springer-Verlag, 1984.

[3] T.A. Cruse and G. Novati. Traction bie formulations and applications to non-planar and multiple cracks. In *22nd ASTM Conference on Fracture Mechanics*, ASTM, 1990.

[4] T.A. Cruse and W. Van Buren. Three-dimensional elastic stress analysis of a fracture specimen with an edge crack. *International Journal of Fracture Mechanics*, 7:1–15, 1971.

[5] N. Ghosh and S. Muhkerjee. A new boundary element method formulation for three dimensional problems in linear elasticity. *Acta Mechanica*, 67:107–119, 1987.

[6] L.J. Gray and E.D. Lutz. On the treatment of corners in the boundary element method. *(In press) Journal of Computational and Applied Mathematics*, 1990.

[7] L.J. Gray, Luiz F. Martha, and A.R. Ingraffea. Hypersingular integrals in boundary element fracture analysis. *International Journal for Numerical Methods in Engineering*, 29:1135–1158, 1990.

[8] L.W. Krishnasamy, L.W. Schmerr, T.J. Rudolphi, and F.J. Rizzo. Hypersingular boundary integral equations: some applications in acoustic and elastic wave scattering. *Journal of Applied Mechanics*, 57:404–414, 1990.

[9] E.D. Lutz, P. Wawrzynek, and A.R. Ingraffea. Implementation of parameterized gaussian quadrature in the 2d bem for elasticity. In A.H.D. Cheng, C.A. Brebbia, and S. Grilli, editors, *Computational Engineering with Boundary Elements: Vol. 2: Solid and Computational Problems*, Computational Mechanics Publications, 1990.

[10] L.F.M. Martha. *Topological and Geometrical Modeling Approach to Numerical Discretization and Arbitrary Fracture Simulations in Three-Dimensions*. PhD thesis, Cornell University, Ithaca, NY, USA, 1989.

[11] H. Okada, H. Rajiyah, and S.N. Atluri. A novel displacemetn gradient boundary element method for elastic stress analysis with high accuracy. *Transactions of the ASME, Journal of Applied Mechanics*, 55:786–794, 1988.

[12] E.Z. Polch, T.A. Cruse, and C.-J. Huang. Traction bie solutions for flat cracks. *Computational Mechanics*, 2:253–267, 1987.

[13] J.L. Sousa, L.F. Martha, P.A. Wawrzynek, and A.R. Ingraffea. Simulation of non-planar crack propagation in three-dimensional structures in concrete and rock. In *Fracture of Concrete and Rock: Recent Developments*, Elsevier Applied Science, 1989.

[14] H. Tada, P. Paris, and G. Irwin. *The Stress Analysis of Cracks Handbook*. Del Research Corporation, 1973.

[15] J. Weaver. Three dimensionial crack analysis. *International Journal of Solids and Structures*, 13:321–330, 1977.

An Integral Equation Method for Geometrically Nonlinear Bending Problem of Elastic Circular Arch

S. MIYAKE*, M. NONAKA** and N. TOSAKA**

*Department of Business Administration,

Faculty of Economics, Kanto Gakuen University, Japan

**Department of Mathematical Engineering,

College of Industrial Technology, Nihon University, Japan

SUMMARY

Integral equation method for geometrically nonlinear bending problem of elastic arches is presented from the two viewpoints. The first nonlinear problem is based on the theory in which axial displacement due to stretching is negligible in comparison with the normal displacement and the axial stress resultant is constant. The second nonlinear problem of arch is taken into considertion of the effect of axial displacement due to stretching. Integral equation formulations and its discretized expressions are also given for both problems. To show the efficiency and validity of our approach, some numerical examples for the first approach are also given.

INTRODUCTION

The geometrically nonlinear problem of thin elastic bodies is one of important problems in structural mechanics. The finite element method[1, 2] based on various schemes seems to be a powerful means. In recent years, the boundary element method for linear problems have been well developed and successfully applied to various kinds of problems such as solid and fluid mechanics. However, efficient application of the boundary element method to nonlinear problems, especially geometrically nonlinear problems is very few at present stage.

It is shown that the so-called boundary-domain element method in the integral equation method can be used efficiently to obtain numerical solutions of geometrically nonlinear problem of shallow spherical shell[3, 4, 5] and shallow sinusoidal arch[6, 7] in our recent papers. And we pursued the nonlinear behaviour of shells with snap-through phenomena. Especially in the analysis of shallow sinusoidal arch[8, 9, 10], we obtained very complicated equilibrium paths including looping, snap-through and bifurcation phenomenon and we verified that our numerical solutions are very close to the Galerkin solutions.

In this paper, we wish to present two integral equation formulations for geometrically nonlinear problem of elastic arch. The first nonlinear arch problem is based on the theory in which axial displacement due to stretching is negligible in comparison with the normal displacement and the axial stress resultant is constant. The second nonlinear problem of arch is taken into consideration of the effect of axial displacement due to stretching. The derived nonlinear integral equations are discretized by using the boundary-domain element approach. To show the efficiency and validity of our approach, some numerical examples for first approach are also given.

INTEGRAL EQUATIONS

(a) Formulation I

Let us consider an elastic isotropic shallow arch of span l, cross-sectional area A, Young's modulus E, moment of inertia I subjected to a normal surface load p^*. We assume this arch as a beam with a small initial curvature described with the initial centroidal locus of the arch, $z = z(x)$. We adopt the well-known nonlinear differential equation which is based on the assumptions such that axial displacement due to stretching is negligible in comparison with the normal displacement and the axial stress resultant is constant. Following our previous papers, the integral equation for this problem is given in terms of the nondimensional transverse deflection:

$$W(Y) = -[Qg^*]_0^\pi + [M\theta^*]_0^\pi - [\theta M^*]_0^\pi + [WQ^*]_0^\pi + \int_0^\pi qg^* dX$$

$$+\frac{1}{2\pi}\int_0^\pi \left(2\frac{d^2\beta^*}{dX^2} - \theta^2\right) dX \begin{bmatrix} \left[\frac{d\beta^*}{dX}g^*\right]_0^\pi & - & \int_0^\pi \frac{d\beta^*}{dX}\theta^* dX \\ -[\theta g^*]_0^\pi & + & \int_0^\pi \theta\theta^* dX \end{bmatrix} \tag{1}$$

in which we introduced the following nondimensional quantities defined by

$$W = \frac{w}{\gamma}\ ,\ \gamma = \sqrt{\frac{I}{A}}\ ,\ X = \frac{\pi}{l}x\ ,\ q = \frac{p^*}{EI\gamma}\left(\frac{l}{\pi}\right)^4\ ,\ \beta^* = \frac{z}{\gamma} \tag{2}$$

Here, we have introduced the following physical quantities, which are the slope angle, the bending moment and the transverse shear in equation (1):

$$\frac{dW}{dX} = \theta, \quad \frac{d^2W}{dX^2} = M, \quad \frac{d^3W}{dX^3} = Q \tag{3}$$

$$\frac{dg^*}{dX} = \theta^*, \quad \frac{d^2g*}{dX^2} = M^*, \quad \frac{d^3g^*}{dX^3} = Q^* \tag{4}$$

The fundamental solution g^* and its derivatives are given by

$$\left.\begin{aligned} g^* &= \tfrac{1}{12}\mid X - Y \mid^3 \\ \theta^* &= \tfrac{dg^*}{dX} = \tfrac{1}{4}\mid X - Y \mid^2 \text{sgn}(X - Y) \\ M^* &= \tfrac{d^2g^*}{dXdY} = \tfrac{1}{2}\mid X - Y \mid \\ Q^* &= \tfrac{d^3g^*}{dX^3} = \tfrac{1}{2}\text{sgn}(X - Y) \end{aligned}\right\} \tag{5}$$

$$\left.\begin{aligned} \tfrac{dg^*}{dY} &= -\tfrac{1}{4}\mid X - Y \mid^2 \text{sgn}(X - Y) \\ \tfrac{d\theta^*}{dY} &= -\tfrac{d^2g^*}{dXdY} = -\tfrac{1}{2}\mid X - Y \mid \\ \tfrac{dM^*}{dY} &= \tfrac{d^3g^*}{dX^2dY} = -\tfrac{1}{2}\text{sgn}(X - Y) \\ \tfrac{dQ^*}{dY} &= \tfrac{d^4g^*}{dX^3dY} = -\delta(X - Y) \end{aligned}\right\} \tag{6}$$

in which sgn denotes the signum function and X and Y denote the observation point and source point, respectively.

Applying the boundary-domain element discretization scheme to the derived nonlinear system of integral equation, we can get finally the nonlinear system of algebraic equations:

$$\left[\ N\left(\frac{d\boldsymbol{W}}{dX}\right)\ \right]\begin{bmatrix} \left\{\frac{d\boldsymbol{W}}{dX}\right\}_\Omega \\ \left\{\frac{d^3\boldsymbol{W}}{dX^3}\right\}_\Gamma \end{bmatrix} = \left\{P\right\} \tag{7}$$

where opetator $\boldsymbol{N}$ represent nonlinear mapping and $\boldsymbol{W}$ and $\boldsymbol{P}$ are nodal and load vectors, respectively.

(b)Formulation II:

The second formulation for nonlinear bending problem of arch, we consider the effect of axial displacement due to stretching. Following our previous paper[10], the fundamental integral equation set is given by

$$\begin{aligned} U_k(Y) &= -\left[Q_\Gamma V_{1k}^*\right]_0^\phi + \left[Q_k^* W\right]_0^\phi + \left[M_\Gamma \Theta_k^*\right]_0^\phi - \left[M_k^* \Theta_\Gamma\right]_0^\phi \\ &\quad - \left[N_\Gamma V_{2k}^*\right]_0^\phi + \left[N_k^* V\right]_0^\phi + \int_0^\phi B_i V_{ik}^* dX \end{aligned} \tag{8}$$

in which we introduce the following nondimensional quantities defined by

$$V = \frac{v}{R}, \quad W = \frac{w}{R}, \quad X = \frac{x}{R}, \quad q = \frac{Rp^*}{EA}, \quad \phi = \frac{l}{R}, \quad \beta = \frac{I}{AR^2} \tag{9}$$

Here, R is the radius of arch, l the length, ϕ the subtended angle. The displacements $W(=U_1)$ and $V(=U_2)$ are the normal and tangential displacements of the mid-surface and B_i is forcing term including the nonlinear terms expressed as

$$\left.\begin{aligned} B_1 &= -q - \frac{d}{dX}\left[\left\{\frac{dV}{dX} - W + \frac{1}{2}\left(\frac{dW}{dX}\right)^2\right\}\frac{dW}{dX}\right] - \frac{1}{2}\left(\frac{dW}{dX}\right)^2, \\ B_2 &= -\left(\frac{dW}{dX}\right)\left(\frac{d^2W}{dX^2}\right) \end{aligned}\right\} \tag{10}$$

In (10), we express components of slope, stress and moment resultants and transverse shear as follows:

$$\left.\begin{aligned} \Theta_\Gamma &= \frac{dW}{dX} + V \\ N_\Gamma &= \frac{dV}{dX} - W \\ M_\Gamma &= -\beta\left(\frac{d^2W}{dX^2} + \frac{dV}{dX}\right) \\ Q_\Gamma &= -\beta\left(\frac{d^3W}{dX^3} + \frac{d^2V}{dX^2}\right) \end{aligned}\right\} \tag{11}$$

We also introduce the quantities expressed in terms of fundamental solution tensor which correspond to the above equation (8), that is,

$$\left.\begin{aligned}
\Theta_k^* &= \frac{dV_{1k}^*}{dX} + V_{2k}^* \\
N_k^* &= \frac{dV_{2k}^*}{dX} - V_{1k}^* \\
M_k^* &= -\beta\left(\frac{d^2V_{1k}^*}{dX^2} + \frac{dV_{2k}^*}{dX}\right) \\
Q_k^* &= -\beta\left(\frac{d^3V_{1k}^*}{dX^3} + \frac{d^2V_{2k}^*}{dX^2}\right)
\end{aligned}\right\} \tag{12}$$

In this case, the explicit form of the fundamental solution tensor V_{ij}^* can be given by

$$\left.\begin{aligned}
V_{11}^* &= \mathcal{M}_{11}\phi^* = (\beta+1)\frac{d^2\phi^*}{dX^2} \\
&= -(\beta+1)\frac{1}{4\beta}(\sin r - r\cos r) \\
V_{12}^* &= \mathcal{M}_{12}\phi^* = -\beta\frac{d^3\phi^*}{dX^3} + \frac{d\phi^*}{dX} \\
&= -\frac{1}{4\beta}\{2\mathrm{sgn}(X-Y) - 2\mathrm{sgn}(X-Y)\cos r \\
&\quad -(\beta+1)r\mathrm{sgn}(X-Y)\sin r\} \\
V_{21}^* &= \mathcal{M}_{21}\phi^* = -V_{12}^* \\
V_{22}^* &= \mathcal{M}_{22}\phi^* = -\beta\frac{d^4\phi^*}{dX^4} - \phi^* \\
&= \frac{1}{4\beta}\{2r + (\beta+1)r\cos r + (\beta-3)\sin r\}
\end{aligned}\right\} \tag{13}$$

Applying the boundary-domain element procedure to the derived nonlinear system of integral equations, we arrive at the matrix expression in terms of nodal vectors $\boldsymbol{V}$ and $\boldsymbol{W}$ and their derivatives, slope $\boldsymbol{\Theta}_\Gamma$, moment resultant $\boldsymbol{M}_\Gamma$, stress resultant $\boldsymbol{N}_\Gamma$ and transverse shear $\boldsymbol{Q}_\Gamma$ expressed as follows:

$$\begin{bmatrix} G \end{bmatrix}\begin{Bmatrix} \boldsymbol{\Theta}_\Gamma \\ \boldsymbol{N}_\Gamma \\ \boldsymbol{M}_\Gamma \\ \boldsymbol{Q}_\Gamma \end{Bmatrix} + \begin{bmatrix} N(\boldsymbol{V},\boldsymbol{W}) \end{bmatrix}\begin{Bmatrix} \boldsymbol{V} \\ \boldsymbol{W} \end{Bmatrix} = \begin{Bmatrix} P \end{Bmatrix} \tag{14}$$

where $\boldsymbol{G}$ and $\boldsymbol{N}$ are coefficient matrices obtained from the boundary terms and domain integral terms due to the nonlinearity of governing equations, respectively, and $\boldsymbol{P}$ is the load vector.

NUMERICAL EXAMPLES

In this place, we show some numerical results for the Formulation I. The initial shape is assumed as the simple sinusoidal one such that

$$\beta^* = H^* \sin X \tag{15}$$

where H^* denotes the dimensionless rise parameter. Boundary conditions prescribed herein are assumed as the simply supported conditions:

$$W(0) = W(\pi) = 0, \qquad M(0) = M(\pi) = 0 \tag{16}$$

As the loading condition, we assumed following three kinds of loading in our previous paper [8]:

(1) the sinusoidal distributed loading $\quad q = \tau \sin X$,

(2) the uniform distributed loading $\quad q = \tau$,

(3) the concentrated loading $\quad q = \tau\delta(X - \pi/2)$.

And we showed numerical results for the case of $H^* = 10.0$ under the above-mentioned three loading conditions. We obtained reasonable numerical solutions through the comparisons with the seven-terms Galerkin solutions. In this paper we wish to show another results for the case of $H^* = 12.0$ under the concentrated load in the following.

Fig.1. illustrates the load-deflection curves through comparisons with the eleven-terms Galerkin solutions. The number of mesh used in this computation is $n = 50$. We can trace the very complicated nonlinear behaviour in which loading and bifurcation phenomenon are observed. In Fig.2., the fundamental (looping) path as well as its deflection modes for each loading point are depicted corresponding to the points, $1, 2, \ldots, 6$ indicated on this fundamental path. And, the bifurcation path and the corresponding unsymmetric modes are also illustrated on the same manner in Fig.3.

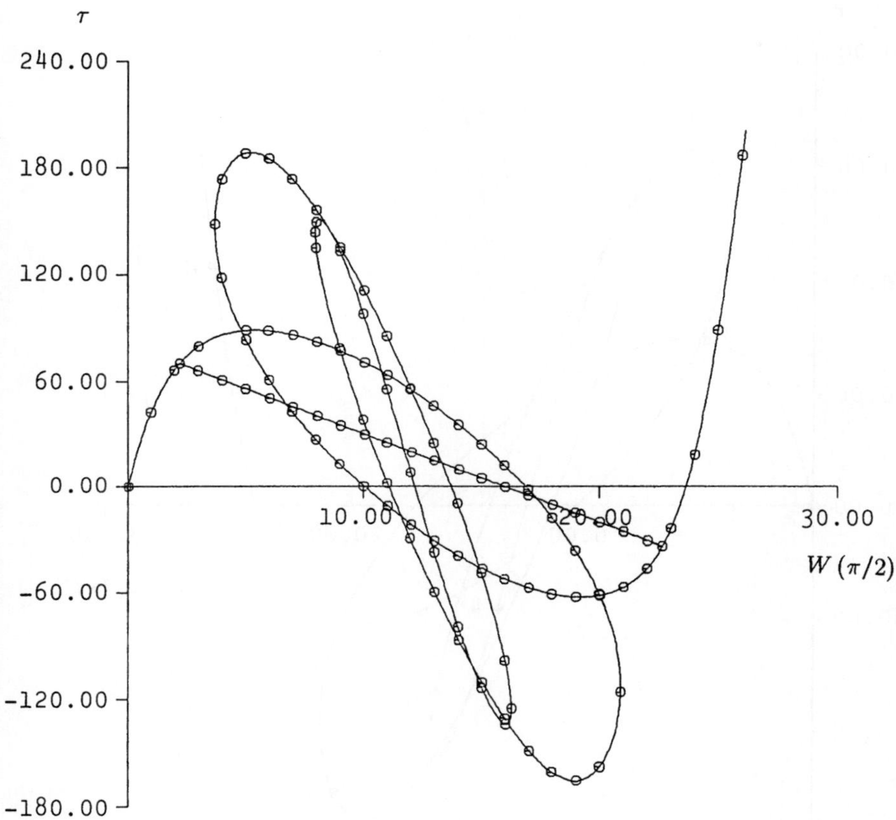

— : Galerkin Method (eleven modes)

o : Present Method (n=50)

Fig.1. Load-deflection curve

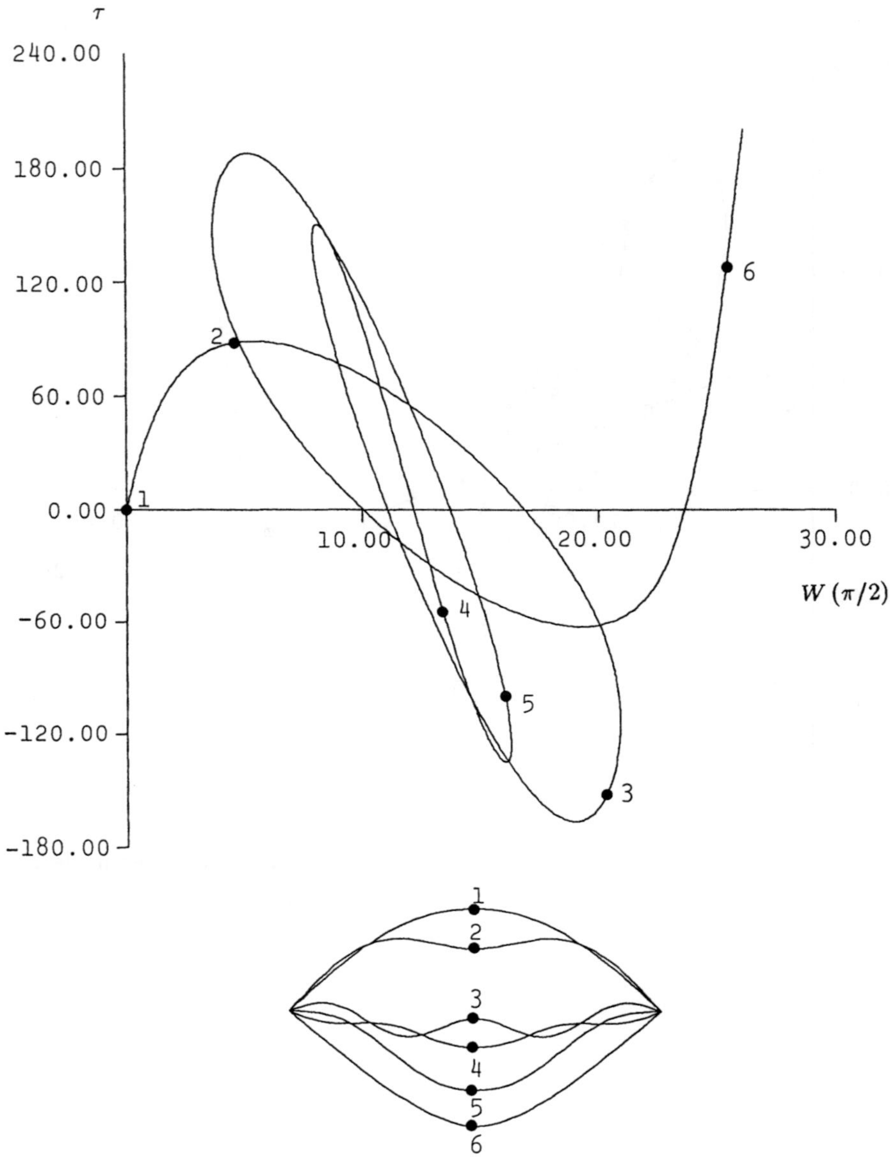

Fig.2. Fundamental path and deflection modes

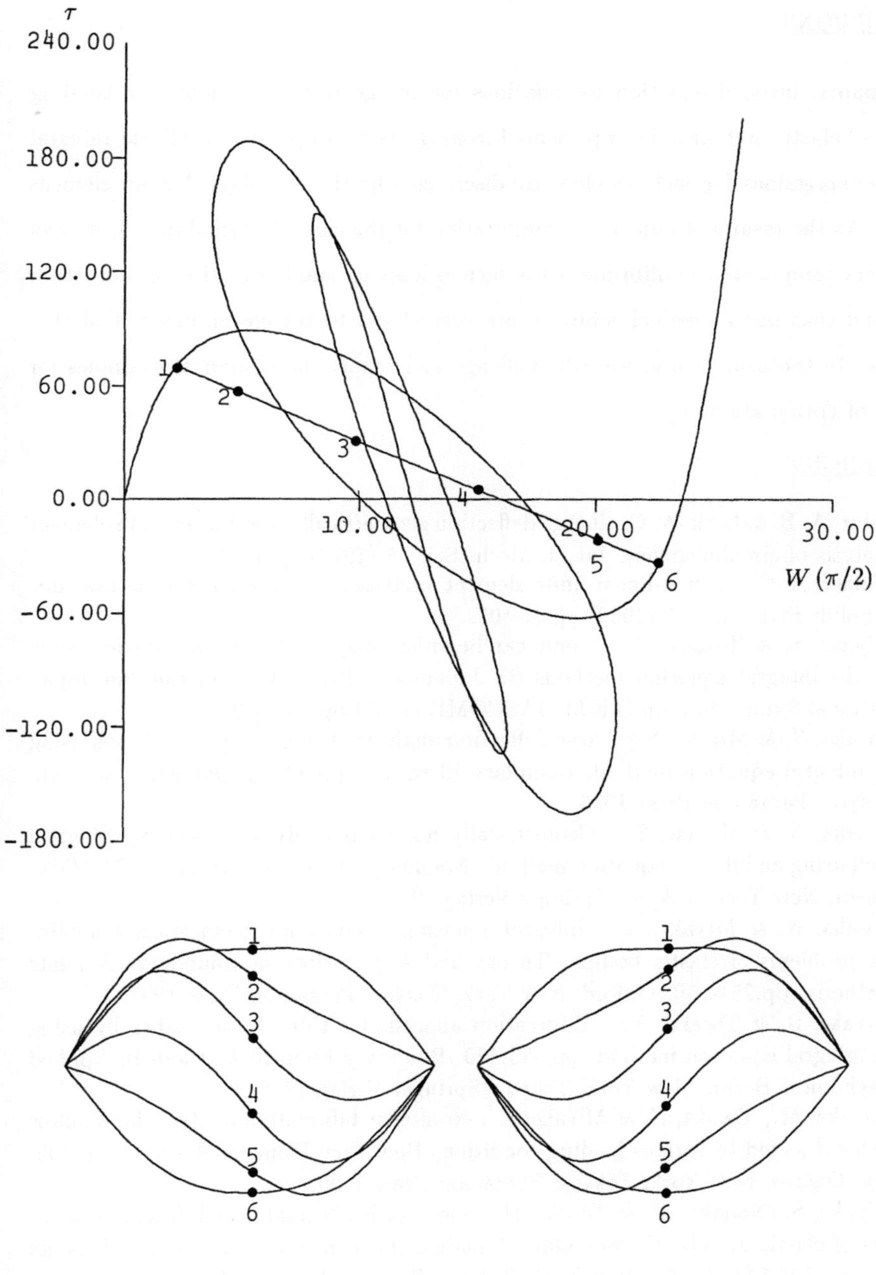

Fig.3. Bifurcation path and deflection modes

CONCLUSIONS

In this paper, integral equation formulations for the geometrically nonlinear bending problem of elastic arch have been presented from the two viewpoints. Nonlinear integral equation expressions for each problem are discretized by the boundary-domain element method. As the results of numerical computation for the case of Formulation I, we can obtain very complicated equilibrium paths such as looping and bifurcation phenomenon. It is found that our numerical solutions are very closed to the eleven-modes Galerkin solutions. In the near future, we will challenge and report the numerical examples for the case of Formulation II.

REFERENCES

[1] Sabir, A. B. & Lock, A. C. : Large deflection geometrically non-linear finite element analysis of circular arches. Int. J. Mech. Sci. 15 (1973) pp.37-47.

[2] Walker, A. C. : A non-linear finite element analysis of shallow circular arches. Int. J. Solids Structures, 5 (1969) pp.97-107.

[3] Miyake, S. & Tosaka, N. : Nonlinear bending analysis of shallow spherical shell by the integral equation methods (in Japanese). Proceedings of the 2nd Japan National Symposium on B.E.M. JASCOME (1985) pp.257-262.

[4] Tosaka, N. & Miyake, S. : Large deflection analysis of shallow spherical shell using an integral equation method, Boundary Elements. pp.59-66. Oxford, New York, Tokyo : Pergamon Press 1986.

[5] Tosaka, N. & Miyake, S. : Geometrically nonlinear analysis of shallow spherical shell using an integral equation method. Boundary Elements VIII. pp.2/573-2/576. Berlin, New York, Tokyo : Springer-Verlag 1986.

[6] Tosaka, N. & Miyake, S. : Integral equation analysis for geometrically nonlinear problems of elastic bodies. Theory and Applications of Boundary Elements Methods. pp.251-260. Oxford, New York, Tokyo : Pergamon Press 1987.

[7] Miyake, S. & Tosaka, N. : Bifurcation analysis for thin elastic bodies by using an integral equation method. pp.483-490. Boundary Element Methods in Applied Mechanics. Berlin, New York, Tokyo : Springer-Verlag 1988.

[8] Nonaka, M., Tosaka, N. & Miyake, S. : Nonlinear bifurcation analysis for shallow arch subjected to various loading condition. Boundary Element Methods. pp.405-414. Oxford, New York, Tokyo : Pergamon Press 1990.

[9] Miyake, S., Nonaka, M. & Tosaka, N. : Geometrically nonlinear bifurcation analysis of elastic arch by the boundary-domain element method. Boundary Elements XII. pp.503-514. Berlin, New York, Tokyo : Springer-Verlag 1990.

[10] Miyake, S., Nonaka, N. & Tosaka, N. : An integral Equation Formulation for Geometrically Nonlinear Problem of Elastic Circular Arch. Boundary Element Methods. pp.289-296. Oxford, New York, Tokyo: Pergamon Press 1990.

A Numerical Method for the Analysis of Nonlinear Sloshing in Circular Cylindrical Containers

Tsukasa NAKAYAMA and Hiroaki TANAKA*

Department of Precision Mechanical Engineering
Chuo University, Tokyo, Japan

Summary

A new computational method has been developed for the analysis of three-dimensional large-amplitude motion of liquids with free surfaces in moving containers. The problem is formulated mathematically as a nonlinear initial-boundary value problem under the assumption of irrotational flow of an inviscid fluid. Basic equations of the problem are discretized spacewise by the boundary element method based on Green's second identity and timewise by a forward-time Taylor series expansion. The size of a time increment is determined every time step so that the remainder of truncated Taylor series should be equal to a given small value. This variable time-stepping technique has made a great contribution to numerically stable computations. As a numerical example, swirl motion of a liquid free surface in a circular cylindrical container undergoing horizontal excitation has been analyzed.

Introduction

It is very important in engineering field to learn the dynamic behavior of liquids in partially filled containers subjected to forced excitation. Such dynamic motion of liquids, called "sloshing", often causes serious technical problems in the design of liquid-propellant spacecrafts, ships carrying liquid cargoes, oil reservoirs and so forth. For small-amplitude liquid oscillations, the linearized theory of sloshing based on the potential flow assumptions is routinely applied in design procedure. However, when excitation is near liquid natural frequencies, this results in large free surface motion which exhibits nonlinearities [1]. For example, in an axisymmetric container undergoing horizontal excitation near resonance, a rotary response of a wave on a liquid free surface occurs. This phenomenon, called "swirl" or "rotary sloshing", arises primarily from the nonlinearities. To learn such nonlinear effects theoretically, we must solve nonlinear initial-boundary value problems, even though governing equations are simplified by assuming a irrotational flow of an inviscid fluid. Then, the help of digital computers and numerical methods are required.

The first-named author has proposed a new computational method for the analysis of two-dimensional unsteady motion of a fluid with a free surface [2,3]. Basic equations

*Present address: Komukai Works, Toshiba Corporation, Kawasaki, Japan

to be solved are discretized spacewise by the boundary element method and timewise by a forward-time Taylor series expansion. The nodal points on the free surface are moved during each time step in a Lagrangian manner. The size of a time increment is variable and is determined every time step so that the remainder of truncated Taylor series should be equal to a given small value. This variable time-stepping method has made a great contribution to numerically stable computations.

In the present paper, the method mentioned above is extended for the analysis of three-dimensional nonlinear sloshing. As a numerical example, swirl motion of a free surface in a circular cylindrical container undergoing horizontal excitation has been simulated.

Mathematical Formulation of Sloshing

We consider oscillatory motion of a liquid in a circular cylindrical container as shown in Fig. 1. The container is partially filled with a liquid and is subjected to a forced horizontal oscillation. A rectangular Cartesian coordinate system, $o - xyz$, has its origin o at the center of the stationary free surface and is fixed to the container in such a manner that the z-axis coincides with the axis of symmetry of the container and is directed upwards. The fluid region V is surrounded by a free surface S_1 and wetted parts of the wall and bottom, S_2. The fluid is assumed to be inviscid and incompressible and the flow to be irrotational. Under these assumptions, the velocity potential $\phi(x, y, z, t)$ can be defined as $\nabla\phi = (u, v, w)$. Here u, v and w are the x-, y- and z-components of the fluid velocity relative to the coordinate system, respectively. Then, the governing field equation and boundary conditions are given as follows:

$$\nabla^2\phi = 0 \quad \text{in } V\,, \tag{1}$$

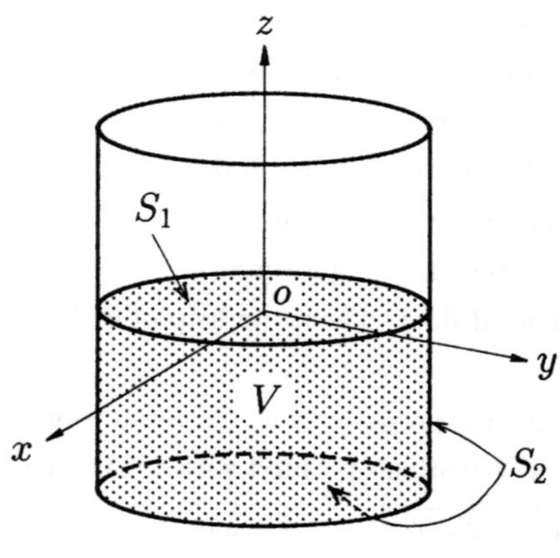

Fig. 1. A circular cylindrical container partially filled with a liquid

$$\frac{D\phi}{Dt} - \frac{1}{2}(u^2 + v^2 + w^2) + \mu\,\phi + a(t)\,\xi + g\zeta = 0 \quad \text{on } S_1\,, \tag{2}$$

$$u = \frac{D\xi}{Dt}, \quad v = \frac{D\eta}{Dt}, \quad w = \frac{D\zeta}{Dt} \quad \text{on } S_1\,, \tag{3}$$

$$\frac{\partial\phi}{\partial n} = 0 \quad \text{on } S_2\,, \tag{4}$$

where t is time; the operator D/Dt denotes the Lagrangian time differentiation and $\partial/\partial n$ means the differentiation along the outward normal drawn on the boundary; μ is a coefficient of the so-called Rayleigh damping; $a(t)$ is the forced acceleration in the x-direction; g is the acceleration due to gravity; (ξ, η, ζ) are the coordinates of fluid particles on the free surface. Those particles are used both to represent the free-surface configuration and to trace its time history.

As for initial conditions, we assume that the liquid is entirely at rest at $t = 0$.

Thus, the problem under consideration has been reduced to the nonlinear initial-boundary value problem containing ξ, η, ζ and ϕ as unknown quantities.

Solution Procedure

We consider two successive time instants, t and $t + \Delta t$, and suppose that a typical fluid particle on the free surface moves from the position (ξ, η, ζ) to the position (ξ', η', ζ') during the time interval Δt. The kinematic boundary condition (3) assures us that the particle remains on the free surface at time $t + \Delta t$. Our task is to evaluate ξ', η' and ζ' and the velocity potential on the new free-surface position using the value of those quantities at time t.

The coordinates ξ', η' and ζ' can be expanded into Taylor series about (ξ, η, ζ, t). The coordinate ξ', for example, is expanded and approximated by truncation as follows:

$$\xi' \approx \xi + \Delta t\frac{D\xi}{Dt} + \frac{(\Delta t)^2}{2}\frac{D^2\xi}{Dt^2} + \cdots + \frac{(\Delta t)^n}{n!}\frac{D^n\xi}{Dt^n}\,. \tag{5}$$

η' and ζ' are expressed similarly. To know the new position of the free surface, each term in the Taylor series must be evaluated. Then, our attention is focused on the computation of the Lagrangian time derivatives of ξ, η and ζ.

First-order Lagrangian derivatives

The first-order Lagrangian derivatives are computed by

$$\frac{D\xi}{Dt} = u = \frac{\partial\phi}{\partial x}\,, \tag{6}$$

$$\frac{D\eta}{Dt} = v = \frac{\partial\phi}{\partial y}\,, \tag{7}$$

$$\frac{D\zeta}{Dt} = w = \frac{\partial\phi}{\partial z} = \frac{1}{n_z}\left(\frac{\partial\phi}{\partial n} - \frac{\partial\phi}{\partial x}n_x - \frac{\partial\phi}{\partial y}n_y\right)\,, \tag{8}$$

where n_x, n_y and n_z are the x-, y- and z-components of the unit normal vector drawn outwardly on the free surface, respectively. The derivatives $\partial\phi/\partial x$, $\partial\phi/\partial y$ are calculated by numerical differentiation.

To obtain the normal derivative $\partial\phi/\partial n$, the boundary value problem

$$\nabla^2\phi = 0 \quad \text{in } V\,, \tag{9}$$

$$\phi = \hat{\phi} \quad \text{on } S_1\,, \tag{10}$$

$$\frac{\partial\phi}{\partial n} = 0 \quad \text{on } S_2 \tag{11}$$

is solved. Here $\hat{\phi}$ is a known quantity. Eqs. (9)-(11) are transformed into the boundary integral equation

$$\alpha_P\phi_P - \iint_{S_1}\frac{\partial\phi}{\partial n}\frac{1}{r}\,dS + \iint_{S_2}\phi\frac{\partial}{\partial n}\left(\frac{1}{r}\right)\,dS = -\iint_{S_1}\hat{\phi}\frac{\partial}{\partial n}\left(\frac{1}{r}\right)\,dS \tag{12}$$

via Green's second identity. In this equation, r is the distance between a source point P on the boundary and an observation point Q which also lies on the boundary. If P is on a smooth part of the boundary, the coefficient α_P takes the value of 2π, and it is the interior solid angle at P if P lies on a corner point. ϕ_P denotes the value of the velocity potential at P. The solution of the integral equation (12) yields $\partial\phi/\partial n$ on the free surface.

Second-order Lagrangian derivatives

The second-order Lagrangian derivatives are expressed as

$$\frac{D^2\xi}{Dt^2} = \frac{Du}{Dt} = \frac{\partial\phi_t}{\partial x} + u\frac{\partial u}{\partial x} + v\frac{\partial u}{\partial y} + w\frac{\partial u}{\partial z}\,, \tag{13}$$

$$\frac{D^2\eta}{Dt^2} = \frac{Dv}{Dt} = \frac{\partial\phi_t}{\partial y} + u\frac{\partial v}{\partial x} + v\frac{\partial v}{\partial y} + w\frac{\partial v}{\partial z}\,, \tag{14}$$

$$\frac{D^2\zeta}{Dt^2} = \frac{Dw}{Dt} = \frac{\partial\phi_t}{\partial z} + u\frac{\partial w}{\partial x} + v\frac{\partial w}{\partial y} + w\frac{\partial w}{\partial z}\,, \tag{15}$$

where $\phi_t \equiv \partial\phi/\partial t$. The derivatives of the velocity components with respect to x or y are evaluated by numerical differentiation. Those with respect to z are given by

$$\frac{\partial u}{\partial z} = \frac{\partial w}{\partial x}, \quad \frac{\partial v}{\partial z} = \frac{\partial w}{\partial y}, \quad \frac{\partial w}{\partial z} = -\left(\frac{\partial u}{\partial x} + \frac{\partial v}{\partial y}\right)\,, \tag{16}$$

where the first and second relations are the irrotational conditions of the fluid and the third is derived from the equation of continuity. Since the value of ϕ_t on the free surface can be calculated using the boundary condition (2) as

$$\begin{aligned}\phi_t &= \frac{D\phi}{Dt} - (u^2 + v^2 + w^2) \\ &= -\left\{\frac{1}{2}(u^2 + v^2 + w^2) + \mu\,\phi + a(t)\,\xi + g\,\zeta\right\}\,,\end{aligned} \tag{17}$$

the derivatives $\partial\phi_t/\partial x$ and $\partial\phi_t/\partial y$ are obtained by numerical differentiation. The derivative $\partial\phi_t/\partial z$ is then evaluated by

$$\frac{\partial\phi_t}{\partial z} = \frac{1}{n_z}\left(\frac{\partial\phi_t}{\partial n} - \frac{\partial\phi_t}{\partial x}n_x - \frac{\partial\phi_t}{\partial y}n_y\right) . \tag{18}$$

The normal derivative $\partial\phi_t/\partial n$ is given as the boundary element solution of the following boundary value problem:

$$\nabla^2\phi_t = 0 \quad \text{in } V , \tag{19}$$

$$\phi_t = \hat{\phi}_t \quad \text{on } S_1 , \tag{20}$$

$$\frac{\partial\phi_t}{\partial n} = 0 \quad \text{on } S_2 , \tag{21}$$

where the value of $\hat{\phi}_t$ is known beforehand by Eq. (17).

Higher-order Lagrangian derivatives

We proceed in the same way to calculate higher-order Lagrangian derivatives of ξ, η and ζ up to the order n. In the numerical examples shown later, n is taken as $n = 3$.

Nodal relocation

The nodal points, which are yielded by the subdivision of the free surface into boundary elements, act as fluid particles in the present method. When those nodes are continuously moved in a Lagrangian manner, the elements on the free surface will be gradually distorted. The distortion of elements causes the decrease of computational accuracy and the numerical instability. Therefore nodal relocation is required to avoid them.

The procedure of nodal relocation takes place after all the nodes on the free surface are moved in a Lagrangian manner. We consider the movement of a typical node P_i. Suppose that the computational procedure mentioned above yields the coordinate (ξ_i', η_i', ζ_i') of the node P_i at time $t + \Delta t$. Then, the node is relocated from the position (ξ_i', η_i', ζ_i') to the position (ξ_i^0, η_i^0, ζ_i''), where ξ_i^0 and η_i^0 are the x- and y-coordinates of P_i at $t = 0$. The value of ζ_i'' is calculated by interpolating nodal coordinates before relocation.

The way mentioned above is essentially equivalent to the height-function method, in which the height ζ of a free surface is the only geometrical unknown quantity which determines the free-surafce position, and as the kinematic boudary condition the equation

$$\frac{\partial\zeta}{\partial t} + u\frac{\partial\zeta}{\partial x} + v\frac{\partial\zeta}{\partial y} - w = 0$$

is often used instead of Eq. (3).

The value of the velocity potential ϕ_i'' at the position (ξ_i^0, η_i^0, ζ_i'') is calculated by the truncated Taylor series

$$\phi_i'' = \phi_i + \sum_{k=1}^{2}\frac{1}{k!}\left(\Delta t\frac{\partial}{\partial t} + \Delta z\frac{\partial}{\partial z}\right)^k \phi_i , \tag{22}$$

where $\Delta z = \zeta_i'' - \zeta_i$, and ϕ_i and ζ_i refer to time t.

The Boundary Element Method

As mentioned in the previous section, the boundary value problems of ϕ and its Eulerian time derivatives are solved by the boundary element method. The boundary element formulation starts from the derivation of the integral equation as Eq. (12).

The boundaries of the solution domain are divided into a large number of triangular elements. When Φ denotes ϕ, ϕ_t or a higher-order Eulerian time derivative of ϕ, Φ and $\partial\Phi/\partial n$ are approximated by linear shape functions in each element. Furthermore, the source point, denoted by P in Eq. (12), is chosen to be a node on the boundary, at which the residual of the integral equation is required to vanish. Thus, the integral equation is discretized into a set of linear algebraic equations with the unknown quantities of $\partial\Phi/\partial n$ on the free surface and Φ on the remaining part of the boundary.

A set of linear algebraic equations thus derived is solved by LU-decomposition method. In order to evaluate the Lagrangian time derivatives of ξ, η and ζ up to the order n, n sets of linear algebraic equations are to be solved in a time step. However, since those equations are assembled at the same time instant, they have the same coefficient matrices. Once the coefficient matrix is decomposed into a lower and an upper triangular matrices, we have only to do the forward and the backward substitutions for individual solutions of the equations. Therefore, the process of the solution of the linear algebraic equations does not require so much increase of computing time.

A Variable Time-Stepping Method

Consider a function of time, $f(t)$. Then we have a well-known expression of a Taylor series expansion

$$f(t+\Delta t) = f(t) + \Delta t f'(t) + \frac{(\Delta t)^2}{2!} f''(t) + \frac{(\Delta t)^3}{3!} f'''(t) + \cdots . \tag{23}$$

When this series is truncated at the term of n-th order derivative, the remainder is given by

$$(\text{Remainder}) = \frac{(\Delta t)^{n+1}}{(n+1)!} f^{(n+1)}(\tau) \qquad t \le \tau \le t + \Delta t . \tag{24}$$

The time increment Δt is calculated so that the remainder of the truncated Taylor series should be equal to some small error limit ε. For given value of ε, Δt is determined by

$$\Delta t = \left[\frac{\varepsilon (n+1)!}{f^{(n+1)}(\tau)} \right]^{\frac{1}{n+1}} . \tag{25}$$

When this formula is used in the present method, $f^{(n+1)}(\tau)$ is approximated as follows:

$$f^{(n+1)}(\tau) \approx \max_{1 \le i \le N} \left\{ \left| \left(\frac{D^{n+1}\xi}{Dt^{n+1}} \right)_i \right| , \left| \left(\frac{D^{n+1}\eta}{Dt^{n+1}} \right)_i \right| , \left| \left(\frac{D^{n+1}\zeta}{Dt^{n+1}} \right)_i \right| \right\} \tag{26}$$

where N is the total number of nodes on the free surface.

A Check of Computational Accuracy

The numerical results computed by the present method are compared with available experimental data. A circular cylindrical container having a radius of 0.092 m is filled with water to a height of 0.092 m. The container is subjected to the forced sinusoidal displacement

$$X(t) = X_0 \sin(2\pi f t) \quad \text{for } t \geq 0 .$$

Here the amplitude X_0 takes the value of 1 mm or 3 mm.

In Fig. 2, the maximum value of the free surface displacement is plotted against the different forced frequency f. In the figure, H and f_1 denote the water depth and the natural frequency of the first antisymmetric slosh mode, respectively. The open circles denote experimental data [4] and the solid circles denote the present numerical results. Qualitative agreements between the two are good. In the present computations, the error limit ε used for determining the size of Δt is taken as $\varepsilon = 10^{-5}$ and the damping coefficient μ is equal to zero.

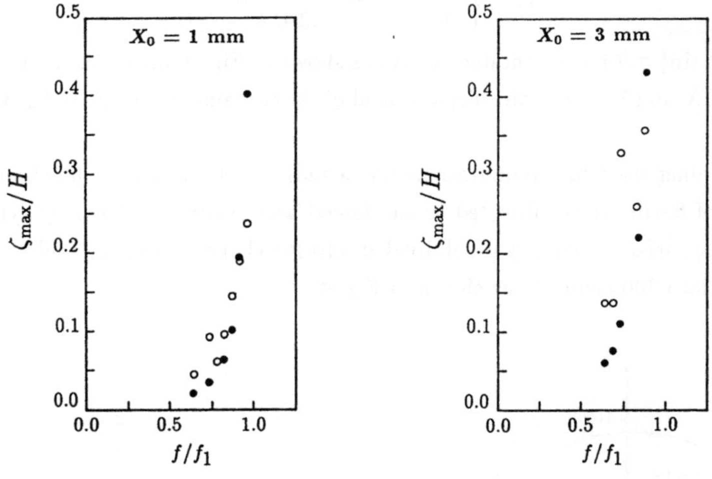

Fig. 2. Maximum free surface displacement at different forced frequency (o : experiment, • : present method)

Numerical Simulation of Swirl Motion

When an axisymmetric container partially filled with a liquid is forced to oscillate horizontally at a frequency considerably below the lowest natural frequency of liquid oscillation, the liquid free surface responds with an antisymmetric mode having a stationary nodal diameter perpendicular to the direction of excitation. At a frequency in the neighborhood of the natural frequency, it is observed that the free surface oscillating with an antisymmetric mode begins to rotate along the side wall of the container. To simulate numerically such a rotary response of a free surface, we should know what causes the transition from an antisymmetric mode to a swirl. It may be the frictional force acting between a liquid and a container wall or some small disturbance initially existing in a fluid. Since the present method is based on inviscid fluid flows, it is difficult to take frictional effects into account. Then, we try to generate a small rotational disturbance at the initial stage of excitation. It follows that, in this approach to the simulation of swirl, our aim of computations is to see if the disturbance causes swirl motion or not.

One of the ways to generate a rotational disturbance is to force a container to circle by oscillating it in the y-direction as well as in the x-direction. For example, when the container is subjected to both the forced acceleration

$$a(t) = A\sin(2\pi f t) \tag{27}$$

with $A/g = -0.0178$ and $f = 0.940\,\mathrm{Hz}$ in the x-direction and that in the y-direction given by

$$b(t) = \begin{cases} B\cos(2\pi f t) & 0 \le t \le 3 \\ 0 & 3 < t \end{cases} \tag{28}$$

where $B = A\sin(\pi t/6)$, the container moves as shown in Fig. 3 during the first 3 seconds. In the figure, X and Y denote the displacement of the container in the x- and y-directions, respectively.

The container used for computations has a radius of 0.5 m and is filled with water to a height of 0.6 m. It is subjected to the forced acceleration of the same type as Eq. (27). The free surface boundary is subdivided into 108 elements and the wall and bottom boundary is into 300 elements as shown in Fig. 4.

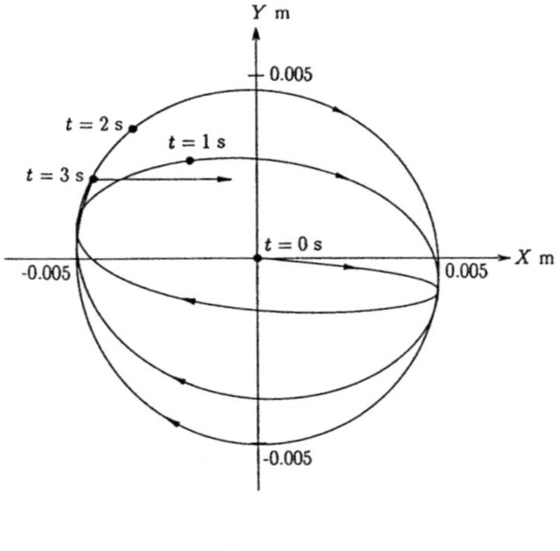

Fig. 3. Circling motion of a container

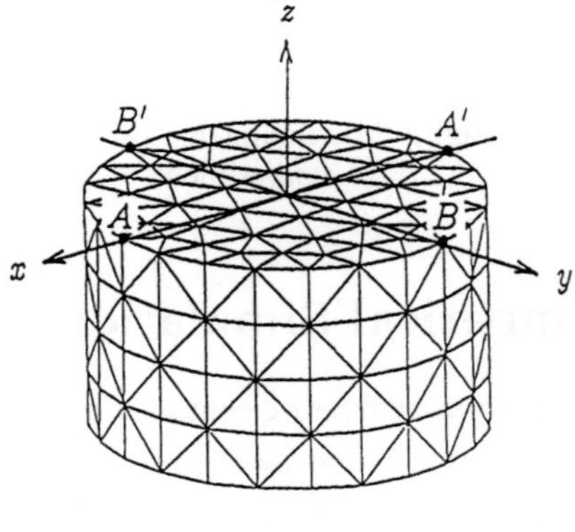

Fig. 4. Subdivision of boundaries

Figures 5(a) and 5(b) show the time histories of the free surface displacement at the stations A, A′ and B, B′ shown in Fig. 4, respectively. The amplitude and frequency of the forced acceleration are taken as $A/g = -0.0178$ and $f = 0.940\,\mathrm{Hz}$. The first natural frequency of the antisymmetric slosh mode is 0.944 Hz. It can be seen in Fig. 5(b) that the amplitude of the displacement at B and B′ begins to increase rapidly about 11 seconds. This means that the free surface begins to rotate and swirl motion occurs. Figure 6 shows the free surface configurations at different time instants. It is observed that the wave on the free surface rotates clockwise along the side wall of the container.

Figure 7 shows the time histories of the free surface displacement in the case of $A/g = -0.0128$ and $f = 0.796\,\mathrm{Hz}$. It is seen that the displacement at B and B′ caused by an initial disturbance decays gradually and swirl motion never occurs.

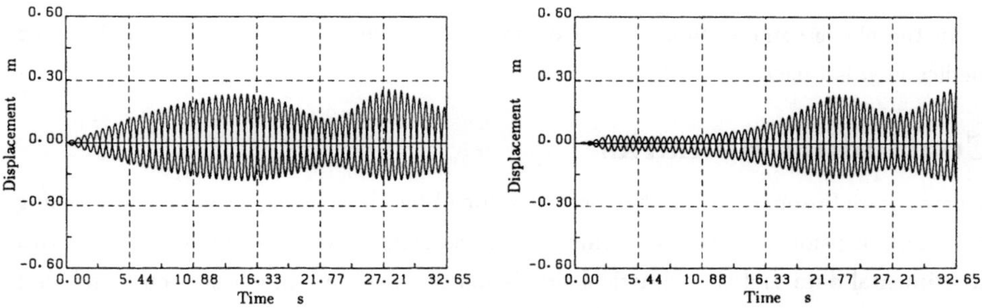

(a) Displacement at A and A′ (b) Displacement at B and B′

Fig. 5. Time histories of free surface displacement ($A/g = -0.0178$, $f = 0.940\,\mathrm{Hz}$)

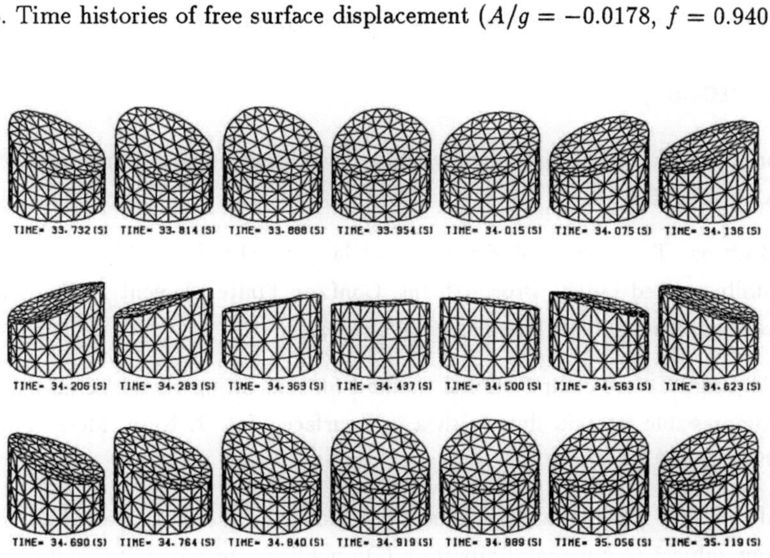

Fig. 6. Swirl motion of a free surface

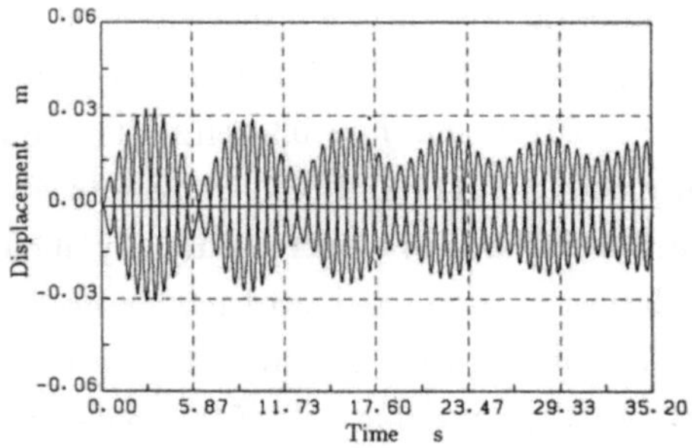

(a) Displacement at A and A′

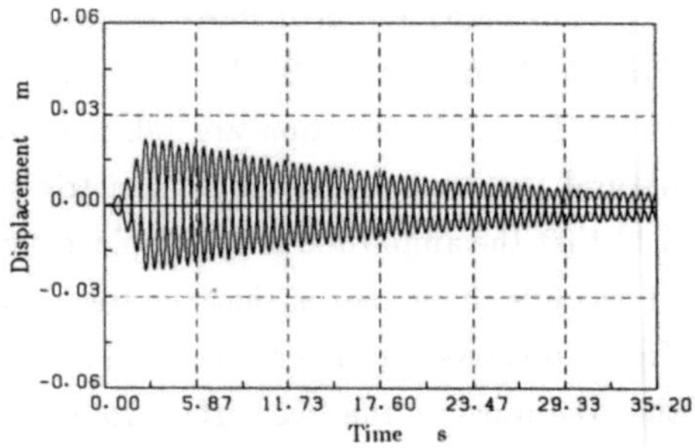

(b) Displacement at B and B′

Fig. 7. Time histories of free surface displacement ($A/g = -0.0128$, $f = 0.796\,\mathrm{Hz}$)

In the above computations, the error limit ε is taken as $\varepsilon = 10^{-6}$ and the damping coefficient μ is taken as $\mu = 0.1\,\mathrm{s}^{-1}$.

Concluding Remarks

A new computational method has been developed for the analysis of nonlinear sloshing in three-dimensional containers. Throughout the comparison of numerical results with experimental data and the numerical simulation of swirl motion, it has been found that the present method is accurate and numerically stable. It will be the next challenging subject to apply the present method to numerical experiments on swirl motion of liquids in spherical or conical containers as well as circular cylindrical containers.

References

1. Abramson, H. N. (editor): The dynamic behavior of liquids in moving containers. NASA SP-106, 1966.

2. Nakayama, T.: Numerical simulation of large-amplitude liquid sloshing in horizontally excited tanks. Proc. 7th Int. Conf. on Finite Element Methods in Flow Problems, Huntsville, Alabama, 1989, 659-664.

3. Nakayama, T.: A computational method for simulating transient motions of an incompressible inviscid fluid with a free surface. Int. J. Num. Meth. Fluids, 10 (1990) 683-695.

4. Sudo, S. and Hashimoto, H.: Dynamic behavior of a liquid in a cylindrical container subject to horizontal vibration (On nonlinear response of liquid surface)(in Japanese). Trans. Japan Soc. Mech. Engng., 52 (1986) 3655-3659.

Coupling of Finite Elements and Consistent Boundary Elements in Structural Analysis

A. NAPPI

Department of Structural Engineering
Politecnico di Milano - P.zza L. da Vinci 32
20133 Milano - Italy

Summary

Boundary elements characterised by continuous displacement fields and by discontinuous tractions, denoted as "consistent" in this paper, are discussed and coupled with finite elements. Some emphasis is given to inelastic problems and to a forced symmetric direct boundary element formulation. Finally, numerical tests are reported which seem to evidence some interesting aspects of the approach based on consistent elements.

A consistent formulation for BE elastic analysis

By following the traditional "direct" boundary element approach [1-3], in the case of a linear elastic analysis we derive a matrix equation such as $\underline{H}\ \underline{u} = \underline{G}\ \underline{p}$, where $\underline{H}$ and $\underline{G}$ are constant matrices, while $\underline{u}$ and $\underline{p}$ are obtained by assembling E subvectors $\underline{u}_e$ and $\underline{p}_e$ (with e=1,..,E and E=number of boundary elements). These subvectors contain displacements and tractions at properly selected points. Thus, by introducing convenient interpolation functions $\underline{\phi}_d(\underline{x})$ and $\underline{\phi}_t(\underline{x})$, the displacements and the tractions along each element are expressed as

$$\underline{u}(\underline{x}) = \underline{\phi}_d(\underline{x})\ \underline{u}_e \quad , \quad \underline{p}(\underline{x}) = \underline{\phi}_t(\underline{x})\ \underline{p}_e \qquad (1a,b)$$

We can now introduce some conditions (essentially related to the interpolation functions) required in order to obtain a formulation which can be denoted as "consistent" according to the definition given in Ref. 4. More specifically, we can impose, for each boundary element, interpolation functions $\underline{\phi}_d(\underline{x})$ and $\underline{\phi}_t(\underline{x})$ whose orders are m and (m-1), respectively (with $m \geq 2$). In addition, for each element we shall define the vector $\underline{p}_e$ at (m-1) points which are never coincident with the element ends, while the vector $\underline{u}_e$ will always be defined at least at the element ends. In the case of plane problems, for m=2 the selection is obvious: the end-points for $\underline{u}_e$ and the mid-point for $\underline{p}_e$. For $m > 2$ a natural choice is m equally spaced points for $\underline{u}_e$ and (m-1) Gauss points for $\underline{p}_e$. Thus displacement continuity is ensured and traction discontinuity is allowed when we pass from one element to a contiguous one. As well known, by collecting the given entries of $\underline{u}$ and $\underline{p}$ into a vector $\underline{X}$, we can write the final equation $\underline{A}\ \underline{X} = \underline{B}$. Of course, the rows of the matrices $\underline{A}$, $\underline{B}$, $\underline{H}$ and $\underline{G}$ are as many as the number of unknowns (say n_x, i.e. the number of terms of the vector $\underline{X}$). Hence, n_x scalar equations must be represented and each equation is obtained through a convenient selection of the so called "collocation points" [1-3]. For instance, let us consider the square panel of Fig. 1 subjected to any set of tractions along C-E or E-G. If the functions $\underline{\phi}_d(\underline{x})$ and $\underline{\phi}_t(\underline{x})$ are of order two and one, respectively, we shall use the points B, C, D, E, F (α, β) as collocation points for the horizontal displacements (tractions) and the points D, E, F, G, H (γ, δ) for the vertical displacements (tractions). In fact, the unknowns are the horizontal displacements at the nodes from B to F, the vertical displacements at the nodes from D to H, the horizontal tractions along A-G and the vertical tractions along A-C (so that n_x=14).

Combination of consistent boundary elements with finite elements

The interpolation functions and the collocation points selected according to the consistent approach followed in the paper, make the boundary elements particularly suitable for their combination with finite elements whose displacement fields are described by interpolation functions of the same order. Indeed, as the number of elements tends to infinite, constant strain fields can be reproduced both at the finite element level and at the boundary element level. In addition, as required by a well known convergence criterion applicable to the finite element method, "if nodal displacements are compatible with a constant strain condition, such constant strain will in fact be obtained" [5].

The coupling procedure is quite obvious and fully analogous to what is currently done with reference to conventional boundary element formulations (see, e.g., Refs. 6-11).

By following a procedure originally proposed for finite elements [12] and subsequently utilised for combined finite and boundary elements [10], we can apply a substructuring technique, which provides some operational advantage. For the sake of brevity, we shall not give details of the intermediate matrix manipulations, which can be found in Ref. 13. In any case we end up with equations such as

$$\underline{K}^{*}\,\underline{u}^{c} = \underline{Q}^{c} + \underline{Q}^{*}\,,\quad \underline{h}\,\underline{u}^{c} = \underline{g}\,\underline{p}^{c} + \underline{b}^{*} \quad\text{or}\quad \underline{p}^{c} = \underline{g}^{-1}\left\{\underline{h}\,\underline{u}^{c} - \underline{b}^{*}\right\} \qquad (2a,b,c)$$

where $\underline{u}^{c}$, $\underline{Q}^{c}$ and $\underline{p}^{c}$ represent displacements, loads and tractions at nodal points along the common interface. $\underline{K}^{*}$, $\underline{h}$ and $\underline{g}$ are constant matrices, while $\underline{Q}^{*}$ and $\underline{b}^{*}$ are constant vectors. The former is due to given loads and displacements in the finite element region, the latter to imposed tractions and displacements in the boundary element region.

The above equations can be combined, since $\underline{Q}^{c}$ must represent a vector of nodal loads equivalent to the tractions defined by the nodal values $\{-\underline{p}^{c}\}$. Thus, we should introduce a relationship such as $\underline{Q}^{c}=\underline{L}\,\underline{p}^{c}$, where $\underline{L}$ is obtained through convenient integration of displacement interpolation functions. Next, by combining Eqs. (2a) and (2b), we obtain

$$\underline{K}\,\underline{u}^{c} = \underline{P} \quad\text{with}\quad \underline{K} \equiv \left[\underline{K}^{*} - \underline{L}\,\underline{g}^{-1}\,\underline{h}\right] = \underline{K}^{*} + \underline{K}_{BE} \quad\text{and}\quad \underline{P} \equiv \underline{Q}^{*} - \underline{L}\,\underline{g}^{-1}\,\underline{b}^{*} \qquad (3a,b,c)$$

Note that the matrix $\underline{K}$ is not symmetric owing to $\underline{K}_{BE}$, but can be inverted if the structure is properly constrained. Hence, we obtain $\underline{u}^{c}$ and, through an obvious back-substitution process, the remaining unknowns.

<u>Extension to inelastic structural analysis</u>

Combined boundary elements and finite elements can be used with potential benefits when inelastic strains occur, since part of the domain is to be discretised in any case. For instance, when damage and/or creep and/or elastic-plastic behaviour should be accounted for, the well-established framework developed within the context of finite elements can be utilised in a straightforward way. In addition, if we apply the substructuring technique suggested in the previous Section, the typical pattern of the finite element approach (including symmetric operators) is preserved to a large extent. However, if we consider the incremental, step-wise holonomic (reversible) problem currently utilised for inelastic structural analysis, the non-symmetric (*and* non-definite) matrix $\underline{K}$ in Eq. (3a) does not allow us to prove some extremal properties which are typical of systems discretised by finite elements [14-16]. As a consequence, some important convergence properties can not be demonstrated. Thus, the coupling procedure discussed in the paper would acquire more interest if a symmetric, positive definite matrix $\underline{K}$ can be introduced into an equation such as (3a).

To this aim we can apply a well known symmetrisation technique based upon the total potential energy $\mathcal{E}$ of the given structural system. In fact, it can be shown that the discrete boundary element model leads to the expression

$$\mathcal{E} \simeq \frac{1}{2}\, \underline{u}^T \left(\int_\Gamma \underline{\Phi}_d^T\, \underline{\Phi}_t \, d\Gamma \right) \underline{p} - \underline{u}^T \int_{\Gamma_p} \underline{\Phi}_d^T\, \hat{\underline{p}} \, d\Gamma \tag{4}$$

where $\underline{\Phi}_d$ and $\underline{\Phi}_t$ are obtained by properly assembling the shape functions which appear in Eqs. (1), $\underline{p}$ is related to $\underline{u}$ through the equation $\underline{H}\ \underline{u}=\underline{G}\ \underline{p}$, $\hat{\underline{p}}$ represents given surface tractions and Γ_p is that part of the boundary along which these tractions are applied. Thus, we get

$$\mathcal{E} \simeq \frac{1}{2}\, \underline{u}^T\, \underline{M}\, \underline{G}^{-1}\, \underline{H}\, \underline{u} - \underline{u}^T\, \underline{P}' \tag{5}$$

where $\underline{M}$ and $\underline{P}'$ are obtained by computing the two integrals in (4). Since the above expression approximates the total potential energy, we can conceive a solution vector $\underline{u}^*$ which minimises $\mathcal{E}$ as defined in Eq. (5). Clearly, this vector also solves the system

$$\frac{1}{2} \left[\underline{M}\, \underline{G}^{-1}\, \underline{H} \right] \underline{u} + \frac{1}{2} \left[\underline{M}\, \underline{G}^{-1}\, \underline{H} \right]^T \underline{u} = \underline{K}'\ \underline{u} = \underline{P}' \tag{6}$$

Here, the matrix $\underline{K}'$ plays the role of a fictitious stiffness matrix. This procedure, based upon a technique originally suggested in Ref. 6, appears to be currently used, although the matrix associated to the quadratic form in Eq. (5) is neither symmetric nor positive definite. Thus, it can not represent (in view of the modelling errors) an elastic strain energy. As a consequence, it is quite arbitrary to look for a solution of the elastic problem in point by minimising $\mathcal{E}$ as given by Eq. (5).

Alternatively, as suggested by Beer [10], we can derive an expression such as (4) with Γ referred to the common interface. Thus, an equation formally identical to (5) is obtained, but $\underline{u}$ and $\hat{\underline{p}}$ are now concerned with the interface only. It is possible to show that this procedure simply implies a forced symmetry of the matrix $\underline{K}_{BE}$ defined in Eq. (3b) when linear displacement fields and constant tractions are assumed along each element. In any case we shall solve a system such as $\underline{K}_s\ \underline{u}^c = \underline{P}$, instead of (3a), where $\underline{K}_s$ is a fictitious symmetric stiffness matrix.

It is worth noting that we might also proceed as follows. First, we compute the surface tractions (say $\underline{p}_k$) which correspond to a component of $\underline{u}$ (say u_k) set equal to one, while all the other components are zero. By updating the non-zero entries of $\underline{u}$, we obtain n vectors of "equivalent nodal loads" (if $\underline{u}$ is a vector of n entries). These are fully analogous to the equivalent nodal loads typical of finite element models and form the columns of the square matrix

$$\underline{K}^{o} \equiv \left[\underline{F}_1 \; \vdots \; \dots \; \vdots \; \underline{F}_n\right] \quad \text{with} \quad \underline{F}_k = \left(\int_{\Gamma} \underline{\Phi}_d^t \, d\Gamma\right) \underline{p}_k \quad \text{and } k=1,\dots,n \qquad (7a,b)$$

Clearly each term K^{o}_{ij} represents the i-th force (or "equivalent nodal load") due to a unit j-th displacement. In view of Betti's theorem $\underline{K}^{o}$ should be symmetric and its symmetry may be enforced. Note that $\underline{K}^{o} = \underline{M}\,\underline{G}^{-1}\,\underline{H}$ when linear interpolation functions are used for displacements. Hence, in this case the symmetric part of $\underline{K}^{o}$ coincides with $\underline{K}'$ in Eq.(6).

The same path of reasoning can be followed by considering the displacements at the nodes along the common interface and the corresponding set of "equivalent nodal loads". Again, in view of Betti's theorem, a fictitious stiffness matrix can be introduced. As before, this matrix coincides with the symmetric part of $\underline{K}_{BE}$ in Eq. (3b) when linear interpolation functions are utilised for the displacements.

Although the problem of symmetric formulations and the above approaches are not new, to the author's knowledge consistent boundary elements have never been considered in this context. Therefore, numerical tests may be of some interest, particularly when referred to the effect of mesh refining. Tests of this kind will be the object of the next Section.

Symmetric Vs. non-symmetric formulation: some numerical tests

We shall discuss numerical results obtained by considering the simple panels of Fig. 2 subjected to plane stress conditions. The panel of Fig. 2b has been obtained by means of a partial distortion of the first panel (Fig. 2a). Such distortion has been introduced with the aim of avoiding non desired effects due to the geometric symmetries.

Linear interpolation functions have always been used for the displacements, while different meshes have been considered, with an increasing number of boundary elements and of finite elements: sixteen (see Fig. 2), thirty-two and sixty-four. Two load distributions have been imposed: a uniform vertical traction p_V and a uniform horizontal traction $p_H=p_V$ acting along the top horizontal edge. The response has been computed by using the original, non-symmetric matrix $\underline{K}$ defined in Eq.

(3b) and by enforcing either $\underline{K}^o$ in Eq. (7a) or $\underline{K}_{BE}$ in Eq. (3b) to be symmetric. As shown in the previous Section, the latter symmetrisation procedure rests on a total potential energy which is function of $\underline{u}^c$ only. The former procedure, on the other hand, implies the definition of a total potential energy as a function of all the nodal displacements. Thus, starting from the fictitious stiffness matrix given by the symmetric part of $\underline{K}^o$, we end up with a system such as $\underline{K}_s\,\underline{u}^c = \underline{P}$, by means of a substructuring technique.

Some results are summarised in the table reported in the next page, where the following information is provided for each case:

a. Matrix whose symmetry has been enforced ($\underline{K}^o$ or $\underline{K}_{BE}$).

b. Panel used (a = panel of Fig. 2a, b = panel of Fig. 2b).

c. Number of boundary elements (n_{BE}) for the discrete model.

d. Applied load (uniform traction p_V or p_V together with p_H).

e. Value of a parameter (which can be denoted as "index of symmetry") related to the pattern of the matrix to be made symmetric. For any $m \cdot m$ matrix $\underline{\mu}$ this index has been defined as

$$i_S \equiv 1 - \frac{2}{m(m-1)} \sum_{1}^{m-1}{}_{i} \sum_{i+1}^{m}{}_{j} \frac{|\mu_{ij} - \mu_{ji}|}{|\mu_{ij}| + |\mu_{ji}|} \tag{12}$$

It ranges between zero (skew-symmetric matrix) and one (symmetric matrix).

f. Minimum eigenvalue (denoted by the symbol β) of the matrix $\underline{K}_s$, obtained by starting from the symmetric part of $\underline{K}^o$, or minimum eigenvalue of the symmetric part of $\underline{K}$ in Eq. (3b).

g. Scalar quantities which quantify the errors introduced by the forced symmetrisation. To this aim we have considered (for each discretisation and each load condition) the vectors $\underline{u}_o$ and $\underline{F}_o$ which represent nodal displacements and loads statically equivalent to the tractions along the constrained elements, as computed without forced symmetrisation. Next, we have determined (for each symmetric formulation) the vectors $\delta\underline{u}$ and $\delta\underline{F}$, which denote the increments of $\underline{u}_o$ and $\underline{F}_o$ due to the forced symmetrisation. Thus we have found the percentage errors e_u=100 $\|\delta\underline{u}\|/\|\underline{u}_o\|$ and e_F=100 $\|\delta\underline{F}\|/\|\underline{F}_o\|$, where $\|\cdot\|$ denotes the Eucledean norm of a vector.

MATRIX MADE SYMMETRIC	PANEL	n_{BE}	APPLIED LOAD	INDEX OF SYMMETRY	β	e_u	e_F
$\underline{K}^o$	a	16	p_v	0.9141	7213	0.45	0.30
	a	32	p_v	0.9357	3625	0.54	0.41
	a	64	p_v	0.9523	1785	0.24	0.21
$\underline{K}_{BE}$	a	16	p_v	0.5659	7415	0.79	0.27
	a	32	p_v	0.6925	3676	0.36	0.19
	a	64	p_v	0.8019	1775	0.15	0.11
$\underline{K}^o$	b	16	p_v	0.7886	5501	12.09	10.04
	b	32	p_v	0.8238	3952	8.89	6.67
	b	64	p_v	0.8535	2021	3.39	3.65
$\underline{K}_{BE}$	b	16	p_v	0.5526	9236	3.06	2.64
	b	32	p_v	0.6195	5193	1.42	0.96
	b	64	p_v	0.6980	2777	0.45	0.46
$\underline{K}^o$	b	16	p_v, p_H	0.7886	5501	48.93	50.23
	b	32	p_v, p_H	0.8238	3952	23.39	27.57
	b	64	p_v, p_H	0.8535	2021	23.25	34.91
$\underline{K}_{BE}$	b	16	p_v, p_H	0.5526	9236	1.17	2.15
	b	32	p_v, p_H	0.6195	5193	0.46	0.63
	b	64	p_v, p_H	0.6980	2777	0.29	0.53

The results obtained by using the second symmetrisation strategy (i.e., enforcement of symmetry for the matrix $\underline{K}_{BE}$) seem to be quite satisfactory. On the contrary, when the symmetric part of $\underline{K}^o$ is considered, large errors often occur, particularly in the presence of shear forces. This is probably due to the following reason: as pointed out in the previous Section, $\underline{K}^o$ basically represents reactions due to convenient sets of imposed displacements. Therefore, these reactions should be reasonably accurate, in order to obtain good results. However, it is well known that the "direct" boundary integral approach tends to give rather poor solutions in terms of tractions. Hence, inaccurate estimates of these tractions (together with the approximation due to forced symmetry) certainly affect the subsequent computations. Of course, when the matrix $\underline{K}_{BE}$ is made symmetric, undesirable effects tend to be concentrated only at the common interface.

It is worth noting that refined meshes generally lead to progressive improvements for the index of symmetry and for the final results. This fact may be typical of a consistent formulation, since convergence towards the correct solution should be obtained as the number of elements tends to infinite.

Closing remarks

Consistent boundary elements have been combined with finite elements and the effectiveness of the coupling procedure has been demonstrated to some extent by means of numerical tests.

Some emphasis has been given to the possibility of using combined boundary elements and finite elements for nonlinear, inelastic structural analysis. In this context the problem of symmetric formulations has been considered, since symmetry and positive definiteness of a particular operator can guarantee the convergence of time integration schemes often utilised for nonlinear analysis (e.g., for elastic plastic systems under certain hypothesis concerning the material behaviour).

As well known, symmetric operators can be obtained with boundary elements by following a Galërkin approach [17]. Here, on the other hand, a classical technique already utilised with traditional boundary elements in the context of the "direct" approach has been applied in order to enforce the symmetry of the operators involved in the elastic analysis.

Numerical tests have been performed to check the consequences of the enforced symmetry, with particular attention given to the eigenvalues of the new (symmetrised) matrices of some structural system. So far the eigenvalue spectra have always shown positive-definiteness of the relevant matrices (so that convegence is ensured for some time-integration schemes under convenient hypotheses on the material behaviour).

In addition, the symmetrisation procedure appears to introduce relatively small errors, at least with refined meshes. In fact, there seems to be a trend towards a more and more symmetric pattern of the original matrices (to be symmetrised) when the boundary elements increase. Indeed, as their number tends to infinite, a given system should be described correctly and the matrices in point should become symmetric, although there is no strict correlation between refined meshes and higher accuracy with boundary elements. This result, however, might be interpreted as a reasonable, characteristic feature of the consistent formulation.

Acknowledgements

A grant from CNR is gratefully acknowledged.

References

1. Banerjee, P.K.; Butterfield, R.:Boundary Element Methods in Engineering Science. London: McGraw-Hill, 1981.

2. Brebbia, C.A.; Telles, J.C.F.; Wrobel, L.: Boundary Element Techniques - Theory and Applications. Berlin: Springer-Verlag, 1984.

3. Stein, E.; Wendland, W.L.: Finite Element and Boundary Element Techniques from Mathematical and Engineering Point of View. Wien: Springer-Verlag, 1988.

4. Nappi, A.: Boundary element elastic analysis on the basis of a consistent formulation, Appl. Math. Mod., 13, 4, (1989) 234-241.

5. Zienkiewicz, O.C.; Taylor, R.L.: The Finite Element Method, Vol. 1. London: McGraw-Hill Book Co., 1989.

6. Zienkiewicz, O.C.; Kelly, D.M.; Bettess, P.: The coupling of the finite element method and boundary solution procedures, Int. J. Numer. Meth. Engng., 11, (1977) 355-376.

7. Brebbia, C.A.; Georgiou, P.: Combination of boundary and finite elements for elastostatics, J. Appl. Math. Modelling, 3, (1979) 212-220.

8. Beer, G.; Meek, J.L.: The coupling of boundary and finite element methods for infinite domain problems in elastoplasticity, in: Boundary Element Method, (Edited by C.A. Brebbia), Springer, Berlin, (1981) 575-591.

9. Tullberg, O.; Bolteus, L.: A critical study of different boundary element stiffness matrices, in: Boundary Element Methods in Engineering, (Edited by C.A. Brebbia), Springer, Berlin, (1982) 625-635.

10. Beer, G.: Finite element, boundary element and coupling analysis of unbounded problems in elastostatics, Internat. J. Numer. Meths. Engrg, 19, (1983) 567-580.

11. Li, H.B.; Han, G.M.; Mang, H.A.; Torzicky P.: A new method for the coupling of finite element and boundary element discretized subdomains of elastic bodies, Comp. Meth. in Appl. Mech. and Engrg., 54, (1986) 161-185.

12. Przemieniecki, J.S.: Matrix structural analysis of substructures, AIAA Journal, 1, (1963) 138-147.

13. Nappi A.: Applications of a consistent boundary element formulation for structural analysis, in: Volume in Memory of Prof. Manfredi Romano, to appear.

14. Martin, J.B.; Reddy, B.D.: Variational principles and solution algorithms for internal variable formulations of problems in plasticity, in: Prof. G. Ceradini Anniversary Volume, Roma, (1988) 465-477.

15. Maier, G.; Nappi A.: Backward difference time integration, nonlinear programming and extremum theorems in elastoplastic analysis, S.M. Archives, 14, 1, (1989) 17-36.

16. Nappi, A.: Application of convex analysis concepts to the numerical solution of elastic plastic problems by using an internal variable approach," Engineering Optimization, to appear

17. Maier, G.; Polizzotto C.: A Galërkin approach to boundary element elastoplastic analysis, Compt. Meth. in Appl. Engrg., 60, (1987) 175-194.

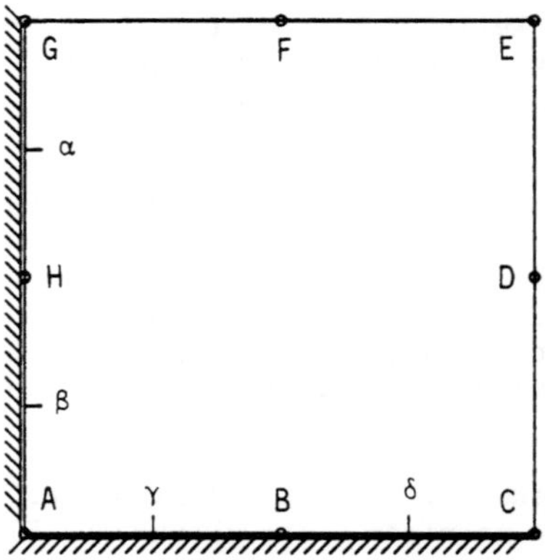

Fig. 1 - Square panel discretised by eight boundary elements

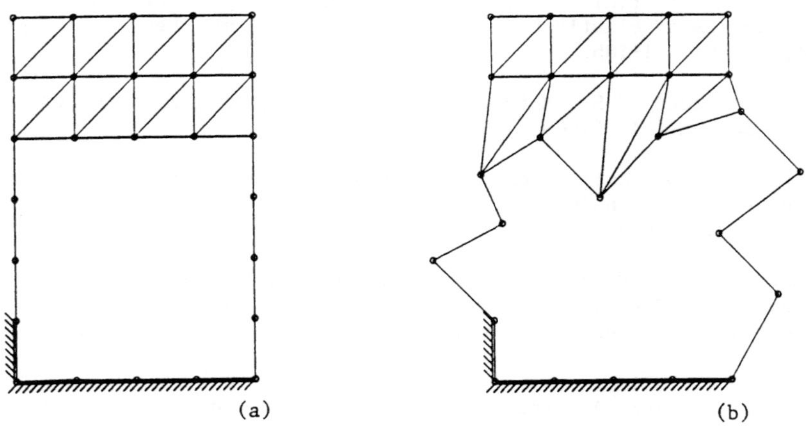

Fig. 2 - Panels used for numerical test as discretised by sixteen boundary elements and sixteen finite elements (lower edge: 100 cm, height: 150 cm, Poisson's ratio: 0.3).

Regularization in 3D for Anisotropic Elastodynamic Crack and Obstacle Problems

Jean-Claude NEDELEC[1]
Eliane BECACHE[1]
Naoshi NISHIMURA[2]

I - INTRODUCTION

The problems of wave scattering by obstacles or cracks appear very often in geophysics and in mechanics. In particular the linearized theory of elastodynamics for 3 dimensional elastic material is used frequently, because this theory keeps the analysis relatively simple. Even with this theory, however, a practical analysis is possible only with the use of some numerical methods. This has been the raison d'être of many numerical experiments carried out in the engineering community. Among those numerical methods tested so far, the boundary integral equation (BIE) method has been accepted favourably by engineers, presumably because it can deal with scattered waves effectively in external problems. In particular the double layer potential representation is considered to be an efficient tool of numerical analysis for wave problems including cracks. The only inconvenience of the double layer potential approach, however, is the hypersingularity of the kernel, which does not permit the use of conventional numerical integration techniques. Hence we can take advantage of this approach only after weakening the hypersingularity of the kernel, or only after 'regularizing' it. As a matter of fact, some of such attemps can be found in the articles by Sladek & Sladek [11], Bui [5], Bonnet [4], Polch et.al [10], Nishimura & Kobayashi [8], [9] who used the collocation method and in Nedelec [7], Bamberger [1] where the variational method has been used.

As the number of the publications on this subject tells, there exist various different possibilities of the regularization. However, not all of these regularizations are universal because some of them work only with collocation methods, and because others may destroy the causality in the time domain. For example the authors of [11], [5], [4], [10] seem to have devised their techniques mainly with collocation methods in their minds. Also the formulation in [7] may not be very useful in time domain because it will produce kernels which violate the causality. In addition, there is no guarantee that the generalizations of the formulations in [1] and [8] preserve the causality in the general anisotropic case, although they do in the isotropic case. In view of this we shall investigate a unified method of generating 'good' regularized

[1] CMAP, Ecole Polytechnique, 91128 PALAISEAU CEDEX, FRANCE

[2] Department of Civil Engineering, KYOTO University, KYOTO 606, JAPAN

integral equations in the double layer potential approach for the general 3 dimensional anisotropic elastodynamics.

This paper begins by recapitulating the governing equations of the 3 dimensional elastodynamics. We then discuss the structure of the hypersingular kernel Σ which appears in the integral equation obtained from the double layer potential in the frequency domain. Specifically, we shall show that Σ allows a decomposition into a sum of a hypersingular kernel 'rot rot rot rot Φ' and a regular kernel R in a way that Φ and R maintain the causality possessed by Σ. Also we shall present an explicit form of Φ for the general 3 dimensional anisotropic case. We then proceed to demonstrate that this decomposition readily produces a variational form in terms of Φ and R. The kernels included in this variational form will be seen to inherit the correct causality possessed by Φ and R. We then discuss the isotropic case. A few remarks concerning collocation conclude this paper.

II - GOVERNING EQUATIONS

Let Ω^- be an open bounded domain of $\mathbf{R}^3$ whose boundary is a regular closed surface Γ and let Ω^+ be the open domain complement to $\overline{\Omega^-}$. We are interested in solving wave propagation problems for anisotropics materials by B.I.E method, in time domain or in frequency domain. The governing equations for the wave scattering problems in the time domain are :

$$\text{(II-1)} \quad \begin{cases} \operatorname{div} \sigma - \rho \dfrac{\partial^2 \vec{u}}{\partial t^2} = 0 & \text{in } \Omega^+ \times \mathbf{R}_+ \\ \vec{u}(x,0^+) = 0 & x \in \Omega^+ \\ \dfrac{\partial \vec{u}}{\partial t}(x,0^+) = 0 & x \in \Omega^+ \\ \sigma \vec{n} = \vec{g} & \text{on } \Gamma \times \mathbf{R}_+ \end{cases}$$

where $\vec{u}$, ρ, σ, $\vec{n}$ are the displacement, mass density, stress tensor, the outward unit normal to the boundary respectively and $\vec{g}$ is a given function.

The stress σ in (II-1) is related to the displacement $\vec{u}$ by Hooke's law given by :

$$\text{(II-2)} \quad \sigma_{ij} = C_{ijkl}\, \varepsilon_{kl} = C_{ijkl}\, u_{k,l}$$

where the summation convention is used, $\varepsilon_{ij}(\vec{u})$ is the strain defined by

$$\text{(II-3)} \quad \varepsilon_{ij}(\vec{u}) = \frac{1}{2}\left(\frac{\partial u_i}{\partial x_j} + \frac{\partial u_j}{\partial x_i}\right),$$

and C is the elasticity tensor which is positive definite and has the symmetry given by

$$C_{ijkl} = C_{jikl} = C_{klij}$$

We also have

(II-4) $\varepsilon_{ij} = A_{ijkl}\, \sigma_{kl}$

where $A = C^{-1}$.

We now take the Fourier Transforms (FT) of equations (II-1) with respect to time to obtain the problem in frequency domain

(II-5)
$$\begin{cases} \operatorname{div} \sigma + \rho\omega^2 \vec{u} = 0 & \text{in } \Omega^+ \\ \sigma\vec{n} = \vec{g} & \text{on } \Gamma \\ \varepsilon(\vec{u}) = A\sigma & \text{in } \Omega^+ \\ \vec{u} \text{ satisfies the radiation condition} & \end{cases}$$

In (II-5) we have used the same symbols $\vec{u}$, σ ... for the time Fourier transforms of $\vec{u}(x,t)$, $\sigma(x,t)$ ... because we will be mainly concerned with the frequency domain.

In order to solve this problem we introduce the double layer potential which satisfies the following equations in addition to the radiation condition :

(II-6)
$$\begin{cases} \operatorname{div} \sigma + \rho\omega^2 \vec{u} = 0 & \text{in } \Omega^- \cup \Omega^+ \\ [\sigma\vec{n}] = 0 & \\ [\vec{u}] = \vec{\varphi} & \\ \varepsilon(\vec{u}) = A\sigma & \text{in } \Omega^- \cup \Omega^+ \end{cases}$$

where $[f] = f_- - f_+$ is the difference between the interior limit f_- and the exterior limit f_+ of a function f regular in $\overline{\Omega^-}$ and $\overline{\Omega^+}$. If we consider the derivatives in (II-2) in the distributional sense we obtain

(II-7)
$$\varepsilon = \tilde{\varepsilon} - t\,\delta_\Gamma$$

where t is defined as

(II-8)
$$t_{ij} = \frac{1}{2}(\varphi_i n_j + \varphi_j n_i)$$

δ_Γ is the surface Dirac measure associated with Γ and $\tilde{\varepsilon}$ is the function part of ε defined as

(II-9) $$\tilde{\varepsilon} = \varepsilon_{/\Omega^+} + \varepsilon_{/\Omega^-}$$

In (II-9) $\varepsilon_{/\Omega}$ stands for the extension by 0 to $\mathbf{R}^3$ of the restriction of ε to Ω. Equation (II-6) can then be rewritten in the distributional sense as

(II-10) $$\begin{cases} \operatorname{div} \sigma + \rho\omega^2\vec{u} = 0 \\ \varepsilon(\vec{u}) - A\sigma = -t\,\delta_\Gamma \end{cases}$$

We now introduce the fundamental solution U, Σ defined by

(II-11) $$\begin{cases} \operatorname{div} \Sigma + \rho\omega^2 U = 0 \\ \varepsilon(U) - A\Sigma = \delta(x)\,\mathbb{I} \end{cases}$$

where

$$(\mathbb{I})_{ij}^{kl} = \frac{1}{2}(\delta_{ki}\delta_{lj} + \delta_{kj}\delta_{li})$$

With U and Σ we can write the double layer potential in (II-6) as

(II-12) $$\begin{cases} \vec{u} = -U * t\,\delta_\Gamma = -\int_\Gamma U(x-y)\,t(y)\,d\gamma_y \\ \sigma = -\Sigma * t\,\delta_\Gamma = -\int_\Gamma \Sigma(x-y)\,t(y)\,d\gamma_y \end{cases}$$

The original boundary value problem is then reduced to an integral equation on Γ given by

(II-13) $$\lim_{\substack{x = x_0 + \varepsilon\vec{n}(x_0) \\ \varepsilon \to 0}} \left\{ -\frac{1}{2}\int_\Gamma \Sigma_{ij}^{kl}(x-y)\,(\varphi_k(y)n_l(y) + \varphi_l(y)n_k(y))\,n_j(x_0)\,d\gamma_y \right\} = g_i(x_0)$$

After solving (II-13) for the unknown φ, we determine $\vec{u}$ and σ by (II-12).

As can be shown, however, the kernel Σ in (II-13) is asymptotically proportional to $1/|x-y|^3$ as $|x-y|$ approaches 0, and we have to give a sense to the limit in (II-13). Because of this strong singularity, we cannot solve (II-13) directly by using conventional numerical methods. In fact, one usually uses integration by parts, or "regularization", to reduce the singularity to an integrable one.and it allows us to define $\sigma(x)\vec{n}(x)$ in a distributional sense as we will see in (IV-1). As we have seen in the introduction, however, the existing regularization techniques may not always be very convenient because some of them are useful only with collocation methods, while others may destroy the causality in time domain. For example, Nedelec [7] proposed to regularize Σ by using the following identity :

$$\hat{\Sigma}_{ij}^{kl} = \frac{1}{|\xi|^4} e_{ipq}\xi_p e_{lab}\xi_a e_{qrs}\xi_r e_{bcd}\xi_c \hat{\Sigma}_{sj}^{kd}$$

$$- \frac{1}{|\xi|^4}\left(e_{ipq}\xi_p e_{qrs}\xi_r \xi_l \hat{\Sigma}_{sj}^{kd} \xi_d + e_{lab}\xi_a e_{bcd}\xi_c \xi_i \hat{\Sigma}_{sj}^{kd} \xi_s - \xi_i \xi_l \hat{\Sigma}_{sj}^{kd} \xi_s \xi_d\right)$$

where

$$e_{imr} = \begin{cases} 1 & \text{if (i,m,r) is obtained by a circular permutation of (1,2,3)} \\ 0 & \text{if 2 indices are equal} \\ -1 & \text{otherwise} \end{cases}$$

He then obtained a variational formulation in a manner analogous to the one to be given in IV. His method works in frequency domain. Unfortunately, however, it destroys the causality in time domain. Indeed, we notice that the Fourier inverse transform of the above expression is given in terms of the convolution of $|x|$ and the derivatives of Σ. Since the former (the fundamental solution of the double laplacian) violates the causality, so does the resulting convolution. Hence we propose in the next chapter a formulation which does not have this inconvenience.

Before closing this chapter we notice that the same integral equation (II-13) solves crack problems, where Γ is a surface in $\mathbf{R}^3$ which is identified as a crack. Bearing this application in mind we shall henceforth drop the assumption that Γ is the boundary of Ω^-.

III - COMPUTATION OF Σ

In this chapter we shall discuss the structure of Σ. The analysis will be in frequency domain unless stated otherwise.To begin with we compute Σ explicitly. From the FT in space of (II-11)(denoted by ^) we deduce that

(III-1) $$\hat{\Sigma}_{ij}^{kl} = \frac{i}{2} C_{ijmn} (\hat{U}_m^{kl} \xi_n + \hat{U}_n^{kl} \xi_m) - C_{ijkl},$$

(III-2) $$-\frac{\xi_j}{2} C_{ijmn} (\hat{U}_m^{kl} \xi_n + \hat{U}_n^{kl} \xi_m) + \rho\omega^2 \hat{U}_i^{kl} = i\xi_j C_{ijkl},$$

where ξ is the parameter of the spatial FT. The inversion of (III-2) leads to :

(III-3) $$\hat{U}_i^{kl} = -\Lambda_{im}^{-1} i\xi_j C_{mjkl}$$

where

(III-4) $$\Lambda = \hat{\Gamma} - \rho\omega^2 I$$

(III-5) $$\hat{\Gamma}_{ik} = C_{ijkl}\xi_j\xi_l$$

(III-6) $$I_{ik} = \delta_{ik}$$

Finally, we substitute U in (III-1) to get the expression of $\hat{\Sigma}$:

(III-7)
$$\begin{cases} \hat{\Sigma}^{kl}_{ij} = C_{ijmn}\, \xi_m\, \Lambda^{-1}_{nr}\, \xi_s\, C_{rskl} - C_{ijkl} \\ \quad = \dfrac{C_{ijmn}\, \xi_m\, (\mathrm{cof}\, \Lambda)_{rn}\, \xi_s\, C_{rskl} - C_{ijkl} \det \Lambda}{\det \Lambda} \end{cases}$$

which satisfies : $\hat{\Sigma}^{kl}_{ij} = \hat{\Sigma}^{kl}_{ji} = \hat{\Sigma}^{ij}_{kl}$ because of the symmetries of C and Λ.

From (III-7) we see that the causality of Σ in time domain is determined by the inverse Fourier transform of $L = (\det \Lambda)^{-1}$. In other words L is linked only to the wave velocities of the material. In fact the numerator in (III-7) is a polynomial in ξ and ω, which implies that Σ is given in terms of certain derivatives of L. Conversely, we see that a function whose FT is written as

(polynomials of ξ and ω).L

possesses the same causality as Σ.

We now proceed to show the existence of a decomposition of Σ which facilitates the regularization without destroying the causality :

THEOREM 1 : *There exists a stress function Φ and a kernel R such that :*

(III-8) $$\Sigma^{kl}_{ij} = rot_i rot_j rot_k rot_l \Phi^{\cdot\cdot}_{\cdot\cdot} + \rho\omega^2 R^{kl}_{ij}$$

holds. These kernels have the same symmetry as Σ and take the following forms:

$$\hat{\Phi} = \frac{P(\xi)}{\det \Lambda} \quad ; \quad \hat{R} = \frac{Q(\omega,\xi)}{\det \Lambda}$$

where P and Q are polynomials of the respective arguments homogeneous of degree 2 and 4, respectively . Hence $rot_i rot_k \Phi^{\cdot l}_{j}$ and R are locally integrable. P is not determined uniquely, although Q is. More precisely, P is written as

$$P^{kl}_{ij} = \overset{o}{P}{}^{kl}_{ij} + \xi_i\, E_{mjkl}\, \xi_m + \xi_j\, E_{mikl}\, \xi_m + \xi_k\, E_{mlij}\, \xi_m + \xi_l\, E_{mkij}\, \xi_m$$

where $\overset{o}{P}{}^{kl}_{ij}$ is a 'particular solution' and E_{mjkl} is a constant tensor symmetric with respect to k and l.

Remark : We define $\mathrm{rot}_i\Phi^{kl}_{.j}$ as the rotation of Φ^{kl}_{ij} considered as a vector in i, with j,k,l being fixed. We therefore have

$$(\mathrm{rot}_i\Phi^{kl}_{.j})_i = e_{imr}\partial_m\Phi^{kl}_{rj}$$

Proof : see [3]

As a matter of fact we can write down the stress function explicitly, as shown in the following

THEOREM 2 *: An expression for $\hat{\Phi}$ is :*

$$\hat{\Phi}^{kl}_{ij} = \frac{A_{ktib}\,A_{lnjd} + A_{ltib}\,A_{knjd}}{\det\,(K-\rho\omega^2 A)/(\rho\omega^2)^3}\, e_{tun}\, e_{bvd}\, \xi_u\, \xi_v$$

where

$$K = \begin{pmatrix} \xi_1^2 & 0 & 0 & 0 & \xi_1\xi_3 & \xi_1\xi_2 \\ 0 & \xi_2^2 & 0 & \xi_3\xi_2 & 0 & \xi_1\xi_2 \\ 0 & 0 & \xi_3^2 & \xi_3\xi_2 & \xi_1\xi_3 & 0 \\ 0 & \xi_3\xi_2 & \xi_3\xi_2 & \xi_2^2+\xi_3^2 & \xi_1\xi_2 & \xi_1\xi_3 \\ \xi_1\xi_3 & 0 & \xi_1\xi_3 & \xi_1\xi_2 & \xi_1^2+\xi_3^2 & \xi_3\xi_2 \\ \xi_1\xi_2 & \xi_1\xi_2 & 0 & \xi_1\xi_3 & \xi_3\xi_2 & \xi_1^2+\xi_2^2 \end{pmatrix}$$

Proof : see [3]

IV - VARATIONAL FORM

We rewrite the integral equation (II-13) as

(II-13) $$\Sigma^{kl}_{ij} * (\varphi_k n_l \delta_\Gamma)\, n_j \delta_\Gamma = -g_i \delta_\Gamma$$

The variational formulation corresponding to (II-13) is

$$b(\varphi,\psi) = -\langle g, \psi\delta_\Gamma\rangle$$

where

$$b(\varphi,\psi) = \left\langle \Sigma_{ij}^{kl} * (\varphi_k n_l \delta_\Gamma), \psi_i n_j \delta_\Gamma \right\rangle,$$

We now use the decomposition of Σ in theorem 1 to get

$$b(\varphi,\psi) = \left\langle (\mathrm{rot}_i \mathrm{rot}_j \mathrm{rot}_k \mathrm{rot}_l)\Phi_{ij}^{kl} * (\varphi_k n_l \delta_\Gamma), \psi_i n_j \delta_\Gamma \right\rangle + \rho\omega^2 \left\langle R_{ij}^{kl} * (\varphi_k n_l \delta_\Gamma), \psi_i n_j \delta_\Gamma \right\rangle$$

Repeated use of integration by parts removes the singularity in this form to yield

$$\text{(IV-1)} \qquad b(\varphi,\psi) = \left\langle F_{ij}^{kl} * (\overrightarrow{\mathrm{rot}}_\Gamma \varphi_k)_l \delta_\Gamma, (\overrightarrow{\mathrm{rot}}_\Gamma \psi_i)_j \delta_\Gamma \right\rangle + \rho\omega^2 \left\langle R_{ij}^{kl} * (\varphi_k n_l \delta_\Gamma), \psi_i n_j \delta_\Gamma \right\rangle$$

with

$$F_{ij}^{kl} = - e_{ima} e_{knb}\, \partial_{mn}^2 \Phi_{aj}^{bl}$$
$$\left(\overrightarrow{\mathrm{rot}}_\Gamma \varphi_k\right) \delta_\Gamma = \overrightarrow{\mathrm{rot}}\, (\varphi_k\, \vec{n}\, \delta_\Gamma)\ .$$

Equation (IV-1) gives a regularized bilinear form for (II-13) in frequency domain because F and R are locally integrable by virtue of theorem 1. We then obtain a variational form in time domain by taking the inverse Fourier transform of (IV-1). The factor ω^2 will give rise to some time derivatives of φ and ψ. For example a bilinear form symmetric, with respect to φ and ψ, is obtained as

$$\text{(IV-2)} \qquad b(\varphi,\psi) = \left\langle F_{ij}^{kl} * (\overrightarrow{\mathrm{rot}}_\Gamma \varphi_k)_l\, \delta_\Gamma, (\overrightarrow{\mathrm{rot}}_\Gamma \psi_i)_j\, \delta_\Gamma \right\rangle - \rho \left\langle R_{ij}^{kl} * (\dot{\varphi}_k n_l \delta_\Gamma), \dot{\psi}_i n_j \delta_\Gamma \right\rangle$$

Again, we have used the same symbols F and R for the Fourier inverse transforms with respect to ω of the frequency domain versions of F and R. Of course $\langle\, ,\, \rangle$ and $*$ in (IV-2) are those for $\mathbf{R}^4 = \mathbf{R} \times \mathbf{R}^3$. For more details about the variational form in the isotropic case, one can see [2].

V - ISOTROPIC CASE

In this case the compliance tensor A is given in terms of the Lamé 's constants (λ,μ) as

$$A_{ijkl} = \frac{1}{4\mu}\left(\delta_{ik}\delta_{jl} + \delta_{il}\delta_{jk} - \frac{2\lambda}{3\lambda + 2\mu}\,\delta_{ij}\delta_{kl}\right)$$

With this formula one easily shows

(V-1)
$$\begin{cases} (A_{ktib}\, A_{lnjd} + A_{ltib}\, A_{knjd})\, e_{tun}\, e_{bvd}\, \xi_u\, \xi_v \sim \dfrac{|\xi|^2 \widetilde{\Phi}_{klij}}{4\mu^2(3\lambda + 2\mu)} \\ \dfrac{\det(K - \rho\omega^2\, A)}{(\rho\omega^2)^3} = \dfrac{D_P D_S^2}{4\mu^5(3\lambda + 2\mu)} \end{cases}$$

where

(V-2)
$$\begin{aligned} \widetilde{\Phi}_{klij} &= (\lambda + 2\mu)(\delta_{ik}\delta_{jl} + \delta_{il}\delta_{jk}) + 2\lambda\, \delta_{kl}\delta_{ij} \\ D_S &= \mu|\xi|^2 - \rho\omega^2 ;\ D_P = (\lambda + 2\mu)|\xi|^2 - \rho\omega^2 \end{aligned}$$

and ~ indicates an equality modulo $\xi_{i,j,k,l}$. These formulae and Theorem 2 determine Φ. Because of the isotropy, however, one can obtain a simpler expression for the stress function :

(V-3)
$$\begin{cases} \Phi_{ij}^{kl}(x) = -\mu^2\, \widetilde{\Phi}_{ijkl}\, h(x) \\ R_{ij}^{kl}(x) = -\, \widetilde{R}_{ij}^{kl}(\partial)\, h(x) \end{cases}$$

where $\widetilde{\Phi}$ is given in (V-2) and

(V-4)
$$\begin{cases} \widetilde{R}_{ij}^{kl}(\xi) = \mu\Big\{|\xi|^2\big(3\mu\delta_{ij}\delta_{kl} + (\lambda + 3\mu)(\delta_{ik}\delta_{jl} + \delta_{il}\delta_{jk})\big) \\ -\, 2\lambda(\delta_{ij}\xi_k\xi_l + \delta_{kl}\xi_i\xi_j) - \mu\,(\delta_{jk}\xi_i\xi_l + \delta_{jl}\xi_i\xi_k + \delta_{ik}\xi_j\xi_l + \delta_{il}\xi_j\xi_k)\Big\} \\ -\, \rho\omega^2\,\big(\lambda\delta_{ij}\delta_{kl} + \mu(\delta_{ik}\delta_{jl} + \delta_{il}\delta_{jk})\big) \\ h(x) = \dfrac{1}{4\pi(\lambda + \mu)\rho\omega^2}\left(\dfrac{e^{ik_T|x|}}{|x|} - \dfrac{e^{ik_L|x|}}{|x|}\right) \\ k_T = \sqrt{\dfrac{\rho\omega^2}{\mu}}\ ;\ k_L = \sqrt{\dfrac{\rho\omega^2}{\lambda + 2\mu}} \end{cases}$$

These results coincide with the expressions obtained by Nishimura & Kobayashi [8]. Notice that the stress function given here has the same symmetry as possessed by the elasticity constant. A lengthy but straightforward calculation shows, however, that

$$\begin{aligned} &e_{i_1 j_1 k_1} \ldots e_{i_4 j_4 k_4}\, \xi_{j_1} \ldots \xi_{j_4}\, a_{k_1 \ldots k_4} = 0 \\ &a_{ijkl} = a_{jikl} = a_{klij} \qquad a_{ijkl} : \text{constant} \end{aligned}$$

imply $a_{ijkl} = 0$. In this sense the symmetric decomposition given above is unique. This decomposition and the general formulae in (IV-1,2) yield a variational form for the present case.

VI - CONCLUDING REMARKS

The decomposition of Σ into a sum of a stress function part and a weakly singular part R is useful with the variational method as well as with the collocation one. It is noted that the collocation requires a C^1 element, while the variational formulation works with a finite element of class C^0.

In the isotropic case, all the kernels are explicit. An application of the present formulation, in this case, to the collocation method in the frequency domain and in the time domain can be found in [8], [9]. A variational formulation in the time domain has been derived from this decomposition, again in the isotropic case, in [2].

There are attempts at the use of non-regularized kernels with a numerical integration formula for hypersingular functions. For references see Martin & Rizzo [6].

An explicit expression for $\hat{\Phi}$ is given in [3]. However in the general anisotropic case we need some extra works to obtain explicitly Φ and R.

REFERENCES

[1] Bamberger, A. (1983) Approximation de la diffraction d'ondes élastiques : une nouvelle approche (I), (II), (III), Internal report n°91, 96, 98 of Centre de Mathématiques appliquées, Ecole Polytechnique, France.

[2] Becache, E & Ha Duong, T (1989) Formulation Variationnelle Espace - Temps Associée au Potentiel de Double Couche des Ondes Elastiques, Internal report n°199 of Centre de Mathématiques appliquées, Ecole Polytechnique, France.

[3] Becache, E , Nedelec , J-C & Nishimura, N (1989) Regularization in 3D for anisotropic elastodynamic crack and obstacle problems, Internal report n°205 of Centre de Mathématiques appliquées, Ecole Polytechnique, France.

[4] Bonnet, M. (1987) Méthode des équations intégrales régularisées en élastodynamique, Bulletin de la direction des études et recherches, EDF, France.

[5] Bui, H.D.(1977) An integral equations method for solving the problems of a plane crack of arbitrary shape, J. Mech. Phys. Solids, 25, 29-39.

[6] Martin, P.A. & Rizzo, F.J. (1989) On boundary integral equations for crack problems, Proc. Roy. Soc. London (A) 421, 341-355.

[7] Nedelec, J.-C. (1983) Le potentiel de double couche pour les ondes élastiques, Internal report n°99 of Centre de Mathématiques appliquées, Ecole Polytechnique, France.

[8] Nishimura, N. & Kobayashi, S. (1989) A regularized boundary integral integral equation method for elastodynamic crack problems, Comp. Mech. ,4, 319-328.

[9] Nishimura, N., Guo, C & Kobayashi, S. (1987) Boundary Integral Equation Methods in Elastodynamic Crack Problems , Proc. 9th Int Conf BEM, vol2, ed Brebbia, Wendland, Kuhn, Springer Verlag, 279-291.

[10] Polch, E.Z. , Cruse, T.A & Huang, C.-J (1987) Traction BIE solutions for flat cracks, Comp. Mech., 2, 253-267.

[11] Sladek, V. & Sladek, J. (1984) Transient elastodynamic three-dimensional problems in cracked bodies, Appl. Math. Model., 8,2-10.

Boundary Element Analysis of Non-Linear Wave Forces on Buried Pipelines

J.M. Niedzwecki and J.A. Earles

Department of Civil Engineering
Texas A&M University
College Station, Texas 77843 USA

Summary

In water depths up to 61m offshore pipelines are buried in excavated trenches to minimize their impact on the local environment and to reduce the possibility of accidental damage. For many practical reasons it is not unusual to have several pipelines buried in very close proximity to one another. A two-dimensional boundary element method which can be used to predict the seabed pressure field and the surface wave induced forces on a single buried pipe or any cluster configuration is presented. Of particular interest for design practice is the introduction of non-linear stream function wave theory into the formulation. A comparison of the boundary element model predictions with the analytical solution for a single buried pipeline, and a finite element model, which was validated using small scale laboratory tests is presented and discussed. The use of linear and non-linear ocean surface wave theories is examined and predictions of wave induced forces on several cluster configurations is presented. The two-dimensional boundary element approach is shown to be efficient and very suitable for this type of offshore application.

Introduction

Offshore pipeline systems are used for many applications including the transport of oil and gas to onshore facilities from offshore platforms or from ships too large to put into port, municipal storm water drainage and sewerage out falls to name a few. They can be found in an about port and harbor facilities and sometimes near recreational areas. In water depths less than 61m (200ft), the pipelines are laid in excavated trenches and then back filled with local fill material unless another cover material has been specified, e.g. crushed stone. Beyond that water depth pipelines are laid in open trenches and any back filling is a result of natural ocean processes [4]. The pipe sizes used offshore vary significantly depending upon the particular application. Pipe diameters on the order of 30cm (1ft) to 60cm (2ft) are used for oil and gas lines, while pipe diameters on the order of 5m (16.4ft) are used for municipal out fall systems. The actual depth of cover, i.e. the depth of the fill over the pipeline again depends upon the particular application. However, from an engineering design viewpoint, it is desirable to minimize the cyclic wave induced forces on the buried pipeline so that it will not work its way to the seafloor and break apart. Various localized anchoring schemes and a variety of cover materials have been used to try and eliminate these types of problems [2,4]. In congested port and harbor areas pipelines are often laid in very close proximity to one another. This only further complicates the design process and possibility of inadvertent damage.

In the past offshore engineers have employed finite difference or finite element methods together with linear design waves to predict the wave induced loads on buried pipelines. The design wave approach is a quasi-static procedure which allows the engineer to use a range of anticipated worst case storm conditions as a basis for design. The seaway in shallow and moderate water depths is often inaccurately modeled for design computations using linear wave theory, which is in fact valid for only a very limited range of wave conditions. Linear waves are sinusoidal in form, but real waves are characterized by steeper crests and shallower troughs than those predicted by linear theory. The use of numerical wave theories which more accurately satisfy the kinematic boundary conditions at the free surface offer a more accurate means to describe the free surface, the sub-surface kinematics and dynamic pressure component at the seafloor. Stream function wave theory has become quite popular with the offshore industry for deterministic wave simulations [3]. This study presents a two-dimensional boundary element formulation that addresses the use of stream function wave theory and allows the specification of multiple pipes closely spaced within the seabed.

Statement of the Boundary Value Problem

The problem domain of interest is shown in Figure 1. It consists of a fluid layer directly over a porous seabed. As a surface wave, classified as either an intermediate or shallow water wave, moves along the free surface of the fluid it induces orbital motion of the fluid particles and a dynamic pressure component. The velocity and pressure fields generated by the passing wave also penetrate the seabed. In the seabed the buried impermeable pipe or pipes influence the local velocity and pressure fields. In this process the wave is damped since this interaction with the seabed dissipates energy. The orbital motion of the fluid is quickly damped as it progresses into the seabed [8] and it is assumed that no pore pressure build up occurs in the seabed. An accurate prediction of the pressure field around the buried pipe or pipes is required in order to estimate the surface wave induced forces and other engineering quantities of interest.

Reid and Kajiura [11] derived the governing equation based upon Darcy's equation for low Reynold's number flow in a porous seabed. In a similar fashion, Darcy's equation can be expressed as

$$\begin{aligned} \frac{1}{n}\frac{\partial u}{\partial t} &= -\frac{\mu}{\rho K_x}u - \frac{1}{\rho}\frac{\partial p}{\partial x} \\ \frac{1}{n}\frac{\partial w}{\partial t} &= -\frac{\mu}{\rho K_z}w - \frac{1}{\rho}\frac{\partial p}{\partial z} \end{aligned} \tag{1}$$

where, n is the isotropic porosity of the seabed, u is the horizontal fluid velocity component, w is the vertical velocity component, μ is the dynamic viscosity, ρ is the fluid mass density, K is the seabed permeability, and p is the dynamic pressure due to the passing wave. Assuming incompressible flow, the corresponding form of the continuity equation is

$$\frac{\partial u}{\partial x} + \frac{\partial w}{\partial z} = 0 \tag{2}$$

Further assuming that the flow is un accelerated and that $K_x = K_z$, Eq. 1 can be reduced to obtain expressions for each of the velocity components [12]. These can then be differentiated and substituted into Eq.2 to yield the governing equation

$$\frac{\partial^2 p}{\partial x^2} + \frac{\partial^2 p}{\partial z^2} = 0 \tag{3}$$

or more compactly

$$\nabla^2 p = 0 \tag{4}$$

In Figure 1 our attention now moves to the seabed and the definition of the appropriate boundary conditions. At the seafloor the pressure is prescribed using the appropriate wave theory, the lateral boundary conditions are assumed to decay exponentially for the pressure value specified at the seafloor, the bottom boundary condition and the boundary conditions on the impermeable pipe or pipes is that the normal derivative of the pressure is zero. In summary,

$$\begin{aligned}
p(x) &= p(x)|_{\text{wave theory}} \;\; \text{at} \;\; z = -d, \\
p(x,z) &= p(x = X)e^{kz} \quad \text{for} \;\; -d \leq z \leq -Z, \\
p(x,z) &= p(x = -X)e^{kz} \;\; \text{for} \;\; -d \leq z \leq -Z, \\
\frac{\partial p}{\partial n} &= 0 \;\; \text{at} \;\; z = -Z \;\; \text{for} \;\; -X \leq x \leq -X, \;\; \text{and} \\
\frac{\partial p}{\partial n} &= 0 \;\; \text{on each pipe surface.}
\end{aligned} \tag{5}$$

Specification of the Pressure on the Seafloor

The pressure field under a surface wave depends upon its form. Many of the analyses [5–12] assume that the design wave is adequately described using linear wave theory in which case the dynamic pressure at the seafloor is given by the expression

$$p(x) = \frac{\rho g H}{2 \cosh kh} \cos kx \tag{6}$$

where, g is the gravitational acceleration, H is the wave height, k is the wave number $(2\pi/L)$, L is the wave length and x it the horizontal coordinate in Figure 1. The research study Lai, Dominguez and Dunlap [5] present a very complete discussion of the various modifications to the linear form proposed by other researchers and how they can be made collapse to the form of Eq 6. Only the study by Liu and O'Donnell [7] discuss the use of a non-linear wave theory, Solitary wave theory, in place of linear wave theory.

Stream function wave theory is a numerical wave theory [3], that is, it iteratively improves the solution to satisfy the kinematic boundary condition at the free surface. Unlike linear wave theory, the governing equation and the boundary conditions are expressed in terms of a stream function rather than a velocity potential, which is used in linear wave theory. Another difference is that the boundary conditions are transformed so that they move with the wave celerity. In order to fit the wide range of deterministic wave conditions the order of the wave theory is varied and depending on the particular wave conditions the order of the theory may range between two and nineteen. A complete definition of the boundary value problem and its solution can be found in the special report by Dean [3]. The dynamic pressure on the seafloor for stream function wave theory can be expressed as

$$p(x) = \rho g \overline{Q} + \frac{\rho}{2}[\, 2\, C\, u(x) \; - \; u^2(x)\,] \tag{7}$$

where

$$u(x) \;\; = \;\; -k \sum_{n=1}^{N} n\; X(n)\; cos\,(n\, k\, x) \tag{8}$$

and $\overline{Q}$ is an averaged Bernoulli constant [3], C is the wave speed ($C = L/T$), T is the wave period, N is the order of the wave theory, and $X(n)$ are the stream functions coefficients which should be obtained from a stream function computer program. At this point it might be worthwhile to point out that for a given site, h, the water depth, H, the wave height and, T, the wave period must be specified. The wave length is computed from the dispersion equation for linear wave theory and is an output variable in stream function calculations. One must also specify N for the stream function calculations and the output will be $\overline{Q}$, L and $X(n)$. From which one can evaluate the dynamic pressure at any point along the x axis.

Boundary Element Formulation

The development of the boundary element equations follow the typical development of a potential problem with interior holes [1]. The weighted residual statement for the potential field can be expressed as

$$c^i\, p^i + \sum_{j=1}^{n} \int_{\Gamma_j} p\, q^*\, d\Gamma = \sum_{j=1}^{n} \int_{\Gamma_j} q\, p^*\, d\Gamma \tag{9}$$

The numeric value of c^i depends upon whether the point of interest is on the domain boundary Γ, in which case $c^i = \vartheta/2\pi$ and ϑ is defined as an angle between adjacent boundary elements, or if it is in the domain Ω in which case $c^i = 1$. The fundamental solution for the two dimensional case is

$$p^* = \frac{1}{2\,\pi} \ln\left(\frac{1}{r}\right) \tag{10}$$

and the flux is expressed as

$$q^* = \frac{\partial p}{\partial n} \tag{11}$$

The boundary element equation, Eq. 9, can be expressed in a more convenient form by combining the terms on the left-hand side of the equation. After selecting a constant, linear or quadratic element model the resulting equation can be expressed as

$$\sum_{j=1}^{n} H_{ij}\, p_j \;\; = \;\; \sum_{j=1}^{n} G_{ij} q_j \tag{12}$$

In the computer implementation, counterclockwise numbering is used for inward pointing normal vectors and clockwise number is used for outward pointing normal vectors. For the domain representing the seabed the normal vector points into the domain and for the holes, which correspond to the buried pipe or pipes the normal vector points outward from the seabed/pipe interface in towards the center of the pipe. After assembling the equations and specifying either the pressure or its flux at the nodal points the complete set of equations can be represented in matrix form as

$$\mathbf{AX} = \mathbf{F} \tag{13}$$

where, $\mathbf{X}$ is the vector of unknown values of pressures, p, and flux, q. This matrix equation can be solved quite easily and accurately. Once the $\mathbf{X}$ is known one can solve for the values of the pressure field at any point in the domain. However, for this problem the solution of the wave induced force components in the x and z directions on the buried pipe or pipes is adequate.

<u>Analytical Solution for a Single Buried Pipeline</u>

There are two analytical solutions which provide formulae for computing the linear wave induced force on a single buried pipe [9,10]. Thus, it is possible to evaluate the accuracy of the boundary element model. The analytical model developed by MacPherson [9] was selected for this purpose. The dimensionless wave force amplitude, as presented by MacPherson, can be expressed as

$$F_0 = \frac{H\,k}{2\cosh kh}\left[\exp\left(-\frac{k\,a}{\tanh \xi_0}\right) + 4\,\sinh^2\xi_0\,\tanh\xi_0\,e^{-2\xi_0 - ka}\right] \tag{14}$$

where

$$\begin{aligned} \xi_o &= \cosh^{-1}\left(\frac{2\,b}{D}\right) \\ b &= d + \frac{D}{2} \\ a &= b\tanh\xi_0 \end{aligned} \tag{15}$$

and k is the wave number, h is the water depth, d is the cover depth, and D is the pipe outside diameter. The force components are then obtained by evaluating the following expression

$$\left\{\begin{array}{c} F_x \\ F_z \end{array}\right\} = \frac{\rho g \pi D^2}{4} F_0 \left\{\begin{array}{c} \sin kx_c \\ \cos kx_c \end{array}\right\} \tag{16}$$

where, F_x is the horizontal force component, F_z is the vertical force component and x_c is the horizontal distance between the crest of the design wave and the center line of the pipe buried in the seafloor. A graphical presentation of the key parameter necessary to use this equation are presented in Figure 2.

<u>Numerical Results</u>

The four numerical examples which follow are presented in a sequential manner that establishes the accuracy of the boundary element model prior to its use in conjunction with non-linear stream function wave theory and the analysis of a simple pipe cluster. The first example considers the issues of domain size and accuracy by comparing the accuracy of the model with an analytical solution for a linear wave passing over a single buried pipeline [9]. The second example provides a comparison of the boundary element model with finite difference and finite element models. Those models were used to predict the pressures around a single buried pipe resulting from the passage of a linear wave [5]. The third example illustrates the importance of accurately modeling the wave environment for engineering design. The last numerical example is used to illustrate the flexibility of the boundary element model for analyzing pipeline clusters. The increase in vertical wave force on a central pipeline is studied by examining the effects of moving a second pipeline from a distant position up to the point of contact with the central pipeline.

For the new examples the water depth was specified to be 15m (49 ft), the wave height 5.5m (18ft) with a wave period of 8s. The buried pipeline had a diameter of 0.6m (2ft) and the ratio of the depth of cover to the pipe diameter was allowed to vary from 1 to 5. The crest of the wave was located directly over the center line of the single or central buried pipeline. Linear wave theory predicted a wave length of 81.7m (267.9ft) while the non-linear stream function wave theory predicted a wave length of 85.9m (282.1 ft). A second order stream function wave theory, $N = 2$, was selected so that the errors in the maximum velocity and acceleration would be less than one percent. The corresponding stream function coefficient values were found to be $X(1) = -211.91$, $X(2) = -7.115$, for use with the ft-lb-s system of units, and the constant value of the stream function on the free surface, ψ_η, was -10.98m (-36.01 ft).

The sensitivity of the dimensionless wave force amplitude to changes in element type and pipeline depth are presented in Table 1. Two variations in the domain modeling were examined. The first was L by L and the second was L/4 by L. Symmetry of the single pipe system were ignored since the eventual target was the analysis of buried pipeline clusters. The constant element model provides an adequate degree of accuracy an there appears to be no advantage in using linear elements. The constant element model, which will serve as the basis for later comparisons, required 120 elements to model the seabed and 12 to model each buried pipeline. Thus, for the analysis of a single buried pipeline a total of 132 element were used in the boundary element model. Interestingly, it can be observed that the force amplitude does not decay very much for the pipeline burial depths considered.

An earlier study by Lai, Dominguez, and Dunlap [5] was used as the basis for the second example. They developed both a finite difference and a finite element model for their study. For the analysis of a single buried pipeline subject to a linear wave the finite difference model required a 37 by 37 grid and the finite element model was chosen to be 38 by 38 triangular elements. Table 2 presents a comparison of the predictions of pressure around the single buried pipeline by the three discrete element methods. Note that for a local coordinate system centered in the pipe position one is located on the pipe surface at zero degrees and that the remaining positions are located at ninety degree intervals in a counterclockwise direction around the pipe. Surprisingly the finite difference model provides results within 4% of the boundary element model while the finite difference model provides estimates within 5 to 15%.

A comparison of the dimensionless wave force amplitudes based upon linear wave theory and non-linear stream function wave theory are shown in Table 3. The most striking fact for design engineers is that, although linear wave theory is convenient to use, there can be a significant difference in the wave force predictions. In this particular case the results differ by at least a factor of two. Thus, the accurate specification of the design wave can a critical factor for the design of buried pipelines.

The analysis of buried pipelines in close proximity to one another is an important engineering problem and it is in the modeling of this problem that the benefits of using the boundary element approach are quite evident. Each additional pipeline in the two-dimensional model developed in this study requires only 12 additional elements. The orientation and movement of additional pipelines with respect to a central pipeline is easily implemented. In this example a two pipe system, or cluster, is studied under linear and non-linear wave conditions. A depth of cover to pipe diameter ratio of 4 was selected and only relocation of the second pipeline in a horizontal plane was considered. The increase in the wave force on the central pipeline with the inclusion of the second pipeline as compared to a similar single pipeline configuration for various separation distances is presented in Table 4. It is shown in Table 4 that if the dimensionless spacing ration X_0/D is greater that 5, the central pipeline experiences no change in wave force. However, as the two pipelines are moved closer up to the point of contact there is an increase in the wave force. It is interesting to note that the magnitude of the dimensionless force ratio and their rate of change do not appear to depend upon the wave theory.

Closing Remarks

The boundary element approach was shown to be very suitable for the analysis and design of offshore pipelines especially for the analysis of pipeline clusters. This study presents a mathematical formulation that incorporates a powerful non- linear wave theory and allows the specification of pipeline clusters. In the past engineers have extensively used linear wave theory to characterize the design wave and its sub-surface kinematics. The use of an appropriate design wave theory was shown to be an important design consideration together with selection of the appropriate discrete element method.

Acknowledgements.
This study was supported in part by the Offshore Technology Research Center, NSF Engineering Research Centers program grant #CDR-8721512. The senior author would also like to thank Arun S. Duggal for his help in preparing this manuscript.

References

1. Brebbia, C. (1978). *The Boundary Element Method for Engineers*, Halsted Press, John Wiley & Sons, Inc., New York.

2. Brown, R.J. (1971). "Rational Design of Submarine Pipelines," *World Dredging* & Marine Construction, pp. 17–22.

3. Dean, R.G. (1974). *Evaluation And Development of Water Wave Theories For Engineering Applications*, CERC Special Rpt No. 1, Vol. I & II, U.S. Corps of Engineers.

4. Herbich, J.B., C. (1981). *Offshore Pipeline Design Elements*, Marcel Decker, Inc., New York.

5. Lai, N.W., Dominguez, R.F. and Dunlap W.A. (1974). "Numerical Solutions for Determining Wave-Induced Pressure Distributions Around Buried Pipelines," *Sea Grant Report*, TAMU-SG-75-205, pp.92.

6. Lennon, G.P. (1985). "Wave-Induced Forces on Buried Pipelines", *ASCE Journal of the Waterway, Port, Coastal and Ocean Division*, Vol. 111, WW3, pp. 511–524.

7. Liu, P.L.F. and O'Donnell, T. (1983). "Wave-Induced Forces on Buried Pipelines in Permeable Seabeds," *ASCE Civil Engineering in the Oceans*, pp. 111–121.

8. Liu, P.L.F. (1973). "Damping of Water Waves over Porous Bed," *ASCE Journal of the Hydraulics Division*, Vol. 99, HY 12, pp. 2263–2271.

9. MacPherson, H. (1978). "Wave Forces on Pipeline Buried in Permeable Seabed" *ASCE Journal of the Waterway, Port, Coastal and Ocean Division*, Vol. 104, WW4, pp. 407–419.

10. McDougal, Davidson S.H., Monkmeyer P.L. and Sollitt, C.K. (1988). "Wave-Induced Forces on Buried Pipelines", *ASCE Journal of the Waterway, Port, Coastal and Ocean Division*, Vol. 114,WW3, pp. 220–236.

11. Reid, R.O. and Kajiura, K. (1957). "On the Damping of Gravity Waves Over a Permeable Sea Bed," *Transactions of American Geophysical Union*, Vol. 38, pp. 662.

12. Sleath, J.F.A. (1970). "Wave-Induced Pressure in Beds of Sand", *ASCE Journal of the Hydraulics Division*, Vol. 96, HY 2, pp. 367–378.

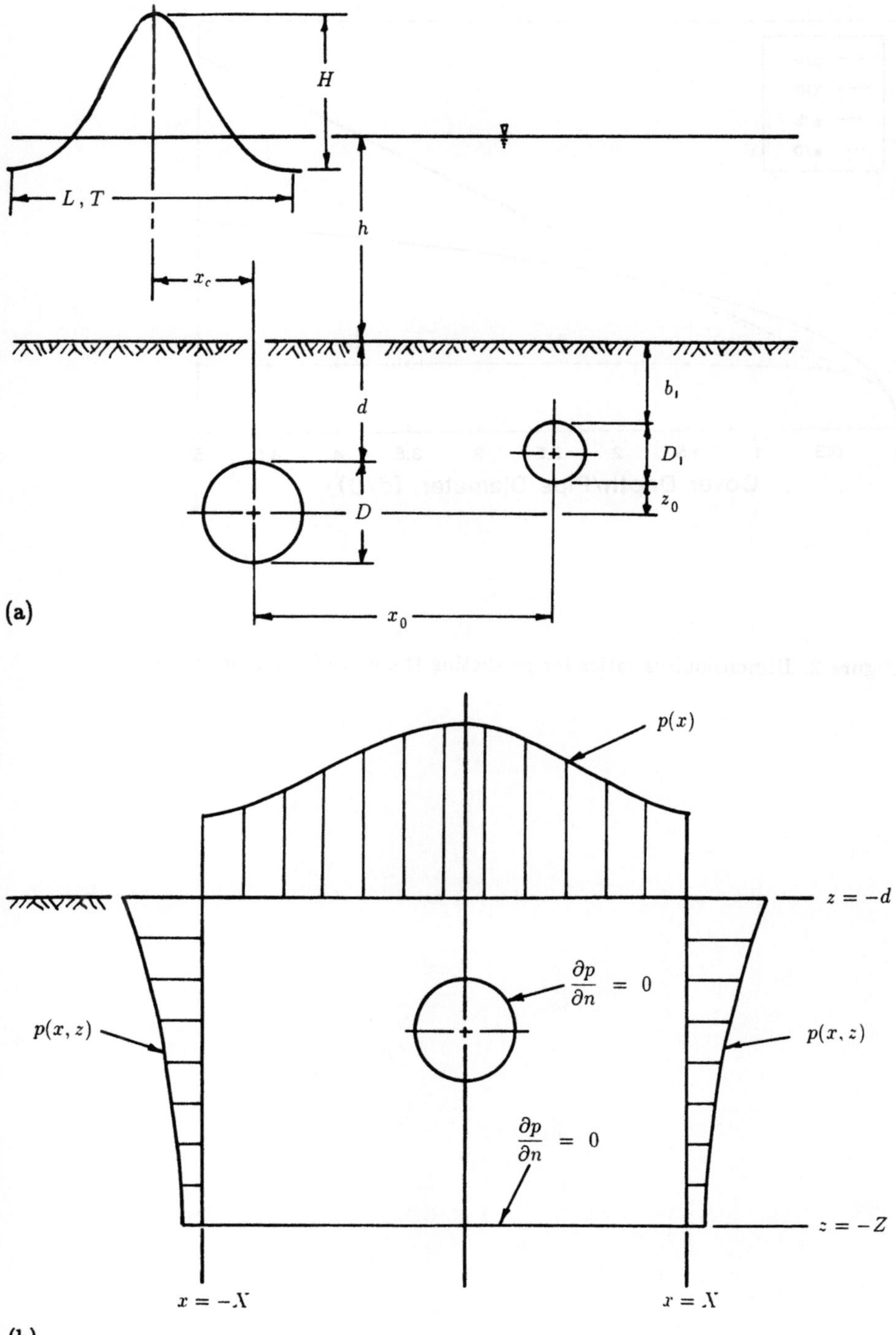

Figure 1. Definition sketch for buried pipelines.

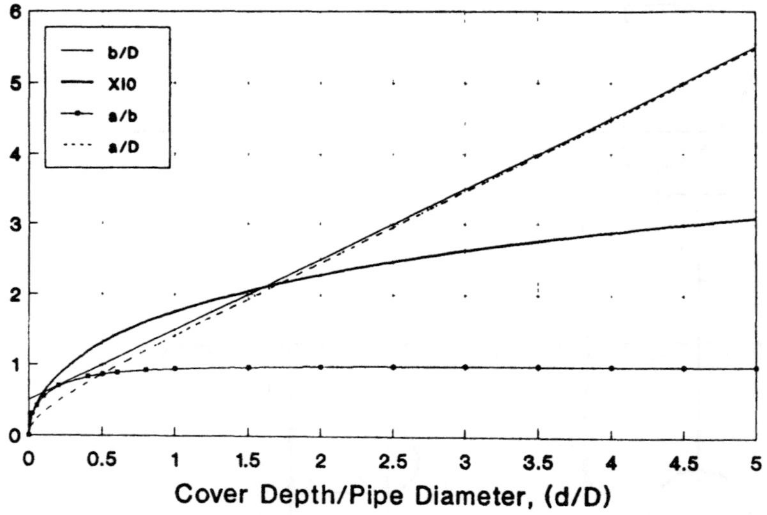

Figure 2. Dimensionless ratios for predicting the wave force amplitude.

Table 1. Sensitivity of Force Predictions to Domain Width.

Element Type	d/D	F_0 Exact Solution	$X = L/2$	$X = L/8$
Constant	1.0	0.214	0.223	0.222
	2.0	0.212	0.214	0.214
	3.0	0.204	0.206	0.205
	4.0	0.196	0.196	0.196
	5.0	0.187	0.188	0.188
Linear	1.0	0.214	0.221	0.220
	2.0	0.212	0.213	0.213
	3.0	0.204	0.205	0.204
	4.0	0.196	0.196	0.195
	5.0	0.187	0.187	0.186

Table 2. Comparison of Discrete Element Models.

Position	Model			% Difference	
	FD	FE	BE	FD	FE
1	0.53	0.48	0.55	3.6	12.7
2	0.78	0.74	0.78	0.0	5.1
3	0.53	0.48	0.55	3.6	12.7
4	0.40	0.33	0.39	2.6	15.4

Table 3. Comparison of Linear and Stream Function Wave Force Predictions.

d/D	Exact Solution	Linear Wave	Non–linear Wave
1.0	0.214	0.222	0.492
2.0	0.212	0.214	0.484
3.0	0.204	0.206	0.475
4.0	0.196	0.196	0.466
5.0	0.187	0.188	0.457

Table 4. Wave forces on a two pipe system.

Wave Theory	Wave Force Ratio	Spacing, X_0/D						
		27	20	15	10	5	2	1
Linear	F_z/F_1	0.995	0.995	0.995	0.997	1.003	1.058	1.460
Non–linear	F_z/F_1	1.001	1.001	1.001	1.002	1.009	1.064	1.470

Further Applications of Regularised Integral Equations in Crack Problems

N. Nishimura and S. Kobayashi

Department of Civil Engineering, Kyoto University,
Kyoto 606, Japan

Summary

This paper discusses an application of BIEM to an inverse problem of determining the geometry of cracks by boundary measurements. The inverse problem considered intends to reconstruct the shape and location of an interior crack from experimental data obtained in certain boundary measurements. The measured physical quantity is assumed to be governed by Laplace's equation. We solve this problem by minimising the error of the direct boundary integral equation (BIE). This minimisation, however, requires solutions of hypersingular integral equations. Regularisation methods suitable for solving these equations are proposed. Several 2D and 3D numerical examples demonstrate the efficiency and robustness of the present method.

Introduction

Assume that a body D is known to contain a crack S whose location and shape are unknown. One is given an instrument to measure the boundary flux associated with a physical quantity u governed by Laplace's equation. Some examples of u are antiplane elastic displacements (2D), electrostatic potentials, temperature etc. With this instrument one carries out experiments, which are interpreted as prescribing several Dirichlet data and measuring the corresponding Neumann data on the exterior boundary. For example one may give elastic antiplane displacements, electrostatic potentials, temperature distribution etc. on the exterior boundary and measure the associated elastic traction, electric current, heat flux etc. We are now interested in determining the geometry of the crack form the obtained data and from the fact that the homogeneous Neumann condition is satisfied on the crack; the last condition means that the crack is traction-free in elasticity, or that the crack is perfectly insulating in electrostatics and in thermostatics. This paper tries to establish an algorithm to solve this problem by BIEM, which was found to be effective in related subjects such as shape optimisation [1] and crack analysis [2].

Inverse problems of this type have so far been considered by several authors. For example the uniqueness of the solution has been established by Friedman & Vogelius [3] in 2D and by Kubo et al. [4] in nD ($n = 2, 3$); they showed that n series of experiments determine S uniquely in nD problems. Also, numerical efforts to solve this and related inverse problems are found in Santosa and Vogelius [5] who used FEM in 2D, and in Kubo et al. [6,7] who used 3D BIEM to solve direct problems for many candidate crack locations and picked up the one which fits experimental data the most. Nishimura and Kobayashi [8] also proposed a complicated 2D BIEM which uses Newton's method. Their approach was simplified considerably in [9] as they replaced Newton's method by a nonlinear programming technique. The purpose of the present paper is to provide more numerical examples to confirm the applicability of this simplified version.

This paper begins by reformulating the original inverse problem into another of minimising the error of boundary integral equations to be satisfied on the exterior boundary of D. This minimisation is carried out with the help of Powell's variable metric method, which needs the gradient of the function to be minimised (cost function). The computation of this cost and its gradient needs solutions of hypersingular integral equations which are solved via Nedelec's variational method [10]. This paper concludes with some 2D and 3D numerical examples to test the efficiency and robustness of the present method. For the purpose of brevity we shall describe our results mainly in the 3D context unless stated otherwise. The 2D counterpart is obtained in a self-evident manner.

Formulation

Let D be a bounded domain in R^3 which has a smooth boundary ∂D. Also let S be a smooth non-self-intersecting curved surface contained in D. The direct crack problem for Laplace's equation is formulated into the following boundary value problem: Find a function $u(\mathbf{x})$ in $D \setminus S$ which satisfies

$$\Delta u = 0 \quad \text{in } D \setminus S, \qquad \left(\frac{\partial u}{\partial n}\right)^+ = \left(\frac{\partial u}{\partial n}\right)^- = 0 \quad \text{on } S,$$

$$\lim_{x(\in S)\to x_0(\in \partial S)} \varphi(\mathbf{x}) = 0, \quad \varphi := u^+ - u^-,$$

subject to a certain boundary condition (Dirichlet, Neumann or mixed) on ∂D, where the superposed $+(-)$ indicates the limit on S from the positive (negative) side and φ is the gap of u across S called crack opening displacement in mechanics. The positive (negative) side in this statement indicates the side of S into which the normal vector $\mathbf{n}$ $(-\mathbf{n})$ points. The solution to this problem has a well-known potential representation given by

$$u(\mathbf{x}) = \int_{\partial D} G(\mathbf{x}-\mathbf{y})\frac{\partial u(\mathbf{y})}{\partial n}dS_y - \int_{\partial D} \frac{\partial G(\mathbf{x}-\mathbf{y})}{\partial n_y}u(\mathbf{y})dS_y + \int_S \frac{\partial G(\mathbf{x}-\mathbf{y})}{\partial n_y}\varphi(\mathbf{y})dS_y, \qquad G(\mathbf{x}-\mathbf{y}) = \frac{1}{4\pi|\mathbf{x}-\mathbf{y}|} \qquad (1),(2)$$

for $\mathbf{x} \in D \setminus S$. The function $G(\mathbf{x}-\mathbf{y})$ is called the fundamental solution of Laplace's equation. The unknown parts of u and $\partial u/\partial n$ on ∂D and φ on S are determined from the boundary integral equations given by

$$0 = \frac{u(\mathbf{x})}{2} - \int_{\partial D} G(\mathbf{x}-\mathbf{y})\frac{\partial u(\mathbf{y})}{\partial n} dS_y + \int_{\partial D} \frac{\partial G(\mathbf{x}-\mathbf{y})}{\partial n_y} u(\mathbf{y}) dS_y$$
$$- \int_S \frac{\partial G(\mathbf{x}-\mathbf{y})}{\partial n_y}\varphi(\mathbf{y})dS_y, \quad \mathbf{x} \in \partial D \tag{3}$$
$$0 = \frac{\partial}{\partial n_x}\left(\int_{\partial D} G(\mathbf{x}-\mathbf{y})\frac{\partial u(\mathbf{y})}{\partial n} dS_y - \int_{\partial D} \frac{\partial G(\mathbf{x}-\mathbf{y})}{\partial n_y} u(\mathbf{y}) dS_y\right)$$
$$+ \,=\!\!\!\!\!\!\int_S \frac{\partial}{\partial n_x}\frac{\partial}{\partial n_y} G(\mathbf{x}-\mathbf{y})\varphi(\mathbf{y})dS_y, \quad \mathbf{x} \in S \tag{4}$$

where the integration sign with a superimposed $=$ indicates that the integral is carried out in the sense of finite part.

We next consider the case where the shape and location of S are unknown. We now prescribe a series of Dirichlet data u^I ($I = 1 \sim N$) on ∂D and measure the corresponding Neumann data $\partial u^I/\partial n$, or vice versa, where $N \geq 3$. Our interest is to find the most plausible shape and location of the crack S from these data. We shall now try to solve this problem by converting it into a constrained minimisation problem. The cost function $J(S)$ to be minimised is defined as follows: For a given crack S, solve

$$g^I_{,i}(\mathbf{x})n_i(\mathbf{x}) + \,=\!\!\!\!\!\!\int_S n_i(\mathbf{x})n_j(\mathbf{y})G_{,ij}(\mathbf{x}-\mathbf{y})\varphi^I(\mathbf{y})dS_y = 0 \quad \text{on } S, \quad I = 1 \sim N \tag{5}$$

for φ^I, where $_{,i} = \partial/\partial x_i$ and (See (4).)

$$g^I(\mathbf{x}) = \int_{\partial D} \frac{\partial G(\mathbf{x}-\mathbf{y})}{\partial n_y} u^I(\mathbf{y}) dS_y - \int_{\partial D} G(\mathbf{x}-\mathbf{y})\frac{\partial u^I}{\partial n}(\mathbf{y}) dS_y \quad \text{on } S. \tag{6}$$

Notice that one can calculate this quantity because both u^I and $\partial u^I/\partial n$ are known on ∂D via measurements. One then defines $J(S)$ by

$$J(S) := \frac{1}{2}\sum_{I=1}^{N}\int_{\partial D}\left(f^I(\mathbf{x}) + \int_S G_{,i}(\mathbf{x}-\mathbf{y})n_i(\mathbf{y})\varphi^I(\mathbf{y})dS_y\right)^2 dS_x, \tag{7}$$

where f^I is a function computed from experimental data by (See (3).)

$$f^I(\mathbf{x}) = \frac{u(\mathbf{x})}{2} + \int_{\partial D} \frac{\partial G(\mathbf{x}-\mathbf{y})}{\partial n_y} u^I(\mathbf{y}) dS_y - \int_{\partial D} G(\mathbf{x}-\mathbf{y})\frac{\partial u^I}{\partial n}(\mathbf{y}) dS_y \quad \text{on } \partial D.$$

The crack S is then obtained as the solution to the following problem:

$$\underset{S}{\text{Minimise}}\, J(S) \quad \text{subject to} \quad S \in D. \tag{8}$$

Obviously the meaning of $J(S)$ is the sum of errors of the integral equations in (3) to be satisfied on ∂D. We remark that the solution to (5) is obtained numerically with the help of Nedelec's variational equation [10] given by

$$\int_S e_{PQ} \frac{\partial \psi(\mathbf{x})}{\partial s_P} \frac{\partial x_p(\mathbf{x})}{\partial s_Q} \left[\int_S G(\mathbf{x}-\mathbf{y}) e_{KL} \frac{\partial \varphi(\mathbf{y})}{\partial s_K} \frac{\partial x_p(\mathbf{y})}{\partial s_L} ds_{1_y} ds_{2_y} \right] ds_{1_x} ds_{2_x}$$

$$= - \int_S \psi g^I_{,i} n_i ds_1 ds_2 \tag{9}$$

where $s_{1,2}$ are arbitrary curvilinear coordinates to describe S, e_{IJ} is the 2D permutation symbol and ψ is a test function which vanishes on ∂S. We henceforth adopt a convention that upper case subscripts run from 1 to 2.

In the present paper we choose to solve the minimisation problem in (8) by using Powell's variable metric method. This method of nonlinear programming, however, needs grad $J(S)$, where 'grad' indicates the derivatives with respect to shape parameters of S denoted by t_i, $(i = 1, \ldots, M)$. We here compute this gradient by differentiating $J(S)$ directly with respect to t_i. To perform this manipulation we introduce an immobile 'reference crack' denoted by S_0, and describe S as the image of S_0 via a mapping $\mathbf{x} = \mathbf{x}(\mathbf{X}, \mathbf{t})$, $\mathbf{x} \in S$, $\mathbf{X} \in S_0$. Also the function $\varphi^I(\mathbf{x})$ is redefined as $\varphi^I(\mathbf{x}) := \varphi^I_0(\mathbf{X}(\mathbf{x}, \mathbf{t}), \mathbf{t})$ in terms of a function φ^I_0 of $\mathbf{X}$ and $\mathbf{t}$, where $\mathbf{X}(\mathbf{x}, \mathbf{t})$ is the inverse of $\mathbf{x}(\mathbf{X}, \mathbf{t})$. With these preliminaries we differentiate J as

$$\dot{J}(S) = \sum_I \int_S \left[e_{jip} e_{klp} \dot{x}_j(\mathbf{x}) n_l(\mathbf{x}) \frac{\partial \varphi^I}{\partial x_k}(\mathbf{x}) - n_i(\mathbf{x}) \dot{\varphi}^I(\mathbf{x}) \right] dS_x \cdot$$

$$\cdot \int_{\partial D} G_{,i}(\mathbf{x}-\mathbf{y}) \left(f^I(\mathbf{y}) + \int_S G_{,m}(\mathbf{y}-\mathbf{z}) n_m(\mathbf{z}) \varphi^I(\mathbf{z}) dS_z \right) dS_y \tag{10}$$

where e_{ijk} is the permutation symbol in 3D and the symbol '$\dot{}$' stands for the 'Lagrangian' derivative with respect to one of t_i. Since φ^I has already been obtained in the computation for J, and the calculation of J always precedes that of $\dot{J}$ in Powell's method, we see that all the quantities in (10) are known except for $\dot{\varphi}^I$. By differentiating (9) with respect to t_i, however, we obtain a variational equation to determine $\dot{\varphi}^I$ given by

$$\int_S e_{PQ} \frac{\partial \psi}{\partial s_P}(\mathbf{x}) \frac{\partial x_p}{\partial s_Q}(\mathbf{x}) ds_{1_x} ds_{2_x} \int_S G(\mathbf{x}-\mathbf{y}) e_{KL} \frac{\partial \dot{\varphi}^I}{\partial s_K}(\mathbf{y}) \frac{\partial x_p}{\partial s_L}(\mathbf{y}) ds_{1_y} ds_{2_y}$$

$$+ \int_S e_{PQ} \frac{\partial \psi}{\partial s_P}(\mathbf{x}) \frac{\partial \dot{x}_p}{\partial s_Q}(\mathbf{x}) ds_{1_x} ds_{2_x} \int_S G(\mathbf{x}-\mathbf{y}) e_{KL} \frac{\partial \varphi^I}{\partial s_K}(\mathbf{y}) \frac{\partial x_p}{\partial s_L}(\mathbf{y}) ds_{1_y} ds_{2_y}$$

$$+ \int_S e_{PQ} \frac{\partial \psi}{\partial s_P}(\mathbf{x}) \frac{\partial x_p}{\partial s_Q}(\mathbf{x}) ds_{1_x} ds_{2_x} \cdot$$

$$
\begin{aligned}
&\cdot \int_S G_{,k}(\mathbf{x}-\mathbf{y})(\dot{x}_k(\mathbf{x})-\dot{x}_k(\mathbf{y}))e_{KL}\frac{\partial \varphi^I}{\partial s_K}(\mathbf{y})\frac{\partial x_p}{\partial s_L}(\mathbf{y})ds_{1_y}ds_{2_y} \\
&+\int_S e_{PQ}\frac{\partial \psi}{\partial s_P}(\mathbf{x})\frac{\partial x_p}{\partial s_Q}(\mathbf{x})ds_{1_x}ds_{2_x}\int_S G(\mathbf{x}-\mathbf{y})e_{KL}\frac{\partial \varphi^I}{\partial s_K}(\mathbf{y})\frac{\partial \dot{x}_p}{\partial s_L}(\mathbf{y})ds_{1_y}ds_{2_y} \\
&-\int_S e_{PQ}\frac{\partial \psi}{\partial s_P}(\mathbf{x})\frac{\partial x_p}{\partial s_Q}(\mathbf{x})e_{ijp}g^I_{,i}(\mathbf{x})\dot{x}_j(\mathbf{x})ds_1 ds_2 = 0,
\end{aligned}
\tag{11}
$$

where we have used an equality $\Delta g^I = 0$ which follows form (2) and (6). Notice that all the terms in (11) except for the one including $\dot{\varphi}^I$ are known since φ^I is. Therefore the discretised version of this equation can be solved for $\dot{\varphi}^I$ numerically.

In the special case of a planar circular crack (9) and (11) simplify to

$$
\begin{aligned}
&\int_{S_0}\frac{\partial \psi}{\partial X_J}(\mathbf{X})dX_1 dX_2\int_{S_0} G(\mathbf{X}-\mathbf{Y})\frac{\partial \varphi^I}{\partial Y_J}(\mathbf{x}(\mathbf{Y}))dY_1 dY_2 \\
&\quad = -a\int_{S_0}\psi(\mathbf{X})g^I_{,i}(\mathbf{x}(\mathbf{X}))n_i dX_1 dX_2,
\end{aligned}
\tag{12}
$$

and

$$
\begin{aligned}
&\int_{S_0}\frac{\partial \psi}{\partial X_J}(\mathbf{X})dX_1 dX_2\int_{S_0} G(\mathbf{X}-\mathbf{Y})\frac{\partial \dot{\varphi}^I}{\partial Y_J}(\mathbf{x}(\mathbf{Y}))dY_1 dY_2 \\
&\quad = \frac{1}{a}\int_{S_0} e_{JK}\frac{\partial \psi}{\partial X_J}(\mathbf{X})\frac{\partial x_p}{\partial X_K}e_{ijp}g^I_{,i}(\mathbf{x}(\mathbf{X}))\dot{x}_j(\mathbf{X})dX_1 dX_2 \\
&\quad + \dot{a}\int_{S_0}\psi(\mathbf{X})g^I_{,i}(\mathbf{x}(\mathbf{X}))\,n_i dX_1 dX_2
\end{aligned}
\tag{13}
$$

where a is the radius of S, S_0 is the reference crack which is taken to be a unit circle, and $\mathbf{X}$ and $\mathbf{Y}$ are points on S_0 which are referred to by a fixed cartesian coordinate system on S_0. For a general S, however, one has to use the general formula in (11).

Numerical Analysis

In this section we briefly describe the numerical procedures used to obtain the results given in the next section. In 3D the variational equation in (9) is solved directly with the help of Galerkin's method. Namely, we discretise (9) by substituting Ω^α for ψ and $\sum_\beta \Omega^\beta \varphi^{I,\beta}$ for φ^I, where Ω^α is a shape function on S and $\varphi^{I,\alpha}$ is the nodal value of φ^I. As Ω^α we use the ordinary three node piecewise linear (isoparametric) element neglecting the near tip singularity of φ^I; we follow Nedelec [10] for this choice. The inner (outer) integral in the discretised version of (9) is then computed analytically (numerically). Notice that the same matrix equation is obtained as one discretises (11) with the same shape functions Ω^α for $\dot{\varphi}^I$. This observation allows us to construct the matrix equation only when we evaluate J, and to reuse it later in the calculation of $\dot{J}$.

In the special case of a circular crack the matrix equation is derived directly from (12) or from (13). A mesh on the reference crack S_0 is used to this end. The matrix equation thus obtained stays invariant regardless of the size and location of S. Hence we need to compute and solve this matrix equation *only once* in the whole process of minimisation.

Notice that our formulation requires several multiple integrations, which are usually prohibitive. It is therefore essential to develop fast methods to compute these integrals. This is the more so considering the iterative nature of the nonlinear programming algorithm. Here are some examples of integration techniques: (a) Analytical integrations are employed as often as possible because numerical integrations are usually slower. (b) We compute integrals on S such as the one in (10) by integrating interpolations of the nodal values of the integrands. Piecewise linear interpolation functions are used to this end. (c) The computation of $g^I_{,i}$ is made faster when the data on ∂D are of Dirichlet type and the corresponding U^I admits an analytical expression, where U^I is the no crack solution of the Dirichlet problem defined by $\Delta U^I = 0$ in D and $U^I = u^I$ on ∂D for $I = 1 \sim N$. Indeed, we then use (1) with $\varphi = 0$ and (6) to have an expression for $g^I_{,i}$ given by

$$g^I_{,i}(\mathbf{x}) = -U^I_{,i}(\mathbf{x}) - \int_{\partial D} G_{,i}(\mathbf{x}-\mathbf{y}) \left(\frac{\partial u^I}{\partial n}(\mathbf{y}) - \frac{\partial U^I}{\partial n}(\mathbf{y}) \right) dS_y \quad \text{on } S. \tag{14}$$

(d) It is advisable to seek possibilities of reusing integration results. In the computation of J, for an example, we calculate the integrals written as

$$\int_{\partial D} G_{,i}(\mathbf{x}-\mathbf{y})\omega(\mathbf{y})dS$$

analytically for all the nodal points $\mathbf{x}$ on S and all shape functions ω on ∂D, and store the results. These integrals and (14) are used immediately for evaluating $g^I_{,i}$, which we need to obtain φ^I and, therefore, to obtain J. We reuse these integrals with (10) later in the evaluation of $\dot{J}$.

Finally we remark that the use of 'integrated' regularised integral equations discussed in [9] is more efficient than Galerkin's method in 2D.

Numerical Examples

In this section we present several 2D and 3D numerical examples. We replace experimental boundary measurements by numerical simulations because our interest is to test the proposed method. Namely, we solve, by BIEM, several direct problems with the given true crack geometry and Dirichlet data on ∂D, and then use $\partial u^I/\partial n - \partial U^I/\partial n$ thus obtained as the input to the inverse problem solver.

1. 2D examples We consider for D a circular domain having a radius of r. On the exterior boundary we give 2 Dirichlet data given by $u^1 = x_1, u^2 = x_2$. The exterior boundary ∂D is modelled by 24 piecewise linear boundary elements, and 5 DOF cubic spline elements are used for the crack which is assumed to be linear. Hypersingular integral equations are solved with the method in [9].

In the first example both the initial guess and the true crack are straight lines shown in Fig. 1(a). We use noisy data produced by giving ±10% of error, with an alternating sign, to $\partial u^I/\partial n - \partial U^I/\partial n$, as shown in Fig. 1(a). The crack configurations after (almost) every 5 iteration steps are also plotted in the same figure. Fig. 1(b) shows the most plausible crack configuration obtained by our method. In Fig. 2 we plotted results for the same problem as in Fig. 1 but with +10% of error given to the data on ∂D. Our analysis converged to the line segment shown in Fig. 2(b). These examples prove the robustness of the present approach. The CPU time for each of these examples was less than 1 sec. on Fujitsu FACOM M780 (scalar processor).

2. 3D example We consider for D a cube having a side length of $2l$. On the exterior boundary we give 3 Dirichlet data given by $u^1 = x_1, u^2 = x_2, u^3 = x_3$. The crack is modelled by 21 DOF piecewise linear elements, and ∂D is discretised into 96 piecewise constant boundary elements (See Fig. 3.). No noise in data is considered.

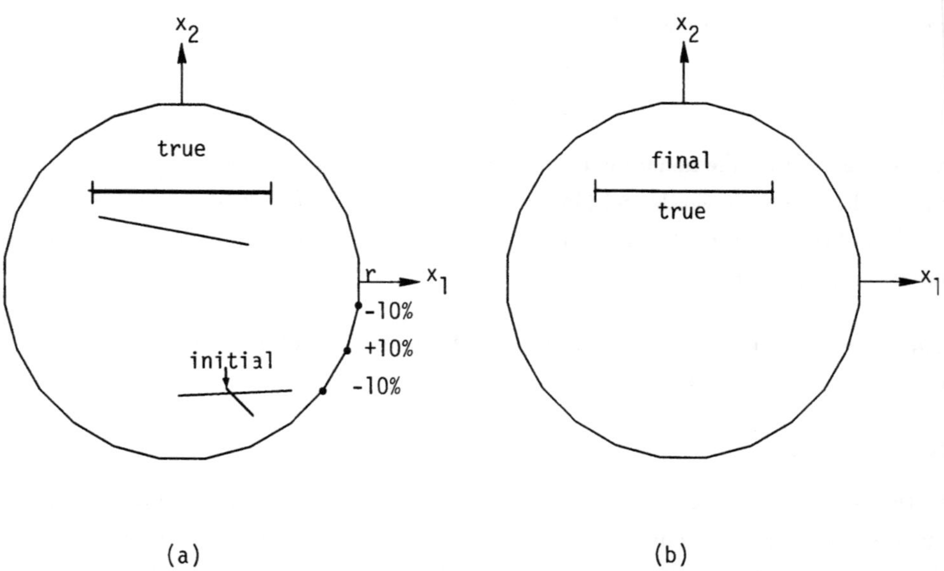

Fig. 1. 2D line crack search with noisy data 1. (a) Mode of convergence (b) final resul

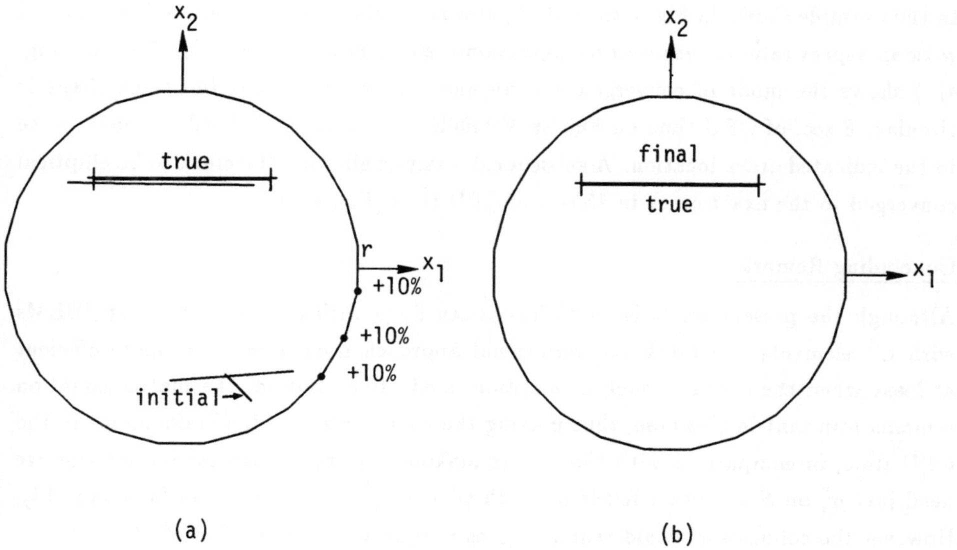

Fig. 2. 2D line crack search with noisy data 2. (a) Mode of convergence (b) final result

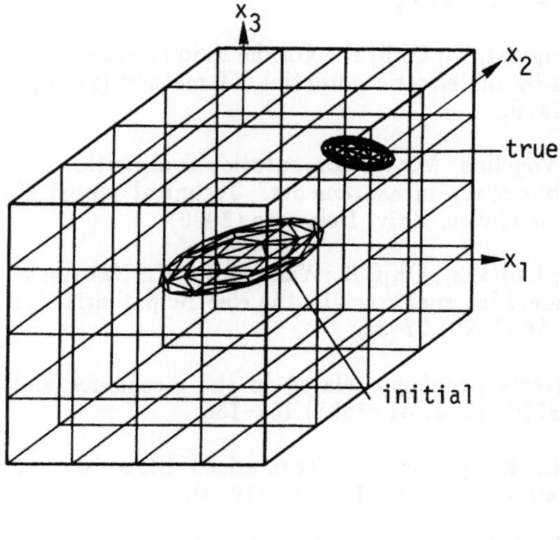

Fig. 3. Mesh, initial guess and true crack for 3D elliptical crack analysis

In the example shown in Fig. 3 the initial guess is circular and the true crack is elliptical with an aspect ratio (=major axis length/minor axis length) of $(5+\sqrt{2})/(5-\sqrt{2})$. Fig. 4(a) shows the mode of convergence in an analysis assuming that the crack shape is circular. 8 sec. of CPU time on Fujitsu VP400E (vector processor) led to convergence to the indicated crack location. A subsequent analysis allowing the crack to be elliptical converged to the exact crack in 15 sec. of CPU time (Fig. 4(b)).

Concluding Remark

Although the present analysis could have been done with certain collocation BIEMs with C^1 elements, we think the variational approach used here to be more efficient at least when the circular crack assumption is effective. Indeed, the matrix equation remains invariant in this case, thus making the computation of RHSs dominant in the CPU time, in comparison with the matrix making. In the variational formulation we need just $g^I_{,i}$ on S to obtain RHSs for both φ^I and $\dot{\varphi}^I$ as we can see in (12) and (13). However the collocation would require $g^I_{,ij}$ as well, in the computation of $\dot{\varphi}^I$.

References

1. Barone, M.R.; Yang, R.-J.: Boundary integral equations for recovery of design sensitivities in shape optimization. AIAA J. 26 (1988) 589–594.

2. Nishimura, N.; Kobayashi, S.: A regularized boundary integral equation method for elastodynamic crack problems. Comp. Mech. 4 (1989) 319–328.

3. A. Friedman, A.; Vogelius, M.: Determining cracks by boundary measurements. IMA preprint series #476 (1989).

4. Kubo, S.; Sakagami, T.; Ohji, K.: On the uniqueness of the inverse solution in crack determination by the electric potential CT method (in Japanese). Trans. JSME 55 (1989) 2316–2319.

5. Santosa, F.; Vogelius, M.: A computational algorithm to determine cracks from electrostatic boundary measurements. Technical report No. 90-3, Center for the Mathematics of Waves, Univ. Delaware (1990).

6. Sakagami, T.; Kubo, S.; Ohji, K.; Yamamoto, K.; Nakatsuka, K.: Identification of a three-dimensional internal crack by the electric potential CT method (in Japanese). Trans. JSME 56 (1990) 27–32.

7. Kubo, S.: Inverse problems related to the mechanics and fracture of solids and structures. JSME Int. J. 31 (1988) 157–166.

8. Nishimura, N.; Kobayashi, S.: Regularised BIEs for crack shape determination problems, to appear in Proc. BEM12 (1990).

9. Nishimura, N.; Kobayashi, S.: A boundary integral equation method for an inverse problem related to crack detection. to appear (1990).

10. Nedelec, J.C.: Integral equations with non integrable kernels. Integral Eq. Operator Th. 5 (1982) 562–572.

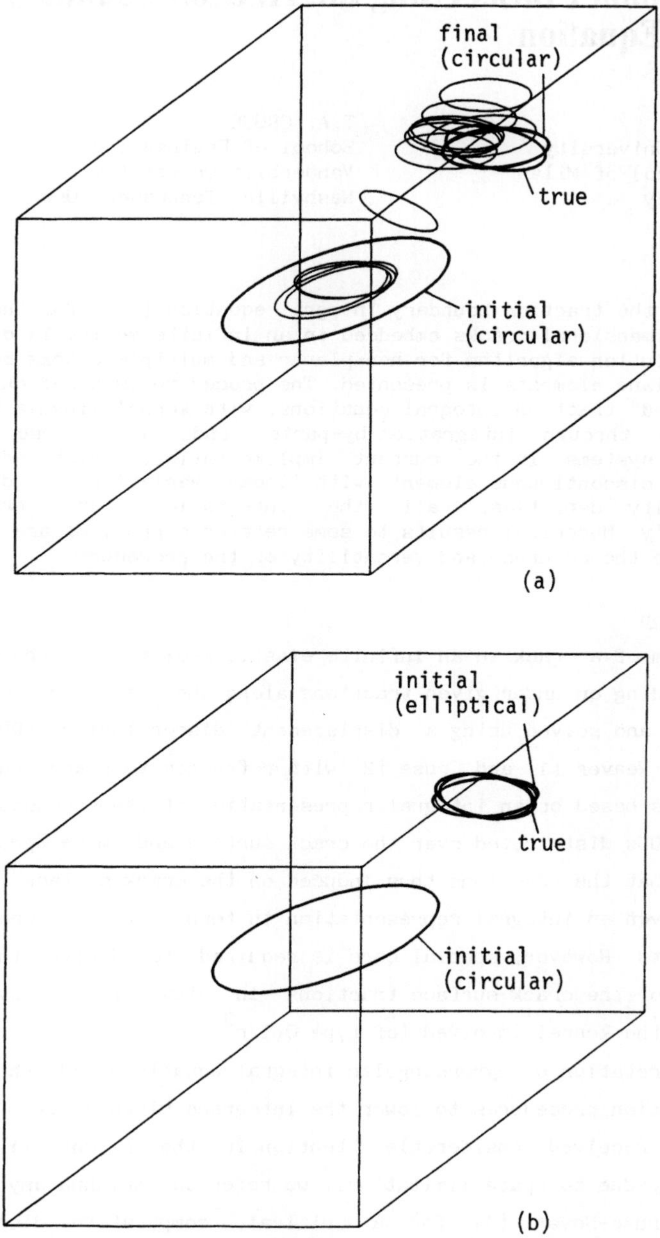

Fig. 4. Mode of convergence. (a) search with circular crack assumption (b) search with elliptical crack assumption

Analysis of Non-Planar Embedded Three-Dimensional Cracks Using the Traction Boundary Integral Equation

G. NOVATI
Technical University
(Politecnico) of Milan
Milan, Italy

T.A. CRUSE
School of Engineering
Vanderbilt University
Nashville, Tennessee, Usa

Summary

The use of the traction boundary integral equation (BIE) for the analysis of three dimensional cracks embedded in an infinite medium is discussed. A general solution algorithm for non-planar and multiple cracks modeled as a patch of plane elements is presented. The procedure is based on a set of "regularized" traction integral equations, with kernel singularity reduced to $O(1/r^2)$ through integration-by-parts, and on the use of local coordinate systems. In the current implementation, which adopts plane triangular discontinuous elements with linear variation of displacement discontinuity densities, all the integrations are carried out analytically. Numerical results to some reference problems are given and demonstrate the accuracy and versatility of the procedure.

Introduction

The problem of a crack in an infinite elastic medium under the action of a remote loading or under given tractions along the crack surface, can be formulated and solved using a displacement discontinuity (DD) approach (see e. g. Weaver [1] and Cruse [2] with reference to plane cracks). This approach is based on an integral representation of the stresses caused in Ω_∞ by the DDs distributed over the crack surface and, more precisely, on the fact that the tractions thus induced on the crack surface itself can also be given an integral representation in terms of the crack opening dispacements. However, special care is required to derive the integral equation for the crack-surface tractions in view of the hypersingular nature of the kernel involved (of type $O(1/r^3)$).

The interpretation of hypersingular integral equations and the study of regularization procedures to lower the integrand singularity, are issues which have received considerable attention in the recent literature on BIEs. Here, due to space limitations, we refer to Krishnasamy et al.[3] and to Cruse-Novati [4] for a critical, comparative discussion on alternative treatments of the hypersingular BIE.

In the frame of the DD methods, this paper presents a new analysis procedure for non-planar and multiple cracks, characterized by the following features: (i) the cracks are modeled as piecewise-flat surfaces; (ii) the hypersingular traction-BIE is regularized through integration-by-parts; (iii) both the regularization process and the integrations are carried out in local coordinate systems; (iv) if the modeled crack is discretized into plane triangular elements over each of which the DD field is assumed to vary linearly, all the needed integrations can be performed analitically in closed form. Note that this discretization strategy is the one adopted in the current implementation. The proposed procedure, only outlined in Ref.[4] and fully developed in Novati-Cruse [5], is concisely described in the next section where the regularized equations are given in their explicit form.

Regularization of the traction-BIE and solution procedure for linear elements

Figure 1 depicts the assumed piecewise-flat crack surface Γ_{cr} made up by plane triangular boundary elements (BEs).

At first focus is on a single triangular BE, denoted by symbol Γ, and on the stress state generated in Ω_∞ when a linear DD takes place across the element Γ itself while no crack opening displacement is active across the remainder of Γ_{cr}; in particular the crack-surface tractions thus induced on Γ are the crucial quantities to be given an integral representation in terms of the modeled DD field pertaining to the same element. Integral representations of tractions due to the DDs on Γ are derived in the current procedure using a local cartesian reference frame, of unit basis vectors ($\underline{e}_1$, $\underline{e}_2$, $\underline{e}_3$), associated to element Γ as shown in Figure 1: $\underline{e}_1$ and $\underline{e}_2$ have in-plane directions and $\underline{e}_3$ is normal to Γ. Let $p_j(\underline{\xi},\underline{n}^\xi)$ denote the traction component, relevant to a point $\underline{\xi}$ and to a surface element through $\underline{\xi}$ of unit normal $\underline{n}^\xi$, due to DDs $\Delta u_i(\underline{\eta}) \equiv u_i^+(\underline{\eta})-u_i^-(\underline{\eta})$ (i = 1,2,3) acting across Γ alone (supescript + refers to the side facing the positive direction of $\underline{e}_3$). The integral equation for the traction $p_j(\underline{\xi},\underline{n}^\xi)$

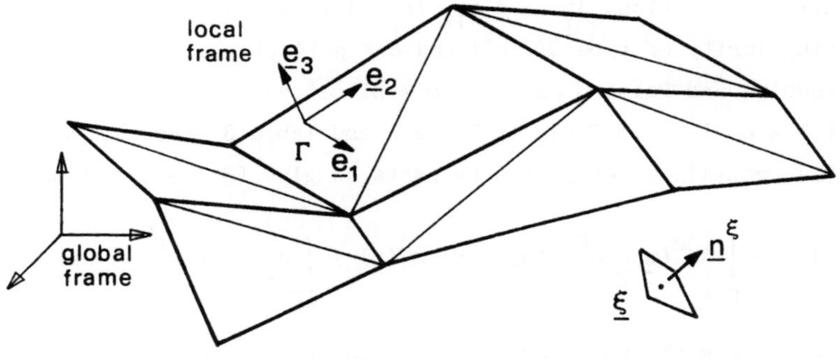

Figure 1. BE crack model and local frame associated to element Γ.

at a point $\underline{\xi}$ off the element Γ, reads, in the local frame:

$$p_j(\underline{\xi},\underline{n}^{\xi}) = \int_{\Gamma} \overset{**}{p}{}_i^j(\underline{\eta},\underline{\xi};\underline{e}_3,\underline{n}^{\xi})\ \Delta\underline{u}_i(\underline{\eta})\ dS(\eta) \tag{1}$$

The kernel function $\overset{**}{p}{}_i^j(\underline{x},\underline{\xi};\underline{n}^x,\underline{n}^{\xi})$, which behaves like $O(1/r^3)$ for $r \to 0$ (r being the distance between field point $\underline{x}$ and source point $\underline{\xi}$), is the influence function for Ω_∞ which gives the i-th traction component at point $\underline{x}$, relevant to an element surface (through $\underline{x}$) of outward normal $\underline{n}^x$, due to a concentrated unit DD acting in the j-th direction at point $\underline{\xi}$ across a surface element of unit normal $\underline{n}^{\xi}$. Equation (1) is obtained as follows: (a) write a Betti's theorem for two elastic states in Ω_∞ whose sources are the DD distribution on the actual crack Γ and a second DD distribution relevant to a fictitious crack through point $\underline{\xi}$ and of normal $\underline{n}^{\xi}$; (b) let the density of the fictitious DD in the j-th direction be represented by a Dirac delta function while the other DD components are made to vanish: this leads to the above traction equation.
It turns out that the kernels $\overset{**}{p}{}_i^j$ of this equation can be cast into an alternative form which makes its r.h.side amenable to integration-by-parts with respect to the in-plane coordinates η_1,η_2 (defined along the axes $\underline{e}_1,\underline{e}_2$), thus leading to a "regularized" traction integral equation. Such kernel transformation has been pursued and carried out in a new and fairly synthetic fashion, described in Ref. [5], through the use of the reciprocity properties linking the displacements $\overset{**}{u}{}_i^j$ (of the fundamental solution due to a concentrated DD) to the Kelvin tractions $\overset{*}{p}{}_k^h$. The alternative expressions obtained for the kernels $\overset{**}{p}{}_i^j$ are:

$$\overset{**}{p}{}_i^j(\underline{\eta},\underline{\xi};\underline{e}_3,\underline{n}^{\xi}) = A_{i,1}^{j1} + A_{i,2}^{j2} = c\ n_k^{\xi}\ [a_{ik,1}^{j1}(\underline{\eta},\underline{\xi}) + a_{ik,2}^{j2}(\underline{\eta},\underline{\xi})] \tag{2}$$

where the commas denote in-plane derivatives with respect to η_1 and η_2, the auxiliary functions $A_i^{j\alpha}$ and $a_{ik}^{j\alpha}$ ($\alpha = 1,2,\ k = 1,2,3$) exhibit a lower order singularity of type $O(1/r^2)$ and $c \equiv \mu/[4\pi(1-\nu)]$, μ and ν being the shear modulus and Poisson ratio; the explicit expressions of $a_{ik}^{j\alpha}(\underline{\eta},\underline{\xi})$ for $j = 1,2,3$ are given in Table 1, Table 2 and Table 3, respectively.
Using (2), eq.(1) is integrated-by-parts to give ($j,i = 1,2,3$; $\alpha = 1,2$):

$$p_j(\underline{\xi},\underline{n}^{\xi}) = -\int_{\Gamma} A_i^{j\alpha}(\underline{\eta},\underline{\xi},\underline{n}^{\xi})\ \Delta u_{i,\alpha}(\underline{\eta})\ dS(\underline{\eta}) + \int_{\partial\Gamma} A_i^{j\alpha}(\underline{\eta},\underline{\xi},\underline{n}^{\xi})\ m_\alpha(s)\ \Delta u_i(\underline{\eta})\ ds \tag{3}$$

where $\partial\Gamma$ denotes the bounding curve for the flat crack-portion Γ, s is an

Table 1. Auxiliary functions $a_{ik}^{1\alpha}$ involved in the expression of the kernels $\overset{**}{p}{}_i^1(\underline{\eta},\underline{\xi};\underline{e}_3,\underline{n}^{\xi})$ (i = 1,2,3) in the local frame.

$a_{11}^{11} = \frac{1}{r^2} r_{,3} \left[+ 1 + 3(r_{,1})^2 \right]$	$a_{11}^{12} = 0$
$a_{12}^{11} = \frac{9}{4 r^2} r_{,1} r_{,2} r_{,3}$	$a_{12}^{12} = \frac{1}{4 r^2} r_{,3} \left[(3-4\nu) + 3(r_{,1})^2\right]$
$a_{13}^{11} = \frac{3}{2 r^2} r_{,1} \left[(r_{,3})^2 - (r_{,1})^2\right]$	$a_{13}^{12} = \frac{-1}{2 r^2} r_{,2} \left[(1-2\nu) + 3(r_{,1})^2\right]$
$a_{21}^{11} = \frac{3}{2 r^2} r_{,1} r_{,2} r_{,3}$	$a_{21}^{12} = \frac{1}{2 r^2} r_{,3} \left[-(1-4\nu) + 3(r_{,1})^2\right]$
$a_{22}^{11} = \frac{1}{4 r^2} r_{,3} \left[(3-4\nu) + 3(r_{,2})^2\right]$	$a_{22}^{12} = \frac{9}{4 r^2} r_{,1} r_{,2} r_{,3}$
$a_{23}^{11} = \frac{1}{2 r^2} r_{,2} \left[(1-2\nu) - 3(r_{,1})^2\right]$	$a_{23}^{12} = \frac{3}{2 r^2} r_{,1} \left[-(r_{,2})^2 + (r_{,3})^2\right]$
$a_{31}^{11} = \frac{1}{r^2} r_{,1} \left[-1 + 3(r_{,3})^2\right]$	$a_{31}^{12} = \frac{1}{r^2} r_{,2} (-2\nu)$
$a_{32}^{11} = \frac{1}{2 r^2} r_{,2} \left[-(1-2\nu) + 3(r_{,3})^2\right]$	$a_{32}^{12} = \frac{1}{2 r^2} r_{,1} \left[-(1-2\nu) + 3(r_{,3})^2\right]$
$a_{33}^{11} = \frac{1}{r^2} r_{,3} \left[1 - 3(r_{,1})^2\right]$	$a_{33}^{12} = \frac{-3}{r^2} r_{,1} r_{,2} r_{,3}$

arc-length parameter running along $\partial\Gamma$, m^{α} (α = 1,2) are the director cosines with respect to the axes $\underline{e}_1, \underline{e}_2$ of the in-plane outward normal to $\partial\Gamma$ and the dependence $\underline{\eta} = \underline{\eta}(s)$ is implicit in the integrand of the line integral. Equation (3) is the regularized version of the traction integral representation, valid for any pont $\underline{\xi}$ off the Γ-element surface. Before considering its limiting form obtained when the source point $\underline{\xi}$ is moved towards a point $\underline{\eta}_0$ interior to element Γ, let us exploit the adopted field modelling. The DD components, assumed linear over each triangle, and their in-plane derivatives are represented on Γ (in the local frame) as:

$$\Delta u_i(\underline{\eta}) = \underline{N}^T(\underline{\eta})\ \Delta \underline{u}_i\ , \qquad \Delta u_{i,\alpha} = \underline{D}_{\alpha}^T\ \Delta \underline{u}_i\ \ (=\text{const.}) \qquad (4a,b)$$

where $\underline{N}(\eta)$ and $\underline{D}_{\alpha}$ are interpolation operators and vector $\Delta\underline{u}_i$ lists the values of the DD components at the three vertices of Γ. The substitution

Table 2. Auxiliary functions $a^{2\alpha}_{ik}$ involved in the expression of the kernels $p^{**2}_{i}(\underline{\eta},\underline{\xi};\underline{e}_3,\underline{n}^{\xi})$ (i = 1,2,3), in the local frame.

$a^{21}_{11} = \frac{9}{4\,r^2}\, r_{,1}\, r_{,2}\, r_{,3}$	$a^{12}_{11} = \frac{1}{4\,r^2}\, r_{,3} \left[(3-4\nu) + 3(r_{,1})^2\right]$
$a^{21}_{12} = \frac{1}{2\,r^2}\, r_{,3} \left[-(1-4\nu) + 3(r_{,2})^2\right]$	$a^{22}_{12} = \frac{3}{2\,r^2}\, r_{,1}\, r_{,2}\, r_{,3}$
$a^{21}_{13} = \frac{3}{2\,r^2}\, r_{,2} \left[-(r_{,1})^2 + (r_{,3})^2\right]$	$a^{22}_{13} = \frac{1}{2\,r^2}\, r_{,1} \left[(1-2\nu) - 3(r_{,2})^2\right]$
$a^{21}_{21} = \frac{1}{4\,r^2}\, r_{,3} \left[(3-4\nu) + 3(r_{,2})^2\right]$	$a^{22}_{21} = \frac{9}{4\,r^2}\, r_{,1}\, r_{,2}\, r_{,3}$
$a^{11}_{22} = 0$	$a^{12}_{22} = \frac{1}{r^2}\, r_{,3} \left[1 + 3(r_{,2})^2\right]$
$a^{11}_{23} = \frac{1}{2\,r^2}\, r_{,1} \left[-(1-2\nu) - 3(r_{,2})^2\right]$	$a^{12}_{23} = \frac{3}{2\,r^2}\, r_{,2} \left[-(r_{,2})^2 + (r_{,3})^2\right]$
$a^{21}_{31} = \frac{1}{2\,r^2}\, r_{,2} \left[-(1-2\nu) + 3(r_{,3})^2\right]$	$a^{22}_{31} = \frac{1}{2\,r^2}\, r_{,1} \left[-(1-2\nu) + 3(r_{,3})^2\right]$
$a^{21}_{32} = \frac{1}{r^2}\, r_{,1}\, (-2\nu)$	$a^{22}_{32} = \frac{1}{r^2}\, r_{,2} \left[-1 + 3(r_{,3})^2\right]$
$a^{21}_{33} = \frac{-3}{r^2}\, r_{,1}\, r_{,2}\, r_{,3}$	$a^{22}_{33} = \frac{1}{r^2}\, r_{,3} \left[1 - 3(r_{,2})^2\right]$

of (4) into the traction equation (3) gives:

$$p_j(\underline{\xi}\,,\underline{n}^{\xi}) = - \int_{\Gamma} A_i^{j\alpha}\, dS\; \underline{D}_{\alpha}^{T}\, \Delta\underline{u}_i \;+\; \int_{\partial\Gamma} A_i^{j\alpha}\, \underline{N}^{T}\, m_{\alpha}\, ds\;\; \Delta\underline{u}_i \tag{5}$$

In Ref.[5] it is shown that all the integrals showing up in (5) for the various possible combinations of indices i,j and α, can be evaluated analytically (in closed form) by carrying over to the present context an earlier algorithm, see Cruse [6].

The limit version of eq. (5) for the case in which $\underline{\xi}$ is taken to a point $\underline{\eta}_o$ on (and interior to) the crack portion Γ and $\underline{n}^{\xi} = \underline{e}_3 \equiv [0,\ 0,\ 1]$, gives the sought discretized integral representation of the crack tractions induced on Γ by the modeled DD across Γ itself. In view of the simple form of their integrands, the strongly singular surface-integrals arising in this case can easily be evaluated by the following two stage process: first, one assumes $\underline{\xi}$ at a finite distance from Γ and transforms

Table 3. Auxiliary functions $a_{ik}^{3\alpha}$ involved in the expression of the kernels $\overset{**}{p}{}_i^{3}(\underline{\eta},\underline{\xi};\underline{e}_3,\underline{n}^{\xi})$ (i = 1,2,3), in the local frame.

$a_{11}^{31} = \frac{3}{2\,r^2}\,r_{,1}\left[-(r_{,1})^2 + (r_{,3})^2\right]$ $a_{12}^{31} = \frac{3}{2\,r^2}\,r_{,2}\left[-(r_{,1})^2 + (r_{,3})^2\right]$ $a_{13}^{31} = \frac{3}{2\,r^2}\,r_{,3}\left[-(r_{,1})^2 + (r_{,3})^2\right]$	$a_{11}^{32} = \frac{1}{2\,r^2}\,r_{,2}\left[-(1-2\nu) - 3(r_{,1})^2\right]$ $a_{12}^{32} = \frac{1}{2\,r^2}\,r_{,1}\left[(1-2\nu) - 3(r_{,2})^2\right]$ $a_{13}^{32} = \frac{-3}{2\,r^2}\,r_{,1}\,r_{,2}\,r_{,3}$
$a_{21}^{31} = \frac{1}{2\,r^2}\,r_{,2}\left[(1-2\nu) - 3(r_{,1})^2\right]$ $a_{22}^{31} = \frac{1}{2\,r^2}\,r_{,1}\left[-(1-2\nu) - 3(r_{,2})^2\right]$ $a_{23}^{31} = \frac{-3}{2\,r^2}\,r_{,1}\,r_{,2}\,r_{,3}$	$a_{21}^{32} = \frac{3}{2\,r^2}\,r_{,1}\left[-(r_{,2})^2 + (r_{,3})^2\right]$ $a_{22}^{32} = \frac{3}{2\,r^2}\,r_{,2}\left[-(r_{,2})^2 + (r_{,3})^2\right]$ $a_{23}^{32} = \frac{3}{2\,r^2}\,r_{,3}\left[-(r_{,2})^2 + (r_{,3})^2\right]$
$a_{31}^{31} = \frac{1}{r^2}\,r_{,3}\left[1 - 3(r_{,1})^2\right]$ $a_{32}^{31} = \frac{-3}{r^2}\,r_{,1}\,r_{,2}\,r_{,3}$ $a_{33}^{31} = \frac{-1}{r^2}\,r_{,1}\left[1 + 3(r_{,3})^2\right]$	$a_{31}^{32} = \frac{-3}{r^2}\,r_{,1}\,r_{,2}\,r_{,3}$ $a_{32}^{32} = \frac{1}{r^2}\,r_{,3}\left[1 - 3(r_{,2})^2\right]$ $a_{33}^{32} = \frac{-1}{r^2}\,r_{,2}\left[1 + 3(r_{,3})^2\right]$

the surface integrals into line integrals along $\partial\Gamma$; secondly, on taking the limit $\underline{\xi} \to \underline{\eta}_o$, one finds that all the integrated contributions relevant to functions $a_{ik}^{j\alpha}$ which contain $r_{,3}$ as a multiplicative factor vanish while all the integrated contributions relevant to the remaining functions $a_{ik}^{j\alpha}$ remain finite and are obtained in closed form.

Besides, it is straightforward to show that the results obtained by computing the surface integrals through the above two-stage process, would also be arrived at by conceiving the load point located right on Γ (so that $r_{,3}$ vanishes identically) and interpreting the surface integrals in the Cauchy Principal Value sense.

The integral equation for $p_j(\underline{\eta},\underline{e}_3)$ (j = 1,2,3) (often referred to by simply saying that it is obtained by collocating (5) at a point $\underline{\eta}$ of Γ without alluding to the underlying limit process) can be written simultaneously for different points of Γ in a compact matrix form; relabeling the crack-surface BE in point by Γ^n and using superscript n to mark the relevant arrays, such matrix equation reads:

$$\underline{p}^{n} = \underline{g}^{nn}\,\Delta\underline{u}^{n} \qquad \text{(no sum on n)} \tag{6}$$

where $\underline{p}^n$ is a $3c^n$-vector collecting traction components at the chosen c^n collocation points on Γ^n, $\Delta\underline{u}^n$ is a $3d^n$-vector of all the unknown DD parameters, d^n being the number of vertices of Γ^n not located on the crack edge (where the DD components have known, zero value), and each coefficient of $\underline{g}^{nn}$ is the sum of the surface- and line-integral contributions subordinated by the limit version of eq.(5).
Introducing the orthogonal matrix $\underline{R}^n$ which transforms vector representations in the Γ^n local frame into the corresponding ones in the global system, equation (6) is expressed in the latter system as:

$$\underline{P}^{n} = \underline{G}^{nn}\,\Delta\underline{U}^{n}, \qquad \text{with} \quad \underline{G}^{nn} \equiv \underline{R}^{n}\,\underline{g}^{nn}\,(\underline{R}^{n})^{T} \qquad \text{(no sum on n)} \tag{7}$$

Considering now the presence of the DDs across the whole BE crack model (instead of across the single element Γ^n, solely), it is obvious that the crack-traction contributions at the c^n collocation points of Γ^n due to the DD on the s-th triangular element, can be expressed first in the Γ^s local frame by the counterpart of eq.(5) (but the first integral on its r.h.side is not singular for r≠s), and then in the global frame through the matrix equation $\underline{P}^n = \underline{G}^{ns}\,\Delta\underline{U}^s$ analogous to (7), where $\underline{G}^{ns} \equiv \underline{R}^s\,\underline{g}^{ns}(\underline{R}^s)^T$. Note that this equations holds unaltered also for multiple cracks, in which case the BEs Γ^n and Γ^s may belong to two separate, coexisting crack-surfaces. Hence, in the global reference frame, the simultaneous presence of DDs on all the elements of the modeled crack is accounted for, through superposition of effects, by simply expressing the resulting tractions at the collocation points of Γ^n with the summation

$$\underline{P}^{n} = \sum_{r=1}^{R} \underline{G}^{nr}\,\Delta\underline{u}^{r} \tag{8}$$

A matrix equation of type (8) is available for all the elements (i.e. for n = 1,..R); a natural way to proceed in order to obtain a final equation system $\underline{G}\,\Delta\underline{U} = \underline{P}$ with square coeff. matrix, is to use discontinuous elements, and guarantee the balance between equations and unknowns by choosing $c^n = d^n$ for each BE Γ^n (i.e. on an element-by-element basis). Then, once the tractions in $\underline{P}$ are identified with the given crack-tractions, the equation system thus generated can be solved for the sought DDs in $\Delta\underline{U}$.

<u>Numerical tests</u>

The two examples considered concern three-dimensional cracks which simulate cracks in plane-strain conditions, subject to a remote loading; they are intended to test numerically the proposed traction-BIE technique for non- planar and multiple cracks. The cracks consist of surfaces parallel to one of the global reference system axes (axis x_2 was chosen) and whose cross-sections, shown in Fig. 2a-b, do not vary along x_2. The cracks are modeled by a set of planar surface strips each of which is discretized by discontinuous linear triangular elements, see Fig.2c, and is indicated by a straight segment in the cross-sections. The crack-surface strips, of variable width, are taken to be symmetric with respect to the plane $x_2 = 0$ and of considerable length in the x_2 direction, so that plane-strain conditions are enforced on the symmetry plane and in its vicinity. Besides the strip mesh, consisting of eight BEs, is kept unchanged along the x_2 direction for both the crack models. Figure 2c also shows the "active" collocation points (i.e. those effectively used according to the criterion illustrated at the end of the previous section) on internal and near-the-edge BEs: they are located along the lines connecting the element centroid to its vertices, at 60% of the distance from the centroid to each vertex. In both the examples, the remote applied stress considered is $\sigma_{33}^{\infty} = 1000$ (stress units) and the adopted material constants are $\nu = 0.3$ and $E = 10^6$ (stress units).

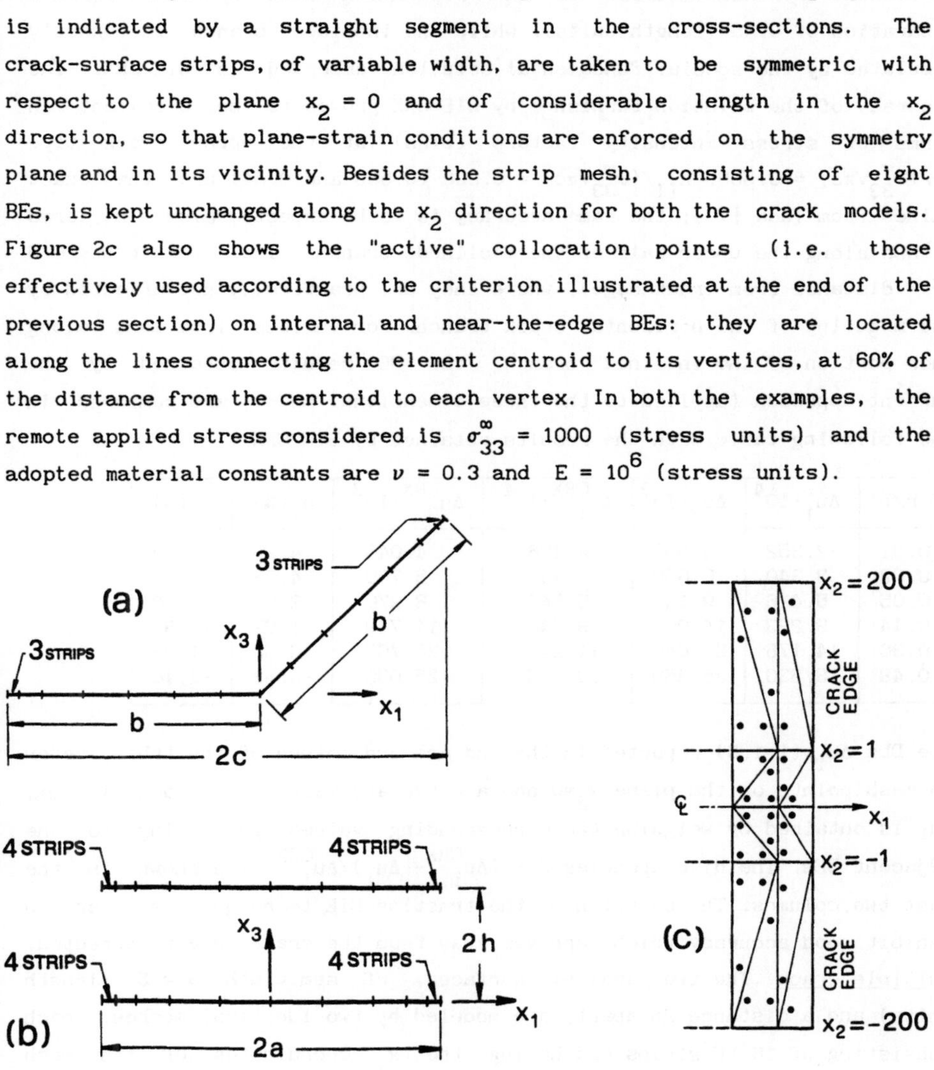

Figure 2. (a) Cross-section of the angled crack and (b) of the multiple crack model; (c) two of the planar BE strips, showing collocation points.

Angled crack. The horizontal and inclined (45 deg.) branches, for which $b = 10$ and $2c = 10+10\sqrt{2}/2$ (length units), are discretized by identical meshes. The x_1 coordinates of the BE vertices (mesh points) lying in the plane $x_3 = 0$ are: -10.0, -9.9, -9.8, -9.5, -8.6, -7.0, -5.2, -3.4, -1.6, -0.0. The complete mesh includes 144 BEs and 95 mesh points. The number of DD unknowns is 1224. A reference solution for the corresponding plane-strain problem has been obtained using the code BIE/CRX, see Ref.[2], which uses a special "implicit-crack" Green's function: the horizontal branch is modeled as an open notch with a notch surface separation of 0.45 (length units), while the inclined branch is exactly simulated by the special fundamental solution embedded in BIE/CRX. The accuracy of the solution furnished by BIE/CRX proves to be excellent in terms of stress intensity factors (SIFs) at the upper crack tip: $K_I/(\sigma_{33}^{\infty}\sqrt{\pi c}) = 0.565$, $K_{II}/(\sigma_{33}^{\infty}\sqrt{\pi c}) = 0.638$ (0.569 and 0.641 are the exact values from Tada [7]); the same accuracy is to be expected also in terms of DDs along the upper half of the inclined branch ,i.e. for $r/b < 0.5$ (r = distance from crack edge), where they are practically not affected by the modeling of the horizontal crack branch. For various locations along such portion of the inclined branch, the DD results obtained by the current approach (applied to the three dimensional model) are compared in the following table with the results obtained by BIE/CRX:

r/b	$\Delta u_1 \cdot 10^4$	$\Delta u_3 \cdot 10^3$	$\Delta u_1^{CRX} \cdot 10^4$	$\Delta u_3^{CRX} \cdot 10^3$	$d_1(\%)$	$d_3(\%)$
0.01	2.352	3.990	2.456	4.046	4.23	1.38
0.02	3.340	5.656	3.480	5.709	4.02	0.93
0.05	5.426	9.140	5.540	8.970	2.06	-1.90
0.14	9.293	14.999	9.446	14.719	1.62	-1.90
0.30	14.176	21.046	14.247	20.763	0.50	-1.36
0.48	18.530	25.359	18.405	25.086	-0.68	-1.10

The DDs Δu_i ($i=1,3$) reported in the 2nd and 3rd column of the table, refer to mesh points on the plane $x_2=0$ and are average values: at each vertex, Δu_i is obtained by weighing the corresponding values pertaining to the adjacent BEs. The discrepancies $d_i \equiv (\Delta u_i^{CRX} - \Delta u_i)/\Delta u_i^{CRX}$ are shown in the last two columns. The solution by the traction-BIE technique is seen to exhibit good accuracy which improves away from the crack edge as expected. Multiple crack. The two parallel surfaces, of semiwidth $a = 5$ (length units) and a distance $2h$ apart, are modeled by two identical meshes, each consisting of 16 BE strips and having the x_1 coordinates of its mesh points equal to 0.0, ∓1.5, ∓3.0, ∓4.0, ∓4.6, ∓4.8, ∓4.9, ∓4.95, ∓5.0. The overall model consists of 256 BEs, 170 mesh points and 2160 DD unknowns.

Two cases have been studied: h/a = 0.2 and h/a = 1.0 . For this crack problem, the accuracy of the traction-BIE solution is assessed in terms of the stress intensity factors evaluated through extrapolation from the DD results; this is because the exact SIFs for the corresponding plane-strain crack problem are available in Ref.[8]. Such evaluation of the SIFs K_I and K_{II} exploits the usual DD asymptotic solution along a portion of the line $x_2 = x_3 = 0$ near the crack front to define the quantities $[\hat{K}_I, \hat{K}_{II}] \equiv \frac{\mu}{4(1-\nu)} \sqrt{2\pi/r}\, [\Delta u_3, \Delta u_1]$ (r = distance from the crack edge); then, by computing $\hat{K}_I$ and $\hat{K}_{II}$ at four points located at $r = 0.01 \cdot a$, $0.02 \cdot a$, $0.04 \cdot a$, $0.08 \cdot a$ and performing a linear extrapolation of these values for $r \to 0$ (using a least-squares fit), the sought estimates of the SIFs are obtained. They are compared in the table below with the exact reference values K_I^{REF}, K_{II}^{REF} $(K_0 = \sigma_{33}^{\infty}\sqrt{\pi a})$. At worst, discrepancies are < 2% and < 9% for the Mode I and Mode II values, respectively.

h/a = 0.2		h/a = 1.0	
$K_I/K_0 = 0.7002$,	$K_{II}/K_0 = 0.1594$	$K_I/K_0 = 0.8188$,	$K_{II}/K_0 = 0.0593$
$K_I^{REF}/K_0 = 0.700$,	$K_{II}^{REF}/K_0 = 0.170$	$K_I^{REF}/K_0 = 0.835$,	$K_{II}^{REF}/K_0 = 0.065$

Acknoledgments. The bulk of this work was performed at the Southwest Research Institute, San Antonio (Texas). The support of this institution during the visiting position of the first author and the employment of the second is gratefully acknoledged. The first author also acknoledges the financial support of a CNR-NATO scholarship during his stay at SwRI.

References

1. Weaver, J.: Three-dimensional crack analysis. Int. J. Solids and Structs. 13 (1977) 321-330.

2. Cruse, T.A.: Boundary Element Analysis in Computational Fracture Mechanics. Kluwer Academic Publishers, Dordrecht, The Netherlands, 1988.

3. Krishnasamy, G., Schmerr, L.W., Rudolphi, T.J., Rizzo, F.J.: Hypersingular boundary integral equations:some applications in acoustic and elastic wave scattering. J. Appl. Mech. 57 (1990) 404-414.

4. Cruse, T.A., Novati, G.: Traction BIE formulations and applications to non-planar and multiple cracks. Presented at the 22nd Natl. Symposium on Fracture Mechanics, June 26-28, 1990, Atlanta, Georgia (Usa).

5. Novati, G., Cruse, T.A.: A regularized integral equation approach for non-planar, piecewise-flat, three-dimensional cracks. In preparation.

6. Cruse, T.A.: An improved boundary integral equation method for three dimensional elastic stress analysis. Comp. & Structs. 4 (1974) 741-754.

7. Tada, H., Paris, P.,and Irwin, G.: The Stress Analysis of Cracks Handbook, Del Research Corporation, St.Louis, Missouri,1985.

8. Rooke, D.P., Cartwright, D.J.: Compendium of Stress Intensity Factors, Her Majesty's Stationary Office, London, England.

Boundary/Field Variational Principles for the Elastic Plastic Rate Problem*

T. PANZECA, C. POLIZZOTTO and M. ZITO

Dipartimento di Ingegneria Strutturale & Geotecnica, DISEG

Università di Palermo, Palermo, Italy.

Summary

An elastic-plastic continuous solid body under quasi-statically variable external actions is herein addressed in the hypoteses of rate-independent material model with dual internal variables and of infinitesimal displacements and strains. The related analysis problem for assigned rate actions is first formulated through a boundary/field integral equation approach, then is shown to be characterized by two variational principles, one of which is a stationarity theorem, the other a min-max one.

Introduction

The elastic-plastic rate problem, that is the analysis problem for infinitely small external actions, can be viewed as a prototype problem in so that it can be transformed into an analogous incremental problem by the aid of some ad-hoc rules, e.g. incremental quantities instead of rate ones. For this reasons the elastic-plastic rate problem has received so much attention in the literature (see e.g. [1]). Variational principles were provided, and these principles turn out to be useful for the applications of discretization procedures by the FEM (finite element method). But, in the recent years, the BEM (boundary element method) has also shown itself to constitute an effective numerical tool within plasticity applications. So, one may whonder whether there exist any variational principles related to the elastic-plastic rate problem, which may be used as a starting point for boundary element (BE) discretizations. How this is possible for elastic-perfectly plastic material model was shown in [12]. The present paper aims at showing the same thing for a more complex material model.

* This paper has been completed with the financial support of the Ministero dell'Università e della Ricerca Scientifica e Tecnologica, Italy.

The Material Model

An elastic-plastic rate-independent material behaviour is herein assumed. The related constitutive equations read [2]:

$$\boldsymbol{\varepsilon} = \boldsymbol{\varepsilon}^e + \boldsymbol{\varepsilon}^p + \bar{\boldsymbol{\varepsilon}}^\theta \tag{1}$$

$$\dot{\boldsymbol{\varepsilon}}^p = \dot{\lambda}\frac{\partial \phi}{\partial \boldsymbol{\sigma}}, \qquad -\dot{\boldsymbol{\eta}} = \dot{\lambda}\frac{\partial \phi}{\partial \boldsymbol{\chi}} \tag{2a, b}$$

$$\phi(\boldsymbol{\sigma}, \boldsymbol{\chi}) \leq 0, \quad \dot{\lambda} \geq 0, \quad \dot{\lambda}\phi(\boldsymbol{\sigma}, \boldsymbol{\chi}) = 0 \tag{3a - c}$$

$$\boldsymbol{\sigma} = \mathbf{E} : \boldsymbol{\varepsilon}^e, \qquad \boldsymbol{\chi} = \frac{\partial \psi(\boldsymbol{\eta})}{\partial \boldsymbol{\eta}}. \tag{4a, b}$$

Here above, the upper dots denote derivatives with respect to the time-like parameter t, the overbar indicates an assigned quantity; $\boldsymbol{\varepsilon}$ is the total strain tensor, split into the elastic, plastic and thermal-like parts; $\boldsymbol{\sigma}$ is the stress tensor; $\boldsymbol{\chi}$ and $\boldsymbol{\eta}$ are dual internal variables (here formally treated as second-order tensors, but they are not necessarily so), which are related to each other through the thermodinamic potential $\psi(\boldsymbol{\eta})$; $\mathbf{E}$ is the usual stiffness fourth-order tensor of linear elasticity; ϕ is the (convex, smooth) yield function and $\dot{\lambda}$ is the plastic activation coefficient; finally, the dot and the colon denote the simple and double index saturation between tensor factors, respectively.

Equations (1)-(4) describe a rather wide class of elastic-plastic nonlinearly hardening material models of associated plasticity, stable in the Drucker sense [3]. They can be integrated over the entire deformation path when an evolutive problem is dealt with, or can be used to determine $\dot{\boldsymbol{\varepsilon}}^p, \dot{\boldsymbol{\eta}}$ in terms of $\dot{\boldsymbol{\sigma}}, \dot{\boldsymbol{\chi}}$ for the rate problem in which the material finds itself in a known state $\boldsymbol{\sigma}, \boldsymbol{\chi}$.

The Structural Rate Problem

A solid body of elastic-plastic material occupying the (finite) domain Ω and sourrounded by the (smooth) surface $\Gamma = \Gamma_1 \cup \Gamma_2$, restrained over Γ_1 in such a way as to prevent rigid motions, is subjected to a specified quasi-static load history. Let the state of the structure be known at the current time t, and let $\boldsymbol{\sigma}, \boldsymbol{\chi}$ denote the relevant stresses and stress internal variables. Let Ω_y be the subregion of Ω where the yield condition is attained, i.e. $\phi(\boldsymbol{\sigma}, \boldsymbol{\chi}) = 0$, at time t, whereas $\phi(\boldsymbol{\sigma}, \boldsymbol{\chi}) < 0$, thus $\dot{\lambda} = 0$, in $\Omega_e = \Omega - \Omega_y$. Within Ω_y one has $\dot{\phi} \leq 0, \dot{\lambda} \geq 0$ with $\dot{\lambda} = 0$ everywhere $\dot{\phi} < 0$ (elastic return). Therefore, introducing the tensors

$$\mathbf{r} = \frac{\partial \phi}{\partial \boldsymbol{\sigma}}, \qquad \mathbf{s} = \frac{\partial \phi}{\partial \boldsymbol{\chi}} \tag{5a, b}$$

$$\mathbf{H} = \frac{\partial^2 \psi}{\partial \boldsymbol{\eta} \partial \boldsymbol{\eta}} \tag{5c}$$

all of which depend on the known body's state at t, eqs. (1)-(4) can be rewritten in the following rate-form:

$$\dot{\boldsymbol{\varepsilon}} = \dot{\boldsymbol{\varepsilon}}^e + \dot{\boldsymbol{\varepsilon}}^p + \dot{\bar{\boldsymbol{\varepsilon}}}^\theta \quad \text{in } \Omega \tag{6}$$

$$\dot{\boldsymbol{\varepsilon}}^p = \mathbf{r}\dot{\lambda}, \qquad -\dot{\boldsymbol{\eta}} = \mathbf{s}\dot{\lambda} \quad \text{in } \Omega \tag{7a, b}$$

$$\dot{\phi} = \mathbf{r} : \dot{\boldsymbol{\sigma}} + \mathbf{s} : \dot{\boldsymbol{\chi}} \leq 0, \quad \dot{\lambda} \geq 0, \quad \dot{\lambda}\dot{\phi} = 0 \quad \text{in } \Omega_y \tag{8a - c}$$

$$\dot{\lambda} = 0 \quad \text{in } \Omega_e = \Omega - \Omega_y, \quad \text{and} \tag{8d}$$

$$\dot{\boldsymbol{\sigma}} = \mathbf{E} : \dot{\boldsymbol{\varepsilon}}, \qquad \dot{\boldsymbol{\chi}} = \mathbf{H} : \dot{\boldsymbol{\eta}} \quad \text{in } \Omega. \tag{9a, b}$$

The tensor $\mathbf{H}(\boldsymbol{\eta})$, the Hessian tensor of ψ, by hypotesis is positive definite. Equations (6)-(9) are thus the constitutive equations for the rate problem, to be supplemented by the compatibility and equilibrium conditions. The latter conditions read:

$$\dot{\boldsymbol{\varepsilon}} = \frac{1}{2}(\nabla \dot{\mathbf{u}} + (\nabla \dot{\mathbf{u}})^T) \quad \text{in } \Omega, \quad \dot{\mathbf{u}} = \dot{\bar{\mathbf{u}}} \quad \text{on } \Gamma_1 \tag{10a, b}$$

$$\nabla \cdot \dot{\boldsymbol{\sigma}} + \dot{\bar{\mathbf{b}}} = \mathbf{0} \quad \text{in } \Omega, \quad \dot{\boldsymbol{\sigma}} \cdot \mathbf{n} = \dot{\bar{\mathbf{t}}} \quad \text{on } \Gamma_2 \tag{11a, b}$$

where $\dot{\bar{\mathbf{b}}}$ denotes body force rates, $\dot{\mathbf{t}}$ traction rates, $\dot{\mathbf{u}}$ displacement rates, ∇ the well known *del* operator and $\mathbf{n}$ the unit external normal to Γ. Equations (6)-(9), (10) and (11) constitute a well posed boundary value problem. Related variational principles, analogous to those given by Capurso [13, 14], can be formulated and utilized for FEM-based solving procedures. But this is not the purpose of the present paper.

Compatibility (10), equilibrium (11) and Hooke's law (9a) can all be enforced within Ω by representing the body's elastic response by Somigliana's formulae [4-6]. The latter apply to the infinite (homogeneous) elastic domain Ω_∞, with the same elastic properties as the embebbed domain Ω and subjected, besides the given load rates, to some unknown rate actions, that is, layered force rates $\dot{\mathbf{g}}$ in Γ_1, layered distortion rates $\dot{\mathbf{v}}$ in Γ_2 and plastic strain rates $\dot{\boldsymbol{\varepsilon}}^p$ in Ω treated as initial strains. The rate-form Somigliana formulae can be written as

$$\dot{\mathbf{u}} = \mathcal{R}_u[\dot{\mathbf{g}}, \dot{\mathbf{v}}, \dot{\boldsymbol{\varepsilon}}^p] + \dot{\mathbf{u}}_I \tag{12a}$$

$$\dot{\mathbf{t}} = \mathcal{R}_t[\dot{\mathbf{g}}, \dot{\mathbf{v}}, \dot{\boldsymbol{\varepsilon}}^p] + \dot{\mathbf{t}}_I \tag{12b}$$

$$\dot{\boldsymbol{\sigma}} = \mathcal{R}_\sigma[\dot{\mathbf{g}}, \dot{\mathbf{v}}, \dot{\boldsymbol{\varepsilon}}^p] + \dot{\boldsymbol{\sigma}}_I \tag{12c}$$

where, by definition,

$$\mathcal{R}_u[\dot{\mathbf{g}}, \dot{\mathbf{v}}, \dot{\boldsymbol{\varepsilon}}^p] = \langle \mathbf{G}_{uu} \cdot \dot{\mathbf{g}} \rangle_{\Gamma_1} + \langle \mathbf{G}_{ut} \cdot \dot{\mathbf{v}} \rangle_{\Gamma_2} + \langle \mathbf{G}_{u\sigma} : \dot{\boldsymbol{\varepsilon}}^p \rangle_{\Omega} \tag{13a}$$

$$\mathcal{R}_t[\dot{\mathbf{g}}, \dot{\mathbf{v}}, \dot{\boldsymbol{\varepsilon}}^p] = \langle \mathbf{G}_{tu} \cdot \dot{\mathbf{g}} \rangle_{\Gamma_1} + \langle \mathbf{G}_{tt} \cdot \dot{\mathbf{v}} \rangle_{\Gamma_2} + \langle \mathbf{G}_{t\sigma} : \dot{\boldsymbol{\varepsilon}}^p \rangle_{\Omega} \tag{13b}$$

$$\mathcal{R}_\sigma[\dot{\mathbf{g}}, \dot{\mathbf{v}}, \dot{\boldsymbol{\varepsilon}}^p] = \langle \mathbf{G}_{\sigma u} \cdot \dot{\mathbf{g}} \rangle_{\Gamma_1} + \langle \mathbf{G}_{\sigma t} \cdot \dot{\mathbf{v}} \rangle_{\Gamma_2} + \langle \mathbf{G}_{\sigma\sigma} : \dot{\boldsymbol{\varepsilon}}^p \rangle_{\Omega} \tag{13c}$$

Here the notation $\langle q \rangle_R$ indicates the integral of q over R and the two point tensor functions $\mathbf{G}_{hk}(\mathbf{x}, \mathbf{y})$ collect fundamental solutions giving effects at $\mathbf{x} \in \Omega_\infty$ due to unit actions applied at $\mathbf{y} \in \Omega_\infty$ through the rule: the first index denotes the effect ($u \rightarrow$ displacement, $t \rightarrow$ traction, $\sigma \rightarrow$ stress), the second index denotes the unit action through the duality relationship ($u \rightarrow$ force, $t \rightarrow$ layered distortion, $\sigma \rightarrow$ volumetric distortion) [7, 8]. Due to Maxwell's theorem, the properties hold true:

$$\mathbf{G}_{hk}(\mathbf{x}, \mathbf{y}) = \mathbf{G}^T_{kh}(\mathbf{y}, \mathbf{x}) \quad \text{for } h, k = u, t, \sigma. \tag{14}$$

The last terms in ($12a-c$) denote the response of Ω_∞ to the given external action rates. They are expressed as [7, 8]:

$$\dot{\mathbf{u}}_I = -\langle \mathbf{G}_{ut} \cdot \dot{\bar{\mathbf{u}}} \rangle_{\Gamma_1} + \langle \mathbf{G}_{uu} \cdot \dot{\bar{\mathbf{t}}} \rangle_{\Gamma_2} + \langle \mathbf{G}_{uu} \cdot \dot{\bar{\mathbf{b}}} \rangle_{\Omega} + \langle \mathbf{G}_{u\sigma} : \dot{\bar{\boldsymbol{\varepsilon}}}^p \rangle_{\Omega} \tag{15a}$$

$$\dot{\mathbf{t}}_I = -\langle \mathbf{G}_{tt} \cdot \dot{\bar{\mathbf{u}}} \rangle_{\Gamma_1} + \langle \mathbf{G}_{tu} \cdot \dot{\bar{\mathbf{t}}} \rangle_{\Gamma_2} + \langle \mathbf{G}_{tu} \cdot \dot{\bar{\mathbf{b}}} \rangle_{\Omega} + \langle \mathbf{G}_{t\sigma} : \dot{\bar{\boldsymbol{\varepsilon}}}^p \rangle_{\Omega} \tag{15b}$$

$$\dot{\boldsymbol{\sigma}}_I = -\langle \mathbf{G}_{\sigma t} \cdot \dot{\bar{\mathbf{u}}} \rangle_{\Gamma_1} + \langle \mathbf{G}_{\sigma u} \cdot \dot{\bar{\mathbf{t}}} \rangle_{\Gamma_2} + \langle \mathbf{G}_{\sigma u} \cdot \dot{\bar{\mathbf{b}}} \rangle_{\Omega} + \langle \mathbf{G}_{\sigma\sigma} : \dot{\bar{\boldsymbol{\varepsilon}}}^p \rangle_{\Omega} \tag{15c}$$

It worths noting that $\dot{\mathbf{u}}_I$ and $\dot{\mathbf{t}}_I$ are discontinuous across Γ_1 and Γ_2, respectively, and there one can write [7, 8]:

$$\dot{\mathbf{u}}_I|_{\Gamma_1^+} - \dot{\mathbf{u}}_I|_{\Gamma_1^-} = -\dot{\bar{\mathbf{u}}} \text{ on } \Gamma_1, \qquad \dot{\mathbf{t}}_I|_{\Gamma_2^+} - \dot{\mathbf{t}}_I|_{\Gamma_2^-} = -\dot{\bar{\mathbf{t}}} \text{ on } \Gamma_2 \tag{16a, b}$$

where Γ^- and Γ^+ denote surfaces infinitely close to Γ from inside and outside, respectively. These discontinuities arise from the singularities of the first integral of ($15a$) and of the second integral of ($15b$).

Compatibility on Γ_1 and equilibrium on Γ_2 for the body, that is eqs.($10b$) and ($11b$), are enforced by writing a set of boundary integral equations, namely

$$\mathcal{R}_u[\dot{\mathbf{g}}, \dot{\mathbf{v}}, \dot{\boldsymbol{\varepsilon}}^p] = \dot{\bar{\mathbf{u}}} - \dot{\mathbf{u}}_I|_{\Gamma_1^-} \text{ on } \Gamma_1 \tag{17a}$$

$$\mathcal{R}_t[\dot{\mathbf{g}}, \dot{\mathbf{v}}, \dot{\boldsymbol{\varepsilon}}^p] = \dot{\bar{\mathbf{t}}} - \dot{\mathbf{t}}_I|_{\Gamma_2^-} \text{ on } \Gamma_2 \tag{17b}$$

where $\dot{\mathbf{v}}$ is assumed continuous on Γ_2. The latter equation set, at least in principle, can be solved for the unknown $\dot{\mathbf{g}}, \dot{\mathbf{v}}$, the plastic strain rate field being arbitrarily assigned in

Ω. Assuming that this has been done, and denoting by $\dot{\mathbf{g}}^*, \dot{\mathbf{v}}^*$ the solution, eqs. (12a, b) give, with the aid of (16a, b):

$$\dot{\mathbf{u}}|_{\Gamma_1^+} = \mathcal{R}_u[\dot{\mathbf{g}}^*, \dot{\mathbf{v}}^*, \dot{\boldsymbol{\varepsilon}}^p]|_{\Gamma_1} + \dot{\mathbf{u}}_I|_{\Gamma_1^+} = \mathbf{0} \tag{18a}$$

$$\dot{\mathbf{t}}\Big|_{\Gamma_2^+} = \mathcal{R}_t[\dot{\mathbf{g}}^*, \dot{\mathbf{v}}^*, \dot{\boldsymbol{\varepsilon}}^p]|_{\Gamma_2} + \dot{\mathbf{t}}_I\Big|_{\Gamma_2^+} = \mathbf{0}; \tag{18b}$$

in other words, whatever the given load rates and the assigned plastic strain rate field $\dot{\boldsymbol{\varepsilon}}^p$, no actions migrate outside the intersurface Γ of Ω_∞ such that the response of the elastic body has been obtained, $\dot{\mathbf{t}}\big|_{\Gamma_1} = \dot{\mathbf{g}}^*$ and $\dot{\mathbf{u}}|_{\Gamma_2} = -\dot{\mathbf{v}}^*$.

Equation (17a) is the classical boundary integral equation (BIE) for displacements, which, enforced over the entire Γ, is used alone as the basis of the direct boundary element method (BEM) [4, 5]. Equation (17b) is the BIE for the tractions, which contains some hypersingular integrals [9, 10].

Equations (17) and (6)-(9) constitute a set of equations enabling one to solve, at least in principle, the body's elastic-plastic rate problem. As pointed out by Zhang and Atluri [11], only mixed boundary/field formulations are allowed for nonlinear problems, with domain integrals containing some unknown fields ($\dot{\boldsymbol{\varepsilon}}^p, \dot{\lambda}$ in the present case), due to the lack of related fundamental solutions. Obviously, the above equation set can only be addressed numerically if one wants to solve it. Though the actual numerical problem is in practice shaped as an incremental problem for a small finite load step, however rate-form elastic-plastic problems are always of interest due the obvious relationships with their incremental conterparts. Therefore, the formulation of variational principles related to the rate-form problems is paramount, primarely because they can be used as tools for boundary elament (BE) discretizations to generate discrete rate problems, secondly because such principles can either be used as the starting points to establish other variational principles related to the corresponding incremental-form problems.

The above equation set can be characterized by variational principles, as shown hereafter.

Stationarity Principle in Terms of $\dot{\mathbf{g}}, \dot{\mathbf{v}}, \dot{\boldsymbol{\varepsilon}}^p, \dot{\boldsymbol{\eta}}, \dot{\boldsymbol{\sigma}}, \dot{\chi}$

Let one introduce the functionals:

$$\begin{aligned} a[\dot{\mathbf{g}}, \dot{\mathbf{v}}, \dot{\boldsymbol{\varepsilon}}^p] = {} & \langle \dot{\mathbf{g}} \cdot \mathbf{G}_{uu} \cdot \dot{\mathbf{g}} \rangle_{\Gamma_1^2} + 2\langle \dot{\mathbf{g}} \cdot \mathbf{G}_{ut} \cdot \dot{\mathbf{v}} \rangle_{\Gamma_1 \times \Gamma_2} + \langle \dot{\mathbf{v}} \cdot \mathbf{G}_{tt} \cdot \dot{\mathbf{v}} \rangle_{\Gamma_2^2} \\ & + 2\langle \dot{\mathbf{g}} \cdot \mathbf{G}_{u\sigma} : \dot{\boldsymbol{\varepsilon}}^p \rangle_{\Gamma_1 \times \Omega} + 2\langle \dot{\mathbf{v}} \cdot \mathbf{G}_{t\sigma} : \dot{\boldsymbol{\varepsilon}}^p \rangle_{\Gamma_2 \times \Omega} + \langle \dot{\boldsymbol{\varepsilon}}^p : \mathbf{G}_{\sigma\sigma} : \dot{\boldsymbol{\varepsilon}}^p \rangle_{\Omega^2} \end{aligned} \tag{19}$$

$$\Pi[\dot{\mathbf{g}},\dot{\mathbf{v}},\dot{\boldsymbol{\varepsilon}}^p,\dot{\boldsymbol{\eta}},\dot{\boldsymbol{\sigma}},\dot{\boldsymbol{\chi}}]=\frac{1}{2}a[\dot{\mathbf{g}},\dot{\mathbf{v}},\dot{\boldsymbol{\varepsilon}}^p]-\langle\dot{\boldsymbol{\sigma}}:\dot{\boldsymbol{\varepsilon}}^p-\dot{\boldsymbol{\chi}}:\dot{\boldsymbol{\eta}}\rangle_\Omega-\frac{1}{2}\langle\dot{\boldsymbol{\eta}}:\mathbf{H}:\dot{\boldsymbol{\eta}}\rangle_\Omega \\ +\langle(\dot{\mathbf{u}}_I-\dot{\bar{\mathbf{u}}})\cdot\dot{\mathbf{g}}\rangle_{\Gamma_1}+\langle(\dot{\mathbf{t}}_I-\dot{\bar{\mathbf{t}}})\cdot\dot{\mathbf{v}}\rangle_{\Gamma_2}+\langle\dot{\boldsymbol{\sigma}}_I:\dot{\boldsymbol{\varepsilon}}^p\rangle_\Omega. \tag{20}$$

The following theorem can be proved:

Theorem 1. The set $(\dot{\mathbf{g}},\dot{\mathbf{v}},\dot{\boldsymbol{\varepsilon}}^p,\dot{\boldsymbol{\eta}},\dot{\boldsymbol{\sigma}},\dot{\boldsymbol{\chi}})$ which makes Π stationary, under the condition $\mathbf{r}:\dot{\boldsymbol{\sigma}}+\mathbf{s}:\dot{\boldsymbol{\chi}}\leq 0$ in Ω_y, can be associated with a scalar function $\dot{\lambda}$ and all together solve the elastic-plastic rate problem. Conversely, the solution to the latter problem makes Π stationary.

Proof. Applying the Lagrange multiplier method, one considers the augmented functional

$$\Pi^a=\Pi[\dot{\mathbf{g}},\dot{\mathbf{v}},\dot{\boldsymbol{\varepsilon}}^p,\dot{\boldsymbol{\eta}},\dot{\boldsymbol{\sigma}},\dot{\boldsymbol{\chi}}]+\langle\dot{\lambda}\,(\mathbf{r}:\dot{\boldsymbol{\sigma}}+\mathbf{s}:\dot{\boldsymbol{\chi}})\rangle_\Omega \tag{21}$$

where the Lagrangian multiplier function $\dot{\lambda}$ satisfies the conditions

$$\dot{\lambda}\geq 0 \text{ in } \Omega_y, \qquad \dot{\lambda}=0 \text{ in } \Omega_e. \tag{22}$$

The first variation of Π^a, after some easy trasformations, reads:

$$\delta\Pi^a=\langle\delta\dot{\mathbf{g}}\cdot(\mathcal{R}_u[\dot{\mathbf{g}},\dot{\mathbf{v}},\dot{\boldsymbol{\varepsilon}}^p]+\dot{\mathbf{u}}_I-\dot{\bar{\mathbf{u}}})\rangle_{\Gamma_1}+\langle\delta\dot{\mathbf{v}}\cdot(\mathcal{R}_t[\dot{\mathbf{g}},\dot{\mathbf{v}},\dot{\boldsymbol{\varepsilon}}^p]+\dot{\mathbf{t}}_I-\dot{\bar{\mathbf{t}}})\rangle_{\Gamma_2} \\ +\langle\delta\dot{\boldsymbol{\varepsilon}}^p:(\mathcal{R}_\sigma[\dot{\mathbf{g}},\dot{\mathbf{v}},\dot{\boldsymbol{\varepsilon}}^p]+\dot{\boldsymbol{\sigma}}_I-\dot{\boldsymbol{\sigma}})\rangle_\Omega+\langle\delta\dot{\boldsymbol{\sigma}}:(\mathbf{r}\dot{\lambda}-\dot{\boldsymbol{\varepsilon}}^p)\rangle_\Omega \\ +\langle\delta\dot{\boldsymbol{\eta}}:(\dot{\boldsymbol{\chi}}-\mathbf{H}:\dot{\boldsymbol{\eta}})\rangle_\Omega+\langle\delta\dot{\boldsymbol{\chi}}:(\mathbf{s}\dot{\lambda}+\dot{\boldsymbol{\eta}})\rangle_\Omega+\langle\delta\dot{\lambda}\,(\mathbf{r}:\dot{\boldsymbol{\sigma}}+\mathbf{s}:\dot{\boldsymbol{\chi}})\rangle_\Omega \tag{23}$$

which shows that the Euler-Lagrange equations related to the above stationarity problem coincide with the equations which govern the elastic-plastic rate problem in point, with $\dot{\lambda}$ having the meaning of plastic activation coefficient. Conversely, the solution to the latter problem makes $\delta\Pi^a$ to vanish for arbitrary $\delta\dot{\mathbf{g}},\delta\dot{\mathbf{v}},\ldots$, with $\delta\dot{\lambda}\geq 0$ in $\Omega_y^p\subset\Omega_y$ where $\dot{\phi}=0$ and $\delta\dot{\lambda}=0$ in $\Omega_y^e\subset\Omega_y$ where $\dot{\phi}<0$, such that Π is stationary.

If the internal variables are dropped and thus the perfectly plastic model is considered, Th. 1 above coincides with one given in [12].

Min-Max Principle in Terms of $\dot{\mathbf{g}}, \dot{\mathbf{v}}, \dot{\lambda}$

The following properties were proved in [12]:

$$
\begin{aligned}
&i)\ \ \langle \dot{\mathbf{g}} \cdot \mathbf{G}_{uu} \cdot \dot{\mathbf{g}} \rangle_{\Gamma_1^2} \quad positive\, definite\, (pd)\\
&ii)\ \ \langle \dot{\mathbf{v}} \cdot \mathbf{G}_{tt} \cdot \dot{\mathbf{v}} \rangle_{\Gamma_2^2} \quad negative\, definite\, (nd)\\
&iii)\ \ \langle \dot{\boldsymbol{\varepsilon}}^p : \mathbf{G}_{\sigma\sigma} : \dot{\boldsymbol{\varepsilon}}^p \rangle_{\Omega^2} \quad negative\, semidefinite\, (nsd)\\
&iv)\ \ \langle \dot{\mathbf{v}} \cdot \mathbf{G}_{tt} \cdot \dot{\mathbf{v}} \rangle_{\Gamma_2^2} + 2\langle \dot{\mathbf{v}} \cdot \mathbf{G}_{t\sigma} : \dot{\boldsymbol{\varepsilon}}^p \rangle_{\Gamma_2 \times \Omega} + \langle \dot{\boldsymbol{\varepsilon}}^p : \mathbf{G}_{\sigma\sigma} : \dot{\boldsymbol{\varepsilon}}^p \rangle_{\Omega^2}\ (nsd)
\end{aligned}
$$

which are reported here for later use. Let one introduce the reduced functional

$$
\begin{aligned}
\hat{\Pi}[\dot{\mathbf{g}}, \dot{\mathbf{v}}, \dot{\lambda}] = \frac{1}{2} a[\dot{\mathbf{g}}, \dot{\mathbf{v}}, \mathbf{r}\dot{\lambda}] - \frac{1}{2} \langle \mathbf{s} : \mathbf{H} : \mathbf{s}\, \dot{\lambda}^2 \rangle_\Omega \\
+ \langle (\dot{\mathbf{u}}_I - \dot{\bar{\mathbf{u}}}) \cdot \dot{\mathbf{g}} \rangle_{\Gamma_1} + \langle (\dot{\mathbf{t}}_I - \dot{\bar{\mathbf{t}}}) \cdot \dot{\mathbf{v}} \rangle_{\Gamma_2} + \langle \dot{\boldsymbol{\sigma}}_I : \mathbf{r}\dot{\lambda} \rangle_\Omega
\end{aligned}
\tag{24}
$$

and let one consider the saddle-point problem:

$$
\min_{(\dot{\mathbf{g}})} \max_{(\dot{\mathbf{v}}, \dot{\lambda})} \hat{\Pi}[\dot{\mathbf{g}}, \dot{\mathbf{v}}, \dot{\lambda}] \quad \text{s. t. } \dot{\lambda} \geq 0 \text{ in } \Omega_y, \quad \dot{\lambda} = 0 \text{ in } \Omega_e \tag{25}
$$

where "s.t." stands for "subject to". With the aid of the above properties, the following theorem can be proved:

Theorem 2. The set $(\dot{\mathbf{g}}^*, \dot{\mathbf{v}}^*, \dot{\lambda}^*)$ solving the saddle-point problem (25) is a solution to the elastic-plastic rate problem. The converse is also true, and the problem (25) admits an unique solution.

Proof. Let the set $(\dot{\mathbf{g}}^*, \dot{\mathbf{v}}^*, \dot{\lambda}^*)$ be a solution to (25). Thus, any feasible solution to (25) can be given the form

$$
\dot{\mathbf{g}} = \dot{\mathbf{g}}^* + \delta\dot{\mathbf{g}} \quad \text{on } \Gamma_1, \qquad \dot{\mathbf{v}} = \dot{\mathbf{v}}^* + \delta\dot{\mathbf{v}} \quad \text{on } \Gamma_2 \tag{26a, b}
$$

$$
\dot{\lambda} = \dot{\lambda}^* + \delta\dot{\lambda} \geq 0 \quad \text{in } \Omega, \quad \dot{\lambda} \geq 0 \quad \text{in } \Omega_y, \quad \delta\dot{\lambda} = 0 \quad \text{in } \Omega_e \tag{26c}
$$

Substitution of the latter expressions in (24) gives

$$
\hat{\Pi}[\dot{\mathbf{g}}, \dot{\mathbf{v}}, \dot{\lambda}] = \hat{\Pi}^* + \delta\hat{\Pi} + \frac{1}{2}\delta^2\hat{\Pi} \tag{27}
$$

where $\hat{\Pi} = \hat{\Pi}[\dot{\mathbf{g}}^*, \dot{\mathbf{v}}^*, \dot{\lambda}^*]$, and moreover

$$
\begin{aligned}
\delta\hat{\Pi} = \langle \delta\dot{\mathbf{g}} \cdot (\mathcal{R}_u[\dot{\mathbf{g}}^*, \dot{\mathbf{v}}^*, \mathbf{r}\dot{\lambda}^*] + \dot{\mathbf{u}}_I - \dot{\bar{\mathbf{u}}}) \rangle_{\Gamma_1} + \langle \delta\dot{\mathbf{v}} \cdot (\mathcal{R}_t[\dot{\mathbf{g}}^*, \dot{\mathbf{v}}^*, \mathbf{r}\dot{\lambda}^*] + \dot{\mathbf{t}}_I - \dot{\bar{\mathbf{t}}}) \rangle_{\Gamma_2} \\
+ \langle \delta\dot{\lambda} (\mathbf{r} : \mathcal{R}_\sigma[\dot{\mathbf{g}}^*, \dot{\mathbf{v}}^*, \mathbf{r}\dot{\lambda}^*] + \dot{\boldsymbol{\sigma}}_I - \mathbf{s} : \mathbf{H} : \mathbf{s}\, \dot{\lambda}^*) \rangle_\Omega
\end{aligned}
\tag{28}
$$

$$\begin{aligned}\delta\hat{\Pi}^2 &= \langle\delta\dot{\mathbf{g}}\cdot\mathbf{G}_{uu}\cdot\delta\dot{\mathbf{g}}\rangle_{\Gamma_1^2} + 2\langle\delta\dot{\mathbf{g}}\cdot\mathbf{G}_{ut}\cdot\delta\dot{\mathbf{v}}\rangle_{\Gamma_1\times\Gamma_2} + \langle\delta\dot{\mathbf{v}}\cdot\mathbf{G}_{tt}\cdot\delta\dot{\mathbf{v}}\rangle_{\Gamma_2^2}\\ &+ 2\langle\delta\dot{\mathbf{g}}\cdot\mathbf{G}_{u\sigma}:\mathbf{r}\delta\dot{\lambda}\rangle_{\Gamma_1\times\Omega} + 2\langle\delta\dot{\mathbf{v}}\cdot\mathbf{G}_{t\sigma}:\mathbf{r}\delta\dot{\lambda}\rangle_{\Gamma_2\times\Omega}\\ &+ \langle\mathbf{r}:\mathbf{G}_{\sigma\sigma}:\mathbf{r}\,(\delta\dot{\lambda})^2\rangle_{\Omega^2} - \langle\mathbf{s}:\mathbf{H}:\mathbf{s}\,(\delta\dot{\lambda})^2\rangle_{\Omega^2}.\end{aligned} \tag{29}$$

Since the first variation $\delta\hat{\Pi}$ must vanish for arbitrary variation functions $\delta\dot{\mathbf{g}}$, $\delta\dot{\mathbf{v}}$ and must be nonpositive for arbitrary $\delta\dot{\lambda}$ satisfying the constraints (26c), eq. (28) gives that the set $(\dot{\mathbf{g}}^*, \dot{\mathbf{v}}^*, \dot{\lambda}^*)$ solves the elastic-plastic rate problem with the following identifications

$$\begin{aligned}&\dot{\boldsymbol{\epsilon}}^{p*} = \mathbf{r}\dot{\lambda}^*,\ \dot{\boldsymbol{\eta}}^* = -\mathbf{s}\dot{\lambda}^* \text{ in } \Omega\\ &\dot{\boldsymbol{\sigma}}^* = \mathcal{R}_\sigma[\dot{\mathbf{g}}^*, \dot{\mathbf{v}}^*, \mathbf{r}\dot{\lambda}^*] + \dot{\boldsymbol{\sigma}}_I \text{ in } \Omega\\ &\dot{\boldsymbol{\chi}}^* = \mathbf{H}:\dot{\boldsymbol{\eta}}^* \text{ in } \Omega.\end{aligned} \tag{30$a-d$}$$

Conversely, let the set $(\dot{\mathbf{g}}^*, \dot{\mathbf{v}}^*, \dot{\lambda}^*)$ solves the elastic-plastic rate problem. Then, the first variation $\delta\hat{\Pi}$ vanishes for any $\delta\dot{\mathbf{g}}, \delta\dot{\mathbf{v}}, \delta\dot{\lambda}$, with $\delta\dot{\lambda} = 0$ in Ω_e and in those points of Ω_y where $\dot{\phi}^* < 0$. Thus eq. (27) reads

$$\hat{\Pi}[\dot{\mathbf{g}}, \dot{\mathbf{v}}, \dot{\lambda}] = \hat{\Pi}^* + \frac{1}{2}\delta^2\hat{\Pi} \tag{31}$$

and the latter equation, by the aid of eq. (27), gives

$$\hat{\Pi}[\dot{\mathbf{g}}, \dot{\mathbf{v}}^*, \dot{\lambda}^*] = \hat{\Pi}^* + \frac{1}{2}\langle\delta\dot{\mathbf{g}}\cdot\mathbf{G}_{uu}\cdot\delta\dot{\mathbf{g}}\rangle_{\Gamma_1^2} > \hat{\Pi}^*, \tag{32a}$$

$$\begin{aligned}\hat{\Pi}[\dot{\mathbf{g}}^*, \dot{\mathbf{v}}, \dot{\lambda}] = \hat{\Pi}^* &+ \frac{1}{2}\Big\{\langle\delta\dot{\mathbf{v}}\cdot\mathbf{G}_{tt}\cdot\delta\dot{\mathbf{v}}\rangle_{\Gamma_2^2} + 2\langle\delta\dot{\mathbf{v}}\cdot\mathbf{G}_{t\sigma}:\mathbf{r}\delta\dot{\lambda}\rangle_{\Gamma_2\times\Omega}\\ &+ \langle\mathbf{r}:\mathbf{G}_{\sigma\sigma}:\mathbf{r}\,(\delta\dot{\lambda})^2\rangle_{\Omega^2} - \langle\mathbf{s}:\mathbf{H}:\mathbf{s}\,(\delta\dot{\lambda})^2\rangle_{\Omega^2}\Big\} < \hat{\Pi}^*\end{aligned} \tag{32b}$$

where the inequalities are a consequence of properties i) and iv) here above, as well as of the positive definiteness of $\mathbf{H}$. These inequalities can be combined in a single continued one, namely

$$\hat{\Pi}[\dot{\mathbf{g}}, \dot{\mathbf{v}}^*, \dot{\lambda}^*] > \hat{\Pi}[\dot{\mathbf{g}}^*, \dot{\mathbf{v}}^*, \dot{\lambda}^*] > \hat{\Pi}[\dot{\mathbf{g}}^*, \dot{\mathbf{v}}, \dot{\lambda}]. \tag{33}$$

As eq.(33) holds for all feasible sets $(\dot{\mathbf{g}}, \dot{\mathbf{v}}, \dot{\lambda})$ different from $(\dot{\mathbf{g}}^*, \dot{\mathbf{v}}^*, \dot{\lambda}^*)$, it follows that $(\dot{\mathbf{g}}^*, \dot{\mathbf{v}}^*, \dot{\lambda}^*)$, solution of the elastic-plastic rate problem, is also a solution of (25). Furthermore, eq.(33) shows that the saddle-point problem admits a unique solution, and thus a unique solution has also the elastic-plastic rate problem.

Theorem 2 can be viewed as a boundary generalization of analogous theorems holding within the field equation approach [13, 14]. If the internal variables are dropped and thus the elastic-perfectly plastic material model is considered, Th. 2 coincides with the analogous given in [12]. In the latter case, however, the uniqueness property is lost as far as $\dot{\lambda}^*$ is concerned. Finally, if plastic strains are also dropped and thus the elastic model is considered, Th. 2 trasforms into the boundary min-max principle of Polizzotto [12, 15, 16].

Conclusion

For a rather wide class of elastic-plastic rate-independent material models, endowed with dual internal variables and related thermodynamic potential, a set of boundary/field integral equations governing the elastic-plastic rate-problem has been established and then characterized by two related variational principles. One of these principles is a simple stationarity theorem, the other is a min-max one. Other principles, not reported here for lack of space, can also be proved for the same rate problem.

Since the governing equations can only be solved via numerical methods, the given variational principles may be used for suitable boundary and interior element discretizations to derive related discrete rate problems. Though the actual numerical problem is in practice shaped as an incremental problem for a small finite load step, and therefore there is a need for related incremental-form variational principles [17, 18], however rate-form variational principles are always of interest, not only because they can be used as tool for suitable discretizations, but also because they can either be used as the starting points to establish increment-form variational principles.

References

1. Hodge, P.G. Jr.: Numerical applications of minimum principles in plasticity. In Heyman, J. and Leckie, F.A. (eds.) Engineering Plasticity, 237-256. Cambridge: University Printing House 1968.
2. Lemaitre, J.; Chaboche, J.L.: Mécanique des materiaux solides. Paris: Dunod 1985.
3. Drucker, D.C.: A definition of stable inelastic material. J. Appl. Mech. 26, Trans. ASME 81 (1959) 101-106.
4. Banarjee, P.K.; Butterfield, R.: Boundary element methods in engineering science. London: Mc Graws-Hill 1981.
5. Brebbia, C.A.; Telles, J.C.F.; Wrobel, L.C.: Boundary element tecniques. Berlin and Heidelberger: Springer-Verlag 1984.
6. Cruse, T.A.: Mathematical foundations of the boundary-integral equation method in solids mechanics. Tech. Rep. AFOSR-TR-77, Pratt & Whitney Aircraft Group, East Hartfort, July 1977.

7. Maier, G.; Polizzotto, C.: A Galerkin approach to boundary element elastoplastic analysis. Comput. Meth. Appl. Mech. Engng. 60 (1987) 175-194.

8. Polizzotto, C.: An energy approach to the boundary element method. Part I: elastic solids. Comput. Meth. Appl. Mech. Engng. 69 (1988) 167-184.

9. Rudolphi, T.J.; Krishnasamy, G.; Schmerr, L.W.; Rizzo, F.J.: On the use of the strengly singular integral equations for crack problems. In: Brebbia, C.A. (ed.), Boundary Elements X, Vol. 3, 249-263. Berlin Heildeberg: Springer- Verlag 1988. Southampton: Computational Mechanics Pubs. 1988.

10. Krishnasamy, G.; Schmerr, L.W.; Rudolphi, T.J.; Rizzo, F.J.: Hypersingular boundary integral equations: some applications in acoustic and elastic wave scattering. J. Appl. Mech. (to appear).

11. Zhang, J.-D.; Atluri, S.N.: A boundary/interior element method for quasi-static and transient response analyses of shallow shalls. Computers & Structures 24 (1986) 213-223.

12. Polizzotto, C.: A symmetric-definite BEM formulation for the elastoplastic rate problem. In: Brebbia, C.A., Wendland, W.L. and Kunh, G. (eds.), Boundary Elements IX, Vol. 2, 315-334. Southamptom and Boston: Computational Mechanics Publications 1987.

13. Capurso, M.: Minimum principles for the incremental solution to elastic-plastic problems, Part I and II (in Italian). Rendiconti Accademia Nazionale dei Lincei, Serie VIII, Vol. XLVI, fascicoli 4-5, April-May 1969, 417-560.

14. Capurso, M.; Maier, G.: Incremental elastoplastic analysis and quadratic optimization. Meccanica 5 (1970), 107-116.

15. Polizzotto, C.: A consistent formulation of the BEM within elastoplasticity. In: Cruse, T.A. (ed.), Advanced Boundary Element Methods, 315-324. Berlin Heildeberg: Springer-Verlag 1988.

16. Polizzotto, C.: A boundary min-max principle as a tool for boundary element formulations Engng. Anal. (to appear).

17. Polizzotto, C.; Zito, M.: A variational formulation of the BEM for the elastic-plastic analysis. In: Kunh, G. and Mang, H. (eds.), Discretization methods in structural mechanics, 201-210. Berlin and Heidelberg: Springer-Verlag 1990.

18. Panzeca, T.; Polizzotto, C.; Zito, M.: A boundary/field element approach to the elastic-plastic structural analysis problem. Proc. of the X Congresso Nazionale AIMETA, 165-168, 1990.

The Inclusion of Shear Deformations in a Plate Bending Boundary Element Algorithm

R. Piltner

Department of Civil Engineering, University of California at Berkeley,

Berkeley, CA 94720, U.S.A.

Summary

A plate bending formulation for thick and thin plates is considered. No ad hoc assumption: are made to derive the plate formulation for the inclusion of shear deformations. The giver representation of the plate displacements ensures a priori the satisfaction of both the three dimensional Navier-equations and the stress boundary conditions on the upper and lower plat faces. The use of the formulation for boundary element calculations is discussed and an exam ple is shown how a symmetric stiffness matrix can be obtained with the aid of boundar integrals. The numerical results are compared with an exact three-dimensional solution.

Three-Dimensional Representation of Displacements and Stresses

The solution of the three-dimensional Navier-equations

$$\mathbf{D}^T \mathbf{E} \mathbf{D} \mathbf{u} = -\bar{\mathbf{f}}, \tag{1}$$

is decomposed into the form

$$\begin{aligned} \mathbf{u} &= \mathbf{u}_h + \mathbf{u}_p \\ &= \mathbf{u}_h^0 + \mathbf{u}_h^p + \mathbf{u}_p , \end{aligned} \tag{2}$$

where $\mathbf{u}_h$ is a solution of the homogeneous system of differential equations and $\mathbf{u}_p$ is a particu lar solution of the nonhomogeneous differential equations. The solution representation is con structed such that the displacement fields $\mathbf{u}_h^0$, $\mathbf{u}_h^p$ and $\mathbf{u}_p$ satisfy the equations

$$\begin{aligned} \mathbf{D}^T \mathbf{E} \mathbf{D} \mathbf{u}_h^0 &= \mathbf{0}, \\ \mathbf{D}^T \mathbf{E} \mathbf{D} \mathbf{u}_h^p &= \mathbf{0}, \\ \mathbf{D}^T \mathbf{E} \mathbf{D} \mathbf{u}_p &= -\bar{\mathbf{f}}. \end{aligned} \tag{3}$$

Moreover, the constructed displacement fields have the following properties: $\mathbf{u}_h^0$ and $\mathbf{u}_p$ ensur the satisfaction of the homogeneous stress boundary conditions on the upper and lower faces o the plate whereas $\mathbf{u}_h^p$ is a particular homogeneous solution ensuring the satisfaction of the loa conditions on the lower and upper plate faces.

For our boundary element procedure we need to construct a series of linearly independent functions for the displacement field $\mathbf{u}_h^0$. Singular or regular functions can be used for $\mathbf{u}_h^0$. In this paper, the use of singular functions for $\mathbf{u}_h^0$ is primarily considered. The singular functions will be constructed with the aid of the Cauchy integral formula which relates harmonic function values inside the solution domain to function values on the boundary. The free parameters of the approximation functions for $\mathbf{u}_h^0$ must be evaluated such that the sum of the solution parts $\mathbf{u}_h^0$, $\mathbf{u}_h^p$ and $\mathbf{u}_p$ satisfy the remaining boundary conditions on the lateral faces of a plate under consideration either exactly or in a defined optimal sense. Details about the three-dimensional plate representation and its derivation are given in references [1,2].

There are three types of solutions for $\mathbf{u}_h^0$. The first type involves powers of the thickness coordinate z, the second has trigonometric functions depending on z, whereas the third type contains hyperbolic functions of z. Here we consider only the first two types, since they contribute the major solution parts for engineering purposes.

The first solution part contributing to the displacement field $\mathbf{u}_h^0$ can be written as

$$
\begin{aligned}
2\mu u &= -z\frac{\partial}{\partial x}G - \frac{1}{4(1-\nu)}\left[\, h^2 z - 2(2-\nu)\frac{z^3}{3} \,\right]\frac{\partial}{\partial x}\Delta G,\\
2\mu v &= -z\frac{\partial}{\partial y}G - \frac{1}{4(1-\nu)}\left[\, h^2 z - 2(2-\nu)\frac{z^3}{3} \,\right]\frac{\partial}{\partial y}\Delta G,\\
2\mu w &= G + \frac{\nu}{2(1-\nu)}z^2\,\Delta G,
\end{aligned}
\tag{4}
$$

$$
\begin{aligned}
\sigma_{xx} &= -\frac{1}{1-\nu}z\left[\, G_{xx} + \nu G_{yy} \,\right] - \frac{1}{4(1-\nu)}\left[\, h^2 z - 2(2-\nu)\frac{z^3}{3} \,\right]\frac{\partial^2}{\partial x^2}\Delta G,\\
\sigma_{yy} &= -\frac{1}{1-\nu}z\left[\, G_{yy} + \nu G_{xx} \,\right] - \frac{1}{4(1-\nu)}\left[\, h^2 z - 2(2-\nu)\frac{z^3}{3} \,\right]\frac{\partial^2}{\partial y^2}\Delta G,\\
\sigma_{zz} &= 0,\\
\tau_{xy} &= -zG_{xy} - \frac{1}{4(1-\nu)}\left[\, h^2 z - 2(2-\nu)\frac{z^3}{3} \,\right]\frac{\partial^2}{\partial x\partial y}\Delta G,\\
\tau_{xz} &= \frac{1}{2(1-\nu)}\left[\, z^2 - \frac{h^2}{4} \,\right]\frac{\partial}{\partial x}\Delta G,\\
\tau_{yz} &= \frac{1}{2(1-\nu)}\left[\, z^2 - \frac{h^2}{4} \,\right]\frac{\partial}{\partial y}\Delta G,
\end{aligned}
\tag{5}
$$

where G(x,y) has to satisfy the biharmonic equation $\Delta\Delta G = 0$ and z represents the thickness coordinate. The solution for G(x,y) can be written in terms of two arbitrary complex functions $\Phi(\zeta)$ and $\chi(\zeta)$ in the form

$$
G = \mathrm{Re}[\, \bar{\zeta}\,\Phi + \chi \,] \tag{6}
$$

where $\zeta = x + iy$. So we can express the displacements and stresses in terms of the derivatives of the functions $\Phi(\zeta)$ and $\chi(\zeta)$. We get for the displacements and stresses the following complex representation:

$$2\mu u = -z\,\mathrm{Re}[\,\Phi + \zeta\,\overline{\Phi'} + \overline{\chi'}\,] - \frac{1}{1-\nu}[\,h^2 z - 2(2-\nu)\frac{z^3}{3}\,]\,\mathrm{Re}[\,\Phi''\,],$$

$$2\mu v = -z\,\mathrm{Im}[\,\Phi + \zeta\,\overline{\Phi'} + \overline{\chi'}\,] + \frac{1}{1-\nu}[\,h^2 z - 2(2-\nu)\frac{z^3}{3}\,]\,\mathrm{Im}[\,\Phi''\,], \tag{7}$$

$$2\mu w = \mathrm{Re}[\,\bar{\zeta}\,\Phi + \chi\,] + \frac{2\nu}{1-\nu}z^2\,\mathrm{Re}[\,\Phi'\,],$$

$$\sigma_{xx} = -\frac{1}{1-\nu}z\,\mathrm{Re}[\,2(1+\nu)\Phi' + (1-\nu)(\,\zeta\overline{\Phi''} + \overline{\chi''}\,)\,]$$

$$-\frac{1}{1-\nu}[\,h^2 z - 2(2-\nu)\frac{z^3}{3}\,]\,\mathrm{Re}[\,\Phi'''\,],$$

$$\sigma_{yy} = -\frac{1}{1-\nu}z\,\mathrm{Re}[\,2(1+\nu)\Phi' - (1-\nu)(\,\zeta\overline{\Phi''} + \overline{\chi''}\,)\,]$$

$$+\frac{1}{1-\nu}[\,h^2 z - 2(2-\nu)\frac{z^3}{3}\,]\,\mathrm{Re}[\,\Phi'''\,],$$

$$\sigma_{zz} = 0, \tag{8}$$

$$\tau_{xy} = -z\,\mathrm{Im}[\,\zeta\,\overline{\Phi''} + \overline{\chi''}\,] + \frac{1}{1-\nu}[\,h^2 z - 2(2-\nu)\frac{z^3}{3}\,]\,\mathrm{Im}[\,\Phi'''\,],$$

$$\tau_{xz} = \frac{2}{1-\nu}[\,z^2 - \frac{h^2}{4}\,]\,\mathrm{Re}[\,\Phi''\,],$$

$$\tau_{yz} = -\frac{2}{1-\nu}[\,z^2 - \frac{h^2}{4}\,]\,\mathrm{Im}[\,\Phi''\,].$$

The displacement field (7) satisfies the three-dimensional Navier-equations for any choice of $\Phi(\zeta)$ and $\chi(\zeta)$. For our boundary element procedure, we can choose the complex functions in the form of a Cauchy integral.

The second plate bending solution part can be written as

$$2\mu u = \frac{\partial g_n}{\partial y}\sin\omega_n z,$$

$$2\mu v = -\frac{\partial g_n}{\partial x}\sin\omega_n z, \tag{9}$$

$$2\mu w = 0,$$

$$\sigma_{xx} = \frac{\partial^2 g_n}{\partial x \partial y}\sin\omega_n z,$$

$$\sigma_{yy} = -\frac{\partial^2 g_n}{\partial x \partial y}\sin\omega_n z,$$

$$\sigma_{zz} = 0, \tag{10}$$

$$\tau_{xy} = \frac{1}{2} \left[\frac{\partial^2 g_n}{\partial y^2} - \frac{\partial^2 g_n}{\partial x^2} \right] \sin \omega_n z,$$

$$\tau_{xz} = \frac{1}{2}\omega_n \frac{\partial g_n}{\partial y} \cos \omega_n z,$$

$$\tau_{yz} = -\frac{1}{2}\omega_n \frac{\partial g_n}{\partial x} \cos \omega_n z,$$

where $g_n(x,y)$ has to satisfy

$$\Delta g_n - \omega_n^2 g_n = 0 \tag{11}$$

and $\omega_n = n\pi/h$ $(n=1,3,5,....)$. For engineering applications it is sufficient to use the solution with the index $n=1$.

A nonsingular approximation function for $g_n(x,y)$ can be written as

$$g_n(x,y) = A_0 I_0(\omega_n r) + \sum_j [A_j \cos j\Theta + B_j \sin j\Theta] I_j(\omega_n r) \tag{12}$$

where $I_j(\omega_n r)$ are modified Bessel functions of the first kind. In complex form the solution of (11) can be be expressed as [3]

$$g_n(x,y) = \mathrm{Re}[\Phi_n(\zeta)] - \int_0^1 \mathrm{Re}[\Phi_n(t\,\zeta)] \frac{\partial}{\partial t} I_0(\omega_n \sqrt{1-t}\,)\, dt. \tag{13}$$

In addition to the homogeneous solution u_h^0, we need particular solutions. The construction of particular solutions of the type u_h^p and u_p is described in reference [1]. For the example of a constant normal load p acting on the upper face of the plate one can construct the following displacement and stress fields:

$$2\mu u = \frac{p\,x}{4h^3} \left[(2-\nu)(4z^3 - 3h^2 z) - 3(1-\nu)(x^2 + y^2)z + \frac{2\nu h^3}{1+\nu} \right],$$

$$2\mu v = \frac{p\,y}{4h^3} \left[(2-\nu)(4z^3 - 3h^2 z) - 3(1-\nu)(x^2 + y^2)z + \frac{2\nu h^3}{1+\nu} \right], \tag{14}$$

$$2\mu w = \frac{p}{16h^3} \left[-8(1+\nu)z^4 + 24\nu(x^2 + y^2)z^2 + 12h^2 (1+\nu)z^2 \right.$$

$$\left. + 3(1-\nu)(x^2 + y^2)^2 - 6h^2 \nu(x^2 + y^2) - \frac{8h^3}{1+\nu} z \right],$$

$$\sigma_{xx} = \frac{p}{4h^3} \left[4(2+\nu)z^3 - (9+3\nu)x^2 z - (3+9\nu)y^2 z - 3h^2(2+\nu)z \right],$$

$$\sigma_{yy} = \frac{p}{4h^3} \left[4(2+\nu)z^3 - (3+9\nu)x^2 z - (9+3\nu)y^2 z - 3h^2(2+\nu)z \right],$$

$$\sigma_{zz} = -\frac{p}{2h^3} (z + h)(2z - h)^2, \tag{15}$$

$$\tau_{xy} = -\frac{3p}{2h^3}(1-\nu)xyz,$$

$$\tau_{xz} = \frac{3p}{4h^3}x(2z-h)(2z+h),$$

$$\tau_{yz} = \frac{3p}{4h^3}y(2z-h)(2z+h).$$

Approximation of the Complex Functions

For the complex functions Φ, χ, Φ_n in the plate bending solution representation, we can use singular or nonsingular functions. Since we do not need the thickness coordinate z for the discussion of the complex functions, the classical notation for complex variables will be used in this paragraph. So we use $z = x + iy$ for a point inside the solution domain Ω and use ζ as a complex boundary coordinate on the boundary Γ. Writing now f instead of Φ, χ, and Φ_n we can construct a nonsingular displacement and stress field by choosing the complex functions in the form

$$f(z) = \sum_j a_j z^j. \tag{16}$$

The coefficients of the complex power series can be calculated from the boundary conditions of the plate with different methods. The collocation, least squares and the Trefftz method are possible solution strategies to get an approximation.

Instead of using nonsingular approximation functions like the one in equation (16), we can use singular functions, which are constructed such that all displacement and stress components remain finite inside the solution domain and on the boundary. For our complex functions, we can use the Cauchy integral formula

$$f(z)=\frac{1}{2\pi i}\int_{\Gamma}\frac{f(\zeta)}{\zeta-z}d\zeta, \tag{17}$$

which relates function values inside the domain Ω to function values on the boundary Γ. In order to get a boundary element algorithm the Cauchy integral formula is discretized along Γ in the form

$$f(z)=\frac{1}{2\pi i}\sum_j\int_{\Gamma_j}\frac{\hat{f}^j(\zeta)}{\zeta-z}d\zeta, \tag{18}$$

where $\hat{f}^j(\zeta)$ is a chosen basis function for the boundary element j with the boundary portion Γ_j. Since the plate solution representation contains third order complex derivatives, seventh order shape functions N_j are used for the basis function

$$\hat{f}^j(s) = N_1(s)\, f_{j-1} + N_2(s)\, f'_{j-1}(z_j - z_{j-1}) + N_3(s)\, f''_{j-1}(z_j - z_{j-1})^2 + N_4(s)\, f'''_{j-1}(z_j - z_{j-1})^3$$
$$+ N_5(s)\, f_j + N_6(s)\, f'_j(z_j - z_{j-1}) + N_7(s)\, f''_j(z_j - z_{j-1})^2 + N_8(s)\, f'''_j(z_j - z_{j-1})^3 \tag{19}$$

where

$$s = \frac{\zeta - z_{j-1}}{z_j - z_{j-1}}. \tag{20}$$

Using straight line boundary elements, exact integration in relationship (18) becomes possible. We obtain an approximation function of the form

$$f(z) = \frac{1}{2\pi i}\Big[\quad f_1\, F_1(z) + f_1'\, G_1(z) + f_1''\, H_1(z) + f_1'''\, I_1(z) + \cdots$$
$$\cdots + f_j\, F_j(z) + f_j'\, G_j(z) + f_j''\, H_j(z) + f_j'''\, I_j(z) + \cdots \tag{21}$$
$$\cdots + f_N\, F_N(z) + f_N'\, G_N(z) + f_N''\, H_N(z) + f_N'''\, I_N(z) \quad \Big],$$

where the functions F_j, G_j, H_j, I_j contain polynomials in z and logarithmic functions of the form

$$N_i(K)\ln\frac{z_j - z}{z_{j-1} - z},\quad \dot{N}_i(K)\ln\frac{z_j - z}{z_{j-1} - z},\quad \ddot{N}_i(K)\ln\frac{z_j - z}{z_{j-1} - z},\quad \dddot{N}_i(K)\ln\frac{z_j - z}{z_{j-1} - z},$$

where

$$K = \frac{z - z_{j-1}}{z_j - z_{j-1}} \tag{22}$$

and $\dot{N}_i(K) = dN_i/dK$.

Although the approximation function f(z) contains logarithmic terms all values on the boundary remain finite since all limit values for the node points exist. More details about the approximation function of the form (21), the calculation of limit values, as well as the correct definition of branch cuts for the complex logarithmic functions are given in references [4,5,6].

The Use of a Hybrid Variational Formulation for the Evaluation of Symmetric Stiffness Matrices for Subdomains

Using the boundary element discretization of the Cauchy integral we are able to construct linearly independent approximation functions of the form (21) with which we get from the representations (7,8) and (9,10,13) the approximation functions for the displacements and stresses. The constructed functions cannot only be used for a collocation process, but we can use them also in a variational formulation. In order to evaluate symmetric stiffness matrices with the aid of boundary integrals [7,8], one can start, for example, with a hybrid functional Π_H^i for the subdomain i in the form

$$\Pi^{i}_{H} = \int_{V^i} [\frac{1}{2}(\mathbf{u}^T \mathbf{D}^T) \mathbf{E} (\mathbf{D}\mathbf{u}) - \mathbf{u}^T\bar{\mathbf{F}}] dV^i - \int_{S^i_T} \mathbf{u}^T \bar{\mathbf{T}} dS^i + \int_{S^i_u} \mathbf{T}^T (\tilde{\mathbf{u}} - \mathbf{u}) dS^i, \qquad (23)$$

where $\mathbf{u}$ is a displacement field for the subdomain V^i. The linearly independent function terms in $\mathbf{u}$ are multiplied with the real and imaginary parts of the discrete complex nodal values Φ_j, χ_j $(j=1,2,\ldots,N)$, where N is the number of node points in the boundary element discretization. $\tilde{\mathbf{u}}$ is an assumed boundary displacement field for the lateral faces S^i_u of the plate, and it contains the final unknowns of the finite element (e.g. w, w_x, w_y). The assumed boundary displacement field $\tilde{\mathbf{u}}$ has the task to couple adjacent subdomains.

Since the constructed approximation displacement field $\mathbf{u}$ ensures the satisfaction of the governing differential equations and the boundary conditions on the upper and lower plate faces, the functional can be simplified to the form

$$\Pi^{i}_{H} = -\frac{1}{2} \int_{\Gamma^i} [\int_{-h/2}^{h/2} \mathbf{T}_h^T \mathbf{u}_h \, dz] \, d\Gamma^i - \int_{\Gamma^i} [\int_{-h/2}^{h/2} \mathbf{T}_h^T \mathbf{u}_p \, dz] \, d\Gamma^i \qquad (24)$$

$$+ \int_{\Gamma^i} [\int_{-h/2}^{h/2} \mathbf{T}_h^T \tilde{\mathbf{u}} \, dz] \, d\Gamma^i + \int_{\Gamma^i} [\int_{-h/2}^{h/2} \mathbf{T}_p^T \tilde{\mathbf{u}} \, dz] \, d\Gamma^i$$

$$+ \text{ terms without } \mathbf{u}_h, \mathbf{T}_h \text{ and } \tilde{\mathbf{u}},$$

where $\mathbf{T}_h = \mathbf{nEDu}_h$ and $\mathbf{T}_p = \mathbf{nEDu}_p$. This means that only integrations on the lateral faces are necessary. Since all functions in thickness direction are known, the integration in (24) with respect to z can be done analytically. So the three-dimensional plate problem can be reduced to the evaluation of line integrals along Γ^i, which is the boundary of the midsurface from the plate subdomain i.

Numerical Example

The boundary element method, based on the discretization of Cauchy integrals, is applied to get a symmetric stiffness matrix for a quadrilateral finite element. The hybrid functional (24) is used so that only boundary integral evaluations are necessary to obtain the symmetric stiffness matrices. With the aid of the considered quadrilateral element, two plate systems are analyzed for which the exact solutions are known.

At the corner nodes of the quadrilateral element we choose the unknowns to be w, w_x, w_y, $(h^2 q_x)$, $(h^2 q_y)$, where h is the plate thickness, and q_x and q_y characterize the magnitude of warping. At each midside node the unknown is w_n.

For a simplification, only the plate solution representation (4,5) is used in connection with the particular solution (14,15). The assumed boundary displacement field $\tilde{u} = [\,\tilde{u}, \tilde{v}, \tilde{w}\,]^T$, which is needed for the coupling of subdomains (finite elements), is assumed in the form

$$\tilde{u} = -z\,\frac{\partial}{\partial x}\tilde{w}(s) - \frac{1}{4(1-\nu)}\,[\,h^2 z - 2(2-\nu)\frac{z^3}{3}\,]\,\tilde{q}_x(s),$$

$$\tilde{v} = -z\,\frac{\partial}{\partial y}\tilde{w}(s) - \frac{1}{4(1-\nu)}\,[\,h^2 z - 2(2-\nu)\frac{z^3}{3}\,]\,\tilde{q}_y(s), \tag{25}$$

$$\tilde{w} = \tilde{w}(s),$$

where

$$\tilde{w}(s) = N_1(s)\,w_i + N_2(s)\left(\frac{\partial w}{\partial s}\right)_i + N_3(s)\,w_k + N_4(s)\left(\frac{\partial w}{\partial s}\right)_k, \tag{26}$$

$$N_1(s) = 1 - 3\xi^2 + 2\xi^3,$$

$$N_2(s) = l\,[\,\xi - 2\xi^2 + \xi^3\,],$$

$$N_3(s) = 3\xi^2 - 2\xi^3, \tag{27}$$

$$N_4(s) = l\,[\,\xi^3 - \xi^2\,],$$

and

$$\xi = \frac{s}{l}. \tag{28}$$

Note that l is the distance between the corner nodes i and k, and s is a boundary coordinate measured from node i. The midside node is denoted with the index j. In order to calculate the derivatives $\partial\tilde{w}(s)/\partial x$ and $\partial\tilde{w}(s)/\partial y$, the normal derivative of the boundary deflection must be chosen. Here we use

$$\frac{\partial\tilde{w}(s)}{\partial n} = N_5(s)\left(\frac{\partial w}{\partial n}\right)_i + N_6(s)\left(\frac{\partial w}{\partial n}\right)_j + N_7(s)\left(\frac{\partial w}{\partial n}\right)_k, \tag{29}$$

where

$$N_5(s) = 1 - 3\xi + 2\xi^2,$$

$$N_6(s) = 4\xi - 4\xi^2, \tag{30}$$

$$N_7(s) = 2\xi^2 - \xi.$$

The warping functions $\tilde{q}_x(s)$ and $\tilde{q}_x(s)$ are chosen in the form

$$\tilde{q}_x(s) = N_8(s)\,(q_x)_i + N_9(s)\,(q_x)_k,$$

$$\tilde{q}_y(s) = N_8(s)\,(q_y)_i + N_9(s)\,(q_y)_k, \tag{31}$$

where

$$N_8(s) = 1 - \xi,$$
$$N_9(s) = \xi. \tag{32}$$

For the numerical examples, every edge of the quadrilateral element was divided into 4 intervals, in which a 7-point Gauss integration formula was used. So 28 integration points were chosen on every edge. In order to eliminate rigid body terms and function terms, which do not contribute linearly independent terms to the displacement field, the following discrete node values are set to zero: Φ_1, Re[Φ_2], χ_1, χ_2.

For the example of a simply supported plate (a=6, b=4, h=0.4, E=1, ν=0.3, 16 elements for one quarter of the plate) under a uniform load p=1, the maximum deflection at the plate center was calculated as w=3.497 (0.3% error). The exact three-dimensional solution for the deflection is w=3.485 [9]. Since the largest errors for the stresses occur at the corner nodes, stresses are only calculated inside the elements. For example, at the center of the element, closest to the plate center, we get the stresses $\sigma_{xx}= -29.59$ (0.01% error), $\sigma_{yy}= -47.90$ (0.7% error), $\tau_{xy}=0.6709$ (0.2% error), $\tau_{xz}=0.3284$ (0.4% error), and $\tau_{yz}=0.7207$ (2.5% error).

In the second example, a clamped plate (a=6, b=4, h=0.01, E=1, ν=0.3, uniform load p=1) is analyzed. The exact deflection at the center of the plate is taken from the book [10] of Timoshenko, and its numerical value is w=12.64. With one finite element we obtain w=13.26 (4.9% error), and with four elements we get w=12.66 (0.2% error).

Acknowledgement

The author gratefully acknowledges his support from the DFG (Deutsche Forschungsgemeinschaft).

References

1. Piltner, R., The derivation of a thick and thin plate formulation without ad hoc assumptions, Report No. UCB/SEMM-89/08, Department of Civil Engineering, University of California at Berkeley, 1989.
2. Piltner, R., Three-dimensional stress and displacement representations for plate problems, Mechanics Research Communications, to appear.
3. Vekua, I.N., New Methods for Solving Elliptic Equations, North-Holland/ John Wiley, Amsterdam, New York, 1967.
4. Piltner, R. and Taylor, R.L. A boundary element algorithm using compatible boundary displacements and tractions, Int. J. Numer. Meth. Eng., 29, 1323 - 1341, (1990).

5. Piltner, R. and Taylor, R.L., A boundary element procedure for plane elasticity based on Cauchy's integral formula, pp. 15 - 25 in: Proceedings of the Eleventh International Conference on Boundary Element Methods in Engineering, Cambridge, Massachusetts, USA, August 1989: Advances in Boundary Elements, Vol. 3: Stress Analysis, (Editors: C.A. Brebbia/ J.J. Connor), Springer, Berlin, Heidelberg, New York, 1989.

6. R. Piltner and R.L. Taylor, A boundary element algorithm for plate bending problems based on Cauchy's integral formula, Proceedings of the International Symposium on Boundary Element Methods, United Technologies Research Center, East Hartford, Connecticut, USA, October 1989, Springer, Berlin, Heidelberg, New York (to appear).

7. Piltner, R. and Taylor, R.L., The evaluation of stiffness matrices for elasticity problems with the aid of boundary integrals, pp. 38 - 45 in: "NUMETA 90: Numerical Methods in Engineering: Theory and Applications", (Eds. G.N. Pande, J. Middleton), Elsevier, London/New York, 1990.

8. Piltner, R. Special finite elements with holes and internal cracks, Int. J. Num. Methods Eng., 21, 1471 - 1485, 1985.

9. Piltner, R., The application of a complex 3-dimensional elasticity solution representation for the analysis of a thick rectangular plate, Acta Mechanica, 75, 77 - 91, 1988.

10. Timoshenko, S., Theory of Plates and Shells, McGraw-Hill, New York/London, 1940.

Application of the Boundary Integral Equation (Boundary Element) Method to Time Domain Transient Heat Conduction Problems

M. A. QAMAR, R. T. FENNER AND A. A. BECKER

Mechanical Engineering Department,
Imperial College of Science, Technology and Medicine
Exhibition Road, London SW7 2BX

Summary

Application of the boundary integral equation (BIE) or boundary element method to two-dimensional transient heat flow problems using higher-order spatial shape functions is presented. Many different formulations have been proposed for the treatment of heat conduction (diffusion) problems by the BIE method, the most efficient of which is the one which employs a time dependent fundamental solution. The formulation adopted for this analysis employs space and time dependent fundamental solutions to derive the boundary integral equation in the time domain. It is an implicit time-domain formulation and is valid for both regular and unbounded domains. A time stepping scheme (time integral method) is then used to solve the boundary initial value problem by marching forward in time. Constant and linear temporal interpolation and quadratic shape functions are used to approximate field quantities in the time and space domains, respectively. Temporal and spatial integrations are carried out to form a system of linear equations. At the end of each time step, these equations are solved to obtain unknown values at that time. Validity of the BIE formulation is demonstrated by solving some test cases whose analytical solutions are known. Two-dimensional heat flow in a cooled turbine rotor blade is carried out as a practical application.

Introduction

The problem of transient heat conduction arises in many engineering problems, particularly in connection with thermoelasticity. The first formulation of the BIE for certain problems of transient heat conduction was presented by Rizzo and Shippy [1]. The governing differential equation is transformed for the time variable into a BIE using Laplace transforms and the resulting integral identity is solved numerically. The subsequent inversion of the Laplace transform is performed numerically to evaluate the field variables. This was followed by Chang et. al. [2] and Shaw [3] who solved the problem in the time domain by employing an integral representation to set up an integral equation, which is discretised in both the space and time domains, using finite element type functions. This approach was further extended [4] to the solution of parabolic problems using boundary integral equations. In this paper, application of the BIE method to two-dimensional transient heat conduction problems using higher-order spatial and temporal shape functions is presented.

The Boundary Integral Equation Formulation

The differential equation governing heat conduction in a two-dimensional homogeneous isotropic solid region R with boundary S, in which heat generation does not take place and material constants are invariant in space and time, is

$$\kappa \phi_{,ii}(q,t) - \frac{\partial \phi}{\partial t}(q,t) = 0 \tag{1}$$

where κ is the thermal diffusivity, $\kappa = k/\rho c$, c is the specific heat, ϕ is the temperature, t is the time, k is the thermal conductivity, and ρ is the density of the material. The usual cartesian tensor notation is used here; the indices range from 1 to 2, a comma indicates partial differentiation and repeated indices imply summation over its range. The time dependent fundamental solution to the heat equation (1), which represents the field of temperature produced by an instantaneous unit point source of heat at position p at time τ, is expressed as

$$\psi(p,q,t,\tau) = \frac{1}{4\pi\kappa(t-\tau)} \, exp\left[\frac{-r^2(p,q)}{4\kappa(t-\tau)}\right] \tag{2}$$

where r represents the distance between the source point p and the field point q. The direct boundary integral equation formulation for heat flow obtained combining the fundamental solution and Green's second identity is [5,6]

$$\begin{aligned} C(P)\,\phi(P,t) = & -\kappa \int_S \int_{t_0}^{t} \psi_{,n}(P,Q,t,\tau)\,\phi(Q,t)\,d\tau\,dS(Q) \\ & + \kappa \int_S \int_{t_0}^{t} \psi(P,Q,t,\tau)\,\phi_{,n}(Q,t)\,d\tau\,dS(Q) \\ & + \int_R \psi(p,q,t,t_0)\,\phi(q,t_0)\,dR(q) \end{aligned} \tag{3}$$

Equation (3) involves integration over the time domain, the boundary S of the region R and the region R. The last integral of the above equation can be transformed into an equivalent boundary integral [5,7].

$$\int_R \psi(p,q,t,t_0)\,\phi(q,t_0)\,dR(q) = \int_S \left[\varphi_{,n}\phi - \varphi\phi_{,n}\right] dS(Q) \tag{4}$$

where

$$\varphi(p,q,t,t_0) = \frac{1}{4\pi} E_1\left[\frac{r^2(p,q)}{4\kappa(t-t_0)}\right] \tag{5}$$

and E_1 is an exponential integral [8]

Numerical Implementation

The BIE under consideration involves spatial and temporal integrations. In the numerical implementation, the geometry of the problem is modelled using three-noded isoparametric quadratic line elements. Time variation of the field quantities is taken into account through constant or linear temporal interpolation functions. Two different time marching schemes designated as Iterative Time Interval method and Time Integral method can be used to solve the boundary integral equation. These methods are developed in such a manner that the integral nature of time is preserved.

The iterative time interval method involves numerical integration over the physical domain, and at the instant t_0=0 the equation is solved using the specified initial and boundary conditions. The temperature is then evaluated at a series of points over the physical domain which are used as initial conditions for the next time step. The process is repeated and the time history is communicated through pseudo-initial conditions. If the time step is chosen too small, the accuracy of the results deteriorates [9,10] and the associated errors tend to grow with time which is an indication of instability of the method.

In the time integral method, which is employed in this work, all time integration processes must be started from t_0=0 to the current time t. It is a summation of solutions of boundary integrals corresponding to the time variants, and time variation of field functions is taken into account in such a manner that evaluation of field variables at internal points of the physical domain at the end of each time step is not required. Hence, the dimensionality of the problem is effectively reduced by one. Since this method assembles a series of integral solutions to develop an approximate formula, the solution at all times depends upon the entire behaviour history of the boundary functions, dating back to t_0=0. It is apparent that either influence coefficients computed at the previous time steps are to be stored for use in successive time steps, or a new set of influence coefficients has to be computed for each time step.

Constant Temporal Interpolation (CTI)

Assume that the boundary functions ϕ and $\phi_{,n}$ are piecewise constant over each time step. In order to obtain the transient response at a time t_k, the time axis is discretised into k equal intervals

$$t_n = n\,\Delta t \qquad n = 1,2,3,\ldots\ldots,k \tag{6}$$

After substitution of CTI, the time integration is performed analytically and spatial integration is performed numerically leading to the following system of equations.

$$[A_{kk}]\{\phi(t_k)\} = [B_{kk}]\{\phi_{,n}(t_k)\} + \{R(t_k)\} + \{C(t_k)\} \tag{7}$$

where

$$\{R(t_k)\} = \sum_{n=1}^{k}\{[B_{kn}]\{\phi_{,n}(t_n)\} - [A_{kn}]\{\phi(t_n)\}\} \tag{8}$$

is the effect of the transient behaviour history of the boundary function and

$$\{C(t_k)\} = [A_{k0}]\{\phi(t_0)\} - [B_{k0}]\{\phi_{,n}(t_0)\} \tag{9}$$

is the result of the contribution of initial conditions up to the current time instant t_k. The sets of coefficient matrices $[A]$ and $[B]$ contain the integrals of temperature and flux kernels and have a triangular form due to the time translation properties of the fundamental solution. $[A_{kk}]$, $[B_{kk}]$ are the matrices related to the unknown and known field variables; the first subscript denotes the collocation point and the second subscript represents the time step at which they are calculated. $\{\phi(t_k)\}$ and $\{\phi_{,n}(t_k)\}$ are the vectors of unknown and known field quantities; the subscripts refer to the time step. In the numerical processing, once the boundary unknowns at the first time step have been determined, the equation can be solved for the second time step, and so on.

Linear Temporal Interpolation (LTI)

Now assume the field variables, ϕ and $\phi_{,n}$, vary linearly during a time step. Using LTI, after numerical integration and the usual assembly process, the resulting system of equations for the kth time step has the form

$$[A^F_{kk}]\{\phi(t_k)\} = [B^F_{kk}]\{\phi_{,n}(t_k)\} + \{C(t_k)\} + \{R(t_k)\} \tag{10}$$

where

$$\{R(t_k)\} = (-1)\sum_{n=2}^{k}\{[A^F_{kn} + A^I_{kn-1}]\{\phi(t_{k-n+1})\} - [B^F_{kn} + B^I_{kn-1}]\{\phi_{,n}(t_{k-n+1})\}\} + [B^I_{kn}]\{\phi_{,n}(t_0)\} - [A^I_{kn}]\{\phi(t_0)\} \tag{11}$$

The vector $\{R(t_k)\}$ represents the effect of the past time history on the current time node, and I, F indicate time nodes on the time co-ordinate. Similar to the constant case, the temporal integration is performed analytically. In order to find the solution to equation (10), the process is started from the first time step. For each time step, a new vector $\{R(t_k)\}$ needs to be formed, and the system of equations can be solved for the unknowns.

The equations can be rearranged for a particular time step such that all the unknowns ϕ and $\phi_{,n}$ are on the left hand side and all known quantities are on the right. Since values are to be computed at a number of time steps, the equations are best solved by an inversion of the coefficient matrix. Due to the time translation properties of the fundamental solution, the inverted matrix computed at the end of the first time step can be stored for use in successive time steps. At the end of each time step, a new vector of known quantities is formed and multiplied with the inverted matrix to obtain the unknowns at that particular time step.

Numerical Results

(a) *Bar subjected to mixed boundary conditions*

This example concerns unsteady heat conduction along a bar of unit side. The boundary of the solution domain is divided into quadratic elements such that the nodal points on the boundary coincide with those generated using the FEM mesh used in [11], (Figure 1). Two sides of the bar, initially at $\phi_0=0$, are maintained at $\phi=1$ and the other two sides are insulated ($\phi_{,n}=0$) as shown in Figure 1. The BIE solution is obtained for a time step $\Delta t=0.05$ at time $t=0.75$ with $\kappa=1$. Figure 2 gives the BIE solution together with the corresponding FEM solution [11] and the analytical solution [12]. Clearly, the BIE solution is in better agreement with the analytical solution. The BIE solution due to LTI is superior to that due to CTI.

(b) Rectangular region subjected to mixed boundary conditions

In this example three sides of the solution domain, initially at $\phi_0=30^{\circ}$F, are maintained at $\phi=0$ and the fourth side is insulated ($\phi_{,n}=0$). The same mesh design as shown in Figure 1, is used in this example. The square domain $0\leq x\leq 3$ m, $0\leq y\leq 3$ m, with $\kappa=1.25$ Btu/(hr m $^{\circ}$F) is considered to find the temperatures at the boundary after 1.2 hr. For time steps $\Delta t=0.1$ hr, the BIE solution is shown in Figure 3 together with the corresponding numerical results obtained using FEM [11] and the analytical solution [12], showing good agreement.

(c) An infinite plane region subjected to Robin-type boundary conditions

This example concerns transient heat conduction in an infinite plane region, initially at constant temperature $\phi_0=1$, with heat removal at a circular hole maintained at zero ambient temperature($\phi_s=0$). The values of heat transfer coefficient used are h=0.2,1.0,5.0. Figure 4 shows the division of the boundary into quadratic elements and Figure 5 shows the BIE solution at $r=1.0$ for $\Delta t=1.0$, together with the analytical solution (11). The results due to LTI show almost negligible oscillation over the first time step. This problem has also been studied using the FEM [13] and the same level of accuracy as offered by the BIE method was obtained using $\Delta t=0.05$. It clearly shows that the BIE method is more suitable and economical than FEM for solving problems which involve infinite solution domains.

(d) Turbine rotor blade subjected to Robin-type boundary conditions

This problem provides a more practical example of the application of the BIE method to a cooled aluminium turbine rotor blade. Gas temperature around the blade was assumed to be 1145 ^{0}C and the heat transfer coefficient to vary from 0.39 to 0.056 Cal/cm^2 sec ^{0}C on the outside surface of the blade. Other numerical values taken from [13] are shown in Figure 6. Fifty four boundary elements with 108 nodes are used to model the blade (Figure 6). In order to draw isotherms 30 interior points are chosen. The BIE solutions for different time values are shown in Figures 6. Although, the transient results obtained by the BIE method show similar pattern to those produced by the FEM analysis [13], their comparison is not presented here because of insufficient details concerning the geometrical data and the variation of the heat transfer coefficient on the outside surface of the blade used in reference [13].

Conclusions

A direct BIE formulation for two-dimensional transient heat conduction has been developed and implemented. The formulation involves temporal and spatial integrals. The time integral method requires spatial discretisation of the boundary only, therefore it effectively reduces the dimensionality of the problem by one. The time integral method is unconditionally stable and demands less labour for input data preparation. If the matrices computed in numerical processing are not stored for use in successive time steps, it becomes less economical than the iterative time interval method. Although LTI is less economical in computing time than CTI, it produces very accurate numerical results. If CTI is used $[2(n\text{-}1)+4]$ matrices are to be computed over a time interval, whereas for LTI case $[4(n+1)]$ matrices are to be evaluated for second or higher time steps, n being the number of the time step. For the first time step, only 6 matrices are involved. Comparison of the BIE solutions with the analytical solutions show their accuracies to be excellent. The BIE method is superior to the FEM. For a typical practical problem of a turbine rotor blade the amount of labour involved in preparing the mesh data is much less than other numerical techniques. A further advantage of the BIE approach over other numerical methods is the reduced computing costs. The BIE method is more economical and suitable for infinite domain for which the domain type methods are unsuitable.

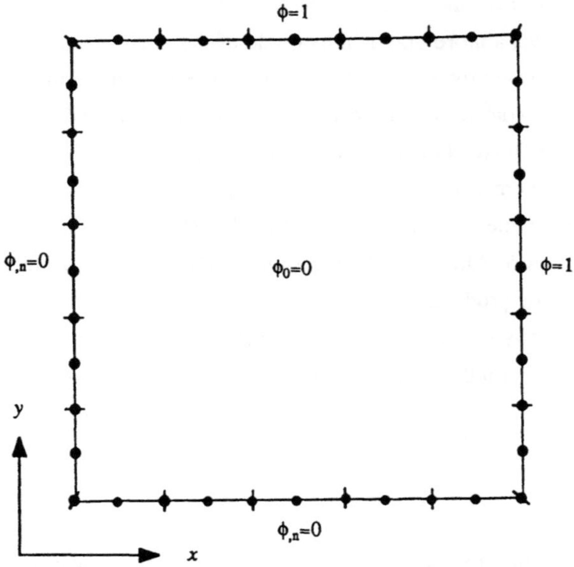

Fig. 1. Mesh design containing 20 quadratic elements and 40 nodal points

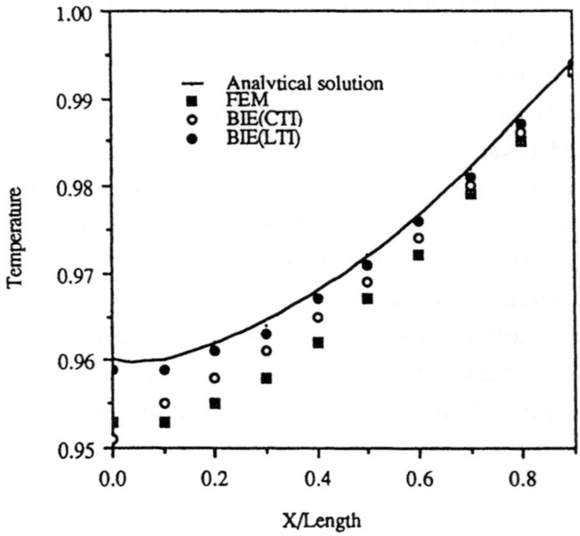

Fig. 2. Surface temperature of bar, initially at zero temperature, at time 0.75 for time step 0.05

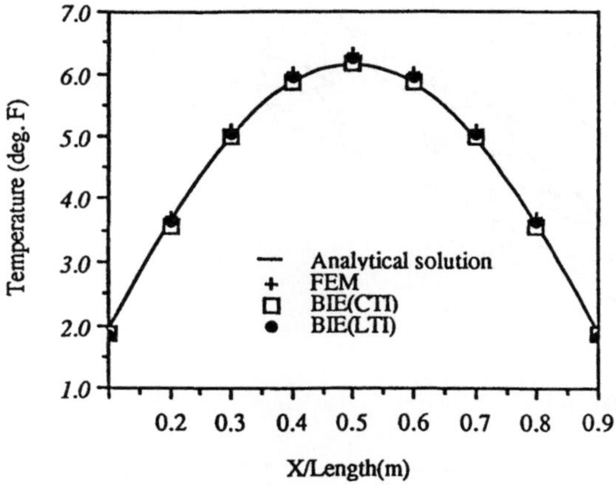

Fig. 3. Surface temperature of rectangular region, initially at constant temperature of 30 °F, at time 1.2 hr for time step 0.05

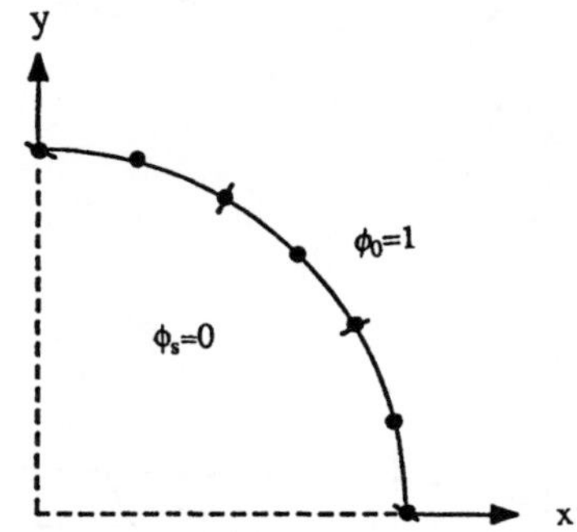

Fig. 4. Mesh design for a quarter of cooling hole in infinite region

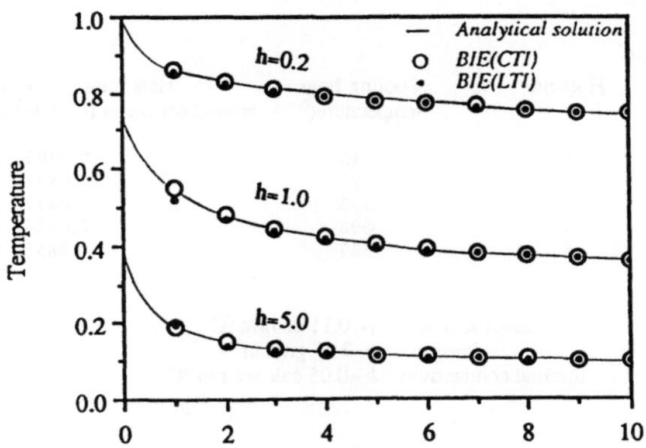

Fig. 5. Temperature variation with time at *r*=1 in an infinite region, initially at unit temperature

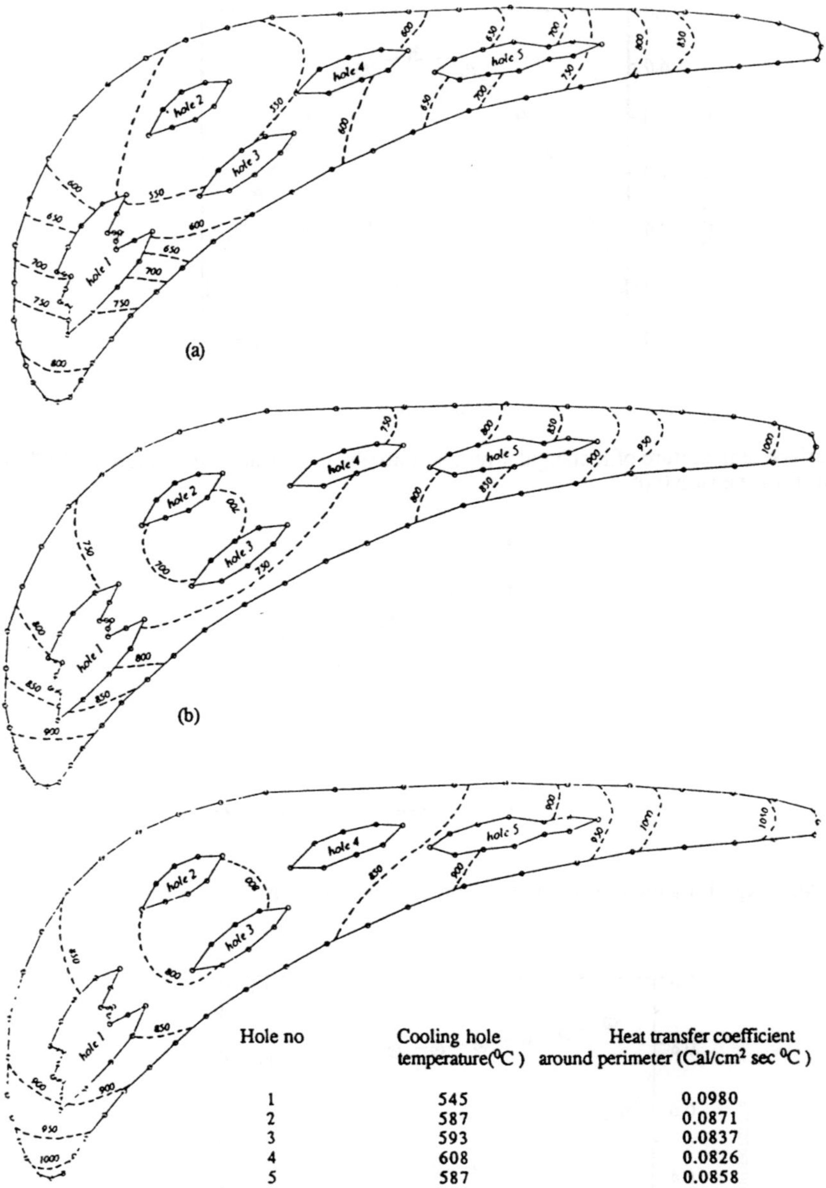

Hole no	Cooling hole temperature(°C)	Heat transfer coefficient around perimeter (Cal/cm² sec °C)
1	545	0.0980
2	587	0.0871
3	593	0.0837
4	608	0.0826
5	587	0.0858

(c)

Specific heat $c = 0.11$ cal/gm °C
density $\rho = 7.99$ gm/cm^3
thermal conductivity $k = 0.05$ cal/ sec cm °C

Fig. 6. Temperature distribution in a cooled turbine rotor blade at time (a) t=0.5 sec, (b) t=1.0 sec, (c) t=10.0 sec

References

1. Rizzo, F. J.; Shippy, D. H.: A method of solution for certain problems of transient heat conduction. AIAAJ 8 (1970) 2004-2009.

2. Chang, Y. P.; Kang, C. S.; Chen, D. J.: The use of fundamental Green's function for the solution of problems of heat conduction in anisotropic media. Int. J. Heat Mass Transfer 16 (1973) 1905-1918.

3. Shaw, R. P.: An integral approach to diffusion. Int. J. Heat Mass Transfer 17 (1974) 693-699.

4. Wrobel, L. C.; Brebbia, C. A.: Time dependent potential problems: in Progress in boundary element methods: vol. 1 London, Pentech Press (1981).

5. Qamar, M. A.: A Boundary Integral Equation Method for Two-dimensional Thermoelastic Analysis. Ph.D. Thesis, University of London, (1990) (in preparation).

6. Banerjee, P.K.; and Butterfield, R.: Boundary Element Methods in Engineering Science. London: McGraw-Hill, 1981.

7. Sharp, S.: A condition for simplifying the forcing term in boundary element solutions of the diffusion equation. Communications in Applied Numerical Methods 4 (1985) 67-70.

8. Abramowitz, M.; Stegun, I.: Handbook of mathematical functions. New York, Dover Publications 1972.

9. Curran, D. A. S; Cross, M.; Lewis, B. A.: Solution of parabolic differential equations by the boundary element method using discretisation in time. Applied Mathematical Modelling 4 (1980) 398-400.

10. Beskos, D. E.: Boundary element methods in mechanics. New York, Elsevier Science Publishers 1987.

11. Bruch Jr., J. C.; Zyvoloski, G.: Transient two-dimensional heat conduction problems solved by the finite element method. Int. J. for Num. Methods in Engng 8 (1974) 481-494.

12. Carslaw, H. S.; Jaeger, J. C.: Conduction of heat in solids. London, Oxford University Press 1959.

13. Zienkiewwicz, O. C.; Parekh, C. J.: Transient field problems: Two-dimensional and Three-dimensional analysis by isoparametric finite elements. Int. J. Num. Methods Engng 2 (1970) 61-71.

Panel Methods for Free Surface Flows

P. D. Sclavounos

Department of Ocean Engineering
Massachusetts Institute of Technology
Cambridge MA 02139

Abstract

Two boundary-element methods are presented for the solution of potential flows arising from tl interaction of floating bodies with the free surface. The first problem studies the radiation ar diffraction of monochromatic surface waves by stationary three-dimensional marine structures ar the second models the generation and propagation of steady surface waves by a ship advancii with forward speed.

Introduction

The boundary element method is widely regarded as the method of choice for the solution of number of potential flows in marine hydrodynamics. Source and dipole singularities are distribut over the domain boundaries and their strengths are determined from the solution of various typ of integral equations obtained from the application the classical Green theorem.

The two types of singularities most often used in marine flows are the *Rankine Source* and tl *Wave Source*. The former will be used in Section 2 for the solution of steady wave flow past ship advancing with forward speed. The wave source, is a solution of the Laplace equation ar is subject to a linearized condition on the mean plane position of the free surface. Its evaluati is more time consuming than the Rankine source, yet its use allows the discretization only of tl body boundary and the analytical enforcement of the radiation condition at infinity. In Secti 1, a panel method is presented based on the use of wave sources for the solution of the radiati or diffraction of time-harmonic surface waves by stationary offshore structures. The occurence "irregular" frequencies is also addressed and their removal is achieved by the solution of a modifi integral equation.

1. Surface-Wave Radiation and Diffraction

Marine structures, e.g. ships and offshore platforms, operating at a stationary position are fr quently exposed to severe wave environments. The loads and reponses induced by ambient wav are often studied by the spectral decomposition of the sea state into monochromatic wave comp nents and the analysis of the effect of each component upon the structure. For small wave slope the linearization of the free-surface condition leads to the boundary-value problem discussed ne Its solution allows the evaluation of the loads and responses of floating bodies in monochromat waves and by linear synthesis in a sea state. [e.g. [1]].

1.1 *The Boundary Value Problem*

Assuming a time-harmonic ideal flow, we may define the velocity potential as follows

$$\Phi(\vec{x},t) = Re\{\varphi(\vec{x})e^{i\omega t}\}, \tag{1.1}$$

where ω is the radian frequency and $\varphi(\vec{x})$ is a complex velocity potential governing the radiation or diffraction of monochromatic waves by the structure. In either problem, φ is governed by the Laplace equation in the fluid domain ($z < 0$), the linearized free-surface condition on the $z = 0$ plane

$$-\omega^2\varphi + g\varphi_z = 0, \tag{1.2}$$

and a Neumann condition on the mean position of the body boundary S_B

$$\vec{n}\cdot\nabla\varphi = V(\vec{x}). \tag{1.3}$$

Here, the normal velocity $V(\vec{x})$ has known value specified by the radiation or diffraction problem being solved. Solutions to (1.1)-(1.3) are not unique unless a radiation condition of outgoing waves is enforced at infinity. The derivation of the boundary value problem (1.1)-(1.3) and definition of the wave induced loads and body responses in terms of the complex velocity potential φ, are described in [1].

1.2 *The Green Integral Equation*

The three dimensional potential flow governed by (1.1)-(1.3) extends over an infinite domain and must satisfy a radiation condition at infinity. Therefore, boundary-element methods based upon the distribution of wave singularities over the body boundary, emerge as very attractive solution schemes.

The relevant wave source potential, or Green function, is a solution of the Laplace equation, satisfies the free-surface and radiation conditions, and is defined as follows ([2])

$$G(\vec{x};\vec{\xi}) = \frac{1}{r} + \frac{1}{r'} + 2K \fint_0^\infty \frac{dk}{k-K} e^{k(z+\varsigma)} J_0(kR), \tag{1.4}$$

where $R^2 = (x-\xi)^2 + (y-\eta)^2$, $r^2 = R^2 + (z-\varsigma)^2$ and $r'^2 = R^2 + (z+\varsigma)^2$.

A direct application of Green's theorm with $\varphi_1(\vec{x}) = \varphi(\vec{x})$ and $\varphi_2(\vec{x}) = G(\vec{x};\vec{\xi})$, leads to the familiar integral equation

$$2\pi\varphi(\vec{x}) + \iint_{S_B} ds\ \varphi(\vec{\xi})\ \frac{\partial G(\vec{\xi};\vec{x})}{\partial n_\xi} = \iint_{S_B} ds\ V(\vec{\xi})\ G(\vec{\xi};\vec{x}) \tag{1.5}$$

for the unknown velocity potential φ over the mean position of the body boundary S_B.

1.3 *A Panel Method for Wave Body Interactions*

A general purpose panel method has been developed for the solution of the integral equation (1.5) over the surface of marine structures of general geometry. Figure 1 illustrates the discretization of a quarter of the wetted surface of an offshore structure known as the Tension-Leg-Platform. The complicated nature of its geometry requires the use of over 12,000 plane quadrilateral panels in order to achieve a solution of the integral equation (1.5) accurate to within a few percent. Tl details of the numerical solution outlined next are contained in [3,4,5].

The body surface is approximated by a collection of plane quadrilateral panels, the velocity potenti φ is assumed constant over the surface of each panel and the integral equation (1.5) is enforced a collocation points $\vec{x}_i$ located at the panel centroids. These discretization steps lead to the set-u and solution of the $N \times N$ complex matrix equation

$$[\, 2\pi \mathbf{I} + \mathbf{D}\,]\, \vec{\varphi} = \mathbf{S}\, \vec{V}, \qquad (1.6$$

where N may be as large as 12,000. In (1.6) $\mathbf{I}$ is the unit matrix, $\vec{\varphi}$ is the complex vector of tl unknown velocity potential over all N panels and $\vec{V}$ is the corresponding vector of known norm velocities. The typical elements of the $N \times N$ influence matrices $\mathbf{D}, \mathbf{S}$ are defined as follows

$$D_{ij} = \iint_{S_j} ds_\xi \frac{\partial G(\vec{\xi}; \vec{x}_i)}{\partial n_\xi}, \qquad S_{ij} = \iint_{S_j} ds_\xi G(\vec{\xi}; \vec{x}_i), \qquad (1.7$$

where S_j is the surface of the j-th panel. The evaluation of the complex influence matrices $\mathbf{D}$, constitutes one of the principal tasks of the panel method. When the radial distances r, r' betwee the two panels are small, the two Rankine sources in the definition (1.4) of the wave source a integrated analytically over the panel surface S_j using the algorithms derived in [6]. Otherwis the entire wave Green function and its gradients are integrated over the panel S_j by a single noc centroid quadrature. Efficient algorithms for the evaluation of the frequency dependent Fouri integral in the definition (1.4) of the wave Green function have been developed in [7].

The computational effort and memory requirements necessary for the solution of the complex matr equation (1.6) may be substantial, especially when Gaussian elimination is used. For large numbe of panels, an $O(N)$ reduction of the computational effort can achieved if an iterative solution schen is available. An accelerated Gauss-Siedel iterative algorithm has been developed in [8] for compl matrices and has been found to converge in about 20 iterations over most of the frequency ran and for all structures of interest in practice. Computations of a linear hydrodynamic force on tl TLP are illustrated in Figure 2.

1.4 *Irregular Frequencies*

There exists an infinite set of discrete frequencies, known as "irregular" frequencies, at which tl Green integral equation (1.5) possesses multiple solutions. Yet, under some mild restrictions c the body geometry, the boundary value problem (1.1)-(1.3) possesses a unique solution ([9]). Thu the irregular frequencies arise because of our choice of the integral equation (1.5) as the mea for solving the boundary-value problem. In their vicinity, the integral equation becomes poor conditioned and the solution of the matrix equation (1.6) develops a significant error.

A number of analytical techniques have been developed for the removal of irregular frequenci in acoustic and surface-wave body interactions. Their numerical implementation however is n always easy or effective. The method adopted with the present panel method was proposed acoustics in [10] and was subsequently studied in connection with wave-body interactions in [1 It is based on the solution of the modified integral equation

$$2\pi\varphi(x) + \iint_{S_B} d\xi\, \varphi(\xi) \frac{\partial G(\xi; x)}{\partial n_\xi} + \alpha \frac{\partial}{\partial n_x} \iint_{S_B} d\xi\, \varphi(\xi) \frac{\partial G(\xi; x)}{\partial n_\xi}$$

$$= \iint_{S_b} d\xi\, V(\xi) G(\xi;x) - 2\pi\alpha V(x) + \alpha \iint_{S_B} d\xi\, V(\xi) \frac{\partial G(\xi;x)}{\partial n_x} \tag{1.8}$$

obtained from the linear superposition of the Green equation (1.5) and its normal derivative with respect to the x−coordinate. The success of the method hinges upon the value assigned to the coupling constant α. It is shown in [11] that when the imaginary part of α is nonzero, the modified integral equation (1.8) is free of all irregular frequencies.

Numerical experiments carried out in [12] suggest that the method performs best when $\alpha = i\beta$ is purely imaginary, with $\beta > 0$. Moreover, the magnitude of β must be no larger than about 0.15. Figure 3 plots the inverse condition number of the matrix derived from (1.8) as a function of the wavenumber $\nu a = \omega^2 a/g$, for different values of β, for a half-immersed sphere of radius a. The vanishing of the inverse condition number indicates the presence of an irregular frequency which is removed by setting $\beta = 0.11$. Computations of the corresponding hydrodynamic forces on the sphere are shown in Figure 4 where the removal of the irregular frequency effect is clearly illustrated.

2. Steady Ship Waves

The modelling and prediction of the surface wave disturbance generated by ships advancing with a forward velocity in calm water has attracted the attention of the scientific community since Euler. Numerous analytical and numerical studies have since shed significant light into this physical problem which occupies a central position in ship hydrodynamics.

Two are the principal reasons why free-surface problems with forward speed are considerably more difficult to solve relative to the radiation-diffraction problem of Section 1. Viscous effects are important within the ship boundary-layer and wake, especially when the calm water resistance is required. Nonlinear effects near the ship ends are often important and not possible to ignore.

Ship boundary layers are thin over most of the length of a ship, therefore the viscous and ideal parts of the flow around their hulls have traditionally been treated separately. Moreover, free-surface nonlinearities may not be important for streamlined ship forms and small forward speeds. Therefore, the linearized forward-speed potential flow past ships has attracted significant attention and numerical methods for its solution approach the maturity enjoyed by the method described in Section 1. The remainder of the present section describes a Rankine panel method developed for the solution of the linearized steady wave flow past ships advancing with constant forward velocity in calm water.

2.1 *The Boundary-Value Problem*

No consensus exists on the most appropriate linearized free-surface condition governing the steady flow past ship forms. The condition studied here is derived in [13] and its validity is justified for most fine-shaped ships as well as for full-shaped ships, as long as they advance at a small forward speed.

Assume an incompressible and inviscid flow and a Cartesian coordinate system $\vec{x} = (x, y, z)$ fixed on the ship which advances with constant forward velocity U in the positive x−direction. The mean position of the free surface lies on the $z = 0$ plane with the positive z−axis pointing upwards. The flow is governed by the velocity potential $\Psi(\vec{x}, t)$, which satisfies the Laplace equation in the fluid domain and is decomposed as follows

$$\Psi(\vec{x},t) = \Phi(\vec{x}) + \phi(\vec{x}). \tag{2.1}$$

The velocity potential Φ is defined as the *basis flow*, and is required to satisfy the rigid-lid condition on the $z = 0$ plane,

$$\Phi_z = 0, \tag{2.2}$$

and the zero-flux Neumann condition

$$\frac{\partial \Phi}{\partial n} = 0 \tag{2.3}$$

on the mean position S_B of the ship hull.

The velocity potential ϕ is the principal unknown and governs the steady wave disturbance generated by the ship. For thin or slow ships, it is assumed small relative to the basis flow Φ, therefore allowing the linearization of the free surface condition as follows

$$\left\{ \begin{array}{l} \nabla\Phi\cdot\nabla\left(\nabla\Phi\cdot\nabla\phi\right) + \frac{1}{2}\nabla(\nabla\Phi\cdot\nabla\Phi)\cdot\nabla\phi + g\phi_z - \Phi_{zz}(\nabla\Phi\cdot\nabla\phi) \\ = -\frac{1}{2}\nabla(\nabla\Phi\cdot\nabla\Phi)\cdot\nabla\Phi - \frac{1}{2}(U^2 - \nabla\Phi\cdot\nabla\Phi)\Phi_{zz}, \quad \text{on} \quad z = 0 \end{array} \right\} \tag{2.4}$$

The boundary condition satisfied by ϕ on the ship hull is homogenous and takes the form

$$\frac{\partial \phi}{\partial n} = 0. \tag{2.5}$$

Therefore, the forcing of this linear boundary-value problem governing the steady ship wave disturbance comes from the right-hand side of the free-surface condition (2.4).

As in the radiation/diffraction problem of Section 1, the boundary value problem (2.1)-(2.5) accepts no unique solution unless suppplemented by a radiation condition. When forward speed is present, is is sufficientsto request that no waves may appear upstream of the ship.

2.2 *The Boundary-Integral Equation*

The free-surface condition (2.4) is linear but contains spatially variable coefficients functions of the basis flow and its gradients. Therefore, it does not invite the use a wave source potential as the Green function in a boundary integral equation. The solution of the boundary-value problem in Section 2.1 has been obtained by invoking Green's theorem and employing the Rankine source potential

$$G(\vec{x};\vec{x}') = \frac{1}{2\pi}\frac{1}{|\vec{x}-\vec{x}'|} \tag{2.6}$$

as the Green function. The fluid domain is bounded by the hull surface S_B, the mean position of the free surface (FS) and a cylindrical 'control' surface (S_∞). The resulting integral equation for the velocity potential ϕ, takes the form

$$\phi(\vec{x}) - \iint_{(FS)} \frac{\partial\phi(\vec{x}')}{\partial z'} G(\vec{x};\vec{x}')d\vec{x}' + \int\int_{(FS)\cup(S_B)} \phi(\vec{x}')\frac{\partial G(\vec{x};\vec{x}')}{\partial n'}d\vec{x}' =$$

$$= \iint_{(S_B)} \frac{\partial \phi(\vec{x}')}{\partial n'} G(\vec{x};\vec{x}')d\vec{x}' \quad , \quad \vec{x} \in (FS) \cup (S_B) \quad . \tag{2.7}$$

The surface integrals over the control surface (S_∞) can be shown to vanish in the limit as (S_∞) tends to infinity with $|\vec{x}|$ kept finite.

The normal derivative of ϕ on the ship surface vanishes by virtue of (2.5), therefore so does the right-hand side of (2.7). Moreover, the vertical derivative ϕ_z on the free surface (FS), is replaced by the appropriate combination of its value and tangential convective derivatives, as indicated by the free-surface condition (2.4). The radiation condition of no wave disturbance upstream is satisfied by enforcing

$$\frac{\partial \phi}{\partial x} = \frac{\partial^2 \phi}{\partial x^2} = 0, \tag{2.8}$$

at some sufficiently large distance $x = x_{UP}$ upstream of the ship. Conditions (2.8) correspond to the physical requirement that the wave elevation and its $x-$derivative vanish along $x = x_{UP}$. The relation of (2.8) to the radiation condition of no waves upstream of the ship is analysed in [14].

2.3 *A Rankine Panel Method*

Figure 6 illustrates the typical discretization of half of the ship hull and free surface by plane quadrilateral panels. Numerical instabilities are known to arise from the discretization of the integral equation (2.7) over the free-surface panel mesh. In most studies to date, the tangential derivatives of the velocity potential on the free surface are approximated by finite difference schemes and numerical error growth is controlled via the use of upstream (upwind) differencing. The principal drawback of such a numerical algorithm is that it introduces numerical damping which may distort significantly the ship wave pattern.

A systematic analysis of the numerical *dispersion, damping* and *stability* of free-surface Rankine panel methods was carried out in [14] and led to the development of a bi-quadratic spline-collocation scheme of cubic order and zero numerical damping ([13]). This representation allows the enforcement of a continous variation of the velocity potential and its first gradient accross panels and permits the direct evaluation of its double derivative appearing under the integral sign in (2.7). Collocation at the panel centers, leads to a real matrix equation for the spline weights. Its solution determines the velocity potential and its gradients on the ship hull and the free-surface wave elevation. The radiation condition (2.8) is enforced at the upstream truncation boundary, while the transverse and downstream boundaries are left free. This is justified because the ship wave disturbance is convected downstream in very nearly a parabolic manner, and is contained within the Kelvin angle of 19.5 deg.

Essential for the performance of the method is a *stability* condition which must be enforced by the proper selection of the parameters of the free surface discretization. They are the grid Froude number $F_h = U/\sqrt{gh_x}$ and the panel aspect ratio $\alpha = h_x/h_y$, where h_x, h_y are the panel dimensions in the streamwise and transverse directions respectively. A stable domain is established in the (F_h, α) plane with a fixed boundary. The enforcement of this condition is essential for this panel method which is free of numerical damping, and its derivation is detailed in [15]. Computations of steady ship wave patterns are illustrated in Figure 5 for a thin strut and a conventional ship. The absence of numerical damping is evident in the computations which are capable to resolve significant detail in the Kelvin ship wave patterns.

3. Acknowledgements

Financial suuport for these studies has been provided by the Office of Naval Research and a group of industrial sponsors, A.S. Veritas, Norsk Hydro and Statoil.

4. References

[1] Faltinsen, O. M. (1990). *Sea Loads on Ships and Offshore Structures.* Cambridge University Press.

[2] Wehausen, J. V. and Laitone (1960). *Surface Waves.* In *Handbuch der Physik*, Vol. pp. 446-778, Berlin, Springer-Verlag.

[3] Breit, S., Newman, J. N. and Sclavounos P. (1986). A New Generation of Panel Programs for Radiation-Diffraction Problems. BOSS'86, Delft.

[4] Korsmeyer, T., Lee C-H, Newman, J. N. and Sclavounos, P. D. (1988). The Analysis of Wave Effects on Tension-Leg Platforms. OMAE'88, Houston.

[5] Sclavounos, P. D. and Newman, J. N. (1985). The User's Manual of WAMIT. Dept. of Ocean Engineering, MIT.

[6] Newman, J. N. (1986). Distributions of Sources and Dipoles over a Quadrilateral Panel. *J. Eng. Maths.* 20, pp. 113-126.

[7] Newman, J. N. (1985). Algorithms for the Free Surface Green Function. *J. Eng. Maths.* 19, pp. 57-67.

[8] Lee, C-H. (1988). Numerical Methods for the Solution of Three-Dimensional Integral Equations in Wave-Body Interactions. PhD Thesis, Dept. of Ocean Enineering, MIT.

[9] John, F. (1950). On the Motion of Floating Bodies. II. Simple Harmonic Motions. *Commums Pure Appl. Maths*, 3, pp. 45-101.

[10] Burton, A. J. and Miller, G. F. (1971). The Application of Integral Equation Methods to the Numerical Solution of Some Exterior Problems. *Proc. R. Soc. Lond.*, A 323, pp. 201-220.

[11] Kleinman, R. E. (1982). The Mathematical Theory of the Motion of Floating Bodies - An Update. DTNSRDC Report 82/074.

[12] Lee, C-H and Sclavounos P. D. (1989). Removing the Irregular Frequencies from Integral Equations in Wave-Body Interactions. *J. Fluid Mechs*, 207, pp. 393-418.

[13] Nakos, D. and Sclavounos, P. D. (1990). Ship Motions by a Three-Dimensional Rankine Panel Method. Proc. 18th Naval Hydrodyn. Conf., Ann Arbor, Michigan.

[14] Sclavounos, P. D. and Nakos, D. (1988). Stability Analysis of Panel Methods for Free Surface Flows with Forward Speed. Proc. 17th Naval Hydrodyn. Conf., The Hague, The Netherlands.

[15] Nakos, D. and Sclavounos, P. D. (1990). Steady and Unsteady Ship Wave Patterns. *J. Fluid Mechs*, 215, pp. 265-288.

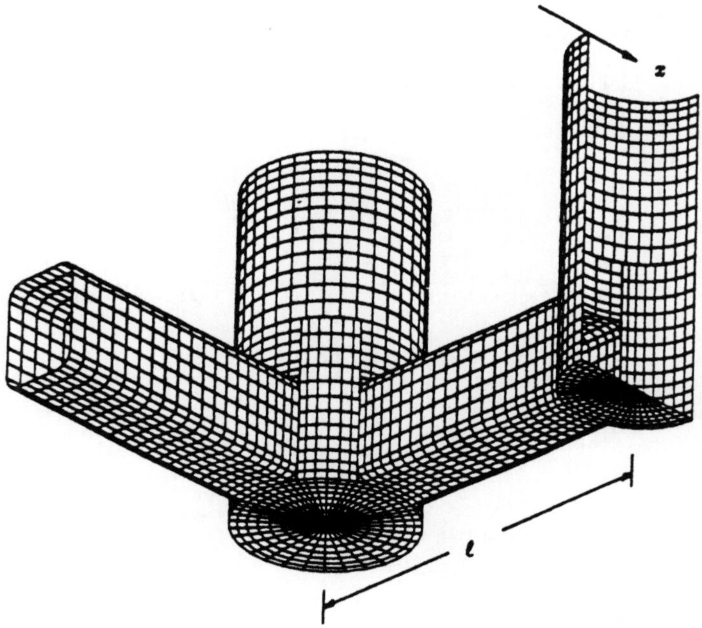

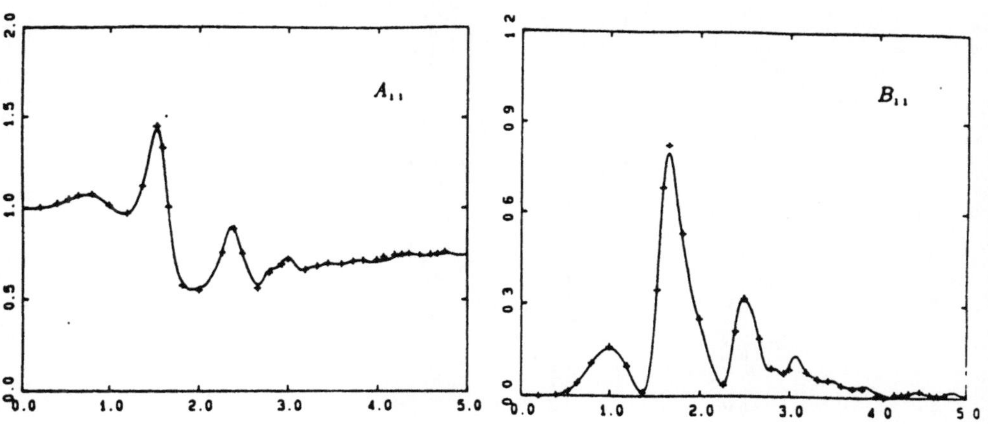

Figures 1, 2: Discretization of the TLP with 12,608 panels (Top). Surge Added-Mass and Damping Coefficients for the TLP. Symbols + are calculated in the frequency domain and the solid curve is obtained from the Fourier transform of an independent time-domain solution ([4]) (Bottom)

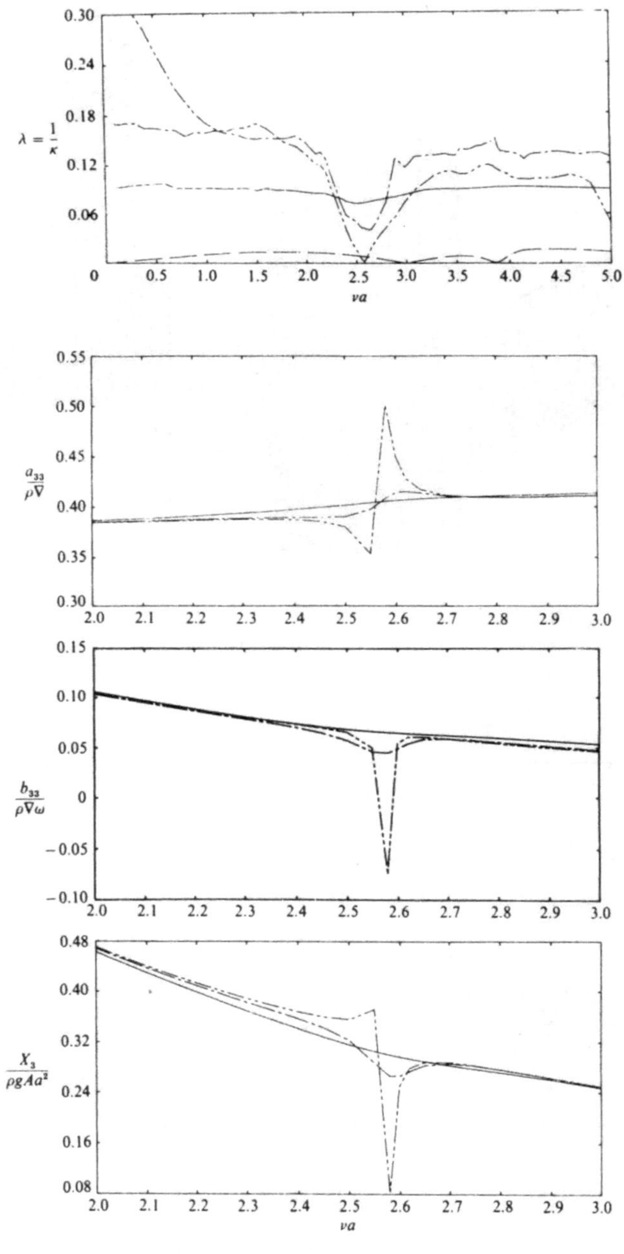

Figures 3, 4: **Inverse of the condition number of the solution matrix for a semisubmerged spher oscillating in heave. (Top). Heave added-Mass, damping coefficient and modulus of the excitin force for the heaving sphere. — – – —, $\beta = 0$; —–—, $\beta = 0.02$; ————, $\beta = 0.11$. (Bottom ([12]).**

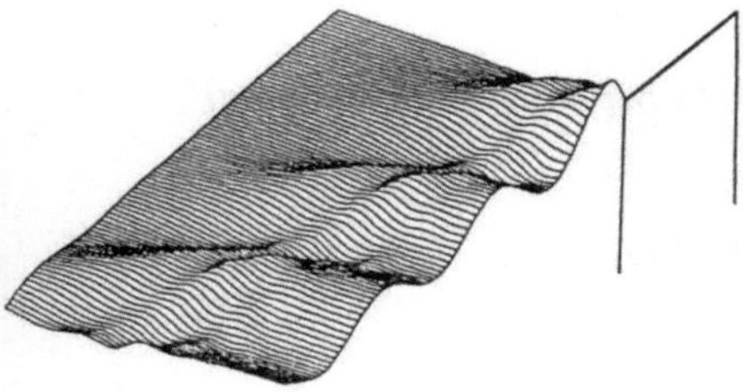

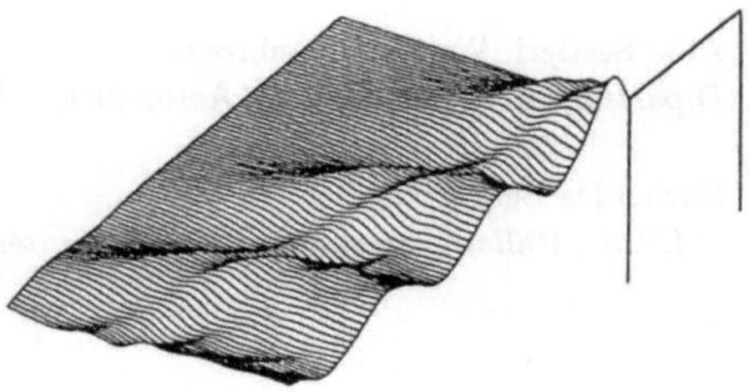

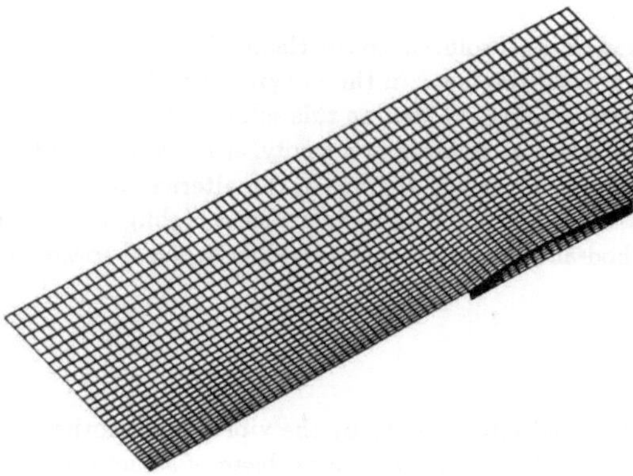

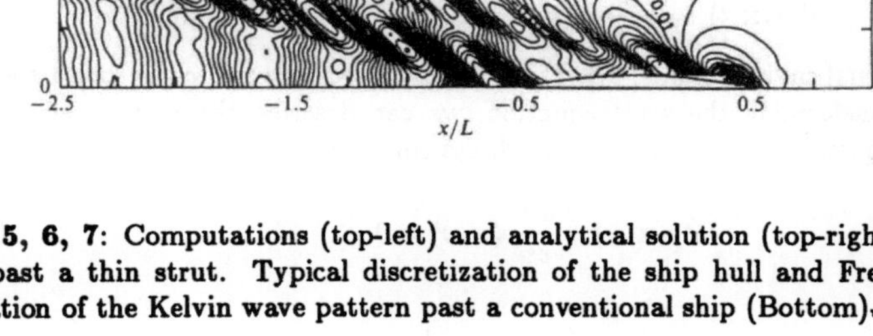

Figures 5, 6, 7: Computations (top-left) and analytical solution (top-right) of the Kelvin wave pattern past a thin strut. Typical discretization of the ship hull and Free Surface. (Middle). Computation of the Kelvin wave pattern past a conventional ship (Bottom), ([14], [15]).

On the Use of Different Fundamental Solutions for the Interior Acoustic Problem

Aldo Sestieri, Walter D'Ambrogio
Dipartimento di Meccanica e Aeronautica, Università "La Sapienza" — Roma (Italy)

Enrico De Bernardis
C.I.R.A., Italian Aerospace Research Center — Capua (Italy)

Abstract

An alternative form of fundamental solution to the Helmholtz equation is presented, which proves to be very effective when employed in the integral formulation for interior acoustic problems. A BEM code was developed based on this solution and gave satisfactory results in dealing with structural-acoustic coupling in a cavity, either with or without absorption walls. Theoretical arguments supporting the use of the alternative fundamental solution are provided; then the discussion of some numerical results highlights the main differences between the present method and the one using the well known free-space Green's function as fundamental solution.

Introduction

The study of the sound radiated into a cavity by the vibrational motion of its walls is an important field of BEM application. The general problem of structural-acoustic coupling calls for the simultaneous solution of the equations governing the vibrations of the walls and the sound field within the cavity, thereby accounting for the feedback of the sound pressure on the wall vibration.

The uncoupled problem does not account for this influence. This simplification is normally adopted because a fully coupled analysis is only necessary in a limited number of cases. In this case the structural problem is originally solved (analytically or by a finite element procedure) and the vibrational response is considered as part of the boundary conditions of the acoustic problem.

The sound induced in a cavity by a vibrating panel is governed by the wave equation:

$$\nabla^2 \bar{p}(\mathbf{x},t) - \frac{1}{c^2}\frac{\partial^2}{\partial t^2}\bar{p}(\mathbf{x},t) = 0 \quad \text{in} \quad V \subset I\!R^3 \tag{1}$$

where V is the internal cavity volume, $\bar{p}(\mathbf{x},t)$ the acoustic pressure and c the sound speed in air. Fourier transforming the wave equation, we can describe the interior acoustic problem, in the frequency domain, by the Helmholtz equation:

$$\nabla^2 p(\mathbf{x},\omega) + \frac{\omega^2}{c^2} p(\mathbf{x},\omega) = 0 \tag{2}$$

where $p(\mathbf{x},\omega)$ is the Fourier transform of $\bar{p}$. The boundary conditions, summarized by the following Neumann and Dirichlet conditions, account for:

- rigid walls : $\partial p/\partial n = 0$
- vibrating panels : $\partial p/\partial n = \rho\omega^2 w$ (ρ : air density, $w = w(\mathbf{x},\omega)$: wall displacement)
- absorbing materials : $p/u = Z_A$
  (u : Fourier transform of the air velocity, normal to the wall, Z_A : point impedance of the absorbing material). Through the momentum equation the previous relation can be also written as follows:

$$\frac{\partial p}{\partial n} = j\frac{\omega\rho}{Z_A}p \tag{3}$$

  This is a very approximate model for absorbing walls since it assumes a point reacting material. In [1], Bliss proposed a bulk reacting model which better describes porous materials, in which the acoustic behaviour of any point depends on the sound field characteristics of the neighbouring points.

The Helmholtz equation can be transformed into an integral equation by applying the Green's theorem, once determined a fundamental solution for the Helmholtz operator. A fundamental solution is a solution of the inhomogeneous Helmholtz equation:

$$\nabla^2 G(\mathbf{x}|\mathbf{x_0}) + k^2 G(\mathbf{x}|\mathbf{x_0}) = \delta(\mathbf{x}-\mathbf{x_0}) \tag{4}$$

G depends on two points: the observation point $\mathbf{x_0}$ and the source point $\mathbf{x}$, with $\mathbf{x},\mathbf{x_0} \in I\!R^3$; δ is the Dirac delta function, and k the wavenumber ω/c.

The fundamental solution satisfying equation (4) is not unique [2]. Uniqueness can be obtained by imposing a radiation condition into the free field: this implies that only outward waves from the source point are possible (Sommerfeld condition). Consequently the fundamental solution takes the form:

$$G(\mathbf{x}|\mathbf{x_0}) = \frac{e^{jkr}}{4\pi r} \tag{5}$$

$r = |\mathbf{x}-\mathbf{x_0}|$ being the distance between source and observation points. $G(\mathbf{x}|\mathbf{x_0})$ is called the "free-space Green's function". This is the common form of Green's solution found in the literature both for the exterior and interior acoustic problems [3,4,5,6].

By using the free-space Green's function, the Helmholtz equation (2) can be turned into the integral equation:

$$\frac{\beta}{4\pi}p(\mathbf{x_0}) + \int_S \left[p(\mathbf{x})\left(\frac{\partial}{\partial n}G(\mathbf{x}|\mathbf{x_0})\right) - G(\mathbf{x}|\mathbf{x_0})\left(\frac{\partial}{\partial n}p(\mathbf{x})\right)\right] dS(\mathbf{x}) = 0 \tag{6}$$

where S is the frontier of V and β the solid angle swept by S at $\mathbf{x_0}$. The problem is then solved in two steps: first the pressure on S is determined by considering both the source and the observation points on the frontier; then the internal acoustic pressure is computed with the source point on the frontier and the observation point internally in the cavity. Only the first step requires the solution of an actual integral equation, the second one involving a simple integral expression.

Though the commonly used fundamental solution is the one shown in equation (5), responding to the Sommerfeld condition, for the interior problem a simpler fundamental solution can be employed:

$$G_R(\mathbf{x}|\mathbf{x_0}) = \frac{\cos(kr)}{4\pi r} \tag{7}$$

i.e. the real part of the free-space Green's function. Note that, since both G_R and G satisfy the inhomogeneous Helmholtz equation (4), the imaginary part $G_I = \sin(kr)/4\pi r$ of the free-space Green's function is a solution of the homogeneous Helmholtz equation. The fundamental solution G_R was used in [7,8] and gave very satisfactory results when dealing with structural-acoustic coupling, either with or without absorption walls. However neither a mathematical demonstration of this form nor the implications of this choice were ever presented compared to the classical form. In this paper we aim to provide the missing link, showing advantages and limits of the above solution.

Theoretical considerations

In this section an outline of the theory upon which the calculations are based is presented.

Fundamental solutions versus free-space Green's function

As we have seen before, among the solutions of the inhomogeneous Helmholtz equation the free-space Green's function satisfies the Sommerfeld radiation condition, that concerns the behaviour of the solution at infinite distance from the source points. This issue is meaningless when an interior problem is solved. Then, part of the free-space Green's function might have no influence on the solution of a problem defined in a domain bounded by a close finite surface.

To see this let us rewrite equation (6) for the exterior problem, by explicitly considering a limit closed surface S_∞, at infinite distance from the source points:

$$\begin{aligned} \frac{\beta}{4\pi}p(\mathbf{x_0}) \quad &+ \quad \int_S \left[p(\mathbf{x})\left(\frac{\partial}{\partial n}G(\mathbf{x}|\mathbf{x_0})\right) - G(\mathbf{x}|\mathbf{x_0})\left(\frac{\partial}{\partial n}p(\mathbf{x})\right)\right] dS(\mathbf{x}) \\ &+ \quad \int_{S_\infty} \left[p(\mathbf{x})\left(\frac{\partial}{\partial n}G(\mathbf{x}|\mathbf{x_0})\right) - G(\mathbf{x}|\mathbf{x_0})\left(\frac{\partial}{\partial n}p(\mathbf{x})\right)\right] dS(\mathbf{x}) = 0 \end{aligned} \tag{8}$$

The pressure field is required to meet the Sommerfeld radiation condition; this can be stated by imposing that only outward waves are permitted at infinity:

$$\frac{\partial p}{\partial n} - jkrp = 0 \quad \text{on} \quad S_\infty$$

The last integral on the left-hand side of equation (8) now reads:

$$\int_{S_\infty} \left[p\frac{\partial G}{\partial n} - jkrGp\right] dS = \int_{S_\infty} p\left[\frac{\partial G}{\partial n} - jkrG\right] dS \tag{9}$$

The complex structure of the free-space Green's function is needed in order to let this integral vanish (what make it possible to find a solution to the exterior problem by solving an integral equation on S), *i.e.* to meet the condition:

$$\frac{\partial G}{\partial n} - jkrG = 0$$

The integral on S_∞ would not appear in the interior problem and, as it is easily seen, only the real part of the free-space Green's function (satisfying the inhomogeneous

Helmholtz equation (4)) is needed in this case. In order to show this point, let us write the integral equation for the interior problem by separating contributions from the real and imaginary parts of the free-space Green's function:

$$\frac{\beta}{4\pi}p(\mathbf{x_0}) + \int_S \left[p(\mathbf{x})\left(\frac{\partial}{\partial n}G_R(\mathbf{x}|\mathbf{x_0})\right) - G_R(\mathbf{x}|\mathbf{x_0})\left(\frac{\partial}{\partial n}p(\mathbf{x})\right)\right] dS(\mathbf{x})$$
$$+ j\int_S \left[p(\mathbf{x})\left(\frac{\partial}{\partial n}G_I(\mathbf{x}|\mathbf{x_0})\right) - G_I(\mathbf{x}|\mathbf{x_0})\left(\frac{\partial}{\partial n}p(\mathbf{x})\right)\right] dS(\mathbf{x}) = 0 \quad (10)$$

By the Green's identity the last integral can be manipulated as follows:

$$\int_S \left[p\frac{\partial G_I}{\partial n} - G_I\frac{\partial p}{\partial n}\right] dS = \int_V \left[p\nabla^2 G_I - G_I\nabla^2 p\right] dV$$

and, since both G_I and p satisfy the homogeneous Helmholtz equation, we finally have:

$$\int_V \left[p\nabla^2 G_I - G_I\nabla^2 p\right] dV = \int_V \left[pk^2 G_I - G_I k^2 p\right] = 0$$

which shows that no contribution comes from the imaginary part of the fundamental solution. Consequently for the interior problem G_I is uninfluential and the integral formulation using G_R is equivalent to the one obtained with the free-space Green's function G.

Uniqueness

In [7], when using the real part of the free-space Green's function and zeroth order rectangular boundary elements (piecewise constant variation of the unknown), singular behaviour (*i.e.* spurious peaks in the pressure level) was observed in the solution, not corresponding either to acoustic or structural eigenvalues. Singular solutions are normally met for the exterior acoustic problem, at wavenumbers corresponding to resonant frequencies of the related interior problem. On the contrary the interior problem should not present any singular behaviour [2], and therefore we concluded that the observed singularities were induced by the numerical scheme adopted. In fact, using triangular instead of rectangular boundary elements, any previuos problem disappeared (at least for the geometries investigated) [8].

Later on it was suggested that the uniqueness for the interior problem was demonstrated for the classical free-space Green's function and not for the cosine one. However, rereading carefully Filippi's basic work [2], we realized that there is no relationship involving the specific form of fundamental solution. On the other side, since uniqueness is ensured for the interior problem when using the free-space Green's function, the same condition holds *a fortiori* for G_R, which is just the real part of G. Consequently, even in absence of further numerical tests, we may assess that the real part of the free-space Green's function can not cause any particular trouble and that the observed singularities were provoked by the numerical procedure. (We never observed numerical singularities when using zeroth order triangular elements.)

Numerical considerations

Let us now discuss whether the procedure employing the real part of the free-space Green's function presents any advantage on the numerical result compared to that involving G. As we have seen, the imaginary part of the free-space Green's function is uninfluential.

However, when discretizing the integral equation through boundary elements, this last part may not vanish, thus inducing significant errors on the final result, especially with a rather coarse mesh. The major error, as we will see discussing numerical results, is obtained on the pressure phase rather than on its magnitude.

Let us consider the case of a complex cavity discretized through flat triangular elements of a given size (depending on the maximum investigated frequency), on which the problem variables (pressure and displacements) are constant. In this way the integral equation becomes an algebraic system:

$$p_k + 2\sum_{i=1}^{N}\left[p_i \int_{S_i} \frac{\partial G}{\partial n_i} dS_i - \frac{\partial p_i}{\partial n_i}\int_{S_i} G dS_i\right] = 0 \qquad k = 1,2,...,N \qquad (11)$$

Expliciting the real and imaginary part of G, we can write the above system as:

$$\begin{aligned} p_k \;&+\; 2\sum_{i=1}^{N}\left[p_i \int_{S_i} \frac{\partial G_R}{\partial n_i} dS_i - \frac{\partial p_i}{\partial n_i}\int_{S_i} G_R dS_i\right] \\ &+\; j\left\{2\sum_{i=1}^{N}\left[p_i \int_{S_i} \frac{\partial G_I}{\partial n_i} dS_i - \frac{\partial p_i}{\partial n_i}\int_{S_i} G_I dS_i\right]\right\} = 0 \qquad k = 1,2,...,N \qquad (12) \end{aligned}$$

Whilst theoretically the last term in square brackets vanishes identically, numerically it does not, leading to some amount of error which depends on the discretization mesh.

Moreover in [7] it had been shown that, when using the real part of the free-space Green's function, the harmonic parts of G_R and $\partial G_R/\partial n$ do not change appreciably on each element: therefore it is possible to carry the harmonic parts ($\sin(kr)$ and $\cos(kr)$) out from the integrals and compute the integrals only once as they are now independent on frequency. Unfortunately this operation does not work when using the free-space Green's function, *i.e.* each integral must be computed any time we change the frequency of analysis, thus heavily increasing the computational burden.

Numerical results

In order to check advantages and limits of the cosine fundamental solution with respect to the free-space Green's function, initially a plane wave propagation problem in a prismatic cavity with the longitudinal dimension much larger than the two transverse dimensions was considered ($1m \times 0.1m \times 0.1m$). For this problem an analytical solution is known. The cavity had rigid walls and the wave perturbation was generated by a rigid displacement of one of the minor surfaces.

Three numerical tests were performed. In the first one the free-space Green's function was used, and the harmonic parts of G and $\partial G/\partial n$ were left in the integrals. The second one used the same Green's function, but the harmonic parts were carried out from the integrals. Finally in the third the cosine fundamental solution was used, with the harmonic functions outside the integrals. The three tests were developed with different boundary discretizations, all sufficiently fine according to the frequency of analysis (100 Hz): 164, 204 and 416 equal triangular elements. Magnitude and phase pressure values were computed at three different points along the axis of the cavity: $x = 0.245m$, $x = 0.49m$, $x = 0.735m$ from the vibrating wall. Since no absorbing wall was considered, the phase value must be theoretically zero. In fig. 1 the difference in the value of pressure magnitude with respect to the analytical solution is shown for the three tests. A minimal difference is observed between the results obtained using the free-space Green's function with harmonic parts inside the integrals and the cosine solution (without the harmonic parts). The free-space Green's function with harmonic part outside the integrals yielded larger errors. The error on the pressure phase is shown in fig. 2 for the two tests on the free-space Green's function. (The error on the cosine solution is not considered, because identically zero). A large error is observed when the harmonic parts are extracted, whilst this is much smaller when they are left in. This result confirms that it is not suitable to extract the harmonic parts from the integrals when using the free-space Green's function. Moreover it emphasizes that, even when the harmonic parts are left in, the error on phases always exists, thus yielding in any case a wrong result with respect to the solution involving G_R.

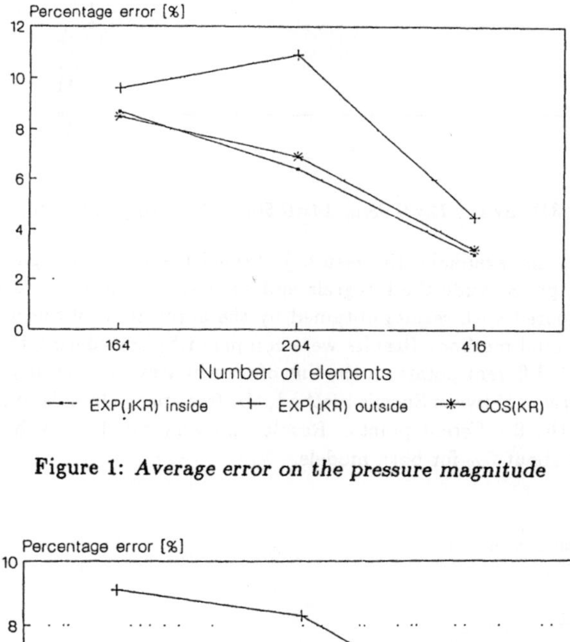

Figure 1: *Average error on the pressure magnitude*

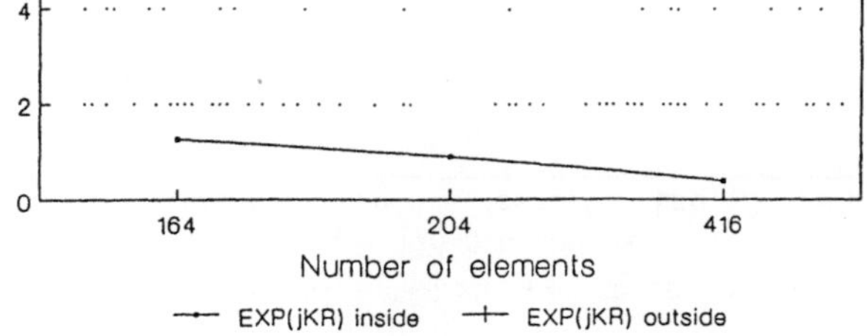

Figure 2: *Average error on the pressure phase*

A second test was performed on a more complex three dimensional cavity (maximum dimension: size $1-2=.98m$), shown in fig. 3. The vibrating panel (2,3,9,10) is an aluminum plate, hinged along the four sides, loaded by a uniform unit pressure. The lower wall (1,2,3,4) is an absorbing wall, modelled as point reacting. Since in this case

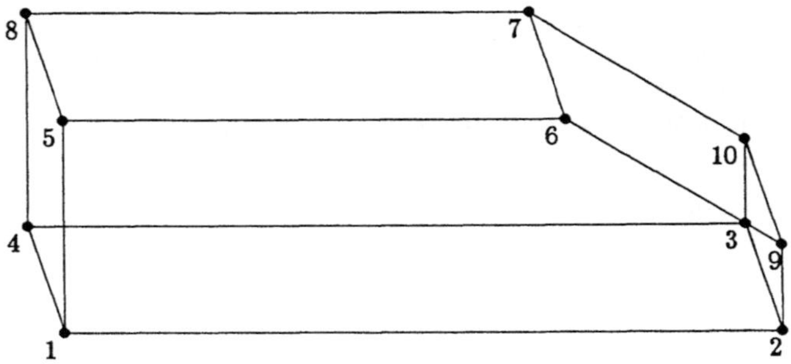

Figure 3: *Sketch of the 3D cavity: 12=0.98m, 14=0.56m, 15=0.7m, 56=0.7m, 29=0.28m*

an analytical solution is not available, the results obtained from the free-space Green's function with harmonic parts inside the integrals and the cosine solution with harmonic parts outside were compared with results obtained by the application of Succi's method [9], which is akin of a modal method. Results were compared by considering the value of pressure magnitudes at 5 different points on the line of coordinates $z = 0.56m, y = 0.28m$. Fig. 4 shows the comparison between Succi's method, the free-space Green's function and the cosine solution, at the 6 different points. Results are very satisfactory because the percentage difference is about 5% for both models.

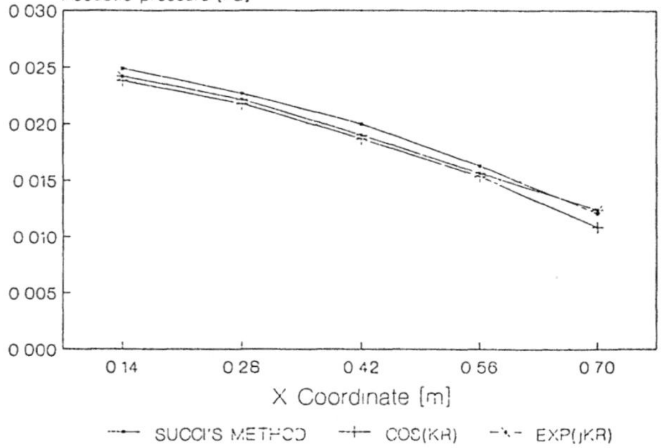

Figure 4: *Comparison between results from three different methods*

Computer burden

As already mentioned, the code using the free-space Green's function requires a larger computation time compared to the one using the cosine solution. The reason of that is twofold.

- When using the free-space Green's function, the number of integrals to compute is twice the correspective required by the cosine solution.
- Integrations can be made independent on frequency only when using G_R.

In consideration of that, a computer time quantification for both solutions is reported in the following diagrams. Times are normalized with respect to the time required by the code using the cosine solution to find the pressure level for the first frequency, in order to make results almost independent on the computer used.

Comparisons refer to the three-dimensional cavity previously considered. Figs. 5 and 6 show the normalized time required by the two models to solve a number of frequencies for two different discretizations (84 and 254 elements, respectively). As it can be observed, the difference is quite relevant not only for the further frequencies, where the integrals are not computed when using the cosine solution, but even for the first frequency, due to the different number of integrals involved. It is worth to point out that the difference slightly decreases for the finer discretization model. This depends on how the of computation time is shared among the different operations. The most of time is required by the following two tasks:

- integration
- construction and solution of the linear algebraic system.

The difference between the two codes lies in the first task, the second one requiring the same time for both codes. This sharing is shown in figs. 7 and 8 for the two models and the two different discretizations. When the number of elements increases, so does the weight of the second task. In fact the number of operations to solve a linear system is proportional to N^3 whilst the number of operations required to solve the integrals is proportional to N^2. Therefore when the number of elements increases, the difference between computation times decreases.

However, since the accuracy of results is very similar, the use of the cosine solution should be preferable, because the computer time is much lower, especially when analysing a consistent number of frequencies.

Conclusions

The alternative use of a simpler fundamental solution for the interior acoustic problem has been presented. With respect to the classical free-space Green's function it shows several advantages that can be briefly summarized as follows.

- The real part of the free-space Green's function is numerically more efficient because it does not introduce a possible source of error, related to the discretization of the integral equation. Since it has been shown that the imaginary part of the free-space Green's function is theoretically uninfluential, it is appropriate to exclude it from the numerical computations and use only the cosine solution. In fact a coarse discretization introduces severe phase errors when using the free-space Green's function instead of the cosine solution.

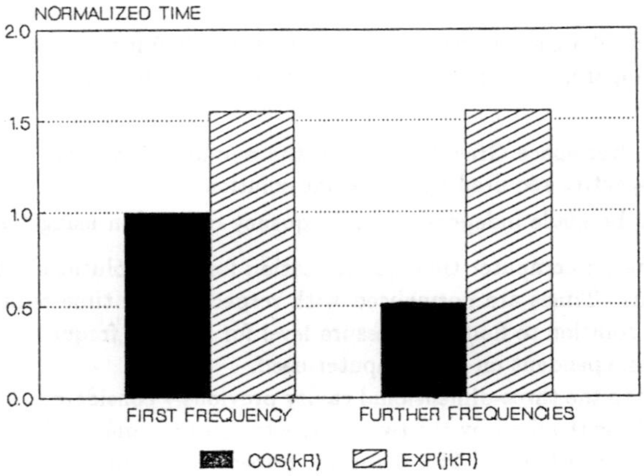

Figure 5: *Normalized times for the 84 elements model*

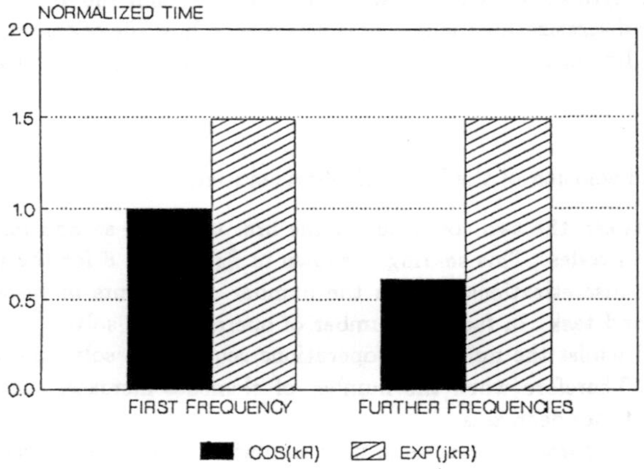

Figure 6: *Normalized times for the 254 elements model*

- When using the cosine solution, the harmonic parts of G_R and $\partial G_R/\partial n$ can be carried out from the integrals. This operation makes the computation faster when many frequencies are analysed because integrals may be computed only once, rather than at each frequency. This procedure is not permitted with the free space Green's function because phases (and moduli) are highly influenced by this approximation.

No drawbacks are legitimately expected, though the experience acquired with this sort of fundamental solution is not so wide as the one developed with the classical free-space Green's function.

References

[1] Bliss, D.B.: A study of bulk reacting porous sound absorbers and a new boundary condition for thin porous layers. J. of the Acoustical Society of America 71 (1982) 533–545.

[2] Filippi, P.J.T.: Layer potential and acoustic diffraction. J. of Sound and Vibration 54 (1977) 473–500.

[3] Schenk, H.A.: Improved integral formulation for acoustic radiation problems. J. of the Acoustical Society of America 44 (1968) 41–58.

[4] Meyer, W.L., Bell, W.A., Zinn, B.T., Stallybrass, M.P.: Boundary integral solutions of three-dimensional acoustic radiation problems. J. of Sound and Vibration 59 (1978) 245–262.

[5] Sayhi, M.N., Ousset, Y., Varchery, G.: Solution of radiation problems by collocation of integral formulations in terms of single and double layer potentials. J. of Sound and Vibration 74 (1981) 187–204.

[6] Banerjee, P.K., Ahmed, S., Wang, H.C.: A new BEM formulation for the acoustic eigenfrequency analysis. Int. J. for Numerical Methods in Engineering 26 (1988) 1299–1309.

[7] Sestieri, A., Del Vescovo, D., Lucibello, P.: Structural acoustic coupling in complex shaped cavity. J. of Sound and Vibration 96 (1984) 219–233.

[8] Sestieri, A.: Discretization procedures for the Green formulation of structural acoustic problems. J. of Sound and Vibration 98 (1985) 305–308.

[9] Succi, G.P.: The interior acoustic field of an automobile cabin. J. of the Acoustical Society of America 81 (1987) 1688–1694.

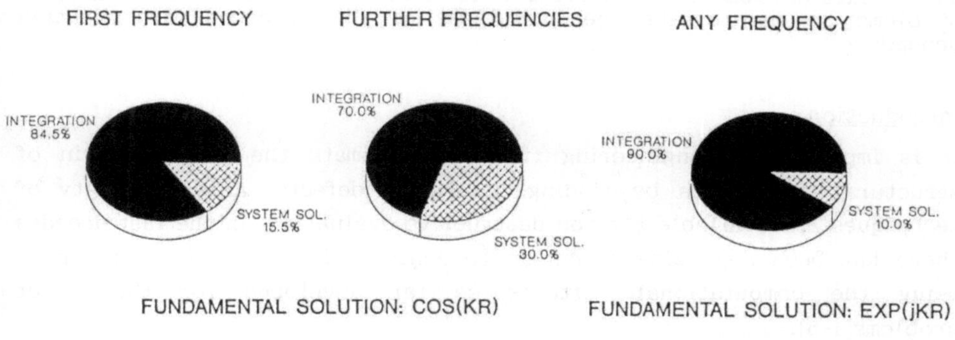

Figure 7: *Sharing of computer time for the 84 elements model*

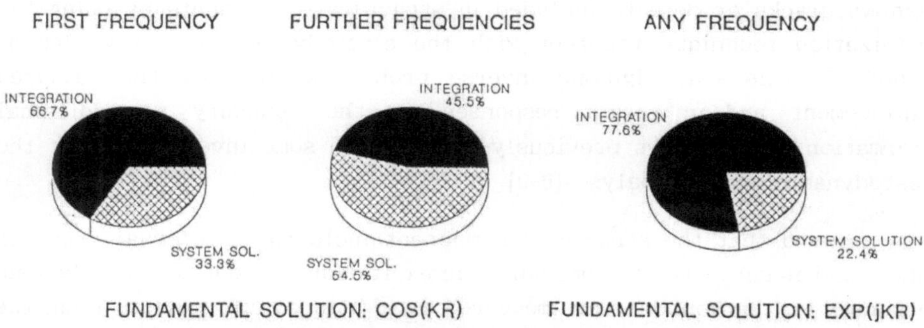

Figure 8: *Sharing of computer time for the 254 elements model*

Identification of Cracks or Defects by Means of the Elastodynamic BEM

Masa. TANAKA and M. NAKAMURA

Department of Mechanical Systems Engineering
Faculty of Engineering, Shinshu University
500 Wakasato, Nagano 380, Japan

T. NAKANO

Graduate School of Shinshu University

Summary

In this paper an attempt is made to identify cracks or defects by means of the optimization technique using the elastodynamic boundary element method. The inverse problem is cast into an optimum problem, in which the objective function is assumed as the square sum of residuals between the reference data measured at selected points on the boundary and the corresponding values computed by the BEM. Assuming that a crack or defect has simple geometry, the optimal set of only several parameters should be determined by the standard optimization technique to estimate the most plausible shape and location. Numerical experiment is carried out for two-dimensional problems to demonstrate the usefulness of the proposed identification procedure.

Introduction

It is important in engineering fields to estimate the safety margin of structural components by finding cracks or defects. A wide variety of techniques are available for non-destructive evaluation. In the last decades there has been a growing attention to analyze these inverse problems by using the computational software so far developed for the direct problems[1-5].

The inverse problem under consideration deals with the identification of unknown cracks or defects included in structural components by using the optimization technique together with the elastodynamic boundary element method. In this elastodynamic inverse problem we may use the measured displacement and/or strain responses on the boundary as additional information. The authors previously reported on some investigation of the elastodynamic inverse analyses[6-9].

It is assumed that the structural component includes an internal crack or defect and is subjected to time-harmonic excitation. In addition, the dynamic responses on the boundary are measured as additional information. Then, the inverse problem can be reduced to an optimal problem solvable by means of the standard optimization technique. Namely, the residuals between the measured data and the computed dynamic responses on the boundary are minimized to find the optimal set of parameters defining the shape and the location of crack or defect. The dynamic responses can be computed by means of the available boundary element software for steady-state elastodynamics. Numerical experiment is carried out for some example problems to demonstrate the usefulness of the proposed method.

Application of Optimization Technique

It is assumed that the material of the structural component under consideration is homogeneous and isotropic, and obeys Hooke's law. The inverse problem can be defined such that the shape and the location of the internal crack or defect in the structural component are not known, while the dynamic responses at some selected points on the boundary are given as additional information. This inverse problem is reduced to an optimum problem which minimizes the square sum of the residuals between the boundary element solutions and the measured data (reference data) given at the selected points on the boundary.

The optimization problem can be stated as follows:

$$\text{Find } \{ y \} = \{ y_1 \ \ y_2 \ \cdots \ y_P \}^T \tag{1}$$

$$\text{which minimizes } W = f(\{ y \}) \text{ under some constraints.} \tag{2}$$

where $\{ y \}$ is an M-dimensional vector which is called here the design vector, and W is the objective function, while P is the total number of design parameters. The superscript T denotes the transpose of a matrix.

In the two-dimensional problems for which numerical experiment will later be carried out, we assume that the unknown shape is a straight line for an internal crack or an ellipse for an internal defect as shown in Fig.1. Then, we should determine the optimal set of only four or five parameters minimizing the objective function. In these cases, the design vectors are defined as

$$\{ y \} = \{ y_1 \ \ y_2 \ \ L \ \ \theta \}^T \quad \text{for a straight line crack} \tag{3}$$

$$\{ y \} = \{ y_1 \ \ y_2 \ \ a \ \ b \ \ \theta \}^T \quad \text{for an elliptical defect} \tag{4}$$

where y_1 and y_2 are the coordinates of centroid of the crack or defect, L is the crack length, a and b are the minor and major axes of the ellipse, while θ is the inclination angle from the coordinate x_1.

When the displacement responses are given as additional information, the

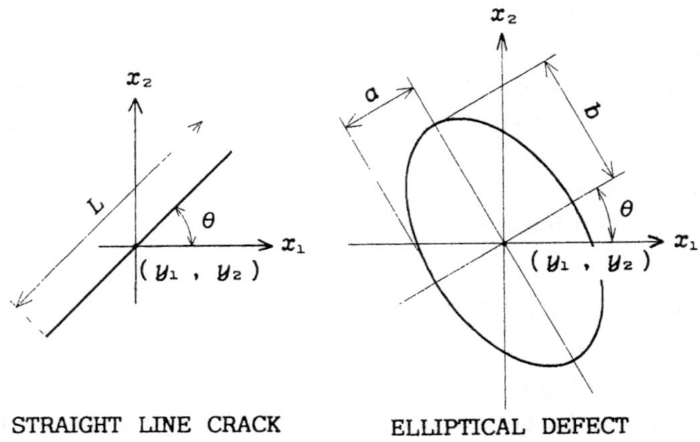

Fig.1. Parameters of straight line crack and elliptical defect

objective function is the square sum of residuals between the measured displacements $\overline{u_i}^n$ at selected points on the boundary and the corresponding displacements u_i^n computed by the boundary element method using the assumed parameter values. For the two-dimensional problems the objective function can be expressed as

$$W = \sum_{n=1}^{N} \sum_{i=1}^{2} (u_i^n - \overline{u_i}^n)^2 \tag{5}$$

where N is the number of measuring points on the boundary. If the strains ε_{ij} are given as additional information, the objective function can be expressed as

$$W = \sum_{n=1}^{N} (\varepsilon_{ij}^n - \overline{\varepsilon_{ij}}^n) (\varepsilon_{ij}^n - \overline{\varepsilon_{ij}}^n) \tag{6}$$

As the non-dimensional expression of equation (5) or (6) we use the following expression:

$$Z = \log (W / W_o)^{1/2} \tag{7}$$

where $W_o = \Sigma \overline{u}_i^n$ for the displacement information, and $W_o = \Sigma (\overline{\varepsilon_{ij}}^n)(\overline{\varepsilon_{ij}}^n)$ for the strain information.

The design vector $\{ y \}$, which is minimizing the objective function W, is modified iteratively by the following equation:

$$\{ y \}^{k+1} = \{ y \}^k + \alpha^* \{ d \}^k \tag{8}$$

where $\{ d \}^k$ is a vector of the search direction at the kth iteration. The coefficient α^* is the optimal step length for the movement along the search direction $\{ d \}^k$. This step length is determined by one-dimensional minimization. We apply in this study the conjugate gradient method[10] to evaluate the vector $\{ d \}^k$. To determine the optimal step length α^* we adopt the quadratic interpolation method[10].

As the convergence criterion of iterative computations, we use either of the following two inequalities:

$$Z^k < \eta_1 < 0 \tag{9}$$

$$| Z^k - Z^{k-1} | < \eta_2 \tag{10}$$

Inequality (9) implies that the non-dimensional residual Z is less than a given negative number η_1, while inequality (10) states that the change in the residual Z is less than a given positive small number η_2.

Selection of Measuring Points

In this study, we locate the measuring points in such a way that the points attain there the largest values of sensitivities of the objective function with respect to the parameters. This method previously proposed by the authors[8,9] can be very efficient for locating the measuring points. In this method we compute the sensitivity S_j^n at the nth measuring point by the following finite difference approximation:

$$S_j^n = \frac{\Delta W}{\Delta y_j} \tag{11}$$

where ΔW is a change in the objective function W, Δy_j is a small change in the parameter y_j. The measuring point is selected in such a way that the absolute value of the sensitivity is maximum for each parameter. This method can be successful, since the maximum sensitivity point includes effective information for modifying the parameter value.

The main flow of the proposed identification procedure using the optimization technique and the boundary element analysis is illustrated in Fig. 2.

Numerical Results and Discussion

Now we show some examples of numerical experiment for the crack identification in two-dimensional problems to demonstrate the usefulness of the above-mentioned procedure. In this numerical experiment we assume that

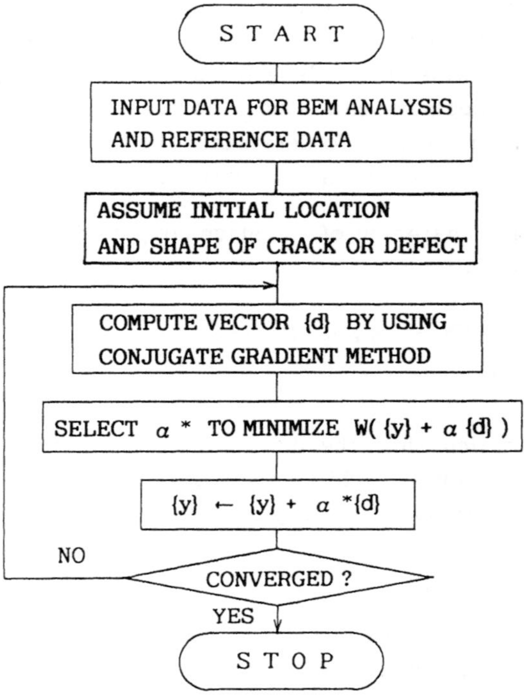

Fig.2. Flow chart of crack or defect identification process

the rectangular plate model includes a straight line crack as shown in Fig.3. The horizontal side of the rectangular plate is 300*mm* long and the vertical one is 200*mm* long. The length and the inclination angle from the axis x_1 of the crack are 80*mm* long and 45°, respectively. The rectangular plate is assumed to be in a plane stress state and to have the following material constants :

Young's modulus : $E = 210\ GPa$

Poisson's ratio : $\nu = 0.3$

Mass density : $\rho = 7.8 \times 10^3\ kg/m^3$

The boundary element discretization used is shown in Fig.4 ; the outer boundary, the crack part, the interface parts 1 and 2 of the rectangular plate are divided into 50, 10, 10 and 10 constant boundary elements, respectively. It is noted that the reference data of this numerical experiment are the displacement responses computed by the elastodynamic boundary element analysis under the above-mentioned computational conditions and the assumption of material constants and crack geometry. At the beginning of iterative computation we assume the 60*mm* long crack with

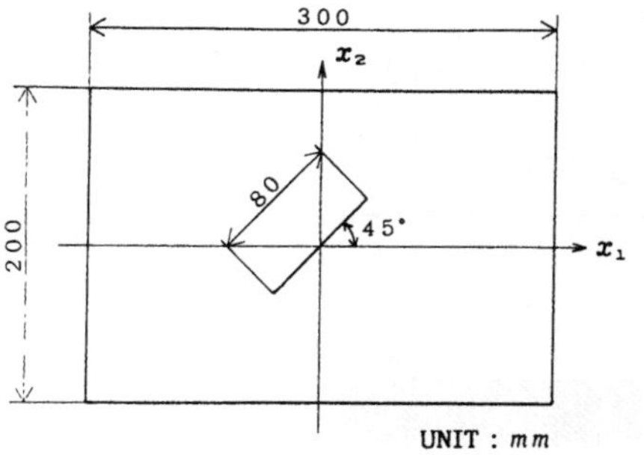

Fig.3. Rectangular plate with a straight line crack

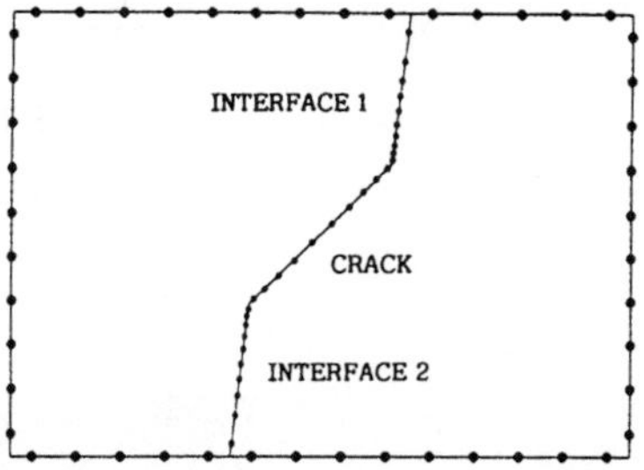

Fig.4. Boundary element discretization

the inclination angle 90° from the axis x_1 located at the center of the rectangular plate. During the iterative computation, the crack part, interface parts 1 and 2 are discretized into 10 constant boundary elements, respectively, which is the same as the computation of the reference data.

The example model is shown in Fig.5. In this example we assume that the center point of real crack is known, while the length and inclination angle are not known. The boundary conditions are also illustrated in Fig.5: The side AD is fixed and the other sides are traction-free. The measuring points of the displacement are arranged at the point M_1 for the crack length parameter and also the point M_2 for the crack inclination angle which is selected by sensitivity analysis. The time-harmonic exciting force ① with angular frequency $\omega=1.0\times10^4$ *rad/sec* (approx. 1600*Hz*) is applied.

Figure 6 illustrates the convergence property with respect to the non-dimensional residual *Z*, while Fig.7 shows how the assumed crack shape is modified during the iterative process. Despite a limited number of the reference data the crack identification is almost successfully carried out by selecting the measuring points with the maximum sensitivity for each parameter.

Next, we examine the identification of the crack location. When the above-mentioned method for selection of the measuring points is used to the crack location identification, the maximum sensitivity points move at each iterative computation. In such cases, to check the convergence of iterative computation we assume that the measuring points to be used only for the

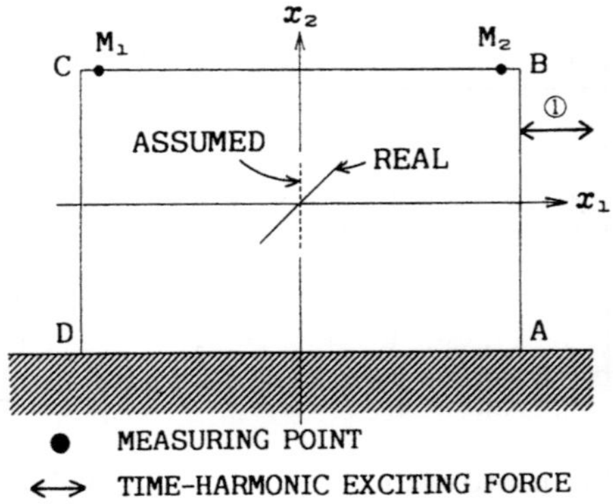

Fig.5. Boundary conditions and selected measuring points

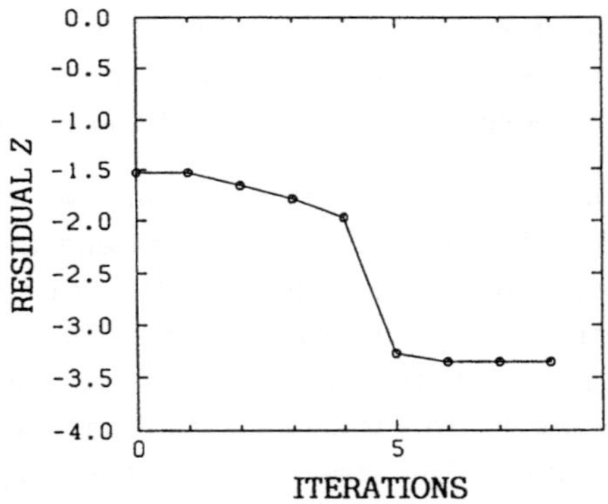

Fig.6. Convergence property with respect to displacement residuals

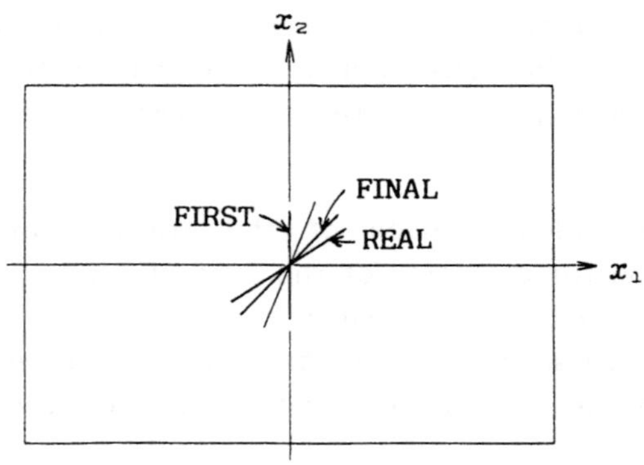

Fig.7. Convergence with respect to crack length and inclination angle

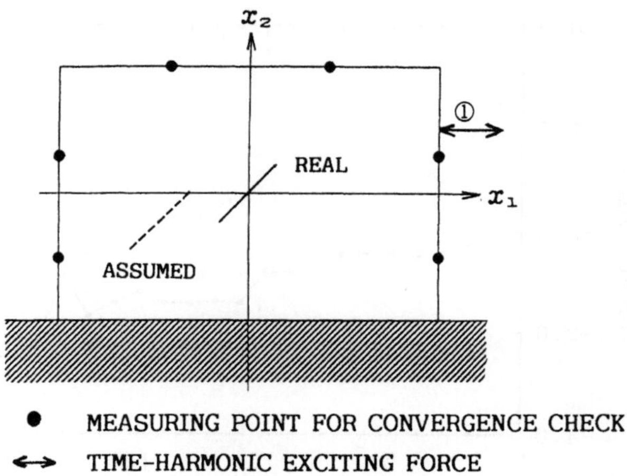

Fig.8. Measuring points to check convergence and boundary conditions

convergence check are distributed uniformly over the boundary. The measuring points with the maximum sensitivity are still used in such cases for modifying the parameter values.

The present model with a straight line crack is shown in Fig.8. The boundary conditions are assumed to be the same as the first model. The measuring points to be used only for the convergence check are arranged as shown in this figure. To modify the parameter values, we use separately another measuring points determined by sensitivity analysis. The time-harmonic exciting force ① with frequency $\omega = 1.0 \times 10^4 rad/sec$ is applied. The real crack is shown by the solid line and the first assumed crack is shown by the dotted line. In this case the length and inclination angle of the real crack are the same as the first assumed crack.

Figure 9 illustrates the convergence property with respect to the displacement residual Z which is calculated by using the data at the six measuring points denoted by the dot in Fig.8. Figure 10 shows how the assumed crack is modified during the iterative process. In viewing the modification process illustrated in Fig.10, we may conclude that the crack identification is successfully carried out.

The numerical results mentioned above were obtained by using the reference data subjected to a single time-harmonic excitation. If we apply the method of multiple force applications previously proposed by the authors[6], more efficient identification could be carried out.

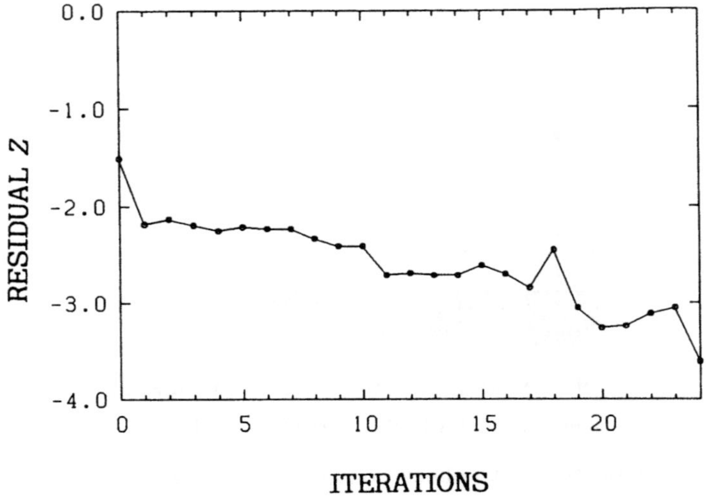

Fig.9. Convergence property with respect to displacement residuals

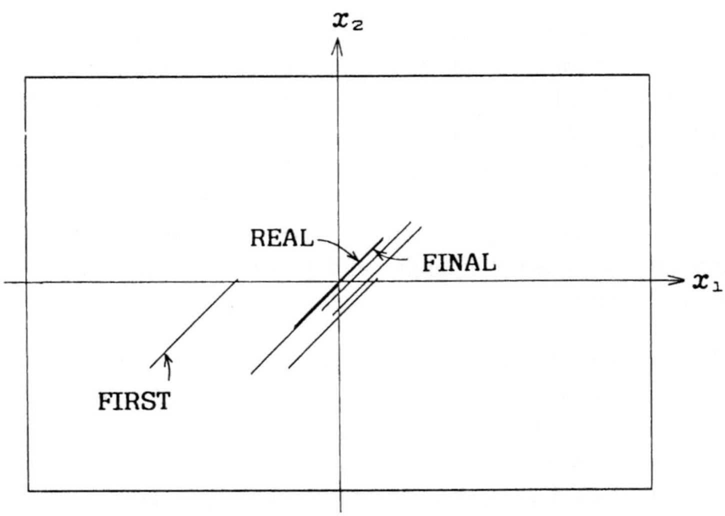

Fig.10. Convergence with respect to crack location

Conclusions

A new computational procedure has been proposed for the inverse problems in steady-state elastodynamics, in which an unknown crack or defect in a structural component should be identified by using some measured data on the boundary. In the proposed computational procedure, the boundary element software and the standard optimization technique were used effectively. Numerical experiments were carried out for a few example problems in the plane stress state to demonstrate the usefulness of the proposed method. In this paper, we presented only two numerical examples for the crack identification. The identification of cavity defects by the same procedure as in this paper is reported on in our previous papers[6,7]. Three-dimensional experiment will be reported in the near future.

References

1. Tanaka, M. : Some recent advances in boundary element research for inverse problems. Boundary Element X, Brebbia, C.A. (ed.), Vol.1 (1988), 567-582, Berlin, Heidelberg, New York, Springer-Verlag.
2. Blakemore, M.; Georgiou, G.A.(eds.) : Mathematical Modeling in Non-destructive Testing. Oxford, New York, Oxford Univ. Press, 1988.
3. Proceedings of the JSME Symposium on Computational Methods and Their Applications to Inverse Problems, (in Japanese). No.890-34, 1989.
4. Kubo, S.; Sakagami, T.; Ohji, K.: Reconstruction of a surface crack by electric potential CT method. Computational Mechanics '88, Atluri, S. N.; Yagawa,G. (eds.), Proc. of Int. Conf. on Computational Engineering Science. Vol.1 (1988), 12.i.1-5, Berlin, New York, Springer-Verlag.
5. Nishimura, N.; Kobayashi, S.: Regularised BIEs for crack shape determination problems. Tanaka, M.; Brebbia, C.A.; Honma, T.(eds.), Boundary Elements XII, Vol.2 (1990), 425-434, Southampton, Boston, Computational Mechanics Publications.
6. Tanaka, M.; Nakamura, M.; Nakano, T.: Defect shape identification by means of elastodynamic boundary element analysis and optimization technique. Advances in Boundary Elements, Brebbia, C.A.; Connor, J.J. (eds.), Proc. of 11th Int. Conf. on Boundary Element Methods, Vol.3 (1989), 183-194, Berlin, New York, Springer-Verlag.
7. Tanaka, M.; Nakamura, M.; Nakano, T.: Defect shape identification by elastodynamic boundary element method using strain responses. Advances in Boundary Element Methods in Japan and USA, Tanaka, M.; Brebbia C.A.; Shaw, R. (eds.), 137-151, Southampton, Boston, Computational Mechanics Publications, 1990.
8. Nakano. T.; Tanaka, M.; Nakamura, M.: Defect identification by elastodynamic BEM - Consideration on selection of additional information, (in Japanese). Proc. of JSME, No.907-1 (1990), 65-66.
9. Tanaka, M.; Nakamura, M.; Nakano. T.: Detection of cracks in structural components by the elastodynamic boundary element method. Tanaka, M.; Brebbia, C.A.; Honma, T.(eds.), Boundary Elements XII, Vol.2 (1990), 413-424, Southampton, Boston, Computational Mechanics Publications.
10. Fox, R.L.: Optimization Methods for Engineering Design. 38-116, Massachusetts, Addison-Wesley Publishing Co., 1971.

Computer Simulation of Duct Noise Control by the Boundary Element Method

Masa. TANAKA

Department of Mechanical Systems Engineering
Faculty of Engineering, Shinshu University
500 Wakasato, Nagano 380, Japan

Y. YAMADA and M. SHIROTORI

Graduate School of Shinshu University

Summary

This paper is concerned with a computer simulation for active control of duct noise by using the boundary element method available for analyzing three-dimensional acoustic field problems. The active noise control under consideration is reduced to an optimum problem to find an optimal set of parameters defining the vibrating state of a secondary noise source to be attached. A computer simulation system is developed, and computation is carried out for typical examples in which the duct is embodied in the infinite plane and the noise through the duct is radiated to the semi-infinite acoustic field. Then, an extension of the developed system is made to the noise source modeling.

Introduction

Control and reduction of noise are very important in engineering, and there are available in the literature a number of investigations. The so-called active noise control, a noise reduction technique by means of the mutual interference of sound waves, has been increasingly attracting attention[1-3]. In particular, the active control of duct noise has been so far studied mainly from the experimental view point.

The authors[4,5] previously treated the noise source identification and the active noise control as an acoustic inverse problem and proposed a procedure in which the boundary element method was combined with the standard optimization technique. A general-purpose computer simulation system has been developed for the active control of noise[5]. In this study, the active control of duct noise is investigated. Computer simulation is carried out for the actual noise consisting of a wide frequency range. Finally, the simulation system developed is extended to a noise source modeling. Numerical results obtained are discussed, whereby the usefulness of the developed simulation system is demonstrated.

Computer Simulation of Active Noise Control

Application of BEM to Acoustic Direct Problems

Under the assumption of a steady state vibration with small amplitude, the acoustic problem is governed by the Helmholtz equation. The corresponding boundary integral equation can be given as follows[6,7]:

$$c(y)p(y) + \int_S q^*(x,y)p(x)dS(x) = -j\omega\rho \int_S p^*(x,y)v(x)dS(x) \tag{1}$$

where $j=\sqrt{-1}$, ω is the angular frequency and ρ is the mass density of the medium. In addition, $c(y)$ is the coefficient depending only on the geometrical property of the boundary surface at point y, $p(x)$ is the sound pressure, and $v(x)$ is the particle velocity, while $p^*(x,y)$ and $q^*(x,y)$ are the well-known fundamental solutions for the Helmholtz equation.

After calculating the sound pressure and the particle velocity on the boundary surface, we can compute the sound pressure at an internal point y by using the following equation:

$$p(y) = -\int_S q^*(x,y)p(x)dS(x) - j\omega\rho \int_S p^*(x,y)v(x)dS(x) \tag{2}$$

Discretization of the boundary integral equation (1) and the integral equation (2) by means of the standard boundary element method can lead to the following system of equations expressed in the matrix form[8]:

$$[H]\{p\}_S = [G]\{v\}_S \tag{3}$$

$$p_V = -[A]\{p\}_S + [B]\{v\}_S \tag{4}$$

where the coefficient matrices $[H]$, $[G]$, $[A]$ and $[B]$ can be calculated from the fundamental solutions, and hence they are all known. The subscripts S and V are used to denote the values on the boundary and in the domain, respectively. Solving equation (3) under the given boundary conditions, we can obtain the nodal values of sound pressure p and particle velocity v on the boundary, and then from equation (4) we compute the sound pressure at an arbitrary internal point in the acoustic field.

Application of Optimization Technique

In this study, the active control of duct noise is treated as an optimum problem in which the boundary element method is used as the central computation technique. The objective function is defined as the sum of absolute values of sound pressures at selected measuring points located at the duct exit. Then, the optimum set of parameters defining the vibrating state of a secondary noise source is searched so that the value of objective function should be minimal. The non-dimensional objective function R can be defined as follows:

$$R = \frac{1}{N}\sum_{i=1}^{N}\left\{\frac{|p_i|}{|p_{0i}|}\right\} \tag{5}$$

where N is the number of measuring points of sound pressure at the duct exit, and p_{0i} is the sound pressure value computed at an ith measuring point where only the primary noise source vibrates. As the convergence criterion of iterative computation, we may use either of the following two expressions:

$$| R^{k} - R^{k-1} | < \varepsilon \ , \quad \sum_{i=1}^{M} | x_i^{k} - x_i^{k-1} | < \eta \tag{6}$$

where k is the number of iterations, and M is the number of parameters, while ε and η are preassigned small positive numbers.

In this study, the conjugate gradient method[9] is employed as an optimization technique. The main flow of the present solution procedure is illustrated in Fig.1.

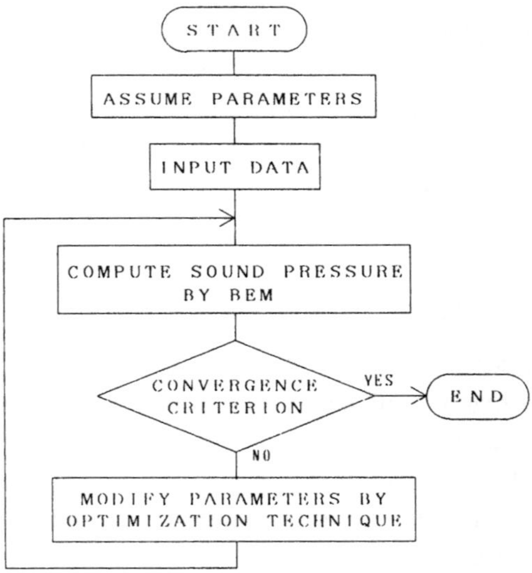

Fig.1. Main flow of solution procedure

Numerical Results and Discussion

We now show the numerical results obtained by carrying out computer simulation of the active noise control in duct by using the computational software developed in this study. It is assumed that a duct with square cross section is embodied in the infinite plane and the noise through the duct is radiated to the semi-infinite three-dimensional acoustic field as shown in Fig.2. We try to cancel this noise at the duct exit by attaching a secondary noise source on the center of duct base. Twenty five measuring points are arranged at the duct exit as shown in Fig.2 where only fifteen points are shown because of symmetry.

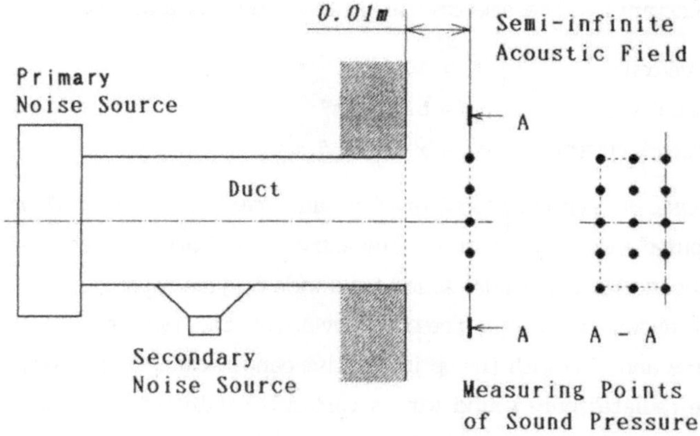

Fig. 2. Duct model

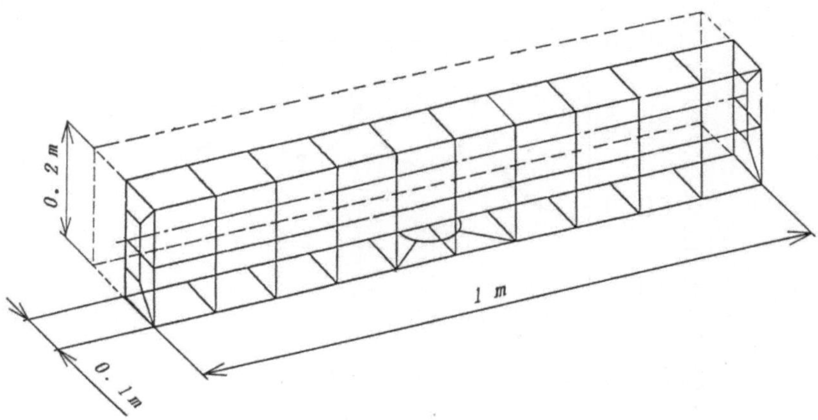

Fig. 3. Boundary element discretization for internal surface of duct with secondary noise source

It is assumed that both the primary and secondary noise sources are excited by uniform pressures with the same frequency and are in a steady state. We further assume that all information on the primary noise source is known and that concerning the secondary noise source only the vibrating state is not known. Under these assumptions the present example can be reduced into an optimum problem to find the optimal set of parameters defining the real and imaginary parts of sound pressure of the secondary noise source.

The secondary noise source is assumed to be circular with the radius *0.05m* and attached on the center of duct base. Because of symmetry only the half of duct surface is discretized into *8*-node isoparametric boundary elements as shown in Fig. 3.

The material constants and parameters used in the computation are assumed as follows:

sound velocity : C_0 = $340\ m/s$
mass density of air : ρ = $1.2\ kg/m^3$
convergence criterion : ε = η = 10^{-5}

Actual noise, in general, consists of a wide frequency range. In order to apply the developed computer simulation system to the active cancellation of such real noise, we carry out the optimal cancellation of pure sound for a wide frequency range.

Figure 4 shows the sound pressure levels at the measuring points without any secondary source and also with the optimal active cancellation. It has been assumed that the primary source radiates pure sound with a particular value of frequency under a uniform sound pressure with the strength *75dB* in the sound pressure level. From Fig.4, it can be seen that a very effective noise reduction can be obtained by the active noise control.

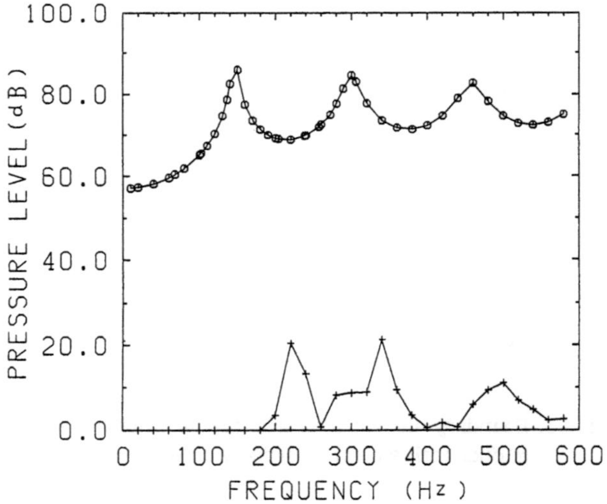

Fig.4. Results obtained for noise with various values of frequency

In Fig.5 is shown the energy ratio of noises radiated from the secondary source to that of the primary source in the optimal noise cancellation. The energy of noise source is defined as the scalar product of the sound pressure and particle velocity multiplied with the area of the noise source. It is clear from Fig.5 that, for the noise with relatively low frequency up to *250Hz*, an effective noise cancellation can be realized by giving the energy of the secondary noise source which corresponds to *1/10* of the primary noise source. However, in the range near the frequency *340Hz*, the energy which is much larger than that of the primary noise

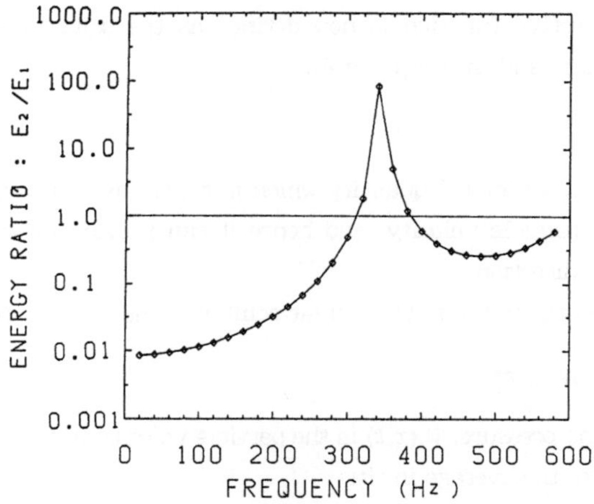

Fig.5. Energy ratio of secondary (E_2) to primary (E_1) noise sources

source is required to realize the optimal noise cancellation. Naturally, it is very effective and useful in practical applications that the noise radiated from the primary source is reduced by a lower energy of the secondary source. In this respect, the results obtained in the vicinity of *340Hz* can not be recommended for practical use.

It is concluded in [5] that the above-mentioned results at *340Hz* occur because the wave length of sound with *340Hz* is approximately equal to the length of the duct used in the present example and hence the position of the secondary noise source coincides with the node of particle velocity. In such a case, an active noise cancellation needs a large amount of energy supplied at this secondary source. By moving the secondary source near the primary source, more effective noise cancellation can be realized; the energy of the secondary source is only *30%* of the primary source energy. It is interesting to note that the active noise cancellation can be performed after a suitable location of the secondary noise source is searched by using the present computer simulation system.

It should be noted that using the present computer simulation system we can estimate the optimal location as well as the optimal shape of the secondary source. Such application is left for future research work, since it needs further study and also consumes much computational time.

Extension of Simulation System to Noise Source Modeling

Now, we shall apply the simulation system developed above to the noise source modeling. It is assumed that the acoustic intensity (*AI*) is given as the reference data at some points in the acoustic field. The objective function is now defined as the square sum of the residuals between the reference data and the computed *AI*.

Acoustic Intensity

The acoustic intensity is a vectoral quantity which is a time-averaged value of the product of sound pressure and particle velocity, and hence it can provide information on both the sound intensity and its direction.

The acoustic intensity in the i-direction at point x in the acoustic field is given by[10]

$$I_{ci}(x) = \overline{P(x,t)V_i(x,t)} \tag{7}$$

where $P(x,t)$ is the sound pressure, $V_i(x,t)$ is the particle velocity in the i-direction, and the superimposed bar denotes the average in time.

Now, we assume that the sound pressure at point x can be expressed in terms of the time harmonic function as follows:

$$P(x,t) = p(x)\exp(j\omega t) \tag{8}$$

Then, evaluation of the time average in equation (7) yields the following expression of the acoustic intensity:

$$\begin{aligned} I_{ci} &= \frac{1}{2} p_v \tilde{v}_{vi} \\ &= I_i(x) + j\,Q_i(x) \end{aligned} \tag{9}$$

where the superimposed tilde denotes the conjugate complex number. In equation (9) $I_i(x)$ is the active intensity usually used, and $Q_i(x)$ is the reactive intensity.

Application of Optimization Technique

The noise source modeling is cast into an optimum problem in a similar way to that mentioned above. It is assumed that the location and shape of the noise source are known and only its vibrating state is not known. The parameters defining the vibrating state of the noise source are first assumed, and then modified by minimizing the objective function. The non-dimensional objective function R in the noise source modeling is defined as follows:

$$R = \log_{10} \left\{ \frac{\sum_{n=1}^{N} \sum_{i=1}^{3} \{ (I_{ni} - I_{oni})^2 + (Q_{ni} - Q_{oni})^2 \}}{\sum_{n=1}^{N} \sum_{i=1}^{3} \{ I_{oni}^2 + Q_{oni}^2 \}} \right\} \tag{10}$$

where N is the number of measuring points, and I_{oni} and Q_{oni} are the reference data of *AI*.

As the convergence criterion, equation (6) is again used and the material constants and the parameters are assumed as the same as in the previous chapter.

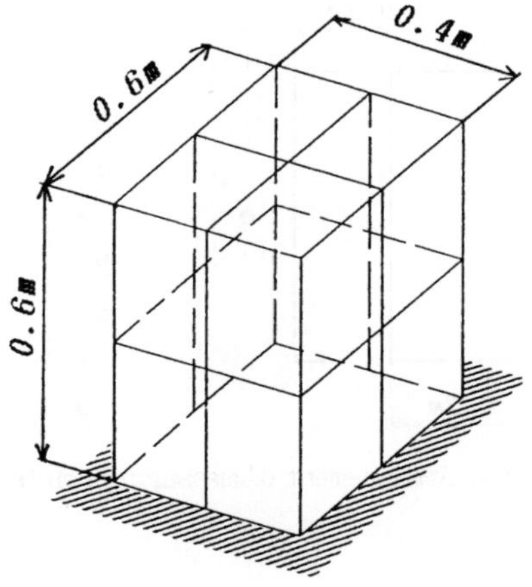

Fig.6. Boundary element discretization for noise source model

Numerical Results and Discussion

It is assumed that the parallelopiped noise source is placed on the infinite plane and a pure sound with *100Hz* is radiated to semi-infinite three-dimensional acoustic field. The noise source surface is discretized into 8-node isoparametric boundary elements as shown in Fig.6. It is assumed that the semi-infinite plane is rigid and that each surface of the noise source vibrates with an identical phase. It is further assumed that the particle velocity V_s of each surface is expressed such that

$$V_s(x_1,x_2) = A_m \cos\left(\frac{x_1}{L_1}\pi\right) \cos\left(\frac{x_2}{L_2}\pi\right) \tag{11}$$

where L_1 and L_2 are the side lengths of the noise source, x_1 and x_2 are rectangular coordinates on each noise source surface. The coefficient A_m of a complex number in equation (11) for each surface is treated as the unknown parameter. In the present numerical experiment, the *AI* values obtained by BEM analysis assuming $A_m=0.05$ for the square surface and $A_m=0.03$ for the rectangular surface are used as the reference data.

Five measuring points are assumed to be on the plane with the distance of *0.1m* from each vibrating surface as shown in Fig.7.

Here we shall consider the two cases: In the first case all the directional components of *AI* are used as the reference data, and in the second case only the *AI* components normal to the noise source surface are used.

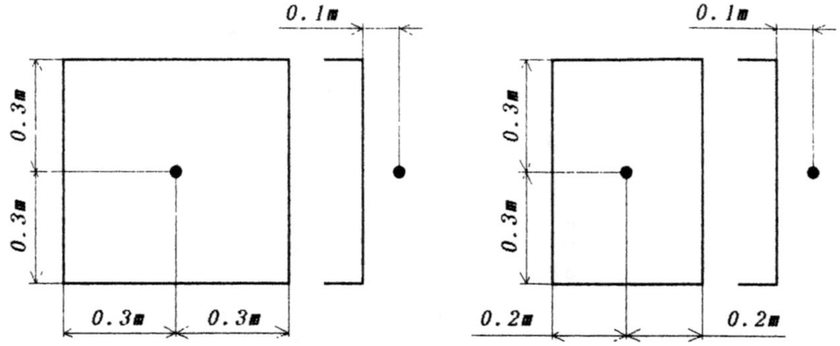

Fig.7. Arrangement of measuring points

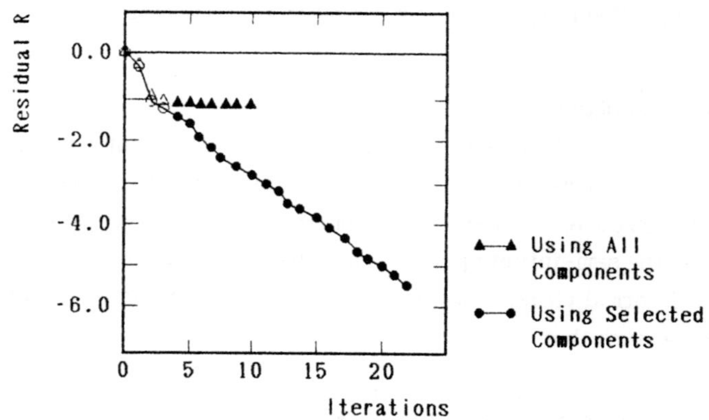

Fig.8. convergence with respect to non-dimensional residuals

Figure 8 shows the convergence property with respect to the non-dimensional residual. In the first case, the converged value of the non-dimensional residual is not acceptable. In the second case, however, the optimal set is obtained after 22 iterations. This implies that attention should be paid to selecting the reference data. If the sensitive reference data are correctly used, successful noise source modeling can be made.

It is interesting to note that once the noise source modeling has been done, computer simulation can be successfully applied to the active noise cancellation as described in the previous chapter or to the far-field estimation of the acoustic field.

Conclusion

In the present paper, the computer simulation system using the BEM software and the standard optimization technique has been developed and applied to active cancellation of duct noise radiated to the semi-infinite three-dimensional acoustic field. It is revealed that the duct noise can be drastically reduced by the active cancellation attaching an optimal secondary noise source. Furthermore, an extension of the simulation system was made to the noise source modeling.

The present computer simulation system can be applied to an arbitrary three-dimensional acoustic field and various acoustic inverse problems. It can be recommended as future research work to apply this system to the practical acoustic inverse problems.

References

1. Kido, K.; Kanai, H.; Abe, M.: Active reduction of noise by additional noise source and its limit. ASME J. of Vibration, Acoustics, Stress, and Reliability in Design, Vol.111 (1989), 480-485.
2. Molo, C.G.; Bernhard, R.J.: Generalized method of predicting optimal performance of active noise controllers. AIAA J., Vol.27 (1989), 1473-1478.
3. Warner, J.V.; Bernhard, R.J.: Digital control of local sound fields in an aircraft passenger compartment. AIAA J., Vol.28 (1990), 284-289.
4. Tanaka, M.; Yazaki, S.; Yamada, Y.: Noise source identification by using the boundary element method. Advances in Boundary Element Methods in Japan and USA, Tanaka, M.; Brebbia, C.A.; Shaw, R. (eds.), 335-349, Southampton, Boston, Computational Mechanics Publications, 1990.
5. Tanaka, M.; Yamada, Y.; Shirotori, M.: Computer simulation of active noise control by the boundary element method. Boundary Elements XII, Vol.2 (1990), 147-158, Tanaka, M.; Brebbia, C.A.; Honma, T. (eds.), Berlin, Heidelberg, New York, Springer-Verlag.
6. Schenck, H.A.: Improved integral formulation for acoustic radiation problems. J. Acoust. Soc. Amer., Vol.44 (1968), 41-58.
7. Seybert, A.F.; Cheng, C.Y.R.: Application of the boundary element method to acoustic cavity response and muffler analysis. ASME J. of Vibration, Acoustics, Stress, and Reliability in Design, Vol.109 (1987), 15-21.
8. Tanaka, M.; Masuda, Y.: A general purpose computer code for acoustic problems, (in Japanese). Trans. Japan Soc. Mech. Engrs., Ser.C, Vol.53 (1987), 387-391.
9. Fox, R.L.: Optimization Methods for Engineering Design. 38-116, Massachusetts, Addison-Wesley Publishing Co., 1971.
10. Hidaka, Y.; Ankyu, H.; Tachibana, H.: Sound field analyses by complex sound intensity, (in Japanese). J. Acoust. Soc. Japan, Vol.43 (1971), 994-1000.

Boundary Element Analysis of Non-Linear Liquid Motion in Two-Dimensional Containers

N.TOSAKA and R.SUGINO

Department of Mathematical Engineering, College of Industrial Technology, Nihon University, Narashino, Chiba, 275, Japan.

Summary

The nonlinear behavior of an incompressible and frictionless liquid with a free surface in two-dimensional containers with various kinds of shapes is analyzed numerically by the boundary element method based on the mixed Eulerian-Lagrangian procedure. The boundary-value problem of Iaplace equation described with the Euler coordinates is solved by means of the direct boundary element method with use of the simplest element (i.e the constant element). On the other hand, the initial-value problem of the free surface conditions given as the evolutional form in terms of the Iagrangian coordinates of a fluid particle on the free surface is solved approximately by using the simples time integration scheme (i.e. Euler scheme). In order to keep on stably numerical performance some effective solution procedures, which are a smoothing technique, an adaptive refinement of mesh and a relocation, are introduced. Applicability and efficiency of the method are examined through sloshing problems in two-dimensional containers subjected to forced acceleration with large amplitude.

Introduction

The complicated phenomenon of a liquid motion in a container which is subjected to forced oscillation is called "sloshing". We have encountered with this phenomenon in various engineering fields, for example, liquid oscillation in a large storage tank or a reservoir due to earthquakes, liquid motion of container of the supertanker caused by swaying and rolling motions during sail, motion of liquid fuel in tanks of air and space crafts[1] and so forth. An estimation of dynamic sloshing loads acted on container has been of even greater concern. The sloshing phenomenon, especially in the case with a large amplitude, can be formulated mathematically as a nonlinear initial-boundary value problem of Laplace's equation in conjunction with a unknown moving boundary called the free surface. This nonlinear problem is one of the difficult mathematical problems to be solved analytically as well as numerically. There exist studies [2]-[7] on the sloshing motions of finite amplitude based on various kinds of numerical solution procedure.

Recently, the boundary element method among various numerical methods has been used widely in solving nonlinear sloshing problems [3]-[7]. Our solution procedure is based on the boundary element method with the mixed Eulerian-Lagrangian approach proposed by Longuet-Higgins and Cokelet[8]. The boundary-value problem of Laplace's equation described with the Euler coordinates can be solved by means of the direct boundary element method. On the other hand the initial-value problem of the free surface conditions described with the Lagrangian coordinates of a fluid particle on free surface is solved approximately by using the time integration scheme. A solution scheme adopted in our study is most elementary one in which we use the constant element in boundary element method and the Euler scheme in the time integration. However, in order to eliminate certain unwanted instabilities in numerical performance some solution procedure, which is composed of a smoothing technique, an adaptive refinement of boundary mesh and a relocation of fluid particle, is proposed.

In this paper, nonlinear sloshing problems in two-dimensional containers with several kinds of shape are analyzed by using the boundary solution procedure. Several numerical computations are carried out in order to verify applicability of the approximate solution procedure developed in this study.

Mathematical Model

We consider the nonlinear motion of a liquid in two-dimensional containers. The fluid occupies the initial region Ω with the boundary $\partial\Omega = \Gamma f \cup \Gamma w$, where Γ_f is the free surface of the container. The unit normal vector drawn outwardly on the boundary is denoted by **n**.

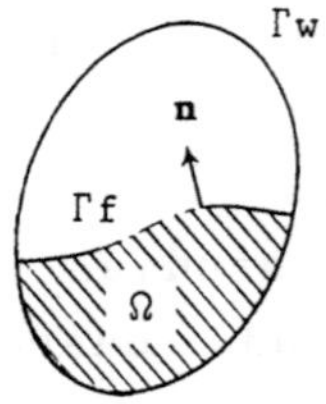

Fig.1 Fluid domain

The liquid contained in a container is assumed to be inviscid and incompressible, and the flow is assumed irrotational. Under these assumptions we can set the following form as a mathematical model of the problem:

$$\mathbf{u} = \nabla\phi \qquad \text{in } \Omega \tag{1}$$

$$\nabla\cdot\mathbf{u} = \nabla^2\phi = \frac{\partial^2\phi}{\partial x^2} + \frac{\partial^2\phi}{\partial y^2} = 0 \qquad \text{in } \Omega \tag{2}$$

$$\frac{\partial\phi}{\partial n} = \mathbf{n}\cdot\nabla\phi \equiv q = 0 \qquad \text{on } \Gamma w \tag{3}$$

$$\frac{D\xi}{Dt} = u = \frac{\partial\phi}{\partial x} \qquad \text{on } \Gamma f \tag{4}$$

$$\frac{D\eta}{Dt} = v = \frac{\partial\phi}{\partial y} \qquad \text{on } \Gamma f \tag{5}$$

$$\frac{D\phi}{Dt} = \frac{1}{2}(\nabla\phi)^2 - g\eta - a\xi \qquad \text{on } \Gamma f \tag{6}$$

where **u** is the velocity vector, ϕ is the velocity potential, ξ and η are the Lagrangian coordinates for a fluid particle on the free surface, D/Dt denotes the substantial differentiation following a given particle, g is the acceleration of gravity and a is the forced horizontal acceleration applied to the container.

In this study, we consider the mathematical model given by Laplace equations (2) and the boundary conditions (3)-(6) as the coupled problem of the boundary-value problem of Laplace equation with the boundary condition (3) and the initial-value problem of the system of evolutional equations (4),(5) and (6) with appropriate initial conditions.

Approximate Solution Method

Let us explain our effective approximate solution method[6] to solve the above mentioned coupled problems.

(i) Boundary element method for boundary-value problem
By using Green's theorem, we can easily transform the Laplace equation(2) as the governing field equation into the following well-known boundary integral equation:

$$\frac{\alpha}{2\pi}\phi = \int_{\Gamma_f}\frac{\partial\phi}{\partial n}\,\omega^* d\Gamma - \int_{\Gamma_w}\phi\,\frac{\partial\omega^*}{\partial n}d\Gamma \tag{7}$$

in which we take the Neumann condition (3) and α denotes the angle between two tangents at a boundary point. The fundamental solution ω^* in (7) is given in terms of the distance $r=\|\boldsymbol{x}-\boldsymbol{y}\|$ by

$$\omega^* = \frac{1}{2\pi}ln\frac{1}{r} \tag{8}$$

If the velocity potential on the free surface is given, then we can determine the value of ϕ on Γw and its derivative q on Γf from the above boundary integral equation. In this paper, we introduce the simplest element (i.e., constant element) in solving approximately (7) by use of the boundary element method.

(ii) Euler time integration scheme for initial-value problem
We can rewrite the kinematical and dynamical conditions into the following evolutional form in terms of the rate of change of the quantities for fluid particles on the free surface :

$$\frac{D}{Dt}\begin{bmatrix}\xi\\ \eta\\ \phi\end{bmatrix} = \begin{bmatrix}\dfrac{\partial\phi}{\partial x}\\ \dfrac{\partial\phi}{\partial y}\\ \frac{1}{2}\{(\frac{\partial\phi}{\partial x})^2+(\frac{\partial\phi}{\partial y})^2\}-g\eta-a\xi\end{bmatrix} \tag{9}$$

This system (9) together with initial conditions may be regarded as the initial-value problem of the first-order ordinary differential equations in time variable t. Although there exist many kinds of time integration schemes to solve the initial-value problem, we adopt the following Euler scheme known as the simplest time integration scheme:

$$\left.\begin{aligned} \xi^{k+1} &= \xi^{k} + \Delta t\left(\frac{\partial\phi}{\partial x}\right)^{k} \\ \eta^{k+1} &= \eta^{k} + \Delta t\left(\frac{\partial\phi}{\partial y}\right)^{k} \\ \phi^{k+1} &= \phi^{k} + \Delta t\left[\frac{1}{2}\left\{\left(\frac{\partial\phi}{\partial x}\right)^{2}+\left(\frac{\partial\phi}{\partial y}\right)^{2}\right\} - g\eta^{k} - a\xi^{k}\right] \end{aligned}\right\} \tag{10}$$

where Δt denotes the short time interval and the superscript " k " indicates the k-th time step. From the above time integration scheme we can determine the time-dependant configuration of the free surface.

Numerical Solution Procedure

In our numerical performance, a saw-toothed wave profile, which was pointed in Longuet-Higgins and Cokelt[8], has also appeared. We cannot continue time integration on account of this phenomenon. Therefore we must introduce effective numerical solution procedures to remove the so-called numerical instability. In this study we adopt the following three effective procedures.

(i) Smoothig technique

In order to smooth a saw-toothed wave profile, we take also the following 5-points smoothing formula shown in [2]:

$$\overline{f}_j = \frac{1}{16}(-f_{j-2}+4f_{j-1}+10f_j+4f_{j+1}-f_{j+2}) \tag{13}$$

This formula is applied for the values of ξ , η and ϕ determined with the time integration scheme at every 30 time-steps.

(ii) Check of accuracy and adaptive mesh

The accuracy of numerical solution generated by our approximation procedure for the problem is checked by calculating, at each time-step, the volume of fluid occupied in container. If we find any change(i.e.,increase or decrease) of the volume at some time-step, we introduce the idea of adaptive refinement of boundary mesh in numerical computation to avoid unreal change of volume.

(iii) Relocation technique

Next, we introduce a relocation technique to remove the failure of our numerical calculation which is caused by concentration of the fluid particle. If we find a very smaller segment than an average-sized segment of the part on free surface, then fluid particles on the edge of segment are relocated to the suitable new positions.

Numerical Examples

In order to examine applicability of our approximation procedure developed in this study. We show several results of numerical examples. In this study we consider the containers subjected to the forced horizontal acceleration given by $a(t)=A\omega^2\sin(\Delta t)$ $(t\geq 0)$ in which A is an amplitude and ω is an angular frequency.

(A) Circular container

First of all, let us show the numerical results of sloshing phenomena in a circular container. In our previous paper[6], we showed numerical results of sloshing problem in circular container subjected to a comparatively small amplitude A=1.0(cm). So we wish to give a few results for the case of large amplitude. In Fig.2(a) we show the profiles of free surface at each time-step in the case of R=0.5(m). And container is filled up 50% with the liquid. Fig.2(b) shows the results of container filled up 12.5% with the liquid.

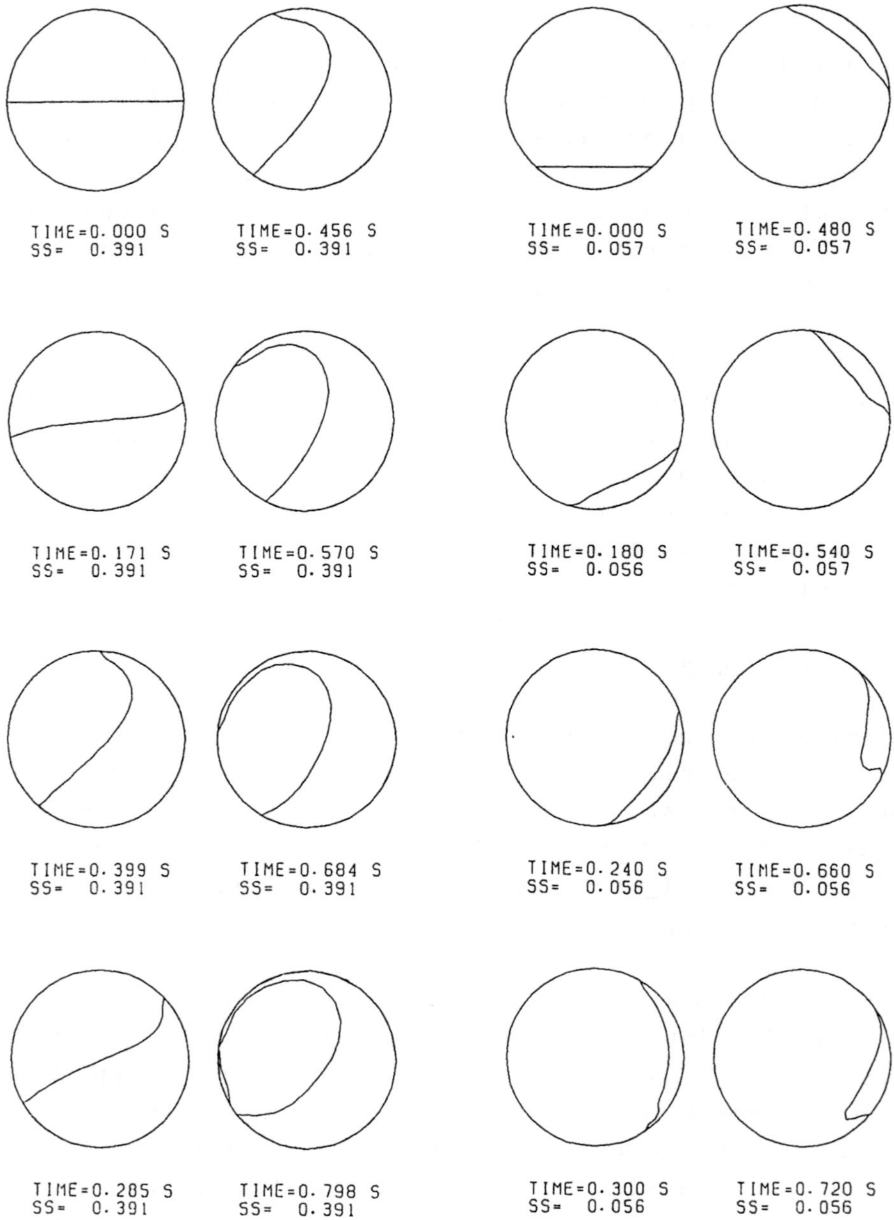

(a) A=0.5(m),ω=5.0(rad/sec) (b) A=2.0(m),ω=5.0(rad/sec)

Fig.2 Large amplitude slosh in circular container

(B) Elliptic container

Next, we show the results in two types of elliptic container. This elliptic container has a long radius a=0.5(m) and a short radius b=0.3(m). Fig.3 shows the profiles of free surface at each time-step in the case of container filled up 25% with the liquid. Next, we indicate in Fig.4 the results in the case of second type elliptic container filled up 25% with the liquid.

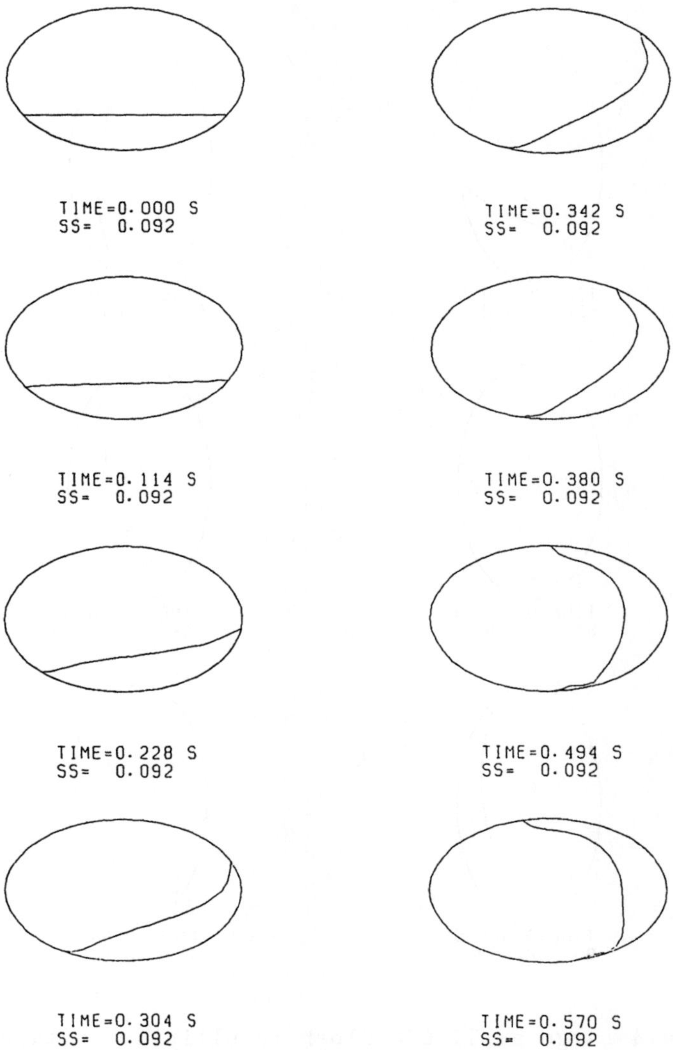

Fig.3 Large amplitude slosh in elliptic container
(A=0.5 m, ω=5.0 rad/sec)

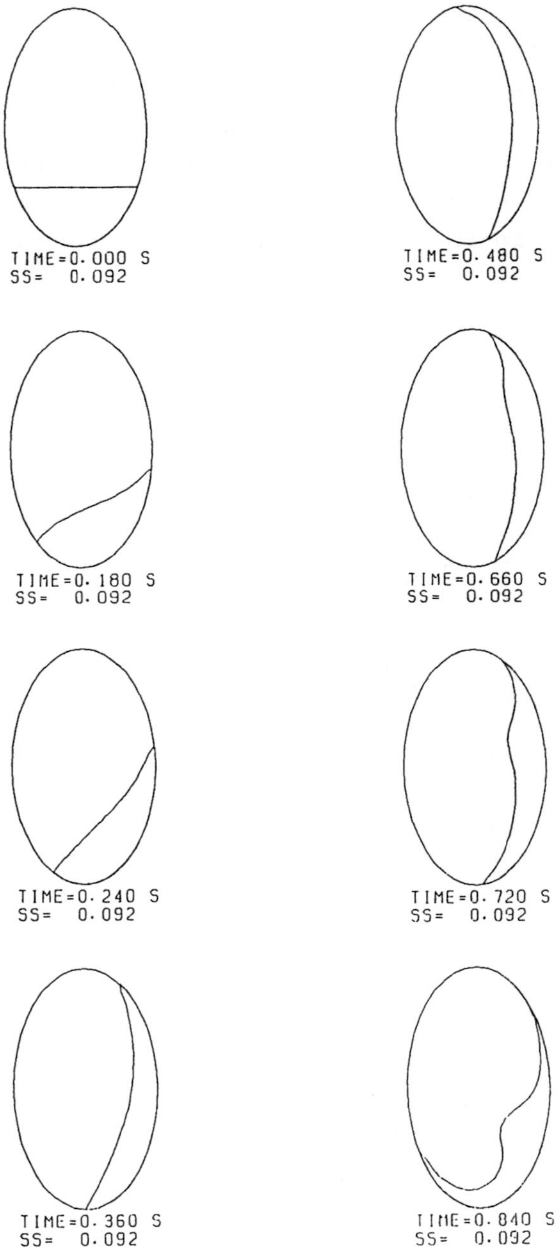

Fig.4 Large amplitude slosh in elliptic container
(A=1.0 m, ω=5.0 rad/sec)

Conclusions

A numerical approximation procedure has been developed in order to simulate large-amplitude liquid motion in two-dimensional containers. The procedure, which is based on the mixed Eulerian-Lagrangian description of a mathematical model of the problem, consists of the boundary element method and the Euler time integration scheme in conjunction with a smoothig technique, adaptive refinement of element mesh and relocation of fluid particle. The solution procedure has been applied to sloshing phenomena in a circular container and an elliptical one. As a consequence of this study the complicated profiles of a free surface subjected to forced acceleration with large amplitude can be simulated numerically.

References

1. Berry,R.L.,L.J.Demchak,J.R.Tegart & M.K.Craig:An analytical tool for simulating large amplitude propellant slosh,AIAA Paper No.81-0500 (1981)
2. Ikegawa,M. : Finite element analysis of fluid motion in a container , Finite Element Methods in Flow Problems (1974) (Eds.J.T.Oden et al), UAH Press,Huntsville,Alabama, 855-860
3. Nakayama,T. & K.Washizu : Nonlinear analysis of liquid motion in a container subjected to forced pitching oscillation, Int.J.Num.Mech.Engng., 15(1980) 1207-1220
4. Nakayama,T. & Washizu,K. : The boundary element method applied to the analysis of two-dimensional nonlinear sloshing problems,Int.J.Num.Mech Engng., 17(1981) 1631-1646
5. Sugino,R. & N,Tosaka :Solution procedure for nonlinear free surface problems by boundary element approach , Theoretical and Applied Mechanics, 38(1989), University of Tokyo Press, 53-59
6. Tosaka,N.,R.Sugino & H.Kawabata:Nonlinear free surface flow problems boundary element-lagrangian solution procedure, Boundary Element Methods,Principles and Applications, Proc. 3th. Japan-China Symposium on Boundary Element Methods., Pergamon Press(1990) 237-246
7. Sugino,R. & N.Tosaka : Large amplitude sloshing analysis in a container with multi-slopped wall by boundary element method , Advances in Boundary Element Methods in Japan-USA (Eds.M.Tanaka,C.A.Brebbia and R.shaw),CMP,(1990) 307-316
8. Longuet-Higgins,M.S & E.D.Cokelet :The deformation of steep surface waves on water I.A numerical method of computation, Proc.R.Soc.Lond.A.350(1976) 1-26

A Combined Finite Element Boundary Element Approach for Elasto-Plastic Analysis

J. L. WEARING, M. A. SHEIKH and M. C. BURSTOW

Department of Mechanical and Process Engineering,
University of Sheffield, U.K.

Summary

A combined Finite Element-Boundary Element approach for elasto-plastic analysis of solids which conform to plane stress or plane strain conditions is presented. Here, a given problem domain is discretised by Finite Element and Boundary Element regions. It is assumed that the response of the Boundary Element regions remains elastic throughout the analysis whilst the elasto-plastic response is captured by the Finite Element regions. Coupling of the two systems of equations is achieved by treating the boundary regions as Finite Element substructures.

Introduction

The Finite Element Method (FEM) is now widely used by industry for the stress analysis of a wide range of components and structures. However the success of the method and the increased power and speed of the current generation of digital computers has led to an ever increasing demand for the analysis of complex three dimensional components. In many cases the Design Engineer is able to use simplified two dimensional models to obtain satisfactory results for such problems. There are many situations, however, where the complexity of the geometry and the type of information required necessitates the use of full three dimensional Finite Element models which are very time consuming both from the modelling and the computational points of view.

Boundary Domain Techniques in which the degrees of freedom during the solution phase are confined to the boundary of the problem, have the potential of improved efficiency at the computational phase and offer the additional benefit of simpler initial models compared with the Finite Element Method. There are however problems which can benefit from part of the domain being modelled by the Finite Element Method (FEM) and the remainder by the Boundary Element Method (BEM). For example in problems involving plasticity, creep or fracture it would be advantageous to model these regions using Finite Elements and the rest of the domain using a Boundary Element Technique [1].

The coupling of the Finite Element Method and the Boundary Element Method has been achieved by using various techniques - [2], [3], [4]. Kelly et al [5] have described various ways of linking Direct and Indirect formulations of the Boundary Element Method (BEM) with the Finite Element Method (FEM). In this paper, a combined approach linking an Indirect Discrete Boundary Method - (IDBM) [6], is discussed for the analysis of planar elasto-plastic problems. The Indirect Discrete Boundary Method (IDBM), which is based on the Indirect Boundary Element Method (IBEM) formulation [7], greatly simplifies the numerical calculation by eliminating the integrations normally associated with Boundary Element Methods and by avoiding the singularities of the fundamental solution of the problem [8]. Its usual derivation gives rise to a system of equations which relate boundary nodal values of displacements and tractions. A technique has been developed which modifies the Indirect Discrete Boundary Method and produces a final system of equations which is analagous to that of the Finite Element Method [9]. It takes the basic energy functional approach used by the Finite Element Method and transforms the energy equations into boundary integral form which are solved by using a boundary technique. The equations for the boundary region are then used by a Finite Element program as a substructure or superelement [10].

Results from two problems of two dimensional elasto-plastic analyses are presented. The plastic regions are modelled with the Finite Element Method and the remainder using the Boundary Element Method.

System of Equations for Boundary Regions

The Indirect Boundary Element Method for linear elastic problems is based on the use of the fundamental (Kelvin) solution which satisfies the governing equilibrium (Navier's) equation [11]. For a homogeneous isotropic domain, this solution relates the displacement field to a fictitious force distribution acting on the boundary of the domain as:

$$u_i(p) = G_{ij}(p,q)\ S_j(q) \qquad (1)$$

Similarly, the surface tractions are given as:

$$t_i(p) = H_{ij}(p,q)\ S_j(q) \qquad (2)$$

In the Indirect Discrete Boundary Method (IDBM), boundary collocation is performed on equations (1) and (2) which result in two matrix equations:

$$\{u\} = [G]\ \{S\} \tag{3}$$

$$\{t\} = [H]\ \{S\} \tag{4}$$

Equations (3) and (4) can be used to make the IDBM compatible with the FEM [12]. From equations (3) and (4):

$$\{t\} = [H]\ [G]^{-1}\ \{u\} \tag{5}$$

Equation (5) can be used in the expression for the total potential energy which is minimised to give [9],

$$\int_\Gamma\ [H]\ [G]^{-1}\ \{u\}\ d\Gamma = \int_\Gamma\ \{p\}\ d\Gamma \tag{6}$$

where {p} are the external boundary loads. Equation (9) can be written in matrix form as:

$$[K]\ \{u\} = \{P\} \tag{7}$$

where {P} is the nodal force vector and [K} is the stiffness matrix given as:

$$[K] = \int_\Gamma\ [H]\ [G]^{-1}\ d\Gamma \tag{8}$$

A system of equations similar to that given by equation (7) is formed for each boundary region. These are then assembled to form an overall system for the boundary region as:

$$[K]_{b.e}\ \{u\} = \{P\}_{b.e} \tag{9}$$

Finite Element Equations

For a discretised system, the Finite Element Method produces a global system of equations:

$$[K]_{f.e}\ \{u\} = \{P\}_{f.e} \tag{10}$$

where the stiffness matrix $[K]_{f.e}$ is given in terms of the strain-displacement matrix [B] and the stress-strain matrix [D] as:

$$[K]_{f.e} = \int_\Omega\ [B]^T\ [D]\ [B]\ d\Omega \tag{11}$$

Global System of Equations

Equation systems (12) and (13) can be merged to produce a global system of equations of the form:

$$[K]\ \{u\} = \{P\} \tag{12}$$

where
$$[K] = [K]_{f.e} + [K]_{b.e} \tag{13}$$

Elasto-Plastic Analysis

Before the onset of any plastic deformation, equation (12) is solved to give the global elastic solution for the problem.

An incremental process is then employed whereby the applied loads are incremented according to the specified load factors. As the elasto-plastic behaviour is confined to the Finite Element regions, only $[K]_{f.e}$ is recalculated in equation (13) after each load increment; $[K]_{b.e}$ remains unaltered. $[K]_{f.e}$ can now be rewritten as:

$$[K]_{f.e} = \int_{\Omega} [B]^T \; [D_{e.p}] \; [B] \; d\Omega \tag{14}$$

where $[D_{e.p}]$ is the elasto-plastic stress-strain matrix, given as:

$$[D_{e.p}] = [D] - \frac{\{d_D\} \; \{d_D\}^T}{H' + \{d_D\}^T \; \{a\}} \tag{15}$$

in which $\{d_D\} = [D] \; \{a\}$ (16)

where $\{a\}$ is the flow vector given as:

$$\{a\}^T = \left(\frac{\partial F}{\partial \sigma}\right) = \left(\frac{\partial F}{\partial \sigma_x} \; , \; \frac{\partial F}{\partial \sigma_y} \; , \; \frac{\partial F}{\partial \tau_{xy}} \; , \; \frac{\partial F}{\partial z}\right) \tag{17}$$

$F(\sigma,K)$ is the yield function; K being the hardening parameter which depends upon the specified yield criterion [13]. Assumption of a work hardening hypothesis and consideration of uniaxial loading conditions results in the scalar term H' given by:

$$H' = \frac{d\sigma}{d\epsilon_p} = \frac{E_T}{1 - E_{T/E}} \tag{18}$$

in which E_T is the elasto-plastic tangent modulus of the uniaxial stress-strain curve; E being the elastic modulus of the material.

For each load increment, the incremental nodal displacements and stresses are calculated. The updated stresses are then brought down to the yield surface and are used to calculate the equivalent nodal forces.

These nodal forces can be compared with the externally applied loads to form a system of residual forces which is brought sufficiently close to zero through an iterative process, before moving on to set the next load increment.

Case Studies

Two problems using the combined Boundary Element-Finite Element Method are presented and the results compared for displacement fields, shape of yield surfaces, and computational time with results obtained using the Finite Element Method.

The first problem is a rectangular plate with a sharp notch, and the second is a rectangular plate with a central hole. These examples are illustrated in figures (1) and (2) respectively. Both plates were assumed to be made of the same material, and the thickness of each plate was sufficient to cause it to behave as a plane strain problem. The material properties were assumed to be:

Youngs Modulus, $E = 7000\ N/mm^2$, Poissons Ratio, $\nu = 0.33$,

Yield Stress, $\sigma_y = 24.3\ N/mm^2$,

Strain Hardening Parameter, $H' = 0.0$ (Perfectly plastic)

1. Notched Plate

Due to symmetry of the problem only one quarter of the plate was analysed and the geometry of the plate and the boundary conditions used in the analysis are shown in figure 1(a). The dimensions of the plate are $w = 10$ mm, $l = 18$ mm, $c = 5$ mm, $a = 45°$. The plate was loaded along its top surface with a uniformly distributed load of $\sigma = 24.3\ N/mm^2$, and seven load increments were applied up to a total load factor of 0.725 times the applied load. The problem was first analysed using two Boundary Element superelements containing a total of 34 three-noded quadratic boundary elements and 24 eight-noded isoparametric finite elements, giving a total of 139 nodes. It was also analysed using 52 eight-noded isoparametric finite elements, with 185 nodes. The combined Boundary Element-Finite Element mesh is illustrated in figure 1(b), and the Finite Element mesh in figure 1(c).

2. Perforated Plate

The geometry of the plate and the boundary conditions used in the analysis are illustrated in figure 2(a), and due to symmetry only one quarter of the plate was analysed. The dimensions of the plate are $w = 10$ mm, $l = 18$ mm, $r = 5$ mm. A distributed load of $\sigma = 24.3\ N/mm^2$ was applied to this top surface of the plate and six load increments were used to take the load factor up to 0.56 times the applied load. The problem was analysed using one Boundary Element superelement containing 16 three-noded quadratic Boundary Elements and 28 eight-noded isoparametric Finite Elements, having a total of 130 nodes. Comparisons were made with a model comprising 42 eight-noded isoparametric Finite Elements, and having a total of 160 nodes. The meshes are illustrated in figures 2(b) and 2(c).

Results

The numerical results showed that for the displacement fields, the results obtained from the combined Boundary Element-Finite Element analysis are within 2% of those obtained from the Finite Element analysis. The shape and rate of growth of the yield surfaces using the combined technique also compared well with results obtained from the Finite Element Method. The growth of the yield surfaces with increase in load factor are shown in figure 1(d) for the notched plate, and in figure 2(d) for the perforated plate. These results compare well with the shape of the yield surfaces obtained by Zienkiewicz [14] for similar problems.

The computer time taken to run each problem for a set of six load increments are presented in Table 1.

Table 1 - Computer Times for Combined and Finite Element Methods

INC.	NOTCHED PLATE			PERFORATED PLATE		
NO.	LOAD FACTOR	TIME (cpu secs) COMB. METHOD	TIME (cpu secs) F.E. METHOD	LOAD FACTOR	TIME (cpu secs) COMB. METHOD	TIME (cpu secs) F.E. METHOD
1	0.3	29	28	0.3	17	17
2	0.4	45	55	0.4	38	45
"	"	"	"	"	"	"
"	"	"	"	"	"	"
"	"	"	"	"	"	"
6	0.7	93	138	0.56	103	133

These results show that a time saving of up to 30% is possible using the combined Boundary Element-Finite Element Method. These time savings result from:

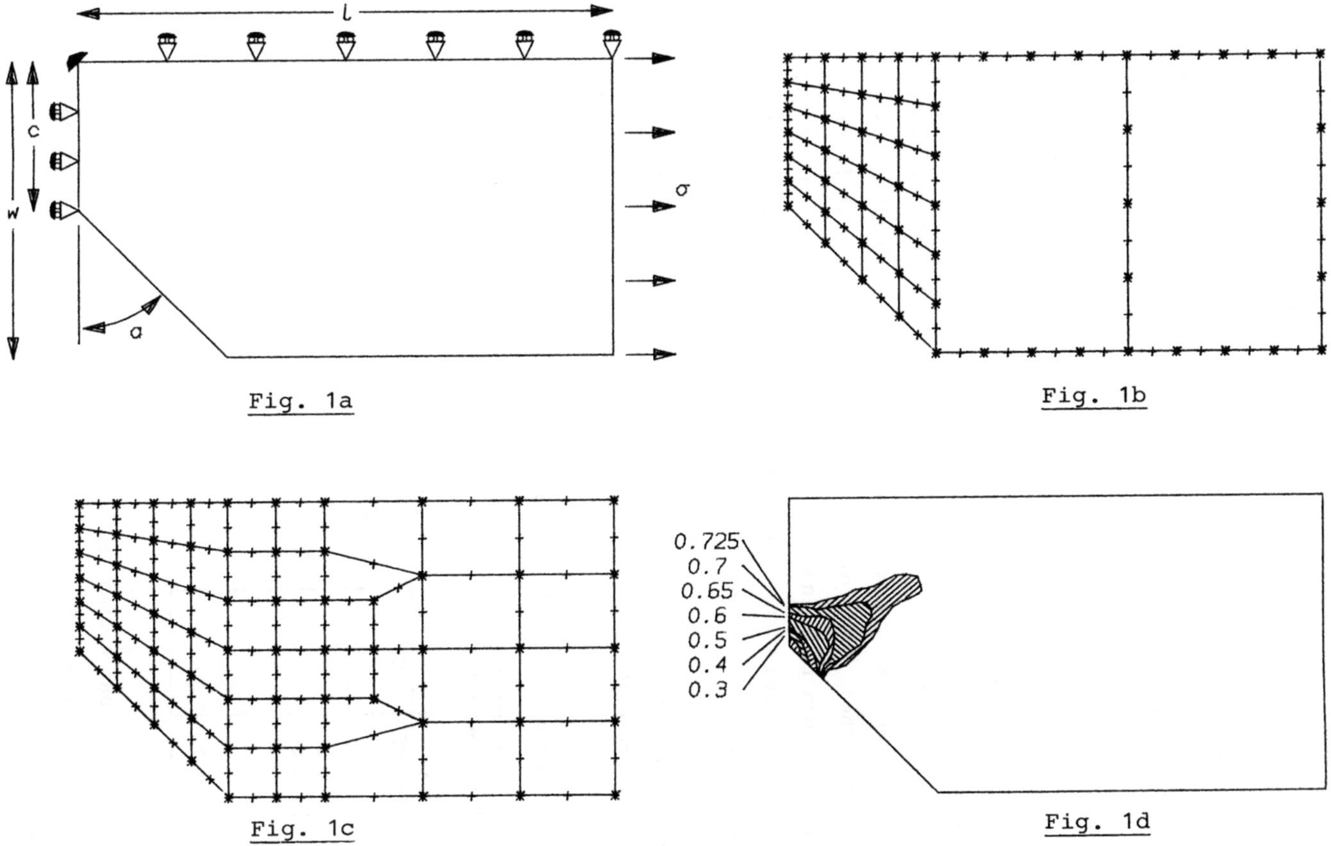

Fig. 1a

Fig. 1b

Fig. 1c

Fig. 1d

FIG. 1 NOTCHED PLATE

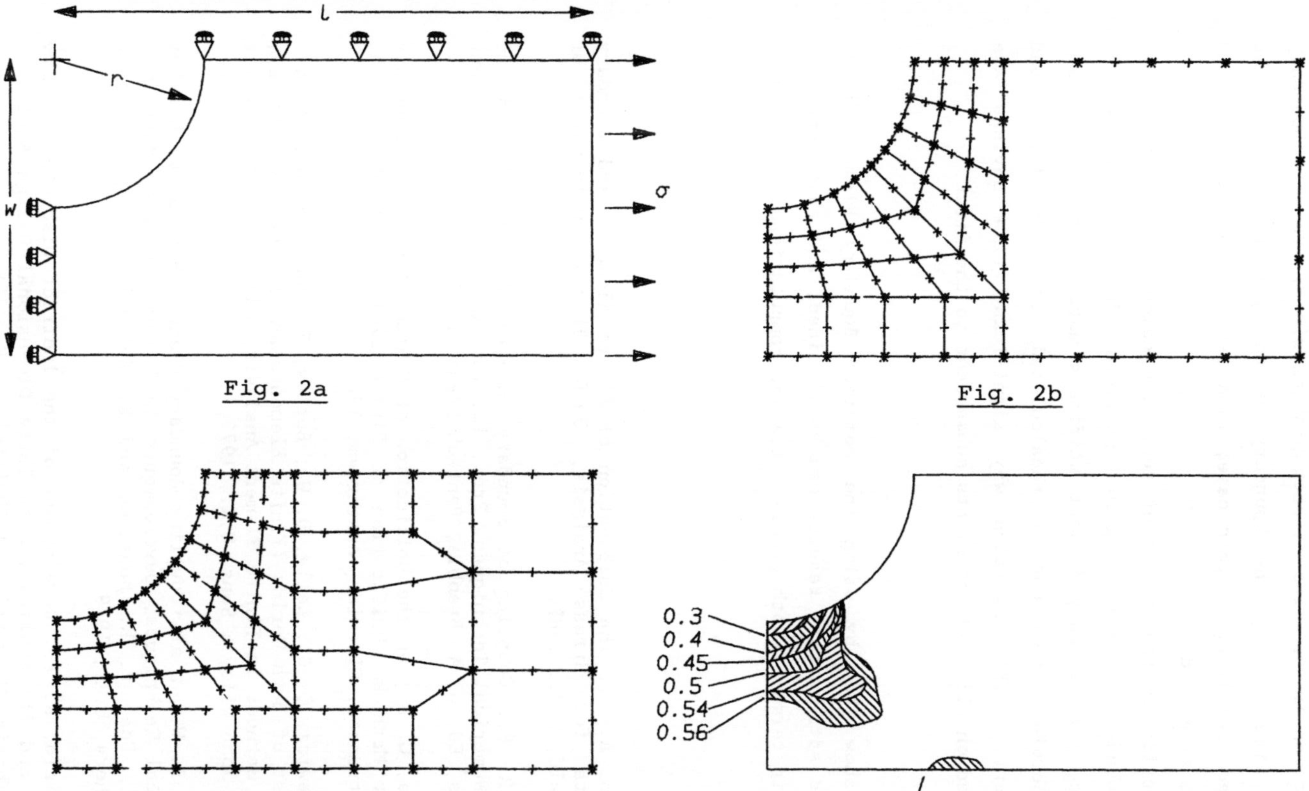

Fig. 2a

Fig. 2b

Fig. 2c

Fig. 2d

FIG. 2 PERFORATED PLATE

1. A reduction in the size of each problem since, over the Boundary Element region only the boundary of the problem needs to be modelled, resulting in this case, in a reduction of up to 25% in problem size, and
2. A reduction in the number of necessary calculations since, over the Boundary Element region, provided the material remains elastic, the Boundary Element stiffness matrix remains the same and therefore need not be recalculated after the first load increment. This explains why significant time savings are achieved on all load increments subsequent to the first.

Conclusions

The results show that by using the combined Boundary Element-Finite Element Method satisfactory results can be obtained, and that considerable savings in terms of both problem size and computational time can be made.

References

1. Hickson, A. J.: The Combination of Finite Elements and Boundary Elements for Stress Analysis, Ph.D. Thesis, University of Sheffield, U.K., 1987.

2. Shaw, R. P.: Coupling of Boundary Integral Equation Methods to Other Numerical Techniques, Proc. 1st Int. Conf. Boundary Element Methods, Ed. C. A. Brebbia, Pentech Press, London, 1978.

3. Brebbia, C. A.: On the Unification of Finite Element and Boundary Element Methods, Unification of Finite Element Methods, Ed. H. Kardestruncer, Elsevier, Amsterdam, 1984.

4. Zienkiewicz, O. C.; Kelly, D. W.; Bettes, P.: Marriage a la Mode - The Best of Both Worlds (Finite Elements and Boundary Integrals), Energy Methods in Finite Element Analysis, Eds. O. C. Zienkiewicz et al, John Wiley & Sons, U.K., 1979.

5. Kelly, D. W. et al: Coupling Boundary Element Methods with Other Numerical Techniques, Developments in Boundary Element Methods - vol. I, Eds. P. K. Banerjee and R. Butterfield, App. Science Publishers, U.K., 1979.

6. Scholfield, R. P.: Development of the Indirect Discrete Boundary Method and its Application to Three Dimensional Design Analysis, Ph.D. Thesis, University of Sheffield, U.K., 1986.

7. Banerjee, P. K.; Butterfield, R.L.: Boundary Element Methods in Engineering Science, McGraw Hill, U.K., 1981.

8. Patterson, C.; Sheikh, M. A.: On the Use of Fundamental Solutions in the Trefftz Method for Potential and Elasticity Problems, Boundary Element Methods in Engineering, Ed. C. A. Brebbia, Springer, Berlin, 1982.

9. Wearing, J. L.; Sheikh, M. A.; Hickson, A. J.: A Combined Finite Element Boundary Element Technique for Stress Analysis, Proc. 10th Int. Conf. Boundary Element Methods, Eds. C. A. Brebbia et al, Springer, Berlin, 1988.

10. MacNeal, R. H.; McCormick, C. W.: Computerized Substructure Analysis, Proc. World Congress Finite Element Methods in Structural Mechanics, 1975.

11. Brebbia, C. A.: The Boundary Element Method for Engineers, Pentech Press, London, 1978.

12. Wearing, J. L.; Sheikh, M. A.: Coupling of Finite Element and Boundary Element Superelement Methods, Proc. Int. Symp. Boundary Element Methods: Advances in Solid and Fluid Mechanics, IABEM, Connecticut, U.S.A., 1989.

13. Owen, D. R. J.; Hinton, E.: Finite Elements in Plasticity: Theory and Practice, Pineridge Press, Swansea, U.K. 1980.

14. Zienkiewicz, O. C.; Valliappan, S.; King, I. P.: Elasto-Plastic Solutions of Engineering Problems 'Initial Stress', Finite Element Approach, Int. J. Num. Meth. in Eng., 1, (1969), 75 - 100.

Boundary Domain Integral Method for the Space Time Dependent Viscous Incompressible Flow

I. ZAGAR, P. SKERGET, A. ALUJEVIC

Faculty of Engineering, University of Maribor, Slovenia

Summary

The partial differential set of equations, governing the laminar or the turbulent motion of viscous incompressible fluid is known as nonlinear Navier-Stokes equations. They constitute the statement of the basic conservation balance of mass, momentum and energy applied to a control volume, i.e. the eulerian description.

The time dependent set of governing equations is handled since it is more stable, and what turns to be very important - the time dependent approach does not presume the existence of steady state solution which may not even exist.

Various approaches exist for the turbulent flow prediction, i.e. full turbulence simulation, large eddy simulation, Reynolds averaged models, etc. The averaged form of the Navier-Stokes equations throught the Reynolds decomposition of instantaneous value of each variable into a time averaged mean value and an instantaneous fluctuation is still the most commonly used approach in the numerical simulation of the turbulence.

The buoyancy force can play an important role in a nonisothermal flow, especially in the case of mixed or pure natural convection. The Boussinesq approximation is used to introduce the buoyancy force.

Boundary-domain integral approach offers some important features. Due to the fundamental solutions, a part of the transport mechanism is transferred to the boundary, producing a very stable and accurate numerical scheme. Diffusion transport part and potential part of the flow is described only by boundary integrals while the convection is described by domain integrals. Boundary vorticity values are included in integral form in kinematic boundary integral formulation, so there is no need to use approximate formulae to determine boundary vorticity conditions. One has to solve implicit systems of equations only for the boundary values, while all internal domain values are computed in an explicit manner.

Existing limitations of boundary-domain integral method in fluid dynamics concerning the computational cost and computer memory demands (matrices are fully populated and non symetric, expensive evaluation of the integrals) can be reduced by using subdomain technique, efficient block solver and clustering. Stability of the method at higher Reynolds numbers can be improved with introducing a part of the convection in to the system matrix.

GOVERNING EQUATIONS

The partial differential equations set, governing the motion of viscous incompressible fluid is known as nonlinear Navier-Stokes equations expressing the basic conservation balance of mass, momentum, and energy.

$$\frac{\partial v_j}{\partial x_j} = 0 \qquad (1)$$

$$\frac{\partial v_i}{\partial t} + \frac{\partial v_j v_i}{\partial x_j} = -\frac{1}{\rho_o}\frac{\partial P}{\partial x_i} + \nu_o \frac{\partial^2 v_i}{\partial x_j \partial x_j} - g_i\, F \qquad (2)$$

Introducing the vorticity vector ω_i and the vector potential Ψ_i of the solenoidal velocity field, the computation of the flow is divided into the kinematics given by the vector Poisson's elliptic equation

$$\frac{\partial^2 \Psi_i}{\partial x_j \partial x_j} + \omega_i = 0 \qquad (3)$$

and into the kinetics described by the vorticity transport equation

$$\frac{\partial \omega_i}{\partial t} + \frac{\partial v_j \omega_i}{\partial x_j} = \frac{\partial \omega_j v_i}{\partial x_j} + \nu_o \frac{\partial^2 \omega_i}{\partial x_j \partial x_j} - \epsilon_{ijk}\, g_k \frac{\partial F}{\partial x_j} \qquad (4)$$

The buoyancy effect is included by Boussinesq approximation and the energy equation is given as follows

$$\frac{\partial T}{\partial t} + \frac{\partial v_j T}{\partial x_j} = a_o \frac{\partial^2 T}{\partial x_j \partial x_j} \qquad (5)$$

The following linear relation between the fluid density and temperature is usually used

$$\frac{\rho - \rho_o}{\rho_o} = F = -\beta\,(T - T_o) \qquad (6)$$

In the region of density anomaly, for example in water, the nonlinear term must be taken into account

$$\frac{\rho - \rho_o}{\rho_o} = F = (0.066576\, T - 0.008322\, T^2)/\rho_o \qquad (7)$$

Direct simulation of turbulence is the solution of the Navier-Stokes equations for the complete details of the turbulent flow. Such simulation is necessarily three-dimensional and time dependent and is too expensive to be widely used. The averaged form of the Navier-Stokes equations is still the most commonly used approach in the practical calculation of the turbulent flows.

When considering the turbulent flow a time averaged form of the Navier-Stokes equations is usually employed through the Reynolds decomposition of the instantaneous value of each variable, e.g. velocity v_i, into a time-averaged mean value $\overline{v_i}$ and an instantaneous deviation or fluctuation v_i' from the mean value ($v_i = \overline{v_i} + v_i'$).

The stress tensor τ_{ij} can be written as a sum of the viscous part written for the time mean values and the Reynolds turbulent stresses ($-\rho\overline{v_i'v_j'}$) in the form

$$\tau_{ij} = -\overline{p}\delta_{ij} + \eta_o\left(\frac{\partial \overline{v_i}}{\partial x_j} + \frac{\partial \overline{v_j}}{\partial x_i}\right) - \rho\overline{v_i'v_j'} \tag{8}$$

The Reynolds mean momentum equations are given as follows

$$\frac{\partial \overline{v_j}}{\partial x_j} = 0 \tag{9}$$

$$\frac{\partial \overline{v_i}}{\partial t} + \frac{\partial \overline{v_j}\,\overline{v_i}}{\partial x_j} = -\frac{1}{\rho_o}\frac{\partial \overline{p}}{\partial x_i} + \nu_o\frac{\partial^2 \overline{v_i}}{\partial x_j \partial x_j} + g_i\overline{F} - \frac{\partial \overline{v_j'v_i'}}{\partial x_j} \tag{10}$$

$$\frac{\partial \overline{T}}{\partial t} + \frac{\partial \overline{v_j}\overline{T}}{\partial x_j} = a_o\frac{\partial^2 \overline{T}}{\partial x_j \partial x_j} - \frac{\partial \overline{v_j'T'}}{\partial x_j} \tag{11}$$

With the exception of the additional Reynolds stress term, the mean velocity of a turbulent flow and the instantaneous velocity of a laminar flow satisfy the same set of differential equations.

The mean flow vorticity transport equation can be derived similary by decomposing the vorticity vector into a time averaged mean value $\overline{\omega_i}$ and an instantaneous deviation ω_i' from the mean value $\omega_i = \overline{\omega_i} + \omega_i'$ and by taking the time averaged form of the instantaneous vorticity equation [6]

$$\frac{\partial \overline{\omega_i}}{\partial t} + \frac{\partial \overline{v_j}\,\overline{\omega_i}}{\partial x_j} = \frac{\partial \overline{\omega_j}\,\overline{v_i}}{\partial x_j} + \nu_o\frac{\partial^2 \overline{\omega_i}}{\partial x_j \partial x_j} - \epsilon_{ijk}g_k\frac{\partial \overline{F}}{\partial x_j} - \frac{\partial \overline{v_j'\omega_i'}}{\partial x_j} + \frac{\partial \overline{\omega_j'v_i'}}{\partial x_j} \tag{12}$$

The turbulent stress terms ($-\rho\overline{v_i'v_j'}$) are usually interpreted in the Boussinesq manner similarly to the viscous stress terms

$$-\rho_o\overline{v_i'v_j'} = \eta_t\left(\frac{\partial \overline{v_i}}{\partial x_j} + \frac{\partial \overline{v_j}}{\partial x_i}\right) - \frac{2}{3}\delta_{ij}\rho_o k \tag{13}$$

in which η_t is turbulent or eddy dynamic viscosity and k mean turbulence kinetic energy. When the temperature is regarded as a passive scalar, the term ($-\overline{T'v_j'}$) is assumed to be related to the mean temperature as follows [3]

$$-\overline{T'v_j'} = a_t \frac{\partial \overline{T}}{\partial x_j} \tag{14}$$

now a_t being the eddy temperature diffusivity.

BOUNDARY-DOMAIN INTEGRAL EQUATIONS

Nonlinear Diffusion-Convective Equation

The mean diffusion-convective transport equations for the momentum, vorticity, temperature, turbulent kinetic energy, dissipation, etc. can be recognized to be formally of the same type as a nonhomogeneous nonlinear parabolic partial differential equation of the form

$$a_o \frac{\partial^2 u}{\partial x_j \partial x_j} - \frac{\partial u}{\partial t} + b = 0 \tag{15}$$

The nonhomogeneous term b represents pseudo-body forces expressing the convection and eddy diffusion term.

Using Green's theorem for the scalar functions and the parabolic fundamental solution u^*, one can derive the following boundary-domain integral statement to eq.(15) in the incremental form for the time step $\tau = t_F - t_{F-1}$ equating $b = \frac{\partial}{\partial x_j}(a_t \frac{\partial u}{\partial x_j} - v_j u)$

$$c(\xi)u(\xi,t_F) + a_o \int_\Gamma \int_{t_{F-1}}^{t_F} u \frac{\partial u^*}{\partial n}\, dt\, d\Gamma = \int_\Gamma \int_{t_{F-1}}^{t_F} a_e \frac{\partial u}{\partial n} u^*\, dt\, d\Gamma$$

$$- \int_\Gamma \int_{t_{F-1}}^{t_F} u\, v_n u^*\, dt\, d\Gamma - \int_\Omega \int_{t_{F-1}}^{t_F} (a_t \frac{\partial u}{\partial x_j} - u v_j) \frac{\partial u^*}{\partial x_j}\, dt\, d\Omega$$

$$+ \int_\Omega u_{F-1}\, u^*{}_{F-1}\, d\Omega \tag{16}$$

when a_e is the effective diffusivity ($a_e = a_o + a_t$).

Velocity-Vorticity Formulation

The boundary domain integral statement for the flow kinematics can be derived from the vector elliptic eq.(3) applying Green's theorem for the vector functions and the elliptic fundamental solution u^*, resulting in the following statement written in the vector notation [4,7]

$$c(\xi)\vec{v}(\xi) + \int_\Gamma (\vec{\nabla} u^* \cdot \vec{n})\vec{v}\, d\Gamma = \int_\Gamma (\vec{\nabla} u^* \mathrm{x} \vec{n}) \mathrm{x} \vec{v}\, d\Gamma + \int_\Omega \vec{\omega} \mathrm{x} \vec{\nabla} u^*\, d\Omega \tag{17}$$

The integral statement (17) represents three scalar equations and only two of them are independent. It is completly equivalent to the continuity equation and vorticity definition, expressing the kinematics of the laminar and turbulent incompressible flow in the integral form. Boundary velocity conditions are included in boundary integrals, while the domain integral gives the contribution of the vorticity field to the development of the velocity field. Equation enables the explicit computation of the velocity vector in the interior of the domain ($c(\xi) = 1$). When the unknowns are the boundary vorticity values or the tangential velocity component to the boundary, one has to use the tangential component of the vector eq.(17) [8]

$$\begin{aligned} c(\xi)\vec{n}(\xi)\mathrm{x}\vec{v}(\xi) + \vec{n}(\xi)\mathrm{x}\int_\Gamma(\vec{\nabla}u^*\cdot\vec{n})\vec{v}\,d\Gamma &= \vec{n}(\xi)\mathrm{x}\int_\Gamma(\vec{\nabla}u^*\mathrm{x}\vec{n})\mathrm{x}\vec{v}\,d\Gamma \\ &+ \vec{n}(\xi)\mathrm{x}\int_\Omega \vec{\omega}\mathrm{x}\vec{\nabla}u^*\,d\Omega \qquad (18) \end{aligned}$$

or when the normal velocity to the boundary is unknown, the normal component of the vector eq.(17) has to be used

$$\begin{aligned} c(\xi)\vec{n}(\xi)\cdot\vec{v}(\xi) + \vec{n}(\xi)\cdot\int_\Gamma(\vec{\nabla}u^*\cdot\vec{n})\vec{v}\,d\Gamma &= \vec{n}(\xi)\cdot\int_\Gamma(\vec{\nabla}u^*\mathrm{x}\vec{n})\mathrm{x}\vec{v}\,d\Gamma \\ &+ \vec{n}(\xi)\cdot\int_\Omega \vec{\omega}\mathrm{x}\vec{\nabla}u^*\,d\Omega \qquad (19) \end{aligned}$$

in order to perform appropriate implicit system of discretized equations. Unknown boundary vorticity values are expressed in the integral form eq.(18) within the domain integral, excluding a need to use approximate formulae for determining boundary vorticity values, which would bring some additional error into the numerical scheme.

Describing the laminar transport of the vorticity and temperature in the integral statement, one has to take into account that each instantaneous component of the vorticity vector and temperature obey a nonhomogenous parabolic equation [5]; the following boundary-domain integral formulations can be derived for the vorticity transfer and temperature transport

$$\begin{aligned} c(\xi)\omega_i(\xi,t_F) + \nu_o\int_\Gamma\int_{t_{F-1}}^{t_F}\omega_i\frac{\partial u^*}{\partial n}\,dt\,d\Gamma &= \nu_o\int_\Gamma\int_{t_{F-1}}^{t_F}\frac{\partial\omega_i}{\partial n}u^*\,dt\,d\Gamma \\ &- \int_\Gamma\int_{t_{F-1}}^{t_F}(\omega_i v_n - v_i\omega_n + \epsilon_{ijk}n_j g_k F)u^*\,dtd\Gamma \\ &+ \int_\Omega\int_{t_{F-1}}^{t_F}(\omega_i v_j - v_i\omega_j + \epsilon_{ijk}g_k F)\frac{\partial u^*}{\partial x_j}\,dtd\Omega + \int_\Omega \omega_{i,F-1}\,{u^*}_{F-1}\,d\Omega \qquad (20) \end{aligned}$$

$$c(\xi)T(\xi,t_F) + a_o\int_\Gamma\int_{t_{F-1}}^{t_F} T\frac{\partial u^*}{\partial n}\,dt\,d\Gamma = a_o\int_\Gamma\int_{t_{F-1}}^{t_F}\frac{\partial T}{\partial n}u^*\,dt\,d\Gamma$$

$$-\int_\Gamma\int_{t_{F-1}}^{t_F} T\,v_n u^*\,dt\,d\Gamma - \int_\Omega\int_{t_{F-1}}^{t_F} Tv_i\frac{\partial u^*}{\partial x_i} + \int_\Omega T_{F-1}\,u^*{}_{F-1}\,d\Omega \tag{21}$$

Turbulent transport equations for the time mean values of the vorticity and temperature eq.(12) and eq.(11) are formally identical to the nonlinear diffusion-convective eq.(15). Following the same idea the integral statements can be written according to eq.(16) for the time mean vorticity and temperature.

$$c(\xi)\overline{\omega}(\xi,t_F) + \nu_o\int_\Gamma\int_{t_{F-1}}^{t_F}\overline{\omega}\frac{\partial u^*}{\partial n}\,dt\,d\Gamma = \int_\Gamma\int_{t_{F-1}}^{t_F}\nu_e\frac{\partial\overline{\omega}}{\partial n}u^*\,dt\,d\Gamma$$

$$-\int_\Gamma\int_{t_{F-1}}^{t_F}(\overline{\omega}\,\overline{v}_n + \epsilon_{ij}n_i g_j\overline{F})u^*\,dtd\Gamma$$

$$-\int_\Omega\int_{t_{F-1}}^{t_F}(\nu_t\frac{\partial\overline{\omega}}{\partial x_i} - \overline{\omega}\,\overline{v}_i - \epsilon_{ij}g_j\overline{F})\frac{\partial u^*}{\partial x_i}\,dtd\Omega + \int_\Omega \overline{\omega}_{F-1}\,u^*{}_{F-1}\,d\Omega \tag{22}$$

$$c(\xi)\overline{T}(\xi,t_F) + a_o\int_\Gamma\int_{t_{F-1}}^{t_F}\overline{T}\frac{\partial u^*}{\partial n}\,dt\,d\Gamma = \int_\Gamma\int_{t_{F-1}}^{t_F} a_e\frac{\partial\overline{T}}{\partial n}u^*\,dt\,d\Gamma$$

$$-\int_\Gamma\int_{t_{F-1}}^{t_F}\overline{T}\,\overline{v}_n u^*\,dt\,d\Gamma - \int_\Omega\int_{t_{F-1}}^{t_F}(a_t\frac{\partial\overline{T}}{\partial x_i} - \overline{T}\overline{v}_i)\frac{\partial u^*}{\partial x_i}dt\,d\Omega$$

$$+\int_\Omega \overline{T}_{F-1}\,u^*{}_{F-1}\,d\Omega \tag{23}$$

SUB-DOMAIN TECHNIQUE

In general sub-domain technique is used to model various material properties of piecewise homogeneous zones, in order to overcome geometry caused problems, and to significantly reduce the computer time and memory demands, what is specially true in the fluid flow computation. The numerical procedure presented in this paper, can be applied to each of the subregions as they are separated from the each other. The final implicit system of equations for the external boundary and interface boundaries of the whole region is obtained by adding the set of equations for each subregion together considering compatibility and equilibrium conditions between interfaces.

Let us now first consider the kinetics of the flow governed generaly by the eq.(20). The compatibility and equilibrium conditions to be applied at the interface Γ_I between Ω^1 and Ω^2 are respectively

$$\{f\}_I^1 = \{f\}_I^2, \qquad \{\frac{\partial f}{\partial n}\}_I^1 = -\{\frac{\partial f}{\partial n}\}_I^2 \tag{24}$$

In the kinematics the only proper compatibility conditions are applied to the tangential and normal velocity component

$$\{v_n\}_I^1 = -\{v_n\}_I^2, \qquad \{v_t\}_I^1 = -\{v_t\}_I^2 \tag{25}$$

On the internal interfaces the boundary vorticity and boundary vorticity flux values are determined in the kinetics. In the kinematics the only unknowns on the interface are tangential and normal velocity components and for the tangential (18) or normal (19) form of the kinematic equation has to be employed. In general, for all source points lieing on the external or interface boundaries the tangential or normal component of the kinematic statement has to be used, while component kinematic eq.(17) has to be employed to evaluate explicitely unknown values in the interior of the each subdomain.

TEST EXAMPLES

Two examples are given here in order to demonstrate the applicability of the proposed method for laminar fluid flow.

Free convection in cylinders

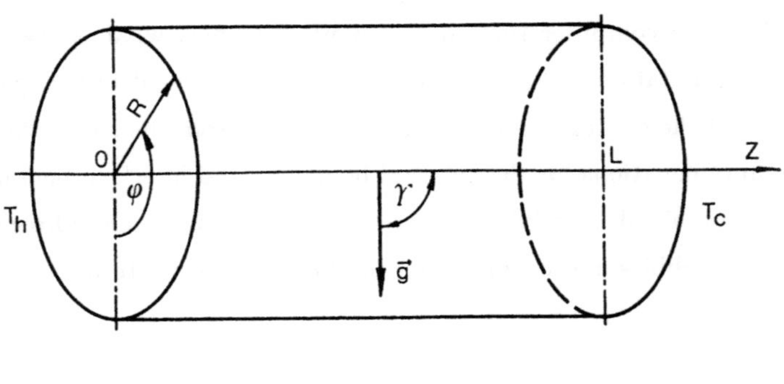

Fig. 1: Geometry of the cylinder

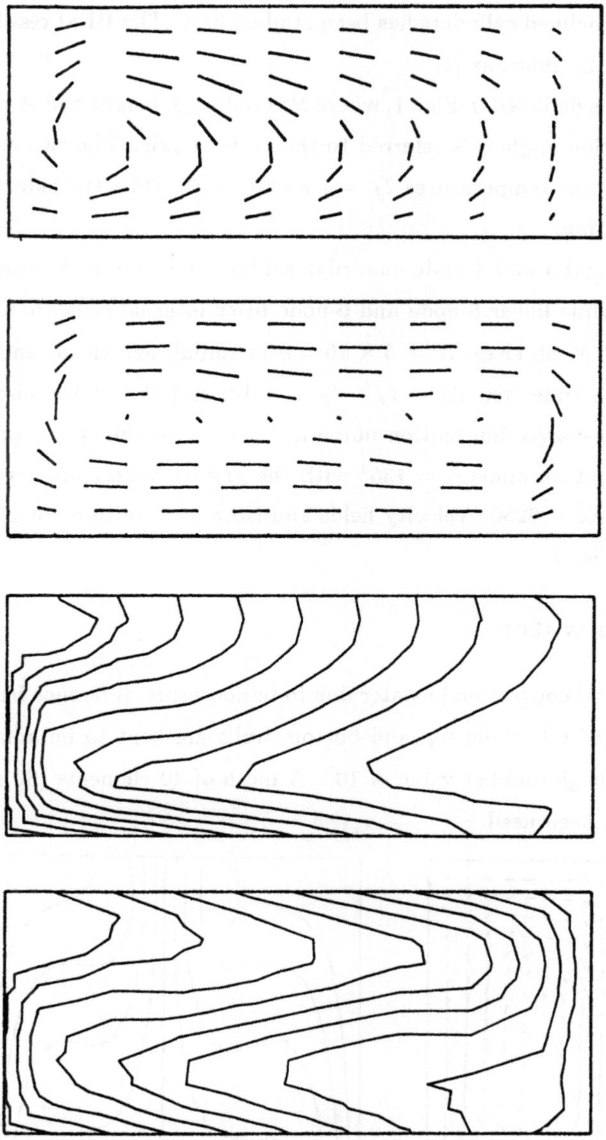

Fig. 2: Velocity fields and isotherm contours in inclined cylinder $\gamma = 135^o$, $\phi = 0^o$ for $Ra = 6250$ at $t = 8$ and 40 s

The free convection in inclined cylinders has been studied first. The BEM results are compared with FDM results, Bontoux [1].

The cylinder geometry is depicted in Fig. 1, where R is radius, L lenght and $A = L/R$ aspect ratio. The inclination angle γ is referred to the vertical axis. The two circular endwalls are kept at constant temperatures $T_h = 1$ and $T_c = 0$, while the side wall is assumed perfectly conducting.

The linear 3-node triangular and 4-node quadrilateral boundary elements were used to model the boundary, while linear 8-node and 6-node brick internal cells are applied to discretise the domain. Mesh sizes $M = 5 \times 16 \times 9$ in radial, azimuthal and axial directions were used. The time step $\Delta t = t_F - t_{F-1} = 1s$, and the under-relaxation factor 0.1 were used. Free-convection motion numerical results for the $A = 5$ cylinder when the axis is inclined at an angle $\gamma = 135^o$ with the gravity vector are presented for the Rayleigh number $Ra = 6250$. Velocity fields and isotherm contours for inclined cylinder are depicted in Fig. 2.

Square cavity with water

A closed cavity with natural convection in water due to temperature difference from the left (8° C) to right side (0° C), while top and bottom walls are kept to be adiabatic, has been studied for Rayleigh number value of 10^5. A mesh of 40 elements (80 nodes) and 400 internal cells has been used.

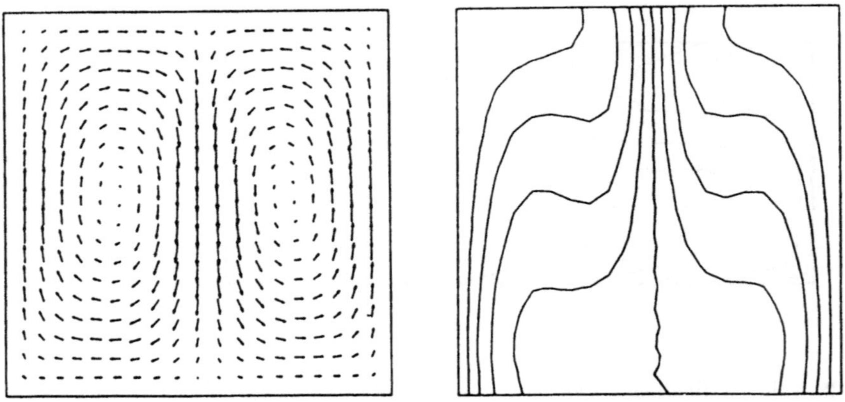

Fig. 3: Velocity and temperature field distribution for steady state ($t = 140$ s)

Figure 3 gives the final steady state ($t = 140$ s) of both velocity and temperature distributions. There are two separated circural zones observerd, while in the middle ($T = 4^o$ C) a symmetry line is developed.

References

[1] Bontoux,P., Smutek,C., Roux,B., Extremet,G.P., Schiroky,G.H., Hurford,A.C., Rosenberg,F.: *Finite Difference Solutions for Three Dimensional Buoyancy Driven Flows in Inclined Cylinder*, Vol.3, 3rd Int. Conf. on Num. Meth. for Nonlinear Problems, Dubrovnik. Pineridge Press, 1986.

[2] Brebbia,C.A., Telles,J.F.C., Wrobel,L.C.:*Boundary Element Methods-Theory and Applications*, Springer-Verlag, New York, 1984.

[3] Nagano,Y., Kim,C.:*A Two-Equation Model for Heat Transport in Wall Turbulent Shear Flows*, Journal of Heat Transfer, Vol.110, 1988.

[4] Skerget,P., Alujevic,A., Zagar,I., Brebbia,C.A., Kuhn,G.:*Time Dependent Three Dimensional Laminar Isochoric Viscous Fluid Flow by BEM*, 10th Int. Conf. on BEM, Southampton, Springer-Verlag, Berlin, 1988.

[5] Skerget,P., Alujevic,A., Brebbia,C.A., Kuhn,G.:*Natural and Forced Convection Simulation Using the Velocity-Vorticity Approach*, Topics in Boundary Element Research (Ed. by Brebbia C.A.), Vol.5, Ch.4, Springer-Verlag , Berlin, 1989.

[6] Tennekes,H., Lumly,J.L.:*A first Course in Turbulence*, The MIT Press, Boston, 1972.

[7] Wu,J.C., Gulcat,U., Wang,C.M., Sankar,N.L.:*A Generalized Formulation for Unsteady Viscous Flow Problems*, Topics in Boundary Element Research (ed. Brebbia C.A.), Vol.5, Ch.3, Springer-Verlag, Berlin, 1989.

[8] Zagar,I., Skerget,P.: *Boundary Elements for Time Dependent 3-D Laminar Viscous Fluid Flow*, Mechanical Engineering Journal, Vol. 10-12, Ljubljana, 1989.

Index

MIX
Papier aus verantwortungsvollen Quellen
Paper from responsible sources
FSC® C105338

If you have any concerns about our products,
you can contact us on
ProductSafety@springernature.com

In case Publisher is established outside the EU,
the EU authorized representative is:
Springer Nature Customer Service Center GmbH
Europaplatz 3, 69115 Heidelberg, Germany

Printed by Libri Plureos GmbH
in Hamburg, Germany